SiP System-in-Package Design and Simulation

SiP System-in-Package Design and Simulation

Mentor EE Flow Advanced Design Guide

Suny Li (Li Yang)
SiP/PCB Technical Specialist
Beijing, China

電子工業出版社
PUBLISHING HOUSE OF ELECTRONICS INDUSTRY
PHEI
http://www.phei.com.cn

WILEY

This edition first published 2017 by John Wiley & Sons Singapore Pte. Ltd under exclusive license granted by Publishing House of Electronics Industry for all media and languages (excluding simplified and traditional Chinese) throughout the world (excluding Mainland China), and with non-exclusive license for electronic versions in Mainland China.

Registered Offices
John Wiley & Sons, Inc., 111 River Street, Hoboken, NJ 07030, USA
John Wiley & Sons Singapore Pte. Ltd, 1 Fusionopolis Walk, #07–01 Solaris South Tower, Singapore 138628

Editorial Office
1 Fusionopolis Walk, #07–01 Solaris South Tower, Singapore 138628

For details of our global editorial offices, customer services, and more information about Wiley products visit us at www.wiley.com.

Wiley also publishes its books in a variety of electronic formats and by print-on-demand. Some content that appears in standard print versions of this book may not be available in other formats.

Library of Congress Cataloging-in-Publication Data
Names: Li (Li Yang), Suny, 1974- author.
Title: SiP system-in-package design and simulation : Mentor EE Flow advanced design guide / Suny Li (Li Yang).
Description: Singapore ; Hoboken, NJ : John Wiley & Sons, 2017. | Includes bibliographical references and index.
Identifiers: LCCN 2017007232 (print) | LCCN 2017011202 (ebook) | ISBN 9781119045939 (cloth) | ISBN 9781119046011 (Adobe PDF) | ISBN 9781119046004 (ePub)
Subjects: LCSH: Integrated circuits--Design and construction. | Multichip modules (Microelectronics)--Design and construction.
Classification: LCC TK7874 .L437 2017 (print) | LCC TK7874 (ebook) | DDC 621.3815--dc23
LC record available at https://lccn.loc.gov/2017007232
John Wiley & Sons Limited is a private limited company registered in England with registered number 641132. Registered office address: The Atrium, Southern Gate, Chichester, West Sussex, United Kingdom. PO19 8SQ.

Cover design by Wiley
Cover Images: (Background) © KTSDESIGN/Gettyimages;
(Gold Chip) Courtesy of the author

Set in 10/12pt Warnock Pro by SPi Global, Chennai, India

10 9 8 7 6 5 4 3 2 1

Contents

About the Author

Mr. Suny Li (Li Yang) is an SiP/PCB Technical Specialist in China; he now works in AcconSys Technology Co., Ltd, (a Mentor Authorized Distributor for China).

Suny has guided and consulted on dozens of SiP projects in China, accumulating plentiful experience in SiP design and simulation.

Suny has 10 years' experience in and knowledge of Application Engineer for Mentor especially in SiP/PCB design and simulation.

Before this, Suny worked in the Chinese Academy of Science and SIEMENS for several years. He has more than seven years' experience in hardware design (HW system design, PCB layout, high-speed signal integrity, power integrity, EMI, etc.).

In the course of his work, Suny has published papers and acquired four patents, and he continues with this work.

Suny is a senior member of the Chinese Institute of Electronics (CIE) and a member of the IEEE.

Suny graduated from Beijing University of Aeronautics & Astronautics (BUAA) in 2000, receiving Master's and Bachelor's degrees in Science and Technology of Aeronautics & Astronautics.

Preface

SiP – System in Package technology is becoming a hotspot of current electronic technology development, receiving attention from many sides, including from traditional package designers, traditional MCM designers and traditional PCB designers, and even SoC designers are beginning to pay close attention to SiP technology.

Comparing SiP to package, SiP is a system-level chip package, with independent system functionality.

In comparison with MCM, SiP is a three-dimensional chip package; its 3D aspect is mainly embodied in the chip stack and substrate cavity; at the same time, the functionality and volume of SiP are greater than those of MCM.

In comparison with PCB, SiP's advantage lies in the aspects of miniaturization, low power consumption and high performance. To achieve the same function as PCB, SiP only needs about 10–20% of PCB area and 40% of PCB power consumption, and system performance will be greatly improved over traditional PCB.

In comparison with SoC, the advantages of SiP technology lie mainly in its short cycle, low cost, ease of use and success rate. To achieve the same result, SiP needs just 10–20% the R&D time required by SoC, and costs in the region of 10–15% of the cost of SoC, and SiP is more likely to be successful.

In China and all over the world, more and more electronic design engineers are beginning to pay close attention to SiP and to study this technology, but since there is no SiP reference book available, designers often do not know how to start; this has to some extent hindered the rapid development of SiP technology.

The author of this book in recent years has been engaged in research into SiP technology and technical support of customers engaged in SiP design and simulation, and has participated in many SiP projects.

During these projects, I realized that there was a more and more pressing need for a comprehensive SiP reference book for users of SiP design and simulation technology. It is for this reason that I wrote this book.

The book is divided into 21 chapters, and systematically describes the whole process of SiP from design to simulation.

Because the SiP projects I have participated in are designed with Mentor SiP tools, the book's description of the SiP design and simulation methods is based on Mentor tools.

The following is a high-level description of the chapters of the book.

- Chapter 1 focuses on the development of the Mentor SiP technology, as well as providing an introduction to the Mentor SiP design and simulation platform.
- Chapter 2 provides a basic description of package and SiP, including details such as package manufacturers and chip suppliers.
- Chapter 3 deals with BGA package, and describes the SiP packaging production process.
- Chapter 4 introduces new packaging technologies and their application, including TSV, IPD, PoP, etc.
- Chapter 5 describes the entire SiP design and simulation process.
- Chapter 6 introduces the establishment and management of the SiP design central library.
- Chapter 7 introduces SiP schematic design, various options and other functions.
- Chapter 8 introduces multi-board project management, concurrent schematic design, and implementation methods.
- Chapter 9 introduces SiP layout creation, setup and operation.
- Chapter 10 describes constraint rules management in SiP design.
- Chapter 11 introduces the bond wire model, parameter setting and design methods.
- Chapter 12 introduces the cavity and chip stack, their definitions and design methods.
- Chapter 13 introduces the concept and the design methods of flip chip and RDL.
- Chapter 14 describes SiP substrate routing and plane shape processing methods.
- Chapter 15 introduces embedded resistor and capacitor technology, materials and design methods.
- Chapter 16 introduces design methods for RF circuits and RF schematic and layout design, as well as the link with simulation tools.
- Chapter 17 describes Xtreme real-time concurrent design technology and operation methods.
- Chapter 18 describes the 3D real-time DRC check, and use of 3D Viewer to simulate SiP production and processing.
- Chapter 19 introduces DRC technology, including online DRC, batch DRC and hazard review.
- Chapter 20 introduces the output of various kinds of production data.
- Chapter 21 describes several SiP simulation technologies.

The chapters provide a complete description of SiP and the problems that may be met in design.

As regards technologies in SiP design that this book doesn't cover, I hope readers will bring these up, and I will deal with them in subsequent editions of the book.

In writing this book, I have striven to create a comprehensive book on SiP. Therefore, in addition to introducing various techniques and methods in SiP design, in the last chapter of this book I introduce SiP simulation technology.

To provide reassurance to beginners in SiP design, in addition to describing SiP design and simulation, the book also introduces SiP manufacturers and bare chip suppliers, and provides other relevant reference information.

Furthermore, the book has three chapters introducing the basics of package and SiP, the production process and emerging SiP technologies, so that engineers who know very little about SiP or package before touching this book can, with the guidance provided, accomplish high-speed, high-quality SiP design.

Of course, although I strove for perfection in writing this book, limitations in my knowledge may lead to faults and errors. I hope that experts and readers will kindly supply suggestions to allow any faults to be corrected in future editions.

I sincerely hope that the publication of this book can play a role in the development of SiP technology and related industries.

Author: Suny Li (Li Yang)
Chinese version, March 2012, Beijing
English version, July 2016, Beijing

1

SiP Design and Simulation Platform

1.1 From Package to SiP

Package is the term commonly used to refer to the protective housing and related accessories of a single Integrated Circuit (IC) bare chip cut down from wafer; it is mainly used to protect the silicon chips. Because silicon chips are very fragile, even very fine dust or water droplets can destroy their functionality, so it is necessary to protect IC chips with package. Another feature of package is to amplify the scale; because the chip itself is usually very small, the scale is increased by package, making it easier to use in subsequent board-level PCB systems. The third function of package is electrical connection: via package, the chip exchanges information with the outside world.

Depending on the process or material, package product is usually divided into three types: plastic package, ceramic package and metal package. Plastic package is mainly used in commercial products and has the advantage of low cost, but the thermal dissipation, stability and air tightness are relatively poor. Ceramic package and metal package are suitable for industrial products as well as in the aerospace, military and other harsh fields, have excellent heat dissipation and air tightness, and high reliability. Meanwhile, ceramic and metal package have the advantage that they can be disassembled easily for fault finding and problem "zeroing". Figure 1.1 shows the three different kinds of package.

IC package typically includes DIP, QFP, BGA, etc. With improvements in technology, packaging technology developed rapidly, along the DIP→QFP→BGA→CSP direction. Package density is becoming higher, scale is increasing quickly, and number of pins is growing fast. Single-chip package has not met the requirements of system design; packaging products are developing from small scale to large scale, from single-chip package to multi-chip package.

Multi-chip package has attracted more and more attention, with most directed to System in Package (SiP).

SiP, as the name suggests, refers to the integration of a system in a package body. Typically, this system requires encapsulating multiple chips able to complete a specific task, such as system-level package-integrated CPU, DRAM, Flash and other IC chips. At present, with the development of package technology, SiP has gradually developed to 3D stacked-chip package.

SiP System-in-Package Design and Simulation: Mentor EE Flow Advanced Design Guide,
First Edition. Suny Li (Li Yang).

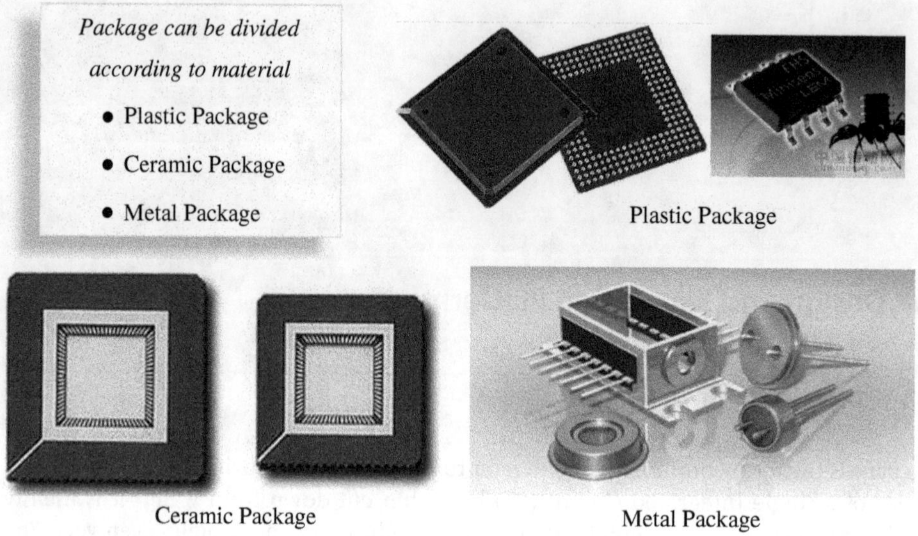

Figure 1.1 Three different kinds of package.

Multi-chip package is not a new concept; MCM (Multi-Chip Module) technology has been popular for many years, and has been widely used in specific areas such as the military, aerospace and aviation.

MCM is a kind of package; as opposed to SiP, small chips are usually used in MCM, which can accomplish relatively simple functions compared to SiP. In MCM the chips are usually in 2D layout.

The bare chips used in MCM usually have a single function, and are smaller and simple. MCM uses 2D flat package, which also means that it is not easy to increase the capability of MCM, it is difficult to shrink in size, and there are other disadvantages.

SiP combines the advantages of both MCM and large-scale IC package.

SiP package is specifically intended for large-scale, multi-chip, 3D packaging. Its stereoscopic 3D nature is mainly reflected in the two aspects of chip stacking and substrate cavity. Figure 1.2 shows the process of IC package and MCM evolving to SiP.

The new trend in SiP technology development is that traditional package design by IC chip manufacturers is being gradually transformed to a new arrangement whereby the

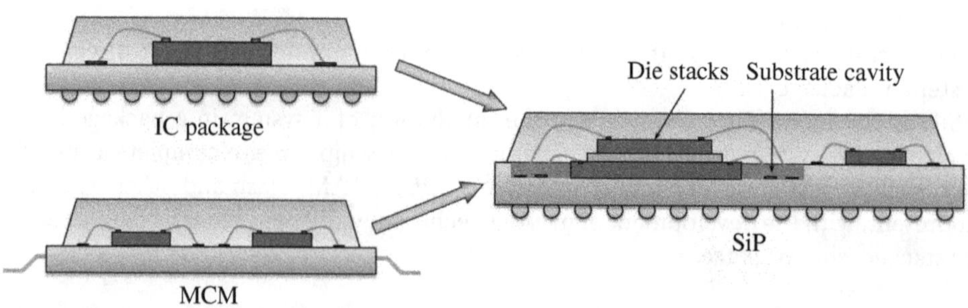

Figure 1.2 Evolution from IC package and MCM to SiP.

system user is beginning to consider and design the package. In the past, chip manufacturers usually packaged chips and then delivered them to the user.

Now, with the development of SiP technology, miniaturization and low-power design requirements, more and more system users want to get the bare dies, and build their system based on bare dies and packaging.

It follows that market demand for bare chips will greatly increase, as more and more designers want to know how to get the bare chips.

As these demands continue to grow, traditional IC agents will continue to expand their bare chips business in order to meet the growing demands of the market.

If the demand does not reach a certain number, orders for bare chips from IC manufacturer are usually made via agents. There will be some delay for the customer. So, the SiP designer, in the early stages of project, should fully consider the order channels and order cycles. When the market demand for bare chips reaches a certain level, and when there is a continuing demand, bare chips agents would consider increasing their inventories to meet customer needs.

The bare chip market is developing, and is driven by the rapidly growing demand for SiP technology. In turn, developments in the bare chip market promote the application and popularity of SiP technology.

Because SiP or package design is gradually shifting from the IC chip manufacturer to the system user and system users are most concerned with system design, the collaboration of package design and system design will also become increasingly important. Package design itself will become a key link to system design, with designers required to realize the function of the entire system in a unified platform. Figure 1.3 shows the relationship between IC bare chips, SiP package and the PCB board-level system.

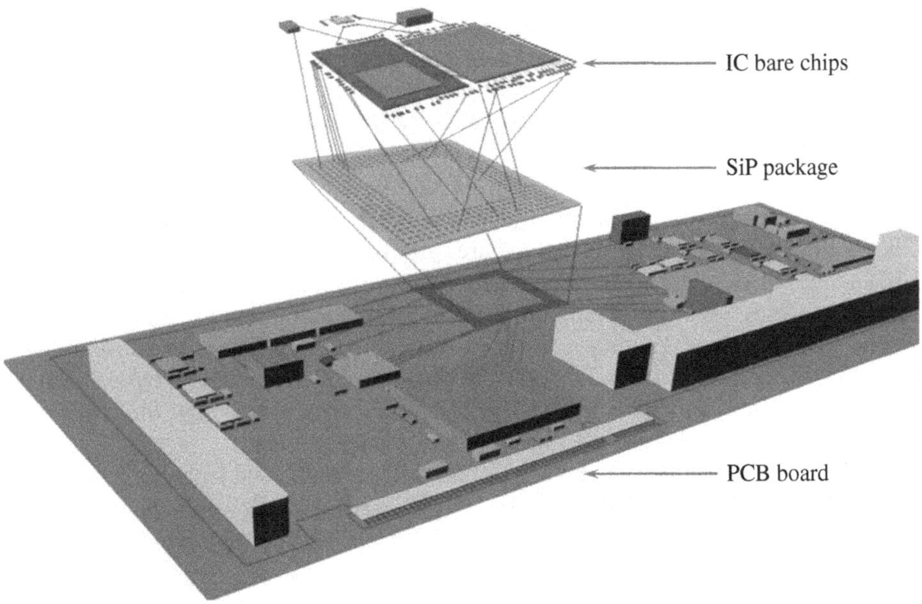

Figure 1.3 The relationship between IC bare chips, SiP package and the PCB board-level system.

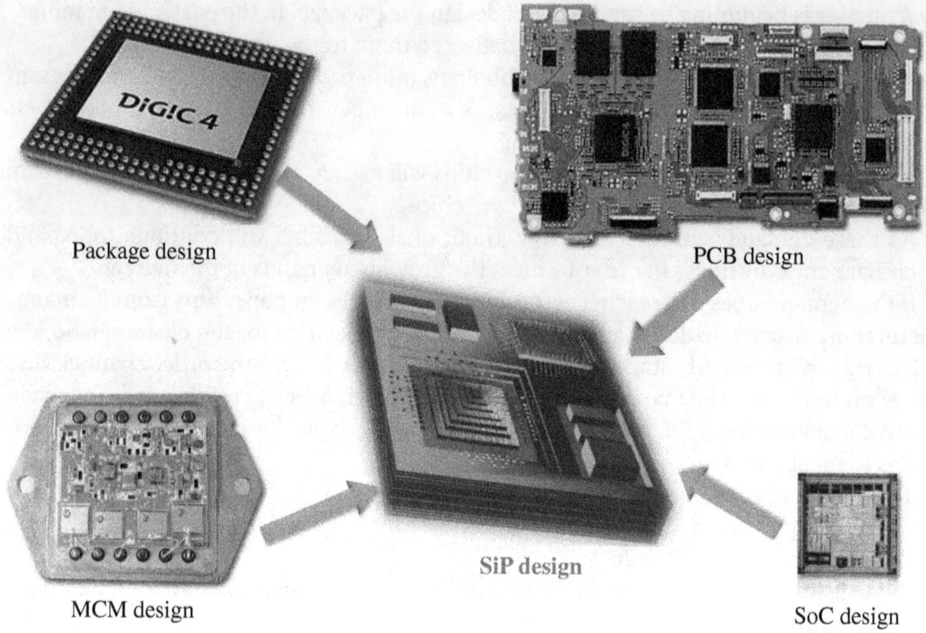

Figure 1.4 SiP is receiving attention from a wide range of areas.

SiP is getting much attention, not only from traditional package designers, but also from traditional MCM designers and PCB designers, and even SoC designers have begun to keep a watchful eye on SiP. Refer to Figure 1.4.

As compared with traditional package, SiP is a system-level package, and can accomplish system functions independently.

Compared with MCM, SiP is a 3D multi-chip package, with the 3D embodied in chip stacks and substrate cavities, while the scale and function of SiP are also greatly increased compared to MCM.

Compared with PCB, the advantage of SiP technology is mainly in terms of miniaturization, low power consumption and high performance. To realize the same function as PCB, SiP only needs about 10–20% of the area and 40% of the power of a corresponding PCB, and also has a relatively large performance improvement over PCB.

Compared with SoC, the advantage of SiP is mainly embodied in short cycles, low cost and ease of success. To achieve the same function, only 10–20% of the development time required for SoC is needed for SiP, the cost of SiP development is typically about 10–15% that of SoC, and SiP is more likely to succeed. Therefore, SiP is often seen as a low-cost, short-term, alternative solution to SoC by many users.

Often, SiP is a forerunner at the start of an SoC project, used to make a rapid and low-cost SiP product. When SiP is successful, giving some initial results on the project, and receiving recognition and support from all, the project then shifts to SoC research and development.

With increasing demand for high-performance, high-speed and versatile design, designers are more concerned about signal integrity, power integrity, crosstalk, EMC/EMI, and functional simulation and verification. A common solution is to use simulation tools to assist in the design flow; HyperLynx SI/PI/Thermal, HyperLynx DRC, HyperLynx 3DEM solver and HyperLynx Analog are commonly used simulation tools.

For high-density, small-size and low-power-consumption design, designers require design rules such as trace width and clearance, smaller passive components, for example, resistors and capacitors, HDI (High Density Interconnection) technology, and buried and blind via technology.

Moreover, design software and product technics support the use of many passive components embedded in the substrate; this is called EP (Embedded Passive) technology, and consists of bare dies mounted directly on the circuit board; this is called COB (Chip On Board) technology.

With the application and maturity of these technologies, PCB system designers began to focus on a new technology, which is the integration of all these technologies: this is SiP technology.

In contrast with PCB, to realize the same function SiP requires only 10%~20% area and 40% power consumption of original PCB; refer to Figure 1.5.

1.2 The Development of Mentor SiP Design Technology

The Mentor company is the largest EDA software supplier of PCB board-level system design in the world. Mentor provides the most advanced solutions for electronics companies and research institutions worldwide.

Mentor has held the absolute leading position in the PCB design market for many years. Statistical data from the global PCB design market (third-party statistics) show that Mentor has more than 50% of the market share, dominating the market and far outstripping competitors.

In May 2009, based on version EE2007.5, Mentor launched the Expedition advanced packaging bundle (Expedition AdvPkg) to support SiP design. Before launching this module, Mentor had no tool for package or SiP design specifically; however, Mentor's related technology can be traced back a few decades.

First, let us recall the processes of the Mentor package and SiP technology accumulation and development.

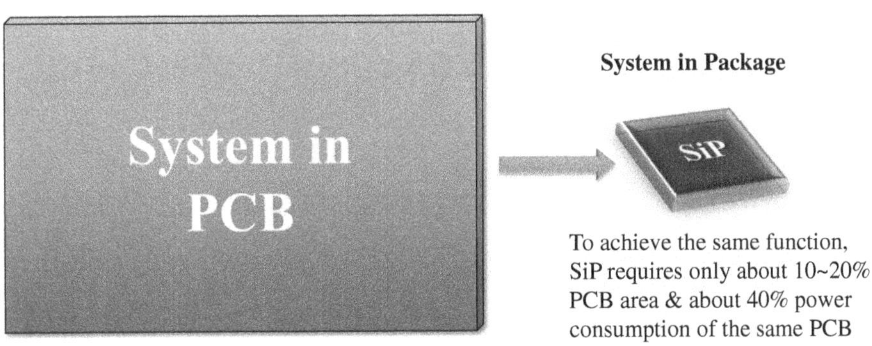

Figure 1.5 Size and power consumption comparison between PCB and SiP.

BoardStation, also called EN (ENTERPRISE), is an excellent Mentor PCB board-level system design platform, mainly used by high-end PCB designers.

In BoardStation there is a module called Hybrid Station, mainly used in hybrid circuits and MCM design, which has a wide range of applications in MCM and mixed-signal circuit design. At present, there are still a number of companies and research institutes worldwide which use this design platform for their projects.

In the 1990s, Mentor acquired DDE's Supermax ECAD. Supermax ECAD is an excellent tool, used mainly for embedded passive, MCM, Hybrid IC package and RF design. As multi-functional layout software, the distinctive feature of Supermax ECAD is the ability to support RF, EP, package and MCM design.

Mentor developed these technologies, and gradually integrated and consolidated them into a new software product: Expedition Enterprise Flow, EE Flow for short.

With the development of technology and design needs, Mentor is gradually shifting its high-end PCB board-level design platform, focused on Expedition Enterprise Flow; the latest version is EE7.9.5.

At the same time, Expedition Enterprise also continues to absorb advanced technology and functionality from other products, covering new areas of design. The Expedition AdvPkg bundle is the new product, which incorporates many of Mentor's advantages in relation to MCM, hybrid circuits, RF, embedded passive components (EP) and other technologies.

Therefore, we can say that while the Expedition AdvPkg bundle is a relatively new product, the technology is very mature, and its technological development can be traced back to the 20th century.

In Figure 1.6, we can see the inheritance and development process of the Mentor SiP design platform and advanced packaging technology.

1.3 The Mentor SiP Design and Simulation Platform

1.3.1 SiP Platform Introduction

Mentor provides a comprehensive SiP/MCM, advanced package and PCB design and simulation platform, as shown in Figure 1.7.

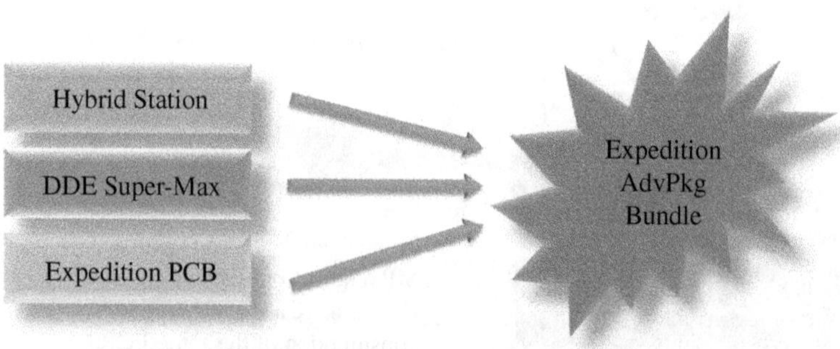

Figure 1.6 Development of the Mentor SiP platform.

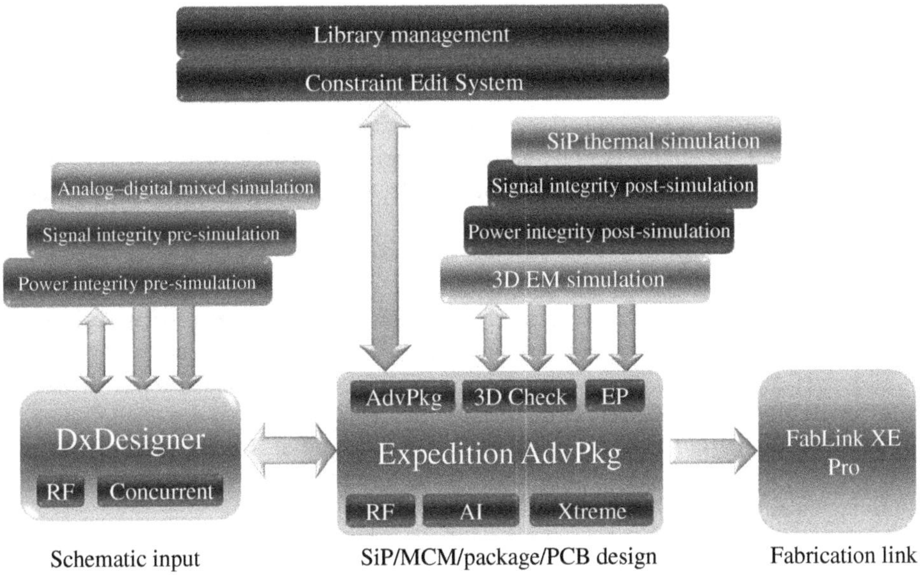

Figure 1.7 Mentor SiP/MCM/advanced package/PCB design and simulation platform.

For SiP design needs, the Expedition AdvPkg bundle comprises AdvPkg, 3D Check, EP, RF and AI (Advanced Interconnection), as well as FabLink XE Pro modules, specialized support for SiP/MCM, advanced packaging design, data output and validation.

Meanwhile, for SiP simulation, Mentor offers the mixed digital–analog simulation tool HyperLynx Analog, the thermal simulation tool HyperLynx Thermal, the signal integrity analysis tool HyperLynx SI, the power integrity analysis tool HyperLynx PI, the 3D electromagnetic simulation tool HyperLynx 3D EM and the EMI/EMC check tool HyperLynx DRC.

Mentor's design and simulation flow for SiP/MCM and advanced package is a Windows-based software platform, compatible with both Linux and Solaris. Mentor EE Flow supported platforms and processors are shown in Figure 1.8.

System Requirements

Release EE 7.9.5 Supported Platforms and Processors

Linux RHEL 4 x86
Linux RHEL 4 x86-64
Linux RHEL 5 x86
Linux RHEL 5 x86-64
Linux RHEL 6 x86/x86-64
Linux SLES 10 x86
Linux SUSE 11 x86/x64
Solaris 10 UltraSPARC
Windows 7 x86/x64
Windows Server 2008 x86/x64
Windows Server2003 x86
Windows Vista x86/x64
Windows XP x86

Figure 1.8 Mentor EE Flow supported platforms and processors.

1.3.2 Schematic Input

DxDesigner is the schematic input tool in the SiP design platform.

In addition to conventional schematic, DxDesigner supports RF input, and its RF components library is synchronized with the Agilent ADS RF library.

In relation to the schematic design phase, DxDesigner is provided with a mixed-circuit simulation tool, to simulate the function of circuits, so as to ensure that "design is correct" from the outset.

Meanwhile, designers can also use the SI pre-simulation tool to do "what if" analysis of the signal integrity of the design, so as to determine the components selection, net topology planning, terminal-matching and substrate laminate structure parameters of the design.

Using the PI pre-simulation tool, designers can plan power plane partition, distribution of power nets, and the number and variety of decoupling capacitors in the schematic design phase, and reasonably control the power plane impedance.

In addition, DxDesigner supports concurrent schematic design for multiple people designing one schematic at the same time; it does not require any partitioning. This technology is especially important for large and complex projects.

1.3.3 Concurrent System Design

Because SiP design and PCB design were eventually carried out by system designers, collaborative design between them became closer and closer. Mentor was sensitive to this new trend, and developed collaborative design capabilities, which also became a key feature of Mentor tools.

As shown in Figure 1.9, multiple designs are managed in one project, including two SiP designs and one PCB board design; each design can be completed independently.

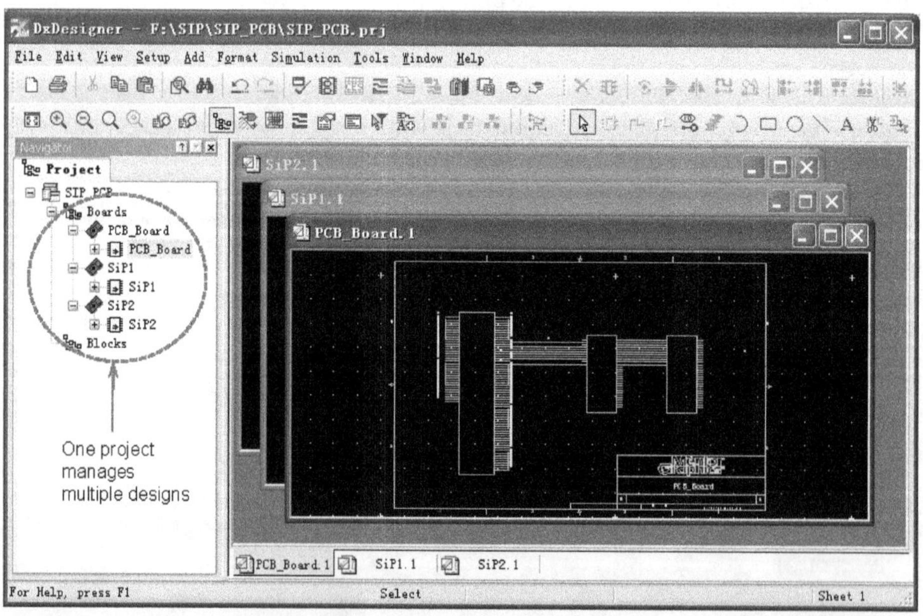

Figure 1.9 One project manages multiple designs, and realizes SiP and PCB collaborative design.

There is a mutual relationship between the three designs, because both SiPs are ultimately installed on the PCB board to work together.

This way of managing projects is more reasonable than having three separate designs, because in considering subsequent PCB system design, the pin assignment of SiP design can be optimized.

1.3.4 SiP Board Design

In the field of PCB board design, Mentor accounted for more than 50% of the global market share, far more than its competitors. Mentor extended its powerful advantages from PCB board design to the SiP design area; all the powerful features of PCB design are also used in SiP design.

Currently, layout is one of the core modules of SiP design. Mentor developed six core design features based on Expedition PCB, features which support the user with all kinds of needs in the SiP design procedure, as shown in Figure 1.10.

1) AdvPkg core design function

   AdvPkg(Advanced Packaging)core design function supports wire bonding and substrate cavity, with no level limit on IC chip stacks, and complex multi-step cavity design, as shown in Figure 1.11.

   AdvPkg also supports both automatic bonding of complex bond wire and automatic generation of power ring.

   Mentor bond wire models can be created using a variety of types of curve fitting, which are more accurate and closer to the actual bonding wire. These models can help designers improve design accuracy, thereby increasing product yield. The Mentor bond wire model is shown in Figure 1.12.

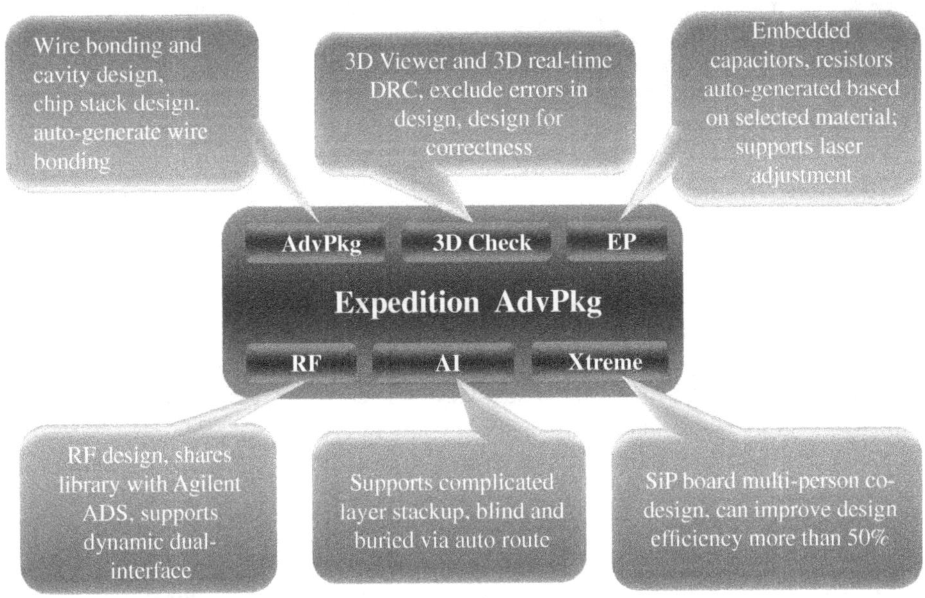

Figure 1.10 The six core design functions of Mentor SiP layout. *(See color plate section for the color representation of this figure.)*

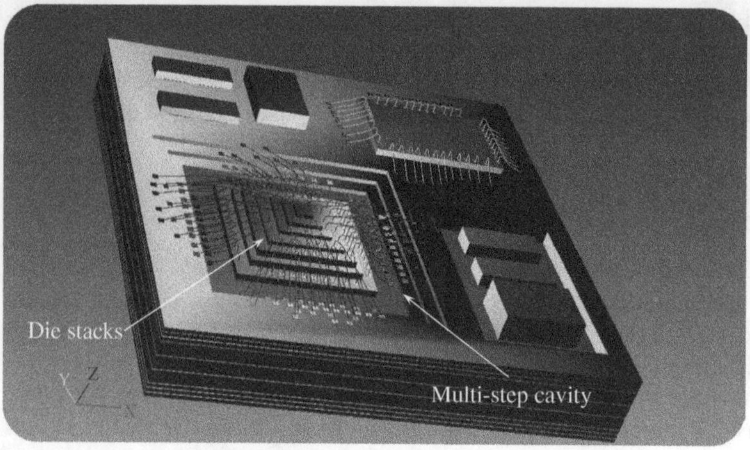

Figure 1.11 Chip stacks and multi-step cavity.

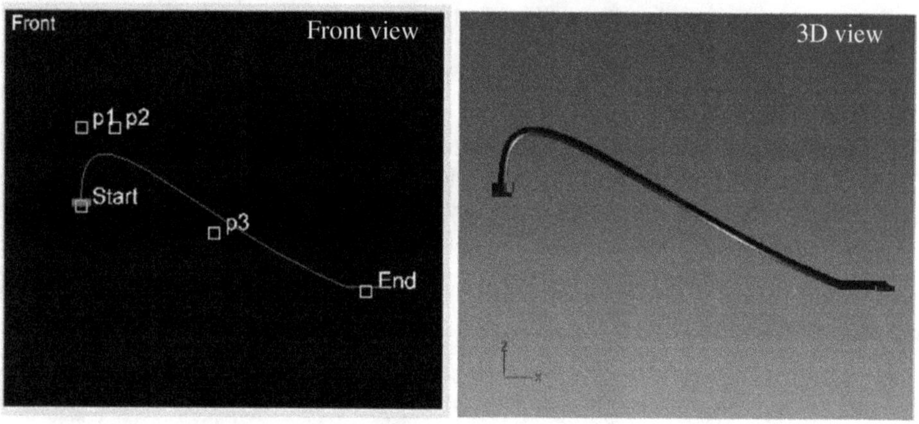

Figure 1.12 Mentor bond wire model.

2) 3D check function

   The 3D check function supports 3D viewing and checking. It supports interactive 3D real-time DRC check of layout design; it allows checks of the chips, including bond wire and cavity, routing, copper, through-hole vias, buried and blind vias, buried resistors and capacitors, etc. By using 3D real-time DRC check, designers can avoid errors at design time, and ensure design is correct. This interactive 3D real-time DRC is shown in Figure 1.13.

   Differing from the traditional method, the Mentor tool is better in relation to understanding the 3D design and check, and realizing 3D real-time inspection during the design process, to ensure design is correct.

   In traditional methods, 3D checking is usually executed after design finish, and the designer needs additional tools for 3D inspection. It is similar to checking routing results in a Gerber browser. Typically, design finish is too late for this kind of post-processing 3D inspection, which can seriously affect the accuracy and efficiency of SiP design.

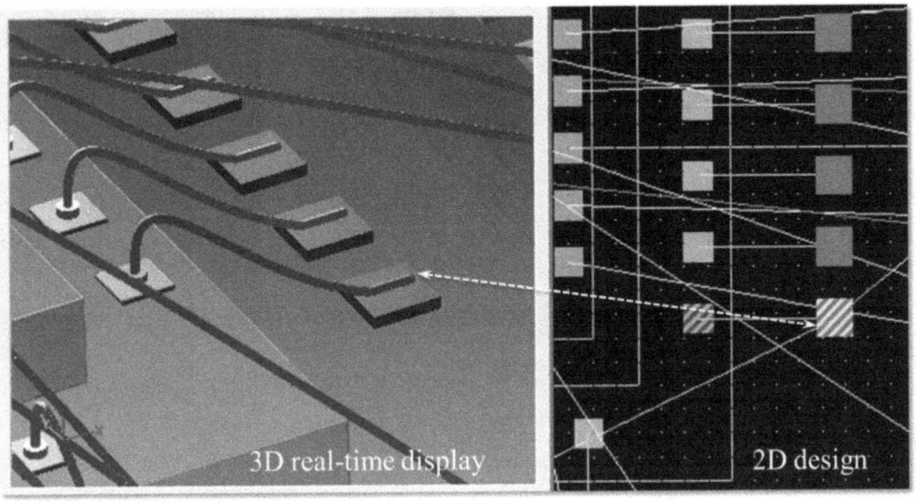

Figure 1.13 Interactive 3D real-time DRC.

3) EP function
   EP function supports embedded passive components, namely buried resistors and buried capacitors. This function supports the automatic integration of embedded resistors and capacitors.

   Depending on the selected material and resistance or capacitance parameters, required resistors or capacitors can be automatically integrated. Laser adjustment is also supported. Expedition supports multiple types of resistor and capacitor, as shown in Figure 1.14; the designer can use resistors and capacitors in any of the substrate layers. Using EP technology, we can greatly save on surface space and reduce the solder joints, thereby increasing the reliability of the design.
4) RF design function
   The RF(Radio Frequency)design function can realize RF circuit design. It supports RF parameter transfer between schematic and layout, and supports RF circuits passing through a dynamic link tool to and from ADS/AWR, as shown in Figure 1.15.

   The RF function can also meet some special design requirements in layout, such as gradient-width trace, stitched via and spiral inductor in power supply.
5) AI function
   The AI feature supports automatic routing of blind and buried vias in complex layer structures, and displays the layer numbers through which the buried or blind vias pass in layout, as shown in Figure 1.16.

   Intelligent AI automatic routing algorithms can greatly improve the routing completion rate and routing efficiency of the design.
6) Xtreme concurrent team design function
   The Xtreme concurrent team design function is suitable for complex SiP projects, with multiple designers concurrently designing one substrate board. It does not require any design segmentation, and provides design data updates to each designer in real-time. It greatly reduces the difficulty of design and alleviates the pressure on designers. Based on statistical data from real projects, Xtreme can improve design efficiency by as much as 50%.

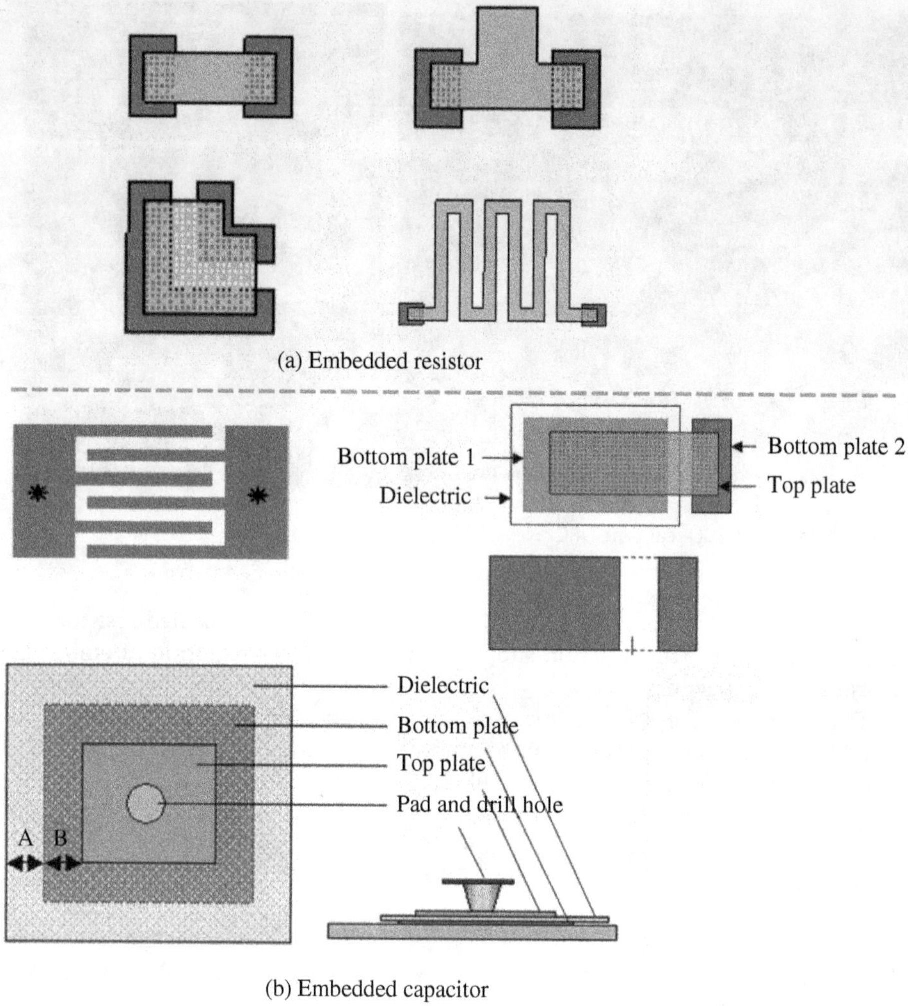

Figure 1.14 Expedition supports four kinds of EP resistor and three kinds of EP capacitor.

It is particularly important for complex SiP design, for which timeframes are very tight. Traditional methods generally do not have such a concurrent design function.

Figure 1.17 shows a diagrammatic sketch of Xtreme concurrent team design.

The SiP six core layout design functions are integrated into the design process of Expedition, belonging to the same environment as PCB design.

Therefore, all the advanced features of Expedition PCB can be used for SiP design, such as bus topology planning and bus routing, intelligent high-speed automatic routing, flexible routing, circuit copy, design reuse, and constraint management and other advanced features can be configured according to the design needs.

In addition to providing powerful layout features, Mentor provides a wide range of simulation and verification tools to support SiP design, including signal integrity, power integrity, thermal analysis, EMI/EMC, etc.

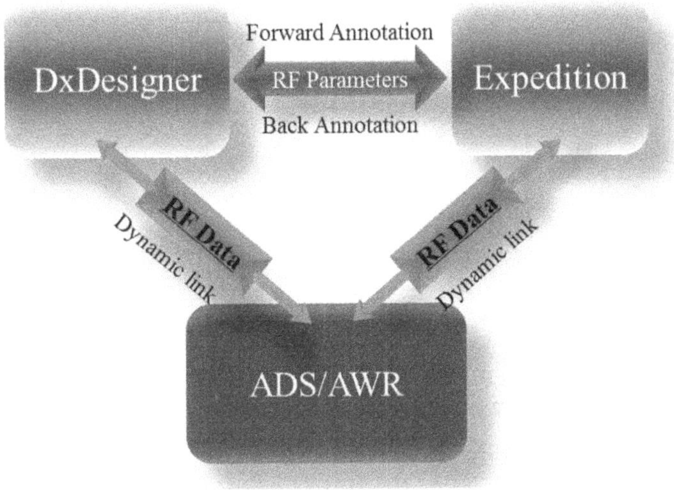

Figure 1.15 RF design data and RF parameter transfer.

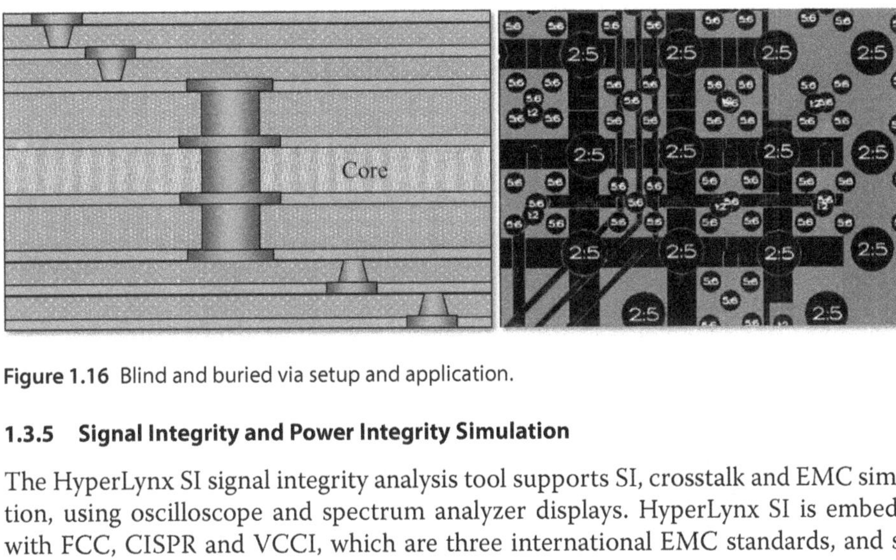

Figure 1.16 Blind and buried via setup and application.

1.3.5 Signal Integrity and Power Integrity Simulation

The HyperLynx SI signal integrity analysis tool supports SI, crosstalk and EMC simulation, using oscilloscope and spectrum analyzer displays. HyperLynx SI is embedded with FCC, CISPR and VCCI, which are three international EMC standards, and supports users in defining their own standards.

HyperLynx PI power integrity analysis tools support DC voltage, current density, AC decoupling and plane noise analysis. Waveform display is provided in two dimension and three dimension. Figure 1.18 shows screenshots of HyperLynx SI and PI.

1.3.6 Thermal Analysis

HyperLynx Thermal is a thermal analysis tool, allowing the designer to diagnose SiP heat-related issues at the early design stage. HyperLynx Thermal analyzes hot issues accurately and reliably, so as to avoid overheating or thermal failure in SiP products. Figure 1.19 shows a snapshot of one MCM's thermal analysis result comprising temperature and thermal gradient.

Figure 1.17 Sketch of Xtreme concurrent team design.

1.3.7 The Advantages of the Mentor SiP Design and Simulation Platform

The Mentor SiP design and simulation platform supports SiP/MCM/package and traditional PCB design, with signal integrity, power integrity, EMC analysis and thermal analysis functions. Completeness of design, simulation and validation ensures high-quality completion of complex SiP projects.

1.3.7.1 Characteristics of Mentor SiP Design and Simulation Platform

1) Comprehensive solution, design and simulation in a unified platform.
2) Real-time 3D DRC checks.
3) Intelligent automatic synthesized embedded passive components, graphical user interface, easy to use.
4) Intelligent blind and buried via automatic routing, no design layer and volume limit.
5) RF design-sharing libraries with ADS/AWR, dynamic data interface, easy to learn and use.
6) Thermal analysis tool for SiP/MCM/package.
7) Supports multi-player concurrent design, greatly improves efficiency.
8) Supports cooperative design with mechanical software, ECAD-MCAD real-time collaborative design.

1.3.7.2 Design Areas of Mentor SiP Design and Simulation Platform

a) Classification according to technology
   1) RF and wireless SiP design
   2) Digital SiP design
   3) General IC package design

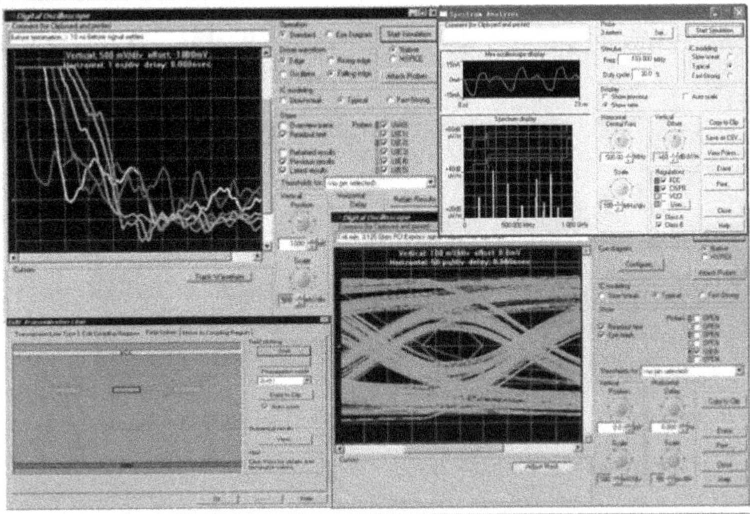

(a) Signal integrity (SI) simulation

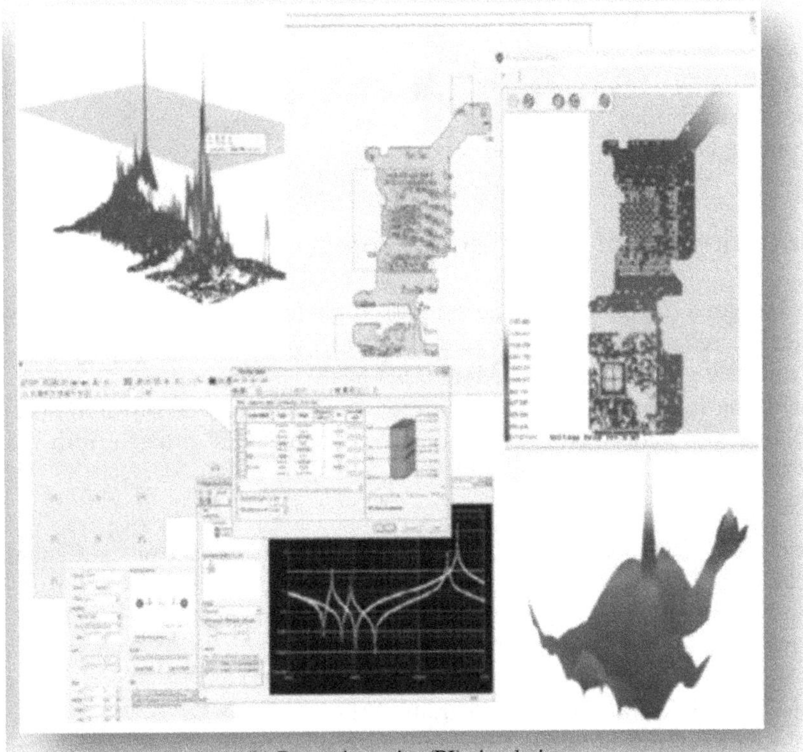

(b) Power integrity (PI) simulation

Figure 1.18 Screenshots of HyperLynx SI and PI.

(a) Temperature (b) Thermal gradient

Figure 1.19 Snapshot of HyperLynx Thermal analysis.

   4) MCM design
   5) Large-scale flip-chip design.
- b) Classification according to industry
   1) Aerospace, aviation, shipping and other military industries
   2) Power system, such as smart power grid
   3) Communication, semiconductor, etc.
   4) Automotive electronics
   5) Intelligent applications such as Internet of things.

1.4 The Introduction of the Finished Project

Since Mentor launched the SiP design and simulation platform, it has attracted many companies and research institutes in China and worldwide. Many of them have designed high-performance SiP products with the Mentor platform.

In China, with the support of Mentor application engineers, several research institutes have completed a variety of complex, high-performance SiP projects with the Mentor platform.

For example:

1) Rudder control MCM design and thermal analysis, Microelectronics Center, Beijing, 2008.
2) RF-baseband mixed SiP design, Institute of Chinese Academy of Sciences, 2009.
3) 5-digital-chip SiP design, research institute in Beijing, 2010.
4) 6-digital-chip SiP design, a Microelectronics Corporation, 2011.
5) 3-SoC-chip SiP design, a Microsystem Institute, 2011.
6) 6-chip dual-cavity ceramic SiP design, research institute in Beijing, 2011.
7) 3-chip design of multi-cavity ceramic SiP, a Beijing company, 2011.
8) 4-chip mixed SiP design, Institute of Chinese Academy of Sciences, 2012.
9) Large-scale flip chip package design, Institute of Chinese Academy of Sciences, 2012.

Figure 1.20 shows one of the user projects completed on the Mentor SiP platform. This project has already successfully progressed to small-batch production, delivered to back-end PCB board-level applications in real projects.

At present, many customers are designing SiP, MCM or package with the Mentor platform, as follows:

1) 5-chip with dual-cavity ceramic SiP design, a research institute in Beijing.
2) 7-chip with multi-cavity ceramic SiP, a technology company in Beijing.
3) 3-chip power system integrated control SiP design, a research center in Nanjing.
4) 6-chip analog and microwave SiP design, an institute in Xian.
5) RF-digital mixed SiP design, a Microsystem research institute in Xian.
6) Flex-circuit SiP package designs, Institute of Chinese Academy of Sciences.

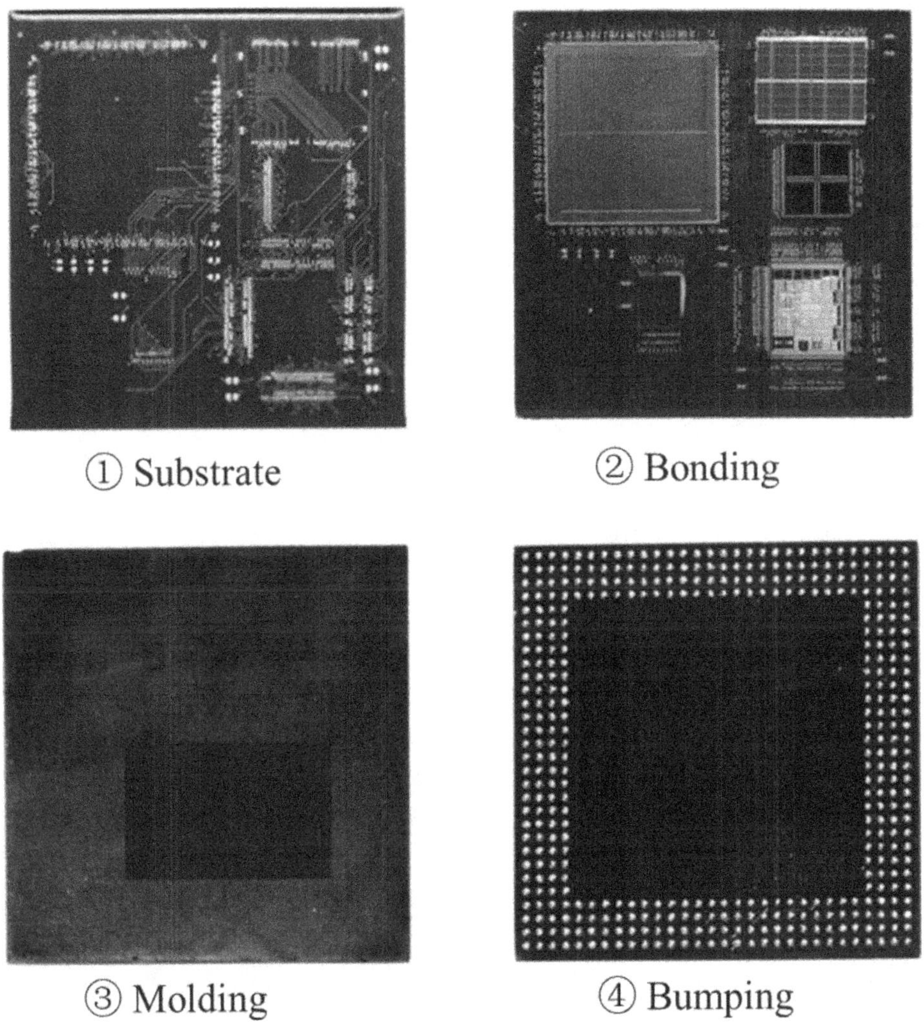

Figure 1.20 One of the finished customer projects using the Mentor SiP platform. *(See color plate section for the color representation of this figure.)*

2

Introduction to Package

2.1 Definition and Function of Package

For the purposes of this book, 'package' can be defined as a sealed shell containing chips placed to form a perfect entity, with the packaging providing protection for chips to operate stably and reliably in required environments and conditions.

The function of semiconductor IC package is to guarantee power supply and input/output signal operation, while dissipating the chip heat to outside environments. Package provides necessary electrical interconnections, mechanical support, environmental protection and thermal structure for semiconductor chips.

Package has the following four main functions:

1) Signal distribution, which mainly involves layout and electromagnetic considerations.
2) Power distribution, including electromagnetic, structural and material considerations.
3) Heat dissipation, including structural and material considerations.
4) Chip interconnection and protection, including mechanical, chemical and electromagnetic considerations.

A typical package structure is shown in Figure 2.1.

Package not only installs, fixes, seals, protects and enhances the heating performance of chips, but also makes electrical connection function available for chips.

After packaging, chips are connected to the outside with package pins; these pins connect to other devices through the trace or copper in the printed circuit board (PCB), so as to make the internal chips connect to external circuits. Meanwhile, packaged chips are isolated from impurities in the air so as to prevent corrosion of the chip circuit, avoiding electrical performance degradation. In addition, the packaged chips are easier to install and transport.

Packaging technology has a direct impact on the performance of the chip itself, and also influences the printed circuit board design, manufacturing and other areas; therefore, it is essential for chips.

If a silicon chip is equivalent to the human brain, packaging is equivalent to the torso and muscle of a human, connecting to the external environment and protecting the key functions of the brain. This vivid metaphor is shown in Figure 2.2.

SiP System-in-Package Design and Simulation: Mentor EE Flow Advanced Design Guide,
First Edition. Suny Li (Li Yang).

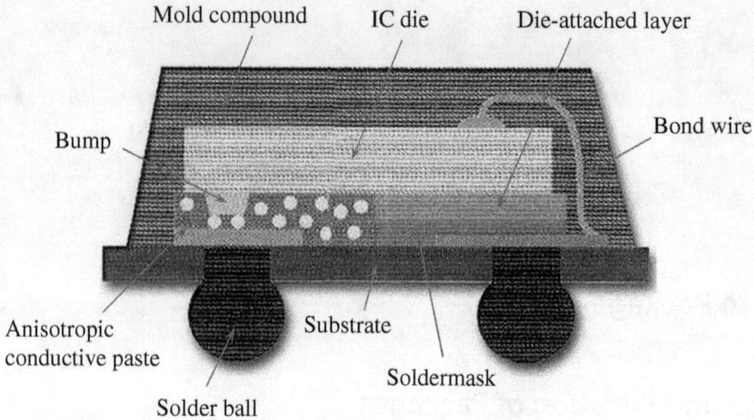

Figure 2.1 A typical package structure.

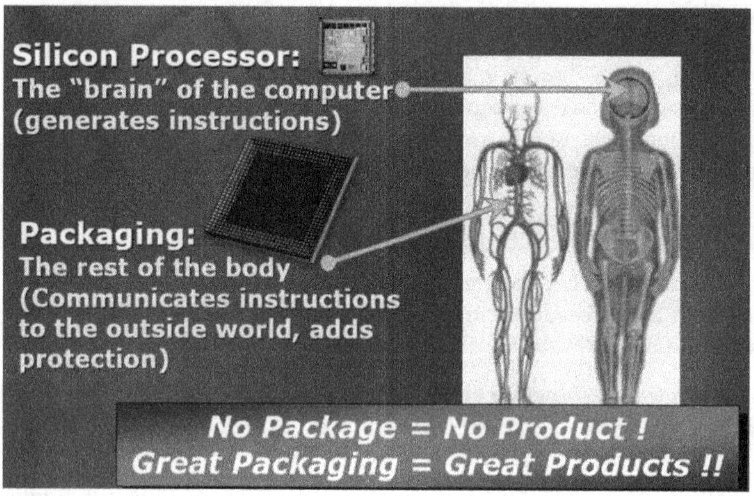

Figure 2.2 The relationship between chip and package is like the human brain and torso.

2.2 Development of Packaging Technology

The evaluation of a packaging technology as advanced or not is based on an important index, which is the ratio of chip area compared to package area; the closer the ratio is to 1, the better.

Currently, in relation to chip package evaluation, we should take the following factors into consideration:

1) Chip area and packaging area ratio. To improve packaging efficiency, the ratio should be as close as possible to 1:1.

2) Pins should be as short as possible in order to reduce delays. In order to ensure non-interference and improve performance, the distance between the pins should extend as far as possible.
3) Based on thermal requirements, the thinner the package, the better.

There are many kinds of classification for package type, based on the packaging material. Package can be divided into ceramic package, metal package and plastic package.

Ceramic package is primarily used in aerospace and military applications, and occasionally for civilian products. Metal package is basically used in the aerospace and military industries. Plastic package is low cost, but reliability is relatively poor; it is mainly used for civilian products.

In relation to sealed class, package can be divided into the airtight package and non-airtight package.

With high-reliability requirements, military products need to be capable of use within a large temperature range, and they also require higher protection against moisture, so they generally use airtight package. Ceramic and metal package can meet such demands. Civilian products use a non-airtight plastic shell, which has an economic advantage.

As regards the connection of the device and mother board, the package can be classified as SMT (Surface Mounted Technology) and THT (Through Hole Technology).

In THT technology, devices are placed at one side of the board, and the pins are soldered on the other side. This technique is known as inserted-welding technology. This welding technique requires the drilling of a hole for each pin, with each pin occupying space on both sides of the motherboard; solder points are also relatively large and the required installation space is large. However, THT has high reliability, and is able to withstand greater stress. For example, cable sockets and other devices that require resistance to high pressure usually use THT package.

In the 1980s, SMT package appeared; the pins of this kind of package are welded on the same side as the parts. Usually SMT parts are higher density and smaller than THT, support reflow soldering, and are more favorable for rapid mass production. Today's PCB mostly uses SMT devices.

Classification of the types of package housings is more numerous: package can be divided into TO (Transistor Outline) Package, DIP (Dual In line Package), QFP (Quad Flat Package) and BGA (Ball Grid Array) package, etc.

Basically, there has been a huge change about every ten years.

In 1955, TO round metal packaging appeared; this type of package has pin numbering 3~12.

In 1965, DIP package appeared, with package pin count 6~64, and pin spacing of 2.54 mm.

TO and DIP package generally belong to the THT type of package. Package size and thickness is relatively large, and pins are easily damaged in the plug process; this has a certain impact on reliability.

In the 1980s, surface mounted technology devices appeared, reflected in the device package form; SMT mainly includes SOT, SOP, SOJ, PLCC, QFP; the pin numbering is 3~300 and the pin spacing is 1.27~3 mm. Among these, QFP has the widest application range.

In order to reduce the parasitic effect of pins, and thus improve RF performance, the lead was omitted, and the lands connect directly to the PCB; this technology evolved out of QFN package (Quad Flat No-lead package).

With the dramatic increase in number of chip pins, BGA package began to be used extensively. The outer pins for soldering BGA are in ball array form, located on the bottom surface of the package; ball spacing is 0.8~1.27 mm. BGA can support more than 2000 pins.

Common package types are shown in Figure 2.3.

With innovations in technology, advanced packaging is developing rapidly; packaging technology has developed from the traditional low-end DIP and QFP forms to BGA, Chip Scale Package (CSP), Multi-Chip Module (MCM), System in Package (SiP) and other high-end forms.

In CSP, the area of the silicon chip and substrate packaging ratio is greater than 80%. MCM is a technology for mounting multiple silicon chips in a high-density multilayer interconnection substrate, composed of a wide variety of electronic modules.

The rapid development of consumer electronics and communication electronics has presented higher requirements for electronic system integration.

In SiP technology, bare chips with different features can be placed or stacked in three-dimensional space, with hybrid technologies, within the same package, constituting a complete system which can work independently. This feature of SiP has attracted more and more attention in recent years, and has become widely used in industry.

With the improvement of IC chip running speed and geometry shrink, package design and manufacturing has become more and more important for system applications.

Electronic packaging design and manufacturing need to start from the very beginning of system design to get the best value.

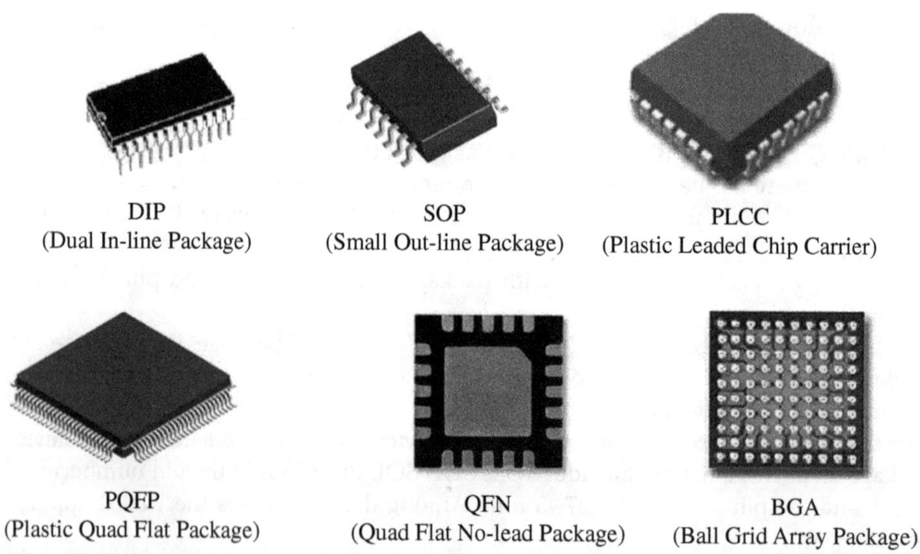

Figure 2.3 Common package types.

The driving forces behind the development of package technology primarily come from semiconductor development and market needs; they can be summarized by the following points:

1) Increasing chip speed and processing ability mean there is a need for more pins, faster clock speed and better power distribution.
2) There is market demand for electronic products with more features, longer battery life and smaller size.
3) The number of electronic devices and electronic products is increasing; new devices are emerging.
4) Demands in chip manufacturing and the electronics market will ultimately determine the developing trend of electronic packaging.
5) Package that is smaller, thinner, lighter, better performing, more powerful, more reliable, more environmentally friendly, more inexpensive and with lower power consumption will be more popular.

Figure 2.4 shows the development trends of electronic packaging technology.

We can see from Figure 2.4 that packaging technology development is moving in the direction of high-density, compact and versatile technology, and its applications are expanding from traditional electronic equipment areas to fields that include aerospace, automotive products, consumer electronics, home appliances, etc.

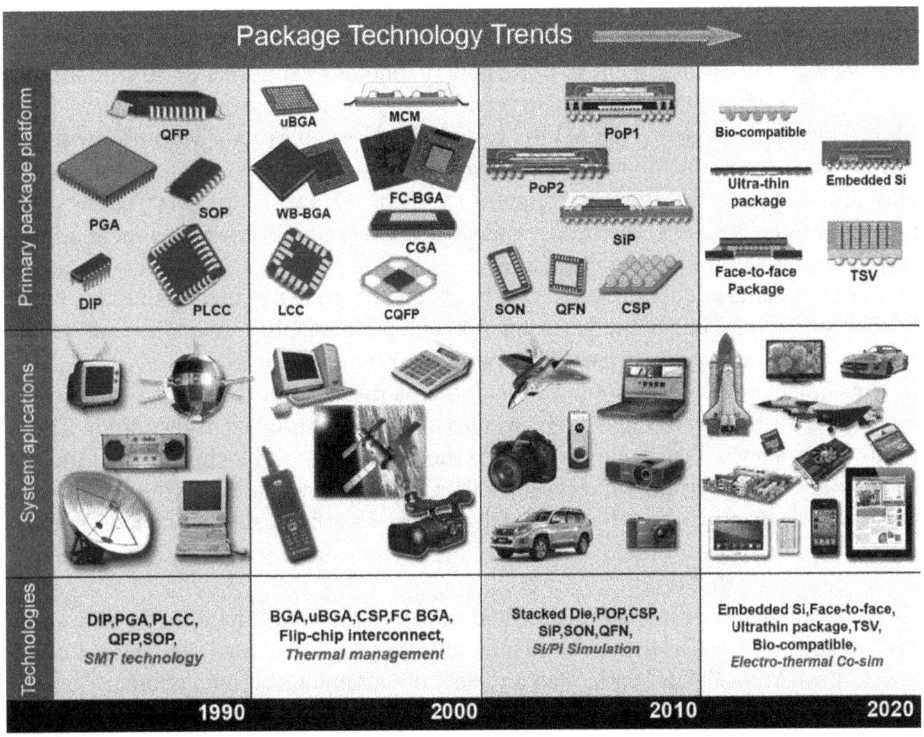

Figure 2.4 Development trends in electronic packaging technology. *(See color plate section for the color representation of this figure.)*

Packaging technology now concerns all areas from initial SMT research to the development of thermal analysis management, signal integrity analysis, power integrity analysis and electro-thermal co-simulation analysis.

For the long term, the following trends in packaging technology development are predicted:

1) Single-chip package will develop to multi-chip package, including MCM and SiP technology.
2) 2D packaging will develop to 3D packaging.
3) SLIM (Single Level Integrated Module) technology will allow adaptation of the current multi-level combination package to one level.
4) The separation between chip, package and PCB manufacture will be significantly reduced, the interconnection between them will further strengthen, and chip-package-PCB collaborative design will become increasingly important.

2.3 SiP and Related Technologies

2.3.1 The Appearance of SiP Technology

The appearance of SiP technology is the result of technology pushing and market pulling.

IC design, semiconductor manufacturing and the packaging industry are three major industries of the microelectronics industry.

Dealing with the development of packaging technology cannot be done without discussion of the microelectronics industry, and we must also mention Moore's law.

Moore's law is the observation of Intel co-founder Gordon Moore. Moore's law shows that the transistor count doubles about every 18–24 months, with transistor performance also doubling, but the price remains the same. Or rather, for every dollar spent on computer performance, the return will double every 18–24 months. Figure 2.5 is a diagram of Moore's law.

Moore's law came out about 50 years. It has been proven to be correct for many years; the semiconductor chip-manufacturing process is developing at a dizzying speed. However, we know that the size of chips cannot shrink indefinitely.

In November 2009, Intel announced that Intel microprocessors based on next-generation 32 nm logic technology had been put into mass production; next-generation 22 nm process technology will soon be used. Once the line width on the chip is up to the nm (10^{-9}m) order of magnitude, which is equivalent to only a few molecules in size, changes in the physical and chemical properties of the material will be qualitative changes, resulting in the current process of semiconductor devices not working correctly. Thus, Moore's law may be coming to its end.

Rapid development of semiconductor technology will still occur, but not as it used to, based on advancement in the wafer-manufacturing process technology; instead it will follow a "post-Moore's law" path, with a variety of technological innovations.

SiP system-level packaging technology is one of the most important future directions; it will also become a more and more important solution in the microelectronics industry.

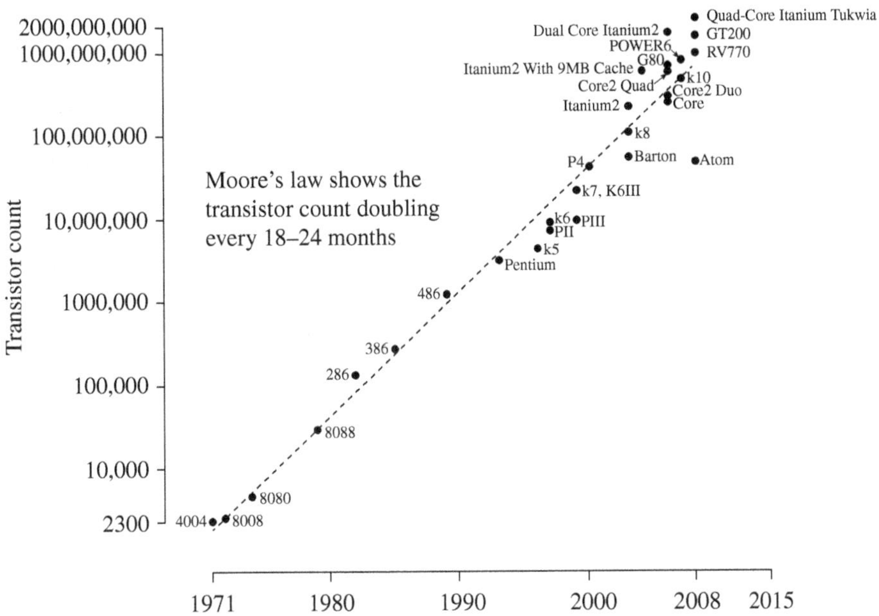

Figure 2.5 Diagram of Moore's law.

2.3.2 SoC and SiP

From the market point of view, for new electronic products, whether communication, consumer, medical or automotive and aerospace electronics, their common characteristics are: high integration, high reliability and shorter product-development time.

There are two solutions in the industry for high integration of electronic systems: one is SoC (System on Chip), the other is SiP.

SoC is an overall realization of CPU, DSP, digital circuits, analog circuits, memory and many other circuits on a single chip, integrating functions such as image processing, speech processing, communication, data processing and others. The greatest advantages of SoC are the small size, better performance and lowest cost in large volumes. However, SoC-based technical features integrating such technologies as digital, analog, RF and microwave on the same chip have technology compatibility issues.

And, due to the system complexity, the SoC product-development cycle is very long, and design errors, delays, repeated chip manufacture and other factors will also lead to costs increasing.

SiP installs chips with different functions in various forms and combinations in three-dimensional space, mixed within the same package, so as to form a complete system package.

Figure 2.6 shows the typical SiP structure.

SiP can not only effectively reduce system volume, improve product performance and accelerate the speed of research and development, but can also speed the product to market, thus becoming the trend that electronic companies are rushing to develop.

Compared to SoC, SiP allows a variety of bare chips with different technologies, processes and materials to be put together, creating a full system and maintaining good

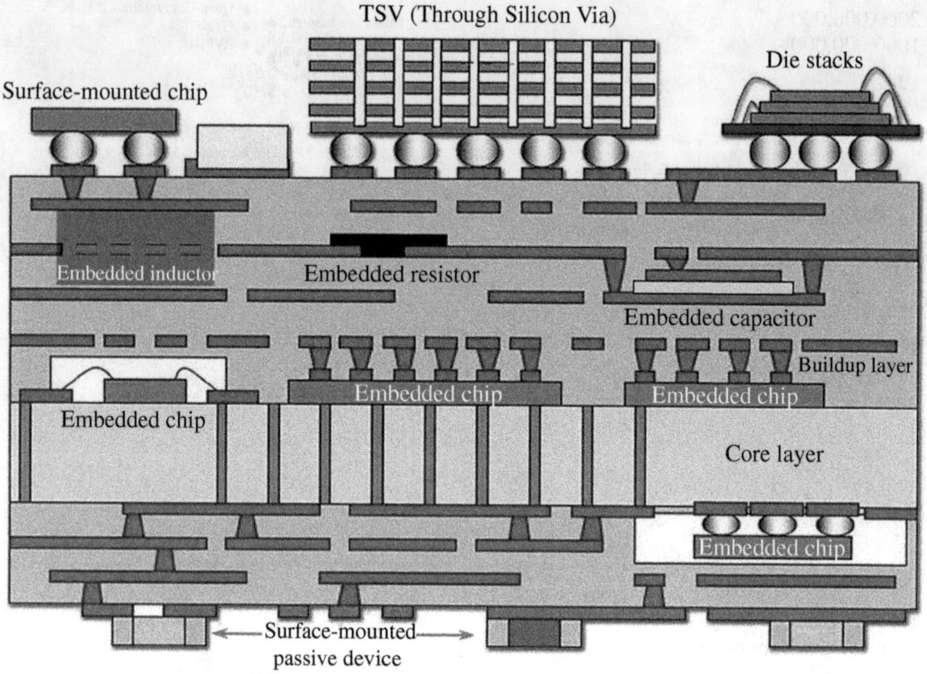

Figure 2.6 Typical SiP structure diagram.

compatibility. Passive components can also be placed on the substrate or embedded in multilayer substrate in order to save space on the surface, reducing the volume of target package.

SiP has a shorter development cycle and lower research and development costs than SoC, and can meet the requirements of new electronic products.

The life cycles of electronic products are now getting shorter, and the chip life cycle has reduced from one year to the current six to nine months.

Using SiP technology, all modules have been certified in advance, so that it is only necessary to put the required modules together to form a multifunction chipset.

So, SiP can not only reduce interference between chips, but can also substantially save required authentication time, and can also, in a relatively short period of time, allow multichip packaging to be put together based on the customer's demand for product features.

Compared to SoC, the SiP development cycle is $^1/_6$ to $^1/_{10}$ shorter than the SoC development cycle, and SiP development costs are $^1/_5$ to $^1/_{10}$ lower than SoC development costs, as shown in Figure 2.7.

2.3.3 SiP-Related Technologies

The prospects for SiP development have been widely recognized. The diversity of market requirements determines the complexity and diversity of SiP technology, and miscellaneous market drivers will push SiP technology to develop in a variety of ways.

At present, the well-known SiP technologies are chip stacking, embedded components, substrate cavity, IPD, TSV and PoP.

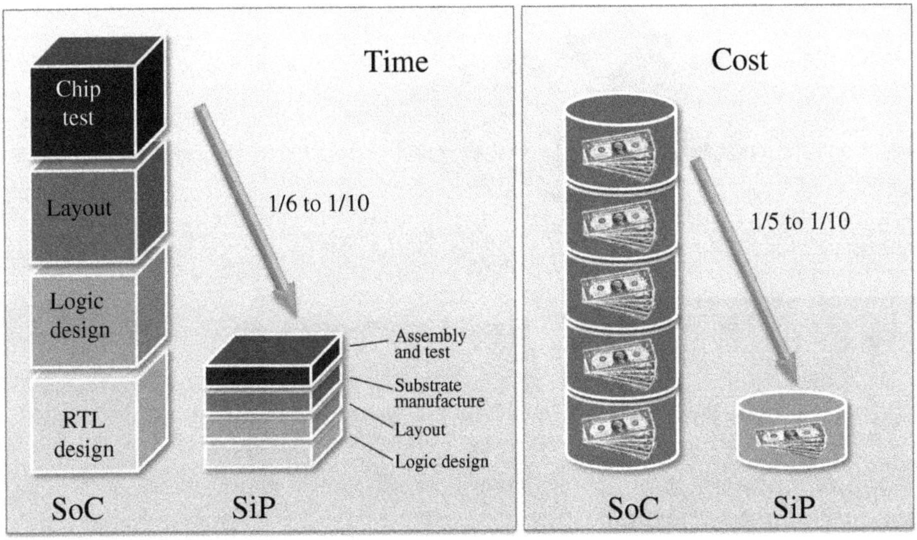

Figure 2.7 Comparison of development time and development cost in SoC and SiP.

1) Chip-stacking technology

   The common chip-stacking method is to stack two or more chips in the vertical direction, use the traditional wire-bonding method for interconnection, then package.

   Chip stacking is usually one of three kinds. First is pyramid stacking, which uses chips of different sizes, with the area of the upper chips less than that of the lower chips, so that the lower surface has enough space for wire bonding.

   Second is cantilever-type stacking, used where there is little difference in chip size, where an interposer is usually inserted between the upper-level chip and the lower-lever chip to facilitate wire bonding. The interposer is usually smaller than the chips, and made of ordinary silicon or other material.

   The third type of chip stack is side-by-side stacking, that is, where multiple pieces of smaller chips are put side by side on a large chip surface.

   Different kinds of stacking modes can be used in combination to increase flexibility.

   Figure 2.8 illustrates the several types of chip stacking.

   Chip stacking is widely used on memory cards. For these products, the key point for increasing capacity is to improve the capacity of flash-memory modules within the memory card.

   Although, with the help of advanced IC-manufacturing technology, we can increase the storage density of single flash-memory chips and store more data in a smaller chip space, the process conversion costs tend to be very large. With the help of chip-stacking technology, we can design multiple flash-memory chip stacks on a storage card, increasing the storage capacity directly and efficiently.

2) Embedded component technology

   "Embedded component" technology refers to the integration or embedding of components in the substrate.

(a) Pyramid stacking

(b) Cantilever stacking

(c) Side-by-side stacking

Figure 2.8 Three kinds of chip stacking.

For IC chips, as regards the general assembly of the substrate, the process is first to create a cavity in the substrate, then embed the IC chip into the substrate, fix, then flush the substrate with epoxy before implementing multilayer routing, finally installing the remaining components on the surface layer.

For passive components such as resistors and capacitors, we can use thick film or thin film technics to integrate with multilayer routing to the substrate surface or to embed the components in the substrate.

For LTCC (Low Temperature Co-fired Ceramic) substrate or alumina ceramic substrate, screen printing can be used directly on the substrate to create graphics and to form a variety of components, such as resistors, capacitors, inductors, etc., by sintering.

For polymeric substrates such as PCBs, there are generally two methods of embedded technics.

The first method is to embed discrete components directly into the substrate, such as 0402 and 0201 resistors, capacitors and inductors. This method of embedding is affected by the size of the discrete components. Its advantage is a discrete device with high precision providing high value.

The second method is printing pulp material or pressing the dielectric film materials into the substrate to make embedded resistors or capacitors, a technique called planar embedding technology. This method of embedding has little effect on the thickness of the substrate, but there are some limitations. For example, the capacitor, limited by the dielectric material itself, can only provide low capacitance. The resistor is also influenced by the size and cannot have too much power. Both kinds of embedded technologies are shown in Figure 2.9.

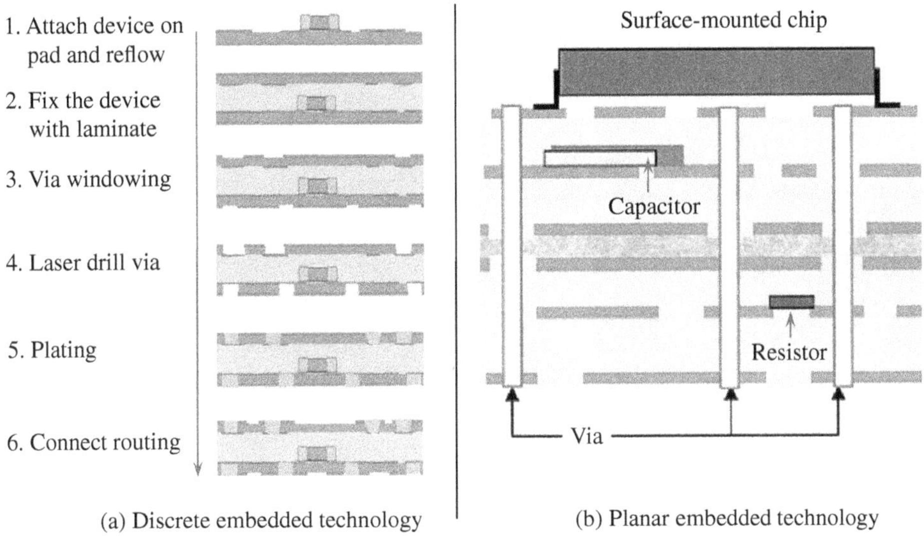

Figure 2.9 Two kinds of embedded technologies.

3) IPD technology

IPD(Integrated Passive Device)technology is also a major concern for the development of 3D integrated SiP package.

Although the chip is the core handling the entire electronic system, in fact a variety of passive components occupy most of the package area.

To increase the integration of passive components and further narrow package volume, buried-device technology developed vigorously, while, in addition, passive-component manufacturers took inspiration from the semiconductor industry, they changed their traditional discrete-device-supplier image, steered products toward high-integration modules, and integrated passive devices in wafer or ceramic substrate to meet demands for reduced volume, improved performance and reduced cost for portable computer and mobile communications equipment, etc.

Figure 2.10 shows IPD technology.

Figure 2.10 IPD technology.

(a) Picture of TSV (b) Diagrammatic of TSV

Figure 2.11 TSV technology.

4) TSV technology
TSV(Through Silicon Via)technology adopts a vertical break over between chip and chip, wafer and wafer; this is the latest technology to achieve interconnection between chips; see Figure 2.11.

Differing from traditional wire-bonding stack technology, TSV enables the highest stacking density, achieving minimum volume with three dimensions of chips, greatly improving chip speed, and lowering power consumption.

TSV is known as the fourth generation of packaging technology, the other three are WB (Wire Bonding), TAB (Tape Automated Bond) and FC (Flip Chip).

TSV has one obvious difference from conventional packaging technologies: TSV technics can be integrated into various stages of the IC-manufacturing process. TSV in the wafer fabrication stage is often called via-first, TSV in the packaging production stage is often referred to as via-last.

Now, some package manufacturers have already begun using via-last technology in the area of high-end Flash and DRAM, that is drilling a through-hole on the edge of the chip, then stacking the chips or wafers; the obvious advantage of this is that the existing process and design of integrated circuits is not changed.

5) PoP technology
PoP(Package on Package)technology has been widely used with the rapid development of portable electronic products such as smart phones, digital cameras and tablet computers such as the iPad.

With storage capacity increasing, and commercial divisions becoming more refined, DRAM, EMMC (Embedded Multi-Media Card) and other semiconductor storage manufacturers such as Micron, Samsung, SanDisk and Toshiba provide standard devices that directly use PoP package.

Handset makers can directly stack storage DRAM on sealed processors to form a complete PoP package.

Now, world-famous packaging makers such as ASE, Amkor and Jiangsu Changjiang Electronics Technology (JCET) have provided PoP products. Currently, the Apple A8 processor used in the iPhone 6 and the Qualcomm AP processor use PoP package.

PoP technology is shown in Figure 2.12.

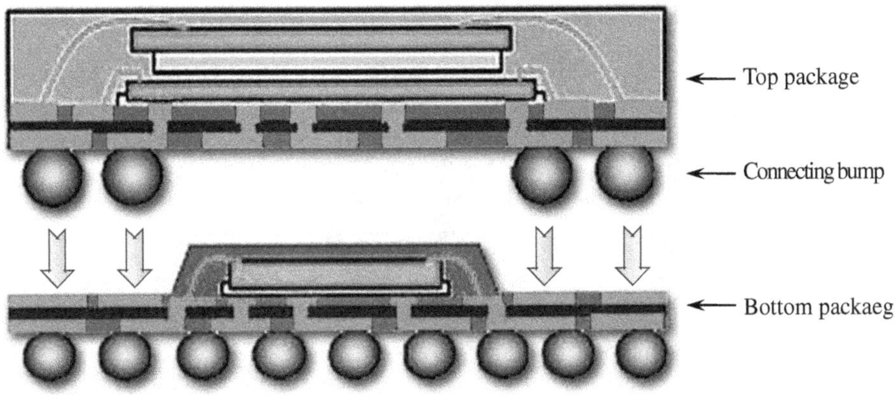

Figure 2.12 PoP technology.

2.4 The Development of the Package Market

Package is a bridge connecting the semiconductor chip and the electronic system. With the rapid development of the semiconductor industry and its quick spread to various other industries, electronic packaging has gradually become a bottleneck for the semiconductor industry, and has attracted more and more attention.

The package industry has greatly developed in the last 20 or 30 years, with considerable progress being made.

China Semiconductor Industry Association Packaging Branch statistics show that in China's three major integrated circuit domains – IC design, manufacturing, and packaging and testing – packaging and testing is still the leader compared with IC design and manufacturing, accounting for more than 50% of the whole industry.

In 2008, the Chinese semiconductor IC industry was worth 124.682 billion yuan, down 0.4% due to the impact of the financial crisis in 2007, which is the first negative growth in the IC industry in China for nearly 20 years. However, the packaging and testing industry grew by 1.4%, and some packaging and testing companies, including JCET, Nantong Fujitsu Microelectronics Limited, TianShui Hua Tian Microelectronics Limited, Guangdong Yuejing High-Tech Company Limited and Wuxi China Resources Micro-Assembly Technology Co. Ltd, grew somewhat.

As regards market share of packaging as a whole, with the development of information electronics and the consumer electronics industry, high-end package has grown rapidly. In 2009, high-end package such as BGA and CSP accounted for 55% of the whole package field, as shown in Figure 2.13.

High-end advanced packaging application areas include mobile phones, digital cameras, personal computers, tablet computers, high-speed switches, routers, game consoles, etc.

In the global electronics market, mobile communication products have become the most important area of SiP application. Mobile phone applications, whether baseband module, wireless module, RF module or camera module, are the main applications using SiP.

In consumer electronics, SiP has a wide range of applications, such as MP3 players, digital cameras, mobile TV, GPS and other products.

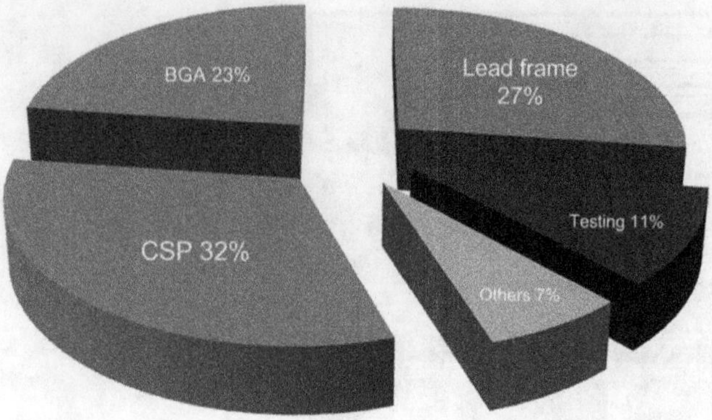

Figure 2.13 Package market share in 2009.

According to the Prismark market survey, in 2005 the market size of SiP was $4.3 billion, RF module accounted for the largest portion with 39% of market share, and the second-largest application was stacked Die-in-Package, which accounted for 29% of market share

By 2010, SiP overall market size exceeded $10 billion, and the proportion of SiP applications in different fields had also changed. Figure 2.14 shows the proportion of SiP applications in different fields in 2005 and 2010.

In the Chinese market, the demand for high-grade packaging products will be very strong over the next ten years. But on a technical level, local packaging and testing companies are at different levels. At present, local packaging and testing enterprises continued to focus on DIP, SOP, QFP and other traditional package forms; there is still a considerable gap between these and the advanced international level.

In recent years, with the increase in Chinese IC design and manufacturing development, a stimulating effect on the Chinese packaging and testing industry has begun to be felt.

Enterprises in China have started to make breakthroughs in BGA, CSP, MCM, SiP, etc., all advanced packaging technologies.

Jiangsu Changjiang Electronics, Nantong Fujitsu, Tian Shui Huatian, etc. have had remarkable achievements in the development and application of advanced package technologies such as BGA, CSP, MCM and SiP.

The gap between the overall Chinese packaging and testing industries and international technical capability is gradually closing.

2.5 Package Manufacturers

2.5.1 Traditional Package Manufacturers

In China, the domestic integrated circuit packaging and testing technology industry, like most domestic industry, focused on low- and mid-end technology; high-end technology was associated more with foreign companies.

In China today, there are more than 60 packaging and testing plants; they can be divided into three major categories.

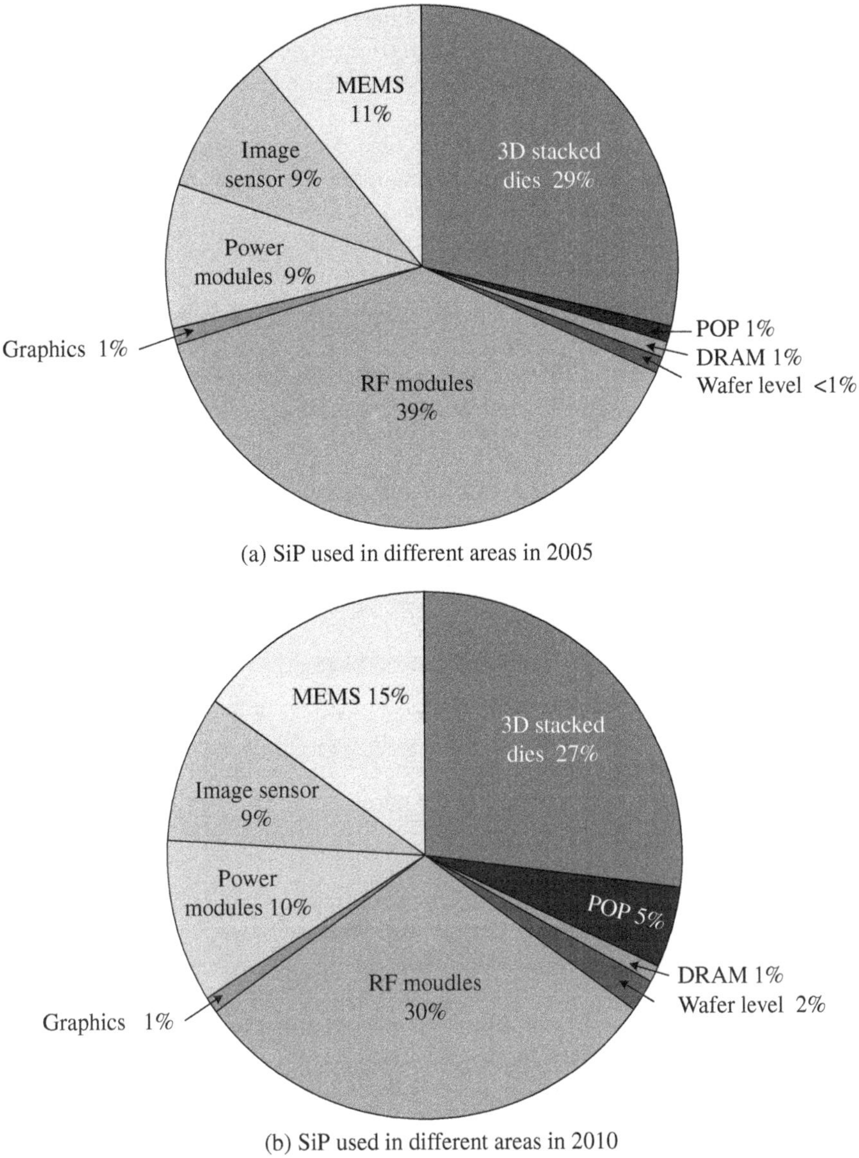

Figure 2.14 The proportion of SiP applications in different fields in 2005 and 2010.

First, international companies operate component-manufacturing packaging and testing plants.

Second, international packaging and testing plants undertake component manufacturing jointly with local operators.

The third category is Taiwan-funded packaging and testing plants.

The first class of packaging and testing plants are foreign companies; currently there are more than ten packaging and testing factories in China, such as Intel, AMD, Samsung, Motorola, Philips, National Semiconductor, Toshiba Semiconductor, General Semiconductor, Amkor, UTAC, Sanyo Semiconductor, ASAT, etc.

These companies are mainly from the United States, Japan and Korea; the purpose of building factories is so that the products of foreign companies in China will gain domestic tax benefits. They are currently concentrated around the Yangtze River Delta cities such as Shanghai and Suzhou.

The second class of packaging and testing plants are joint ventures with local manufacturers, designed to capture the advanced packaging market in China, before China fully opens to this market. This includes such companies as Shenzhen ST Microelectronics, Hitachi Semiconductor, Silicon Semiconductor, Infineon, Panasonic, Mitsubishi Sitong Electronics and Nantong Fujitsu Microelectronics, etc. The funds of the operators mainly come from Taiwan, Japan and Europe. Their locations are more dispersed, but they are concentrated in Jiangsu province and Shenzhen city.

The third category is Taiwan-funded packaging and testing plants. Taiwan-funded manufacturers have gradually joined the Chinese market within the last few years. Their operating mode is more that of the professional foundry.

Almost all Taiwanese packaging and testing companies have a factory in China, including ASE, Hong Sheng Technology, KaiHong Electronic, Jiemin Electronic, Silicon Technology, etc.

These plants are mainly concentrated near Shanghai and Suzhou. The operation scales are still smaller than the operations of the Taiwan parent companies.

2.5.2 New SiP Manufacturers in Different Areas

With SiP technology growing rapidly, and applications growing fast in China, changes and developments are occurring in the SiP chain. In addition to the fact that, as mentioned, chip suppliers have begun adjusting their operation bias to bare dies in place of traditional domestic package such as MCM, PCB manufacturers are turning to SiP substrate and packaging production operations.

1) Plastic packaging

   In the field of plastic packaging, Shennan Circuits and the Wuxi Jiangnan Computer Technology Institute are the leaders in SiP substrate research and development. They have also started SiP packaging operations.

   Shennan Circuits Co. Ltd (SCC), whose main business is high-end PCB research, development and mass production, has a high domestic market share in the area of high-end, multilayer, high-density PCB.

   On this basis, SCC was active in package substrate development and production. They have been an important package substrate supplier in China and all over the world. In addition, SCC has wire bonding and encapsulation for SiP package capability, and can meet the needs of various research organizations for SiP processing and production.

   SCC is constantly improving its SiP packaging and testing technics process; the stability of their products is also increasing; at the same time, SCC products have a certain price advantage.

   Jiangnan Institute of Computing Technology, referred to as Jiangnan Institute, enjoys a high reputation in the field of traditional PCB development and production. As a domestic supplier of high-quality and high-reliability PCB, the quality of its products has been recognized by numerous customers. In the area of HDI package substrate, Jiangnan Institute has unique technics which can support higher-density SiP substrate design and production, and their microvias can be made smaller and more reliable.

2) Ceramic and metal packaging
   In the field of ceramic and metal packaging, Beijing times civil chip corporation, China Electronics Technology Group Corporation and other research institutes are active in SiP substrate- or packaging-related business.

   Beijing times civil chip corporation has a domestic advanced ASIC design, packaging and test center. The corporation is mainly engaged in integrated circuit design, while providing integrated circuit packaging and testing services. It has the ability and experience to design and manufacture complex SiP advanced package, SiP based on multiple large-scale IC chips. It has complex processes such as multi-stage cavity on two substrate surfaces, and chip stacking, and its products are very advanced in the domestic SiP field.

   Hebei Sinopack Electronics Tech Co., Ltd. is a national high-density packaging experimental base for large-scale integrated circuits; it has strong ceramic package development and production capacity, and offers users a wide range of ceramic packages.

   East China Institute of Microelectronics is a key state institute in the area of electronic devices. Its products include DC/DC converters, AC/DC, PWM power circuitry, axis-angle converter circuits, signal processing circuits, amplifier circuits, multi-chip modules, metal housings and thick-film-paste resistive and capacitive components. The institute is the domestic leader in MCM design and manufacture, and its ceramic substrate, LTCC and metal shell are also leaders in the domestic arena. At present, the institute is active in SiP research, development and production processing.

   Nanjing Electronic Devices Institute is a key state institute in the area of electronic devices and applications, mainly in microelectronics, optoelectronics, vacuum electronics and micro electro-mechanical systems research, development and production. The institute has development and production capacity for microsystems packaging and ceramic package substrate and shell.

   Zhongwei High-tech Electronics Co., Ltd. has an IC design, processing, packaging and testing center. Its assembly and test center provide integrated circuit packaging and testing services for domestic customers.

   There are a number of other domestic companies or institutes involved in packaging technology but not all can be mentioned here for reasons of space.

2.6 Bare Chip Suppliers

With the rapid development of SiP technology, design requirements such as miniaturization, low-power consumption and increased reliability, more and more domestic PCB system users want to design and package SiP based on bare chip. When these users design SiP, the first consideration is how to get bare chips.

Typically, when system designers want to find a channel to get bare chips, the first option is the bare chip agents.

Currently, in the Chinese market, the traditional chip agent is well-established, and if users need packaged chips, they will find a variety of traditional chips through these agents.

However, bare chip will be harder to find; the chips that model designers need are often no longer in stock or vendors do not provide such services.

Another scenario is that manufacturers can supply the chip, but the minimum order quantity is so large, frequently thousands of tablets, that the cost runs to hundreds of thousands of dollars or more, and is unacceptable or unaffordable for SiP designers.

However, with the growing market demand for bare chips, traditional agents will continue to expand their bare chip business to meet this.

In cases where demand does not reach a certain number, orders for bare chips are usually made to IC manufacturers, causing some delay.

So, the SiP designer should consider ordering channels and order cycles in a timely manner at the early stages of SiP design, so as not to affect the progress of the project. The author has encountered such delays in many projects, prompting this reminder to designers to think ahead.

When market demand for bare chips reaches a certain level and when there is a continuing demand, agents will consider the adjusting inventories to meet the needs of users in a timely fashion, and the bare chip market will gradually mature.

The bare chip market has developed based on the fast-growing demands of SiP technology, which in turn will promote the rapid application and popularity of SiP technology.

At present, the domestic Chinese bare chip market is far from mature; solving the issue of access to bare chips is a development hurdle for SiP technology.

For the SiP designer's reference, some bare chip vendor websites are listed below:

Beijing Suny Technology Limited, http://www.sunytec.com

Shenzhen Cheer Electronics Co. Ltd, http://www.uesemi.com

Universal Enterprise, http://www.ue.com.hk

State Microelectronics, http://www.ssmec.com

Integrated Silicon Solution Inc., http://www.issi.com

Beijing MXTronics Corporation, http://www.mxtronics.com

These are just some of the bare chip suppliers the author has contacted; it can be seen that the number is still relatively small. However, as we have predicted, as the SiP market matures, there will be more and more bare chip suppliers, and the list will continue to grow.

We believe that the road will not be smooth, but that the future will be bright; SiP technologies will become a hotspot of new technological development.

3

The SiP Production Process

3.1 BGA: The Mainstream SiP Package Form

Most SiP uses BGA package, BGA being an acronym for Ball Grid Array.

BGA package is a milestone in high-pin-count package; this encapsulation technology was developed in the 1960s in IBM mainframe computer assembly. At that time IBM developed the so-called C4 (Controlled Collapse Chip Connection) technology; further development could take advantage of the surface tension of molten bump to support chip weight and control bump height.

In 1993, Motorola was the first to apply BGA in mobile phones. The same year, Compaq also applied BGA to workstations and PCs. Then Intel started to use BGA on computer CPUs (Pentium II, Pentium III, Pentium IV, etc.), as well as the chipset (i850).

At present, BGA is the most popular IC packaging technology. Depending on the chip and substrate connection method, BGA can be divided into Wire Bonding-BGA and Flip Chip-BGA, as shown in Figure 3.1. The former connects bare chip and substrate using bonding, and generally applies to BGA package below 600 pins; the latter connects bare chip and substrate using bumps to support BGA package with over 2000 pins.

BGA package has the following advantages:

1) High-density I/O. With the rapid development of integrated circuits, the number of input/output terminals and pins increased rapidly, and a major advantage of BGA is extremely high I/O density. On the one hand, BGA pins in the array configuration can have pin pitch of 1.27 mm, 1.0 mm, 0.8 mm, 0.65 mm, 0.5 mm, etc. The processing is relatively straightforward, it is easy to weld, gives a higher rate of finished products, and is becoming the best choice of high-pin-count package; on the other hand, the BGA package size is reduced. FC-BGA package I/O leads can be arranged in an array on the chip's surface, providing higher density I/O layout. Compared with traditional package, the area can be reduced by 30~60%.
2) Excellent electrical properties. The pin of BGA package is just a metal ball; compared with QFP package, the parasitic inductance of BGA is much smaller, and Flip Chip-BGA can provide the shortest possible external connection distance. This package not only provides excellent electrical performance, but can also reduce loss and inductance of interconnection, and electromagnetic interference.
3) Higher reliability. In conventional QFP or PLCC package, the lead foot is evenly distributed round the package body; when I/O demand increases, the spacing of pins

SiP System-in-Package Design and Simulation: Mentor EE Flow Advanced Design Guide,
First Edition. Suny Li (Li Yang).

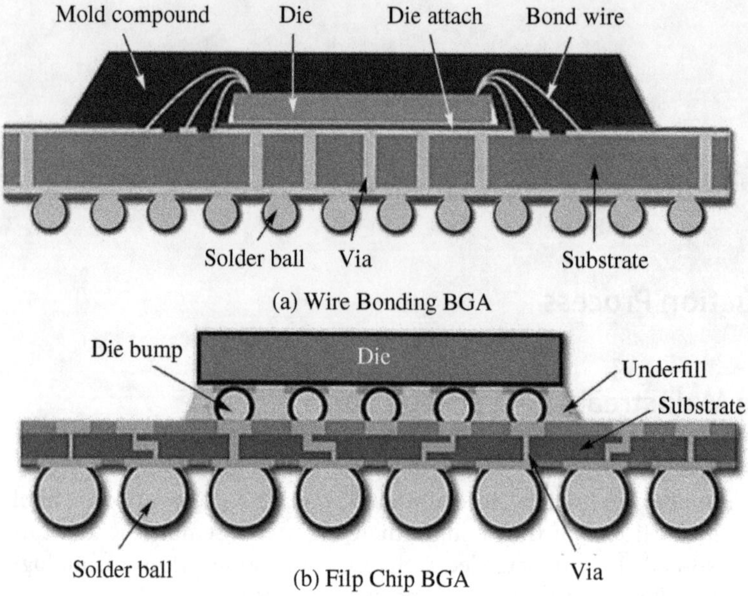

Figure 3.1 Wire Bonding-BGA and Flip Chip-BGA.

must be made smaller. When the spacing is less than 0.4 mm, it is difficult to meet the SMT equipment's accuracy requirements. In addition, lead feet deform easily, resulting in increased mount failure rate. The solder ball for BGA package is distributed in array form in the substrate bottom; it can be arranged for greater I/O; its standard ball pitch is 1.5 mm, 1.27 mm, 1.0 mm. There is also fine-pitch BGA solder ball, with pitch such as 0.8 mm, 0.65 mm, 0.5 mm, 0.4 mm, compatible with existing process equipment in SMT; the mount failure rate is less than 10 ppm.

4) Excellent thermal performance. Metal BGA solder ball grid array provides a sufficient heat dissipation channel; in addition, the substrate can improve cooling efficiency through the metal layer, further enhancing the chip-cooling capacity, and significantly improving the stability of chip running at high speed.
5) Welding auto-alignment. When BGA solder ball melts, the influence of surface tension automatically shortens the distance between the BGA substrate pad and the PCB pad, resulting in a self-alignment effect.

BGA packaging is of various types, usually with a square- or rectangular-shape structure. Depending on the arrangement of the solder ball, it can be divided into peripheral, cross and full array BGA. Depending on the different substrates, it can be divided into three main categories: PBGA (plastic ball grid array), CBGA (ceramic ball grid array) and TBGA (Tape solder ball grid array).

PBGA package uses BT resin/glass laminate as a substrate, plastic (epoxy molding compound) as sealing material, and eutectic solder 63Sn37Pb or quasi-eutectic solder 62Sn36Pb2Ag for solder ball. Some PBGA packages uses a cavity structure; the cavity can be divided in two ways: cavity up and cavity down. This cavity is designed to enhance the thermal performance and is called thermally enhanced BGA, EBGA for short.

The CTE (Coefficient of Thermal Expansion) of BT resin/glass laminate is 14 ppm/°C, the CTE of PCB is approximately 17 ppm/°C; these are close, so heat matching is good. However, PBGA is sensitive to moisture, and does not comply with the air-tightness and high-reliability requirements of device package; therefore, most PBGA applications are used in the business area.

CBGA uses multilayer ceramic substrate sealed by a metal cover. Usually, metal cover-sealed substrates use solder to protect the chip, wire and pad. Its advantages are good air-tightness, good heat dissipation, high moisture-resistance and high reliability in the long-term. CBGA is often used in military and aerospace products.

Due to the difference between the CTE of ceramic substrate and the CTE of PCB, the heat matching is lower, and fatigue of solder ball is the main failure mode. High cost is a fatal flaw of CBGA.

CCGA (Ceramic Column Grid Array) is an improvement on CBGA, using solder columns instead of solder ball to alleviate shear stress caused by the thermal mismatch between ceramic substrate and PCB. This reduces solder joint fatigue, but is difficult to make and complex processes are involved. CBGA and CCGA are shown in Figure 3.2.

TBGA uses a flexible tape package body; the thermal characteristic is good matching with the PCB. This is the most economical BGA package. TBGA has a superior heat dissipation structure to PBGA, but is also sensitive to moisture.

3.2 The SiP Package Production Process

Because BGA is the technology most commonly used in SiP package, the SiP package production process is represented using BGA package, as described below.

Figure 3.3 shows the Wire Bonding-BGA (WB-BGA) and Flip Chip-BGA (FC-BGA) production processes.

1) The Wire Bonding-BGA process includes: wafer thinning→wafer dicing→die attach→wire bond→mold→solder ball attach→solder ball reflow→marking→singulation→final checking→testing and packaging.

   Before packaging, wafer thinning is done, not only making wafer cutting easier, but also reducing the thickness of the final package.

   Die attach uses silver-filled epoxy adhesive to attach the IC chip to the substrate.

   Wire bonding uses pure gold wire to connect chip pins to bond pads on the

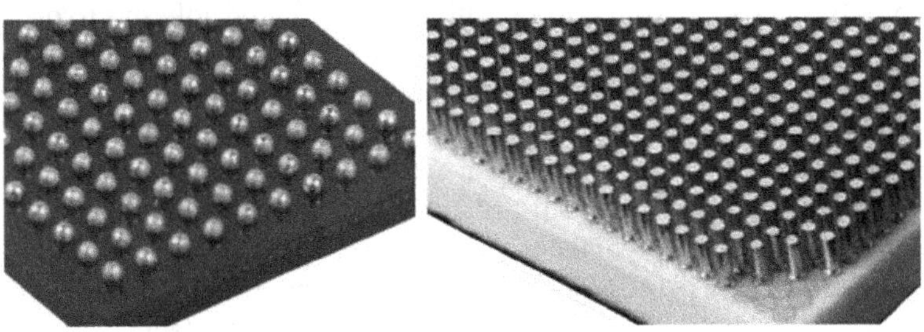

(a) CBGA (b) CCGA

Figure 3.2 CBGA and CCGA.

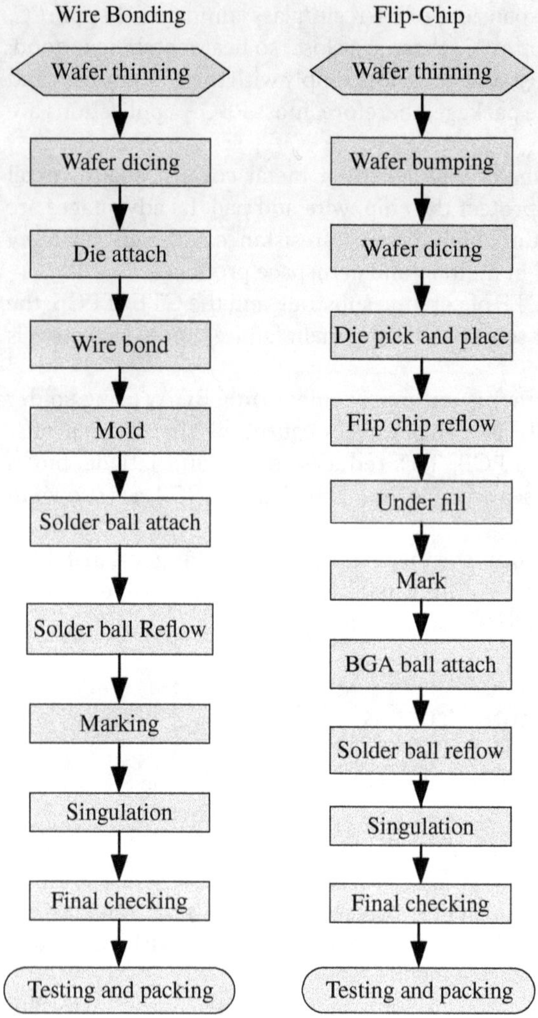

Figure 3.3 WB-BGA and FC-BGA manufacture process.

substrate, then molding encapsulation is done with plastic or liquid adhesive to protect the chips, bonding wires and pads.

BGA substrate bumping involves placing or screening the solder balls on the substrate pads, then reflowing in a soldering reflow furnace, followed by using cleaning agents to clean the substrate to remove residual solder particles and fibers in the package body. Remaining steps include surface marking, cutting, separating, final inspection, testing, packing and warehousing.

The packaging manufacture process for WB-BGA is shown in Figure 3.4.

2) Flip Chip-BGA processes include: wafer thinning→wafer bumping→wafer dicing→die pick and place→flip chip reflow→under fill→mark→BGA ball attach→solder ball reflow→singulation→final checking→test and packing.

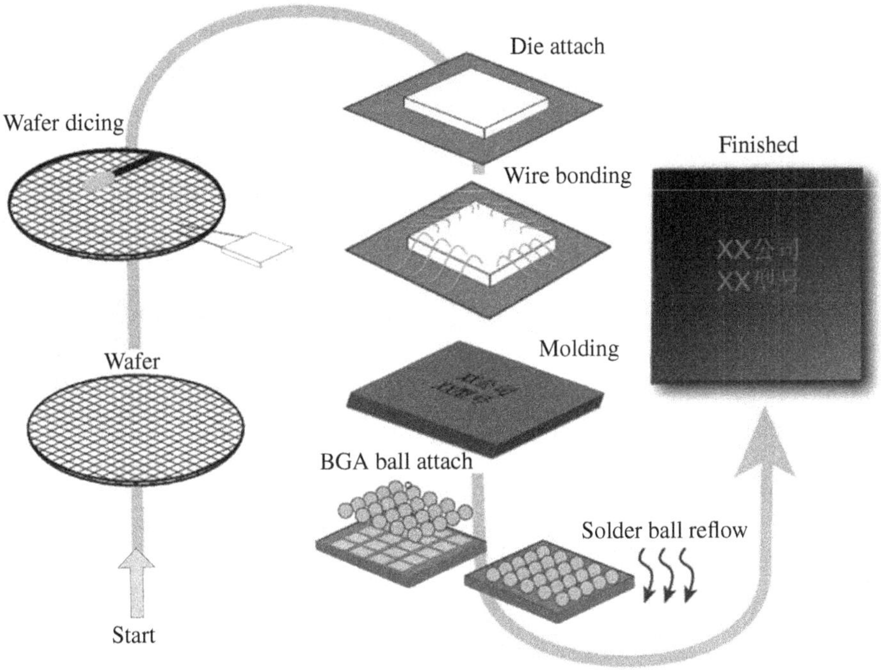

Figure 3.4 WB-BGA packaging manufacture process sketch map.

In this case, wafer thinning is also done before packaging, as is fabricated wafer bumping on silicon. After wafer dicing, flip chip dies are welded to the substrate by reflow. Then chip underfilling is used to glue the chips and to overcome the stresses caused by the CTE inconsistency between chip and substrate. Following this are BGA substrate ball attach, solder ball reflow, cleaning, marking, checking and packing, etc.

The process is generally consistent with Wire Bonding-BGA. Flip Chip-BGA packaging manufacturing processes are shown in Figure 3.5.

Comparing the two processes, it can be seen that the basic process is the same; the main difference lies in the flip chip wafer bump before dicing to discrete chip. The flip chip process requires reflow twice, including wafer reflow and BGA substrate reflow. The wire-bonding process only requires reflow once.

For SiP system-level packaging, because there may be multiple chips, we may encounter a mixed process, namely both WB chip and FC chip in one SiP package.

Although in practice this is rare, it may nevertheless be experienced in design, where some selected bare chips support WB technology and others support FC technology; the need then arises for a hybrid manufacturing technology – SiP.

Figure 3.6 displays the hybrid-technology SiP manufacture process.

3.3 Three Key Elements of SiP

A BGA SiP module can be physically divided into three parts, referred to as the three key elements of SiP package.

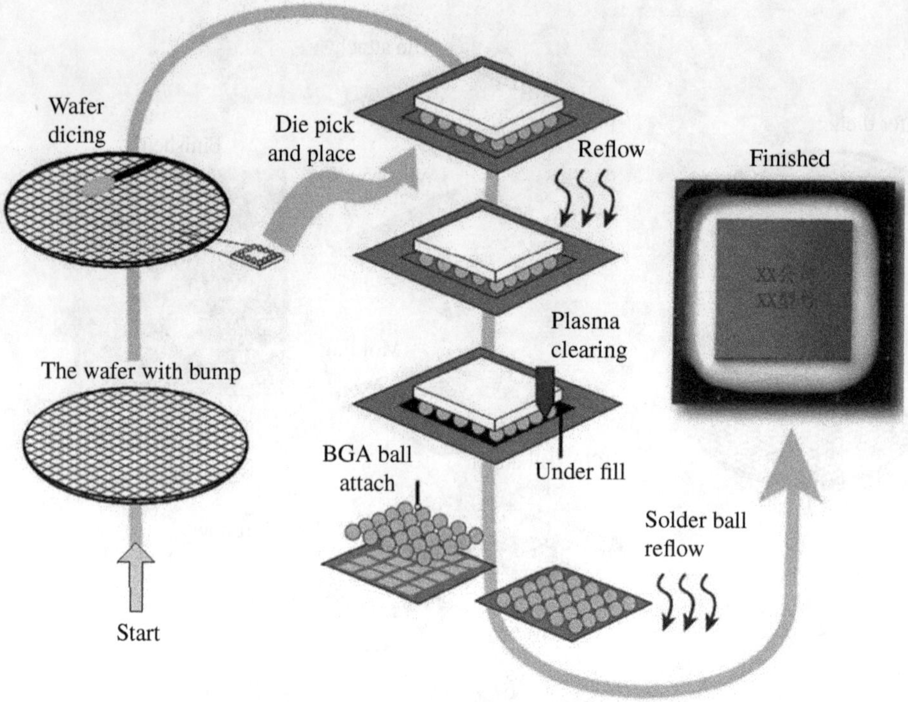

Figure 3.5 Flip Chip-BGA packaging manufacture process sketch map.

1) Bare chips and passive components
   Bare chips come from a wafer factory; they are the core components of SiP package, and, depending on the design needs, two or more bare chips are used in SiP. Bare chip arrangement can take a side-by-side or die-stacking form.
   SiP often necessitates adding a filter circuit to the package, which requires resistors, capacitors, inductors or other passive devices to be included in the design; these passive devices are attached to the substrate, or embedded in the substrate via different processes.
2) Substrate
   Commercial SiP products tend to use organic substrates for cost reasons. This is based on high-density multilayer routing and micro-via technology, which is characterized by lower interconnect resistance and low dielectric constant.
   Limitations of organic substrates are:
   a) High CTE difference between chips and substrate produces thermal mismatch;
   b) In reliability testing, compared with the same type of ceramic package, these are less reliable, mainly due to the adsorption of water vapor.

Currently, a substrate factory will provide appropriate products for the user to select from. Preference depends on the material's CTE, dielectric constant, dielectric loss, resistivity and conductivity. CTE mismatch between substrate and chip or between substrate and PCB is a major cause of product failure.

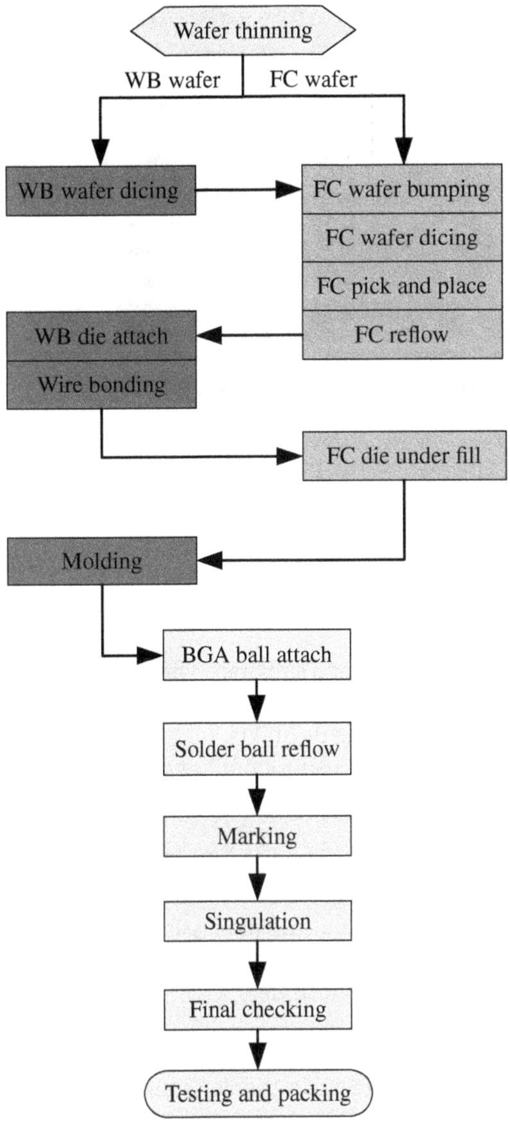

Figure 3.6 Hybrid-technology SiP manufacture process.

Usually dielectric constant and dielectric loss of package substrate have a direct relationship with signal integrity, especially for signal frequency greater than 1GHz.

The factors mentioned above should be taken into account when we select the substrate for SiP package.

For Wire Bonding-BGA and Flip Chip-BGA substrate, because there is a high number of pins and the density is so high, fanout pressure is greater. These technologies typically use HDI (High Density Interconnection) substrate. HDI substrate uses a build-up+laminate structure, usually N+M+N. The M layer stands for laminate layer with mechanical drilling technology and relatively large line width/spacing and

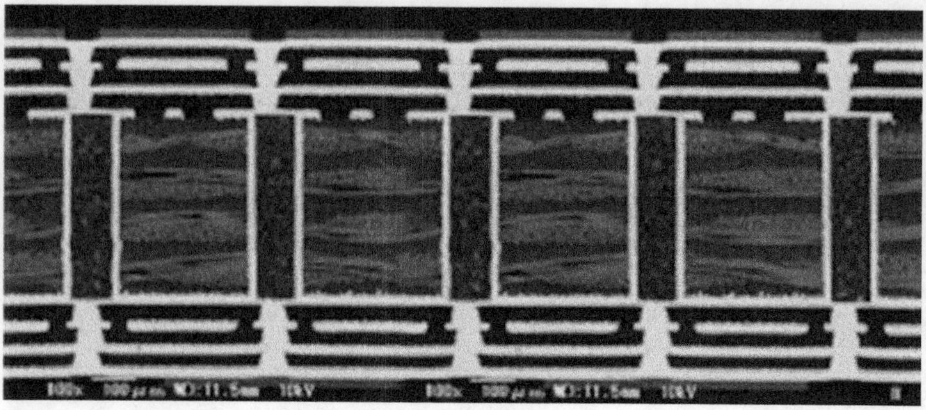

Figure 3.7 HDI substrate cross section.

via diameter; the N layer stands for build-up layer, using laser buried/blind-hole technology, providing higher accuracy of routing and via. However, it is relatively expensive. Figure 3.7 shows a substrate cross section of a typical 3+2+3 HDI structure.

3) The connection of bare chips and substrate
Two methods are usually used to connect bare chips and substrate: wire bonding and flip chip. The wire-bonding method has a long history and a strong technological base, and provides flexible processing. Its machining flexibility, high reliability and material/substrate cost are its major advantages. Its shortcomings are lower efficiency, and the fact that the precision of welding equipment has been developed to the upper limit.

The application of flip chip has increased sharply; it has the advantages of strong joints, short signal-transmission path, wide power/ground net distribution, high-density I/O, small size package, high reliability, etc. But the processing cost is high. Under fill is used in the vast majority of flip chip products; its role is to alleviate shear stress caused by CTE difference between chips and substrate.

Connections also exist between bare chips.

Bare chips placed side by side are usually connected by wire bonding. Stacked bare dies often use wire bonding or TSV (Through Silicon Via) for connection.

4
New Package Technologies

At the end of the 20th century, the upgrading of computers drove the microelectronics industries to develop rapidly, with constantly updated requirements for CPU processors, package pins, performance, frequencies, etc. Power was also gradually increasing. Ball Grid Array (BGA) package spread quickly, and FC-BGA technology became the basic choice for CPU packages.

Then, when the development of microelectronics packaging technology was in a relatively flat stage, the packaging industry quickly moved to China and Taiwan, and other Asian countries and regions. During this period, the industry strove for the development of packaging technology, and various new packaging technologies such as SiP, SoP, PoP, PiP and CSP emerged.

In recent years, consumer electronics markets, especially for smart phones, digital cameras and tablet PCs, developed rapidly. This once again caused high-speed development of the microelectronics industry. Among recent developments, the TSV, IPD and PoP technologies are particularly impressive; they have become the mainstream package for memory, smart phones, processors and other applications.

4.1 TSV (Through Silicon Via) Technology

4.1.1 TSV Introduction

TSV technology is the latest technology to fabricate vertical vias between chip and chip, wafer and wafer, to achieve interconnection between chips, as shown in Figure 4.1.

Different from wire bonding chip-stacking technologies, TSV allows higher-density three-dimensional chip stacking, minimum dimensions, greatly improved chip speed and lower power consumption. TSV is known as the fourth generation of packaging technology after wire bonding, TAB and flip chip.

Continuing past the growth trend of Moore's law, chip technology has come to the so-called "beyond Moore's law" age of 3D integration. From semiconductor fabrication plants (Fab) to packaging and testing plants, and to substrate and assembly operators, the whole supply chain is involved.

TSV technology will accelerate the merging of semiconductor Fab, as well as fabless (design without fabrication) model trends. According to statistics from the market research firm Yole Development, by 2015, TSV wafer shipments will reach millions and may have an impact on 25% of the storage business.

SiP System-in-Package Design and Simulation: Mentor EE Flow Advanced Design Guide,
First Edition. Suny Li (Li Yang).

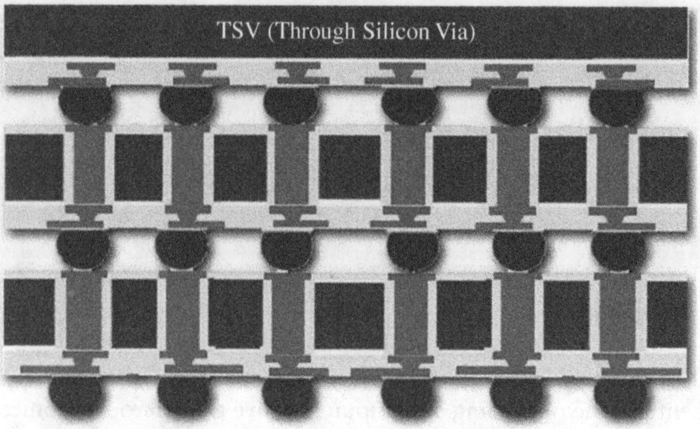

Figure 4.1 TSV sketch map.

By 2015, except in the area of memory, TSV wafers continued to be widely used, constituting more than 6% of the semiconductor industry.

The benefits of TSV integrated technology include smaller dimensions, increased packaging density, ability to meet bandwidth requirements, improved RF performance and power consumption, and reduced costs. Reliability is also a big advantage of TSV, because it can be used to replace wire bonding, TAB and flip chip, using 3D stacked wafers to replace the injection module, with several levels of vertical integration to produce more reliable systems.

In addition, TSV technology can also promote new system applications in automobiles, communications, electric power, etc., and in space-constrained environments.

4.1.2 TSV Technical Characteristics

TSV has an obvious difference from conventional packaging technologies; its production can be integrated into two stages of the manufacturing process.

TSV created before silicon wafer manufacturing completion is usually referred to as via-first. At this point, TSV production can be made prior to front metal interconnection at the Fab plant, to achieve core-to-core connection. This scheme is currently used in the field of high-performance device research such as microprocessors, mainly as an alternative to SoC.

Via-first TSV can also be done in the wafer plant after CMOS production, before the completion of back-end package.

Creating TSV at the packaging production stage is usually referred to as via-last; the obvious advantage of this scheme is that there is no need to change the existing IC production and design processes.

Now, some manufacturers have already begun using via-last technology in the area of high-end Flash and DRAM, that is drilling through-hole vias on the edge of the chips, then stacking the chips or wafers.

Table 4.1 shows a comparison of via-first and via-last technology.

Table 4.1 compares the two basic TSV integration schemes.

Table 4.1 Via-first and via-last technology comparison.

	Via-first	Via-last
Design stage	Before CMOS or BEOL	After BEOL
Intervention time	IC design stage	After wafer manufacture finish
Processing place	IDM Fab	IC packaging and testing plant
Via size	Via size 5~20 μm	Via size 20~50 μm
Critical Dimension	Critical Dimension(CD) control stringent	CD relatively loose
Aspect ratio	Aspect ratio (AR) 3:1 to 10:1	Aspect ratio (AR) 3:1 to 5:1

Note: BEOL (Back End of the Line silicon), IDM (Integrated Design and Manufacture).

As can be seen, via-first technology requires design in the IC design phase, and the CD control requirements are more stringent than for the via-last process.

At this time, the greatest need in TSV process development is for the industry's best technology integration program to reach a consensus as soon as possible.

Because the industry has failed to reach a consensus on TSV technique, it can be said that at present TSV is still in the early stages of development.

While there are many types of TSV process integration, they all face a common problem, that is, in most cases TSV requires drilling through layers of different materials, including silicon, the insulating or conductive layers of the IC.

The etching process is the key to thinned manipulation of wafers, and wafer bonding, measuring and testing are also hot technologies, as shown in Table 4.2.

The best TSV technology must be able to meet the requirements of contour control (including control of profile tilt, taper, sidewall roughness, undercut, etc.) while meeting the need for flexibility in processing capability; be able to cope with multi-level hole size and various layers of materials; have a high yield capability, able to handle 300 mm wafer, with technological repetition, practicality and reliability; and finally, must also meet the IC market's demand for the best price.

A TSV micrograph is shown in Figure 4.2, with 10 μm diameter CD through-silicon vias (via-first), the etching layer including 0.5 μm silicon oxide layer and 60 μm silicon material. The photo shows good uniformity of etching.

Table 4.2 TSV hot technologies.

Shape control	Integration flexibility	Mass production
Scalloping	Big via (20~60 μm) Small via (1~20 μm)	300 mm ability
Profile angle	Stacked layers	Repeatability
Profile tilt	Silicon, oxide layer, metal layer	Long enough life cycle
Taper	Fast switching of technology	Chamber matching
Undercut	Plasma bombardment damage	System reliability
Uniformity between different wafers	Steady State Process	The etching rate

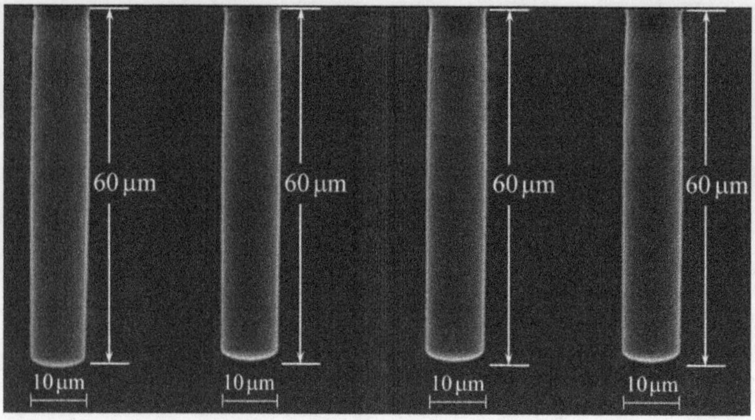

Figure 4.2 TSV micrograph.

As regards choice of etching pattern, the industry is still at the stage of comparing SSP (Steady State Processes) and RAP (Rapid Alternating Processes) technology.

The selectivity of RAP etching is very high, with etching aspect ratio higher than through-hole, and the speed is fast, but surface roughness is a challenge.

SSP etching technology is more close to conventional etching; it is high speed and the production side wall is smooth, but the control of selectivity and undercut is a difficulty.

The real advantage to the user is that the selection and integration of the machine can be done according to the requirements of applications, including the switch realization of SSP and RAP in two modes, and the overall control of etch rate, selectivity, smooth sidewall and aspect ratio.

Of course, this requires the accumulation of large amounts of process knowledge, and knowledge of the manufacturing of components.

The three driving market forces of 3D TSV are performance, small size and low cost; reducing the size is a primary short-term goal of most manufacturers at present.

A number of companies are at the stage of pilot run or already have mature technology; they hope to improve the integrated density of devices using TSV in the short term.

4.1.3 TSV Application and Prospects

From the point of view of mid-term development, up to 2018, TSV manufacture has the capability to put RF, logic, memory, sensor and other different device modules together using TSV technology in 3D IC mode, sidestepping the need to design multiple functional modules in one IC. As regards overall performance, Moore's law continues to apply.

This trend is present in RF design and has been seen as a trend in other applications; further applications will include DRAM and Flash stacking with TSV technology. Many foundries are therefore actively developing this technology, to pave the way for the memory market.

The future of the high-end foundry industry, both wafer manufacturing and packaging and testing production, will require a matching TSV scheme offering for the user.

RAM memory is the first market to use 3D TSV stacking technology; in the future there will be more and more Flash memory integrated in multi-chip package, PoP, SiP, mobile phone card and solid-state hard disk.

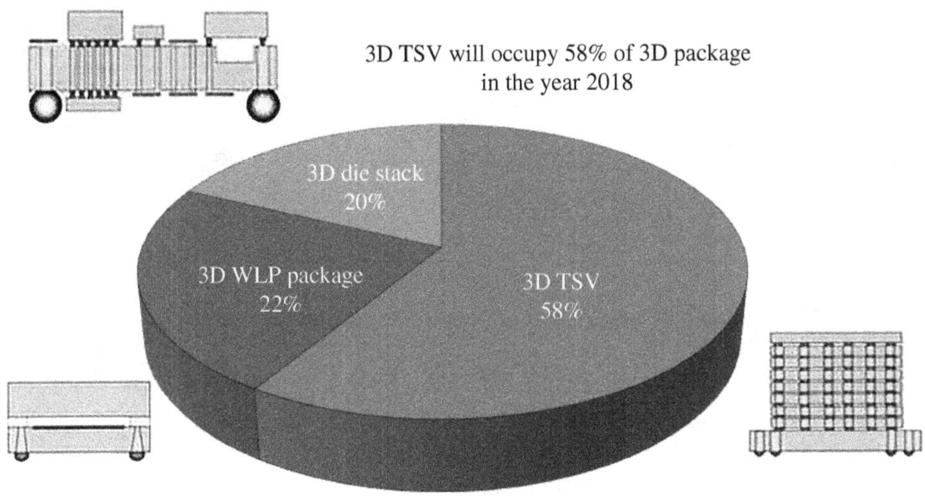

Figure 4.3 By the year 2018, 58% of 3D package will use TSV mode.

A 3D wafer-level packaging platform has been put into use; the CMOS image sensor is available for production with through-hole on the back of the wafer. The scope of this application will be extended to the power amplifier module. Moreover, 3D TSV stacking platforms for memory and logic chips are also in development.

Via-first structure transition is another trend; at present the through-hole diameter of TSV can be as small as 1~5 μm, and the interconnection number can reach 500~2000 per chip.

Figure 4.3 shows TSV as a proportion of total 3D package by 2018.

4.2 Integrated Passive Device (IPD) Technology

4.2.1 IPD Introduction

With the development of semiconductor manufacturing capability, from submicron to nanometer stage, the integration of active electronic components has greatly increased, while the integration of passive components corresponding to the active components has also increased rapidly.

With the mobile phone system becoming more and more powerful, the use of passive components is also increasing rapidly, with this trend likely to continue.

The development trends in relation to electronic products are: light, thin, short and small. With semiconductor manufacturing capacity being enhanced, active chip circuit increases accordingly, while the number of corresponding passive components also increases significantly. There is a need for more space to place these passive components, which will inevitably increase the overall package size.

IPD is an abbreviation of Integrated Passive Device, which is the use of wafer foundry technology to generate different patterns with photo etching in silicon substrate, glass substrate or ceramic substrate to form different devices, so as to realize high-density integrated passive devices such as resistors, capacitors, inductors, baluns, filters, etc.

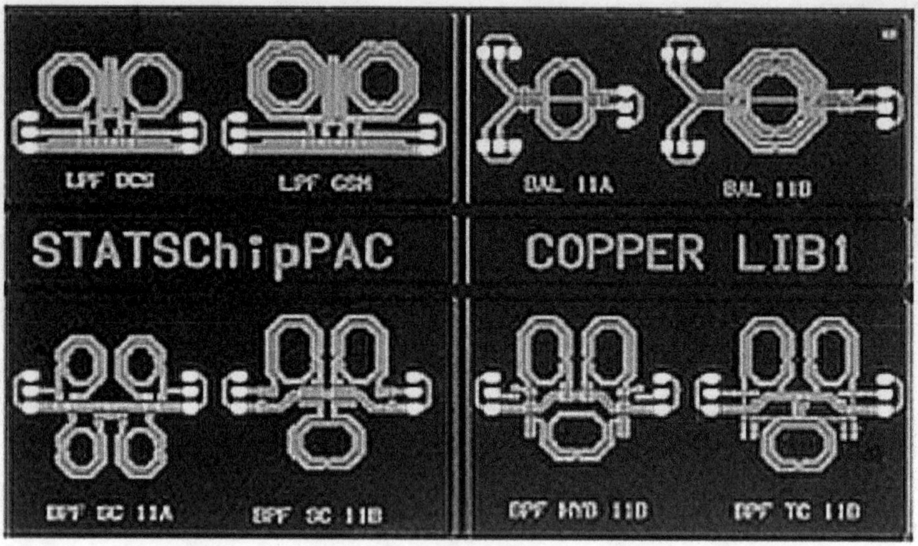

Figure 4.4 IPD sample.

At present, there are a number of companies providing IPD solutions, such as IPDIA, ST Microelectronics, Anson Semiconductor, OnChip Devices, Inc., etc.; some large packaging companies such as ASE, Amkor and STATS ChipPAC also provide IPD chips. Figure 4.4 shows an IPD sample from STATS ChipPAC.

IPD is designed to replace the traditional passive elements; at present it is widely used in high-brightness LED silicon integrated circuits, RF devices, and digital and hybrid circuits.

At present, IPD technology has become the bridge between the front-end and back-end semiconductor processes; it will also become an important part of wafer package and TSV applications in the future.

At the same time, IPD chip itself has excellent electrical performance. In advanced packaging integration, IPD can be used in a variety of stacked packaging with active chip to achieve the shortest possible interconnection and a great reduction in size, so as to improve the electric performance of the whole system.

Figure 4.5 shows four kinds of integrated IPD package.

4.2.2 The Advantages of IPD

The use of IPD technology can bring great benefits to an electronic system:

1) Saves PCB space.
2) Better electric properties.
3) Lower cost.
4) Intellectual property (IP) protection. If IPD is used, it is difficult for competitors to analyze a product's circuit.
5) Improved product supply chain reliability.

Figure 4.6 shows a comparison of one circuit before and after using IPD. Fifty-eight passive elements were replaced by three IPD chips; not only is material supply simpler,

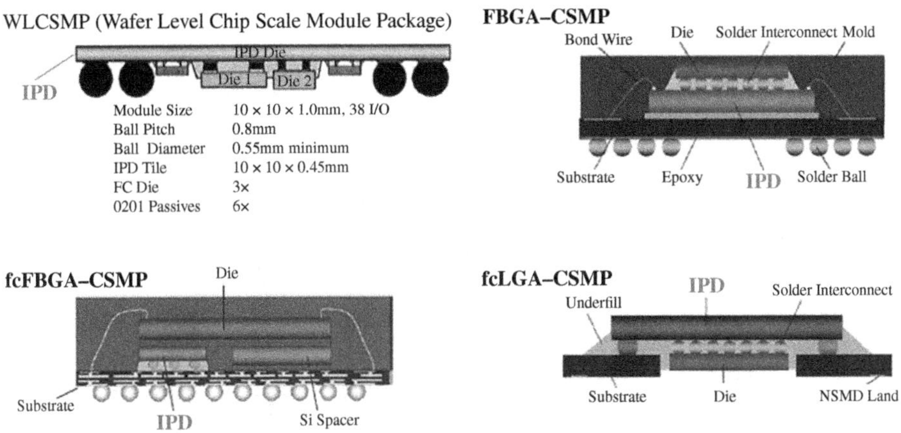

Figure 4.5 Four kinds of integrated IPD package.

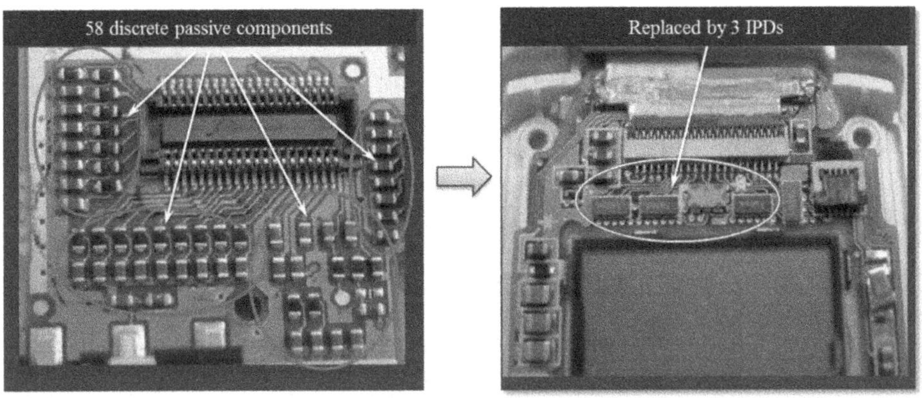

Figure 4.6 Comparison of circuit before and after using IPD.

the number of welding points is also reduced. When circuit change is simpler, reliability will improve rapidly.

Because of the great advantages of IPD technology, it has been widely used in digital cameras, RF modules and LEDs.

At the beginning of 2008, global sales of IPD were more than \$700 million, and are expected to reach \$2100 million in 2018.

Figure 4.7 shows a Yole Development graph of June 2009 illustrating IPD market forecast from 2008 to 2015.

4.3 Package on Package (PoP) Technology

4.3.1 The Limitations of 3D SiP

Three-dimensional dies stacking technology allows for integration of more chips, and has been popular for a long time. Before PoP appeared, chip-stacking SiP techniques

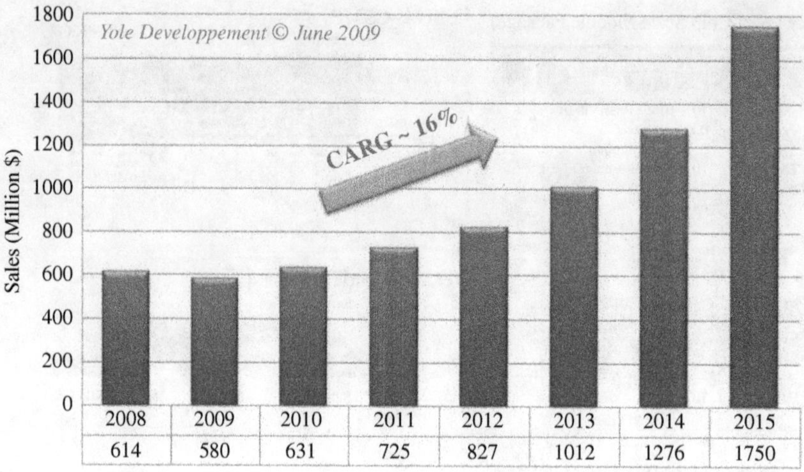

Figure 4.7 IPD market sales data.

were used in integrated memory and logic devices, and were the most popular solutions, with no technology or machining precision problems. However, stacking multiple bare chips makes test and QC (Quality Control) complex, and also reduces packaging flexibility, thereby reducing the flexibility of the product as regards system structure and procurement, and increasing the cost of product.

The key to market success for 3D SiP is not only the size and performance advantage of stacked chips, but also advantages in cost, design and time-to-market constraints.

Using integrated memory, logic-device suppliers can obtain higher profits, but are subject to more risks in relation to design, procurement, inventory and memory product warranty, risks that are costlier than the marginal revenue increase due to memory integration.

Memory suppliers had to develop a new silicon-test method for every new device and special investment, and were also under pressure from SiP pin-arrangement optimization, commodity price and rate of finished products.

The market needs a new and acceptable integrated way of re-dividing the supply chain. This is PoP technology.

4.3.2 The Application of PoP

At present, the multimedia function requirements, large capacity, short life cycle, reduced size and smoother supply chain that make the intelligent mobile phone are becoming the "killer application" pushing PoP development.

PoP involves ultrathin small feet pitch BGA package vertical stacking, with assembly to the PCB surface. These BGA packages realize logic and memory device interconnection with a special design, a flexible yet standard structure.

Figure 4.8 shows a typical PoP package structure.

The earliest PoP dates from the 2003 Electronic Components and Technology Conference (ECTC), when Nokia and Amkor co-published a paper reporting the first joint development of the PoP product.

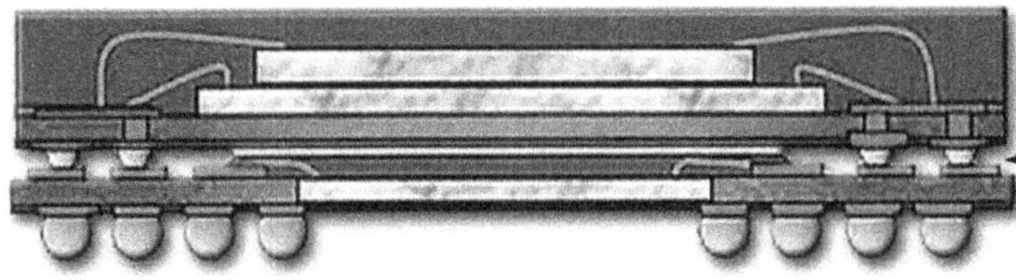

Figure 4.8 The typical structure of PoP package.

In 2005 the market launched ultrathin small feet pitch BGA to support package stacking, which constitutes the basic platform of PoP.

Two years later, all the major suppliers of baseband applications, image-processing and related mobile storage devices had launched or claimed to launch products based on PoP.

The value of the PoP solution lies in the simplicity of the PoP structure.

In PoP, the bottom device, designed according to the JEDEC standard, is usually provided by logic-device suppliers, and has passed testing to ensure the same grade of performance and reliability as for standard components (following the supplier's standard design, packaging and testing process). The top device is usually provided by memory suppliers, and also has the same grade and product flow as standard components.

Original Equipment Manufacturer (OEM) contractors purchase these two kinds of components and stack them up using a modified surface reflow process.

Through the stacking of logic and the memory components, OEM manufacturers not only obtain the benefits of size reduction and integration, but also are unaffected by the high cost limit of stacked SiP products.

The first use of PoP in the mobile phone involved the OEM manufacturer and the logic and memory suppliers acting together, thus solving the issue of the complex technology of integrated high-density logic plus memory.

Since the final product package-stacking technique has successfully solved the problem of high-density integration, PoP provides the best cost benefit, and eliminates the yield rate decline which is inherent in the bare-chip-stack assembly and testing process, and reduces test complexity.

As OEM vendors now have the package-stacking process, while PoP technology allows their logic- and memory-component suppliers to use their existing assembly and testing processes, OEM manufacturers can optimize their cost.

PoP package overcomes the main disadvantages of bare-chip stacking, such as supply chain problems, yield loss, chip and other problems.

PoP packaging requirements are constantly growing, because this technology has the advantage of lower costs, smaller package size, multiple memory-matching logic, assembly flexibility, etc.

The following are the main advantages of PoP package:

1) Package stacking extends functions, which helps to realize more functions in a mobile device.
2) The shorter interconnection achieves a faster data-transfer rate and signal speed between chips.
3) The smaller size reduces the area of the printed circuit board.
4) Less substrate space is taken up, providing design flexibility.

5) Can reduce cost in the manufacturing process.
6) Shortens the time to market.

The achievement of higher performance in a small area for efficient memory architecture, such as storage capacity, and the need for higher scalability and larger data-signal-processing capacity were the main driving forces of PoP applications.

4.3.3 The Emphasis in PoP Design

Before the pins are arranged according to JEDEC standards, collaborative design is a key process for PoP development, with emphasis on memory interface and mechanical optimization.

Now that the industry standard has been established for PoP, the top-level memory device suppliers and bottom-level logic chip device suppliers can follow the standard, with collaborative optimization design focused on the cost and performance of PoP.

Collaborative design is very effective for the optimization of design for mechanical, electrical and thermal requirements, without sacrificing SMT-oriented design, cost-down design, design for testing and design for manufacturability.

The following is a list of the key design parameters in the PoP design flow:

1) Mechanical parameters
   a) The PoP package objective size.
   b) The PoP stack budget height.
   c) The substrate and under fill reliability requirements.
2) Thermal parameters
   a) The maximum power dissipation of all devices.
   b) The working environment temperature.
   c) The maximum junction temperature (Tj) limitation.
3) Electrical parameters
   a) The number of I/O, power supply and ground connection pins between top and bottom devices.
   b) The list of fast-rise time signals and differential pair signals.
   c) The timing budget for all key signals and interfaces.
   d) The mark of any noise-sensitive signal.
   e) The list of specifications of important signals.
4) Design for manufacturability
   a) Design for testing, test interface solutions, consideration of increasing pin numbers in the PoP bottom component, and test coverage for both components.
   b) Design for cost. In PoP applications, the choice of substrate suppliers and the best design rules is key.

PoP design is quite complicated, and must meet design requirements associated with various systems and equipment, with the aim of achieving the best balance in terms of product cost, size, performance and time to market.

PoP generally uses high-density substrate. The bottom package BGA pin pitch is generally 0.5 mm or 0.4 mm.

According to Amkor empirical data collected from a large number of PoP projects, four ring ball arrangements should be able to meet the bottom package I/O pin requirements.

When more solder balls are needed, generally the power and ground pins are concentrated in the inside or center row.

The memory interconnect (the middle interface) ball quantity, the underlying logical bare-chip size, the amount of routing in the underlying substrate, and the bottom package BGA ball are the important factors affecting the reduction of PoP volume.

The optimum PoP size depends on the routing density rating, substrate manufacturing capacity, component cost and performance requirements. Once the best or the smallest PoP package size is selected, the focus will switch to package routing and specific design.

4.4 Apple A8 processor – an Example of a PoP Product

Apple held its 2014 autumn new product release conference in California's Flint Art Center on 10 September 2014. At the conference, Apple introduced its largest iPhones, namely the iPhone 6 and iPhone 6 plus. The dimensions of the new generation iPhone are: iPhone 6: 138.1×67×6.9 mm; iPhone 6 plus: 158.1×77.8×7.1 mm; see Figure 4.9.

The thickness of the iPhone 6 is only 6.9 mm; compared to the iPhone 5 at 7.6 mm, the iPhone 6 is thinner by 10%.

Both the iPhone 6 and iPhone 6 plus use the 64-bit A8 processor.

The highlight of the A8 processor is the manufacturing process. The A8 has created many firsts: it is the first mobile processor batch-manufactured using 20 nm technology, and is the most important processor to date. This is the first time Apple has used a non-Samsung factory OEM; and, of course, this is also the first time Apple has used the

iPhone 6 iPhone 6 plus

Figure 4.9 Apple iPhone 6 and iPhone 6 plus.

Taiwan Semiconductor Manufacturing Company Ltd (TSMC), with TSMC grabbing the biggest volume.

TSMC announced that, compared to 28 nm technology, 20 nm technology can increase chip speed by 30% and integrated degree by 90%, and reduce power consumption by 25%. In A8 the transistor is doubled, but the core area has decreased by 13%.

Apple A8 chip uses PoP technology solutions; the processor and mobile DRAM are integrated in one package with PoP technology.

Amkor Technology and STATS ChipPAC will be responsible for the Apple A8 chip package; each company is responsible for 40%, and in addition, ASE (Advanced Semiconductor Engineering) will be responsible for the remaining 20% of orders.

In fact, unpacking the mainboard from the iPhone 6, it can be seen that the components have been modularized, with each module intended to achieve a particular function, which not only means improved efficiency of the machine at the time of production, but is also conducive to repair and replacement.

The A8 processor and iPhone 6 mainboard are shown in Figure 4.10. The mainboard embodies an ultra-high degree of integration; CPU, memory, communication, audio, touch, gyroscope and other functional modules are integrated in this miniature mainboard; this is the core of iPhone 6 functionality.

The A8 is not in fact only the processor; it uses PoP stacked-package technology, including for the processor core and internal RAM memory. The bottom layer is A8 CPU core, using flip chip to connect to the substrate. The top layer comprises the two layers of RAM made by Samsung, giving the A8 1 GB of memory; see Figure 4.11.

The iPhone 6's powerful functionality and excellent performance are due mainly to the strong performance of the A8 processor, while PoP packaging technology also contributes to a great extent.

Figure 4.10 A8 processor and iPhone 6 mainboard.

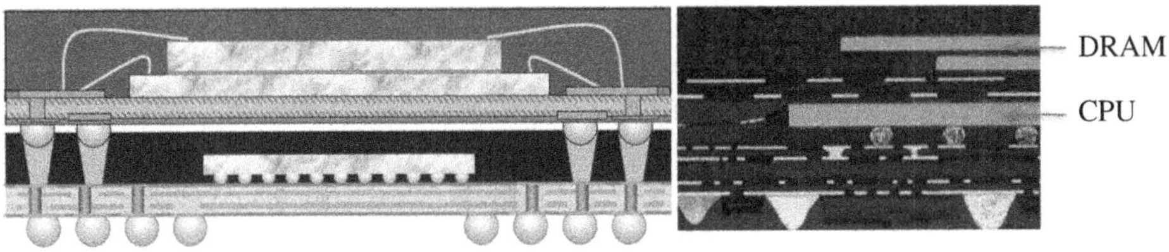

(a) A8 processor structure diagram (b) A8 cross section

Figure 4.11 A8 processor structure diagram and cross section.

With the development of science and technology, product performance will be better, but the price will decrease; this is the benefit of rapid technological development.

Packaging technology will also develop rapidly with the development of science and technology, and have an important impact on people's lives.

5

SiP Design and Simulation Flow

5.1 SiP Design and Simulation Flow

Similar to the design of a PCB board-level system, design and simulation of SiP flow can also be carried out by a process involving library creation, schematic design, layout design, signal integrity analysis, power integrity analysis, thermal analysis and electromagnetic compatibility analysis. Figure 5.1 shows a flow chart for SiP design and simulation.

1) Design preparation
   Design preparation mainly includes:
   a) A variety of data collection: bare chip-related information, pin definition, physical size, whether materials can be purchased or not, etc.
   b) Determining the package type, whether BGA package or some other form of package; determining the package size, the pin pitch, the number of pins.
   c) Using custom pin arrangement or standard package; thinking about later assembly and testing.
   d) Selecting the package technology and materials; selecting plastic, ceramic or metal package according to the field of application and the budget.
2) Library creation and management
   Library creation and management mainly includes the schematic symbols, IC bare chips, BGA or other kinds of package, parts and simulation models.
3) Schematic design
   Schematic design includes schematic entry, RF schematic design, schematic concurrent design, etc.
4) Pre-simulation
   Design pre-simulation involves synchronizing with the schematic design, using "what if" analysis, to determine the key signal topology, impedance matching, layer stack up structure, power plane segmentation, capacitor type and model selection. For analog circuits or analog-digital mixed circuits, functional simulation can be carried out.
5) Process selection
   Process selection mainly involves determining the SiP technics process, such as wire bonding, flip chip, TAB, TSV, etc.; determining whether substrate cavity is needed or not; the selection of one-side cavity or top/bottom double-side cavity;

SiP System-in-Package Design and Simulation: Mentor EE Flow Advanced Design Guide,
First Edition. Suny Li (Li Yang).

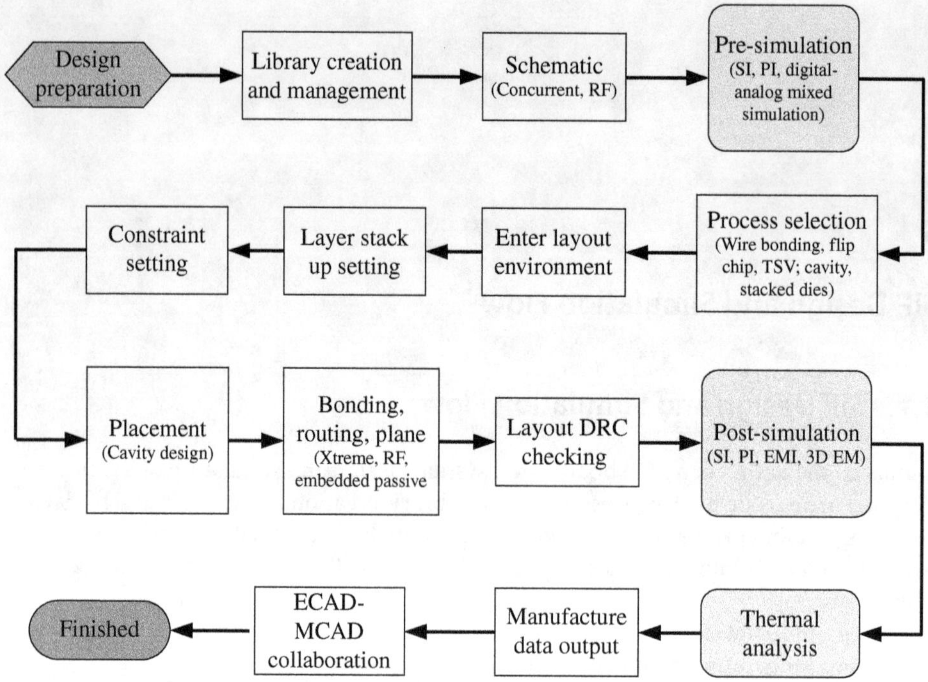

Figure 5.1 SiP design and simulation flow chart.

deciding on cavity depth and cross layers; and making decisions about chip stacking, substrate layers and laminated structure.

6) Enter layout environment
   Using forward annotation, send the schematic net connections and rule definitions to the layout environment; at the same time related cells are called automatically from the central library to the layout design environment.
7) Layer stackup setting
   According to the complexity of the design, set up the layer stackup, including the number of layers and layer structure; select 1+N+1, 2+N+2, m+N+m or ALIVH layer structure.
8) Constraint setting
   Mainly includes net classification, net properties, clearance, electrical constraints, high-speed constraints, differential net constraints, etc.
9) Placement
   Determine the chip placement. If the chips need to be placed into the cavity, determine the depth of the cavity, whether the cavity is single or multi-step, and the cavity shape.
10) Bonding, routing and plane
   Determine the wire bonding, monolayer or multilayer bonding, wire model, power ring setup, interactive routing or automatic routing, plane layer segmentation, RF circuit design, embedded resistor or capacitor synthesis, etc. This step involves quite a large amount of work.

11) Layout DRC check
Using DRC checking, errors can be found in layout and amended, ensuring the correctness of the design.
12) Post-simulation
After design is finished, export design data to the simulation tool, and carry out signal integrity, power integrity and electromagnetic simulation and analysis.
13) Thermal analysis
Import design data to the thermal analysis tool to solve problems of SiP overheating due to excessive power consumption of some chips, ensuring product stability and reliability.
14) Manufacture data output
Including the Gerber file, drill file, BOM, DXF, IDF, GDSII, ODB++ and other files for manufacture.
15) ECAD-MCAD collaboration
ECAD-MCAD collaboration mainly involves collaborative design of electronic software and mechanical software. Expedition mainly supports SiP internal electrical design, including substrate routing, chip assembly, bonding, etc., while the shell data of SiP is usually determined by mechanical design, including design issues such as metal frame, plastic package mold, ceramic package cover plate, metal package shell, etc.
16) Design finished
When the above processes are complete, the design is finished.
Next, using a practical project, we will describe simply the design and simulation process of SiP in Mentor EE (Expedition Enterprise) Flow.

5.2 Design and Simulation Process in Mentor EE Flow

To design SiP in Mentor EE Flow, the basic process is the same as for the PCB design flow; it is very easy to learn and use. At the same time, in addition to the basic design flow, the powerful simulation tools and collaborative design environment supporting this flow enable seamless linking of SiP design with the whole electronic product, such as IC and PCB, as shown in Figure 5.2.

5.2.1 Library Creation

After the necessary preparation for the project, the first step is to create the desired component library, including bare chips, passive components, BGA package, etc. In Mentor EE Flow, this part of the work is accomplished by the Library Manager tool.

In Library Manager, Padstack Editor can be used to create pads and padstacks, and Symbol Editor to create symbols. Symbols can contain the net name and other information used in design. Cell Editor can be used to create a physical cell; a cell contains component manufacturing information such as chip size and pad distribution.

Part Editor can be used to map corresponding symbols and cells; the part contains much information about the design representation of the component. By separating the principle and manufacturing information modes, repeating work for different components with the same cell or same symbol can be avoided. In Library Manager, different parts can reference the same cell or the same symbol.

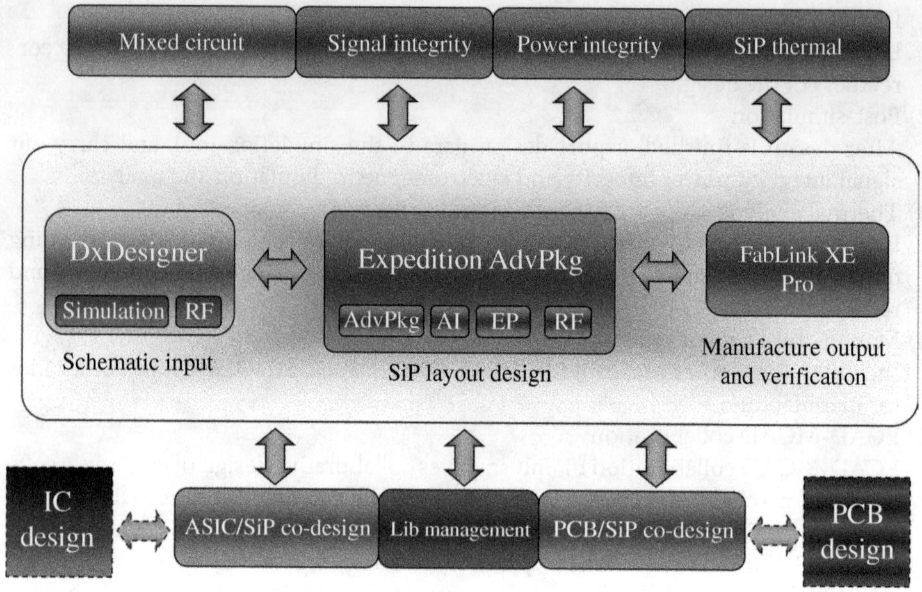

Figure 5.2 SiP design and simulation environment in EE Flow.

5.2.2 Schematic Design

DxDesigner is the schematic editor in EE Flow. For ordinary single-chip BGA package, the net relationship is simply the mapping of die pad to BGA ball. In SiP design, there is a complex net relationship between chip and chip, and chip and BGA package, and this is where DxDesigner has a role to play.

In DxDesigner, select components from the central library, place into schematic pages, and connect the chips according to the logical relationship, as shown in Figure 5.3. If the nets need output in BGA package, connect them to the BGA symbol.

At the same time, whether a single-chip package or SiP is involved, the distribution of ball pins is a very important issue. It not only determines the routing direction in later PCB layout design, but also has an important effect on signal integrity and power integrity.

The principle of distribution of BGA ball pins is usually according to the distribution of bare chip power supply pins or ground pins. Arrange the pins for power:GND:signal close to the proportion 1:1:4.

As shown in Figure 5.4, blue pins are signals, gray pins are GND, pink pins are 3.3VD digital power, yellow pins are 3.3VA analog power, orange pins are 5 V power, green pins are 2.5 V power. The specific power and GND types are related to the chip's function in SiP.

Considering the distribution of power and ground at schematic design stage, we can plan the net connection more fully. Of course, in layout design stage, we can also optimize the net connection and BGA pin assignment, which makes the routing and power distribution more reasonable.

Complete the schematic design, then Package CDB and run Forward Annotation. The net connections and rule settings in the schematic will be automatically transferred

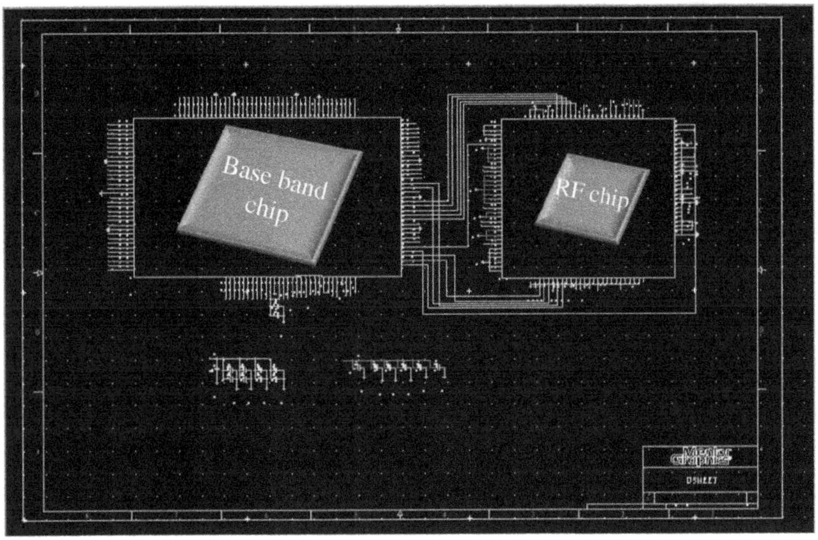

Figure 5.3 SiP schematic design.

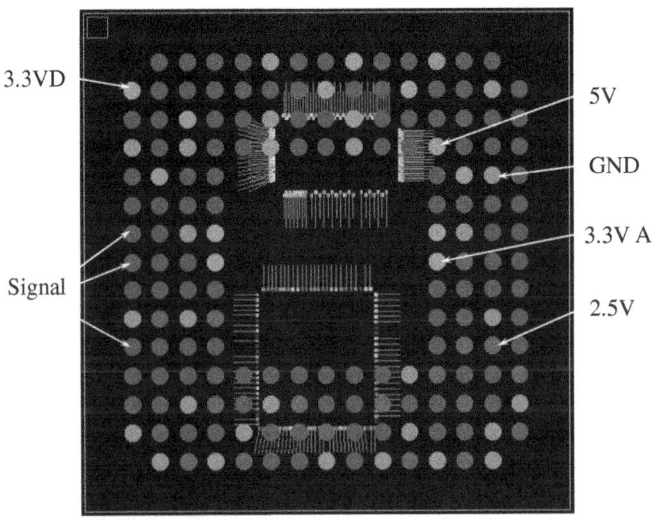

Figure 5.4 Typical distribution of power, GND and signal pins. *(See color plate section for the color representation of this figure.)*

to the Expedition layout tool; call the appropriate cells from the library and enter SiP layout design stage.

5.2.3 Layout Design

In addition to the traditional setting up of line width, clearance and other physical and electrical properties in CES, the main work of SiP layout design in the Expedition environment is to decide on the blind and buried via rules and wire bonding rules, as shown in Figure 5.5.

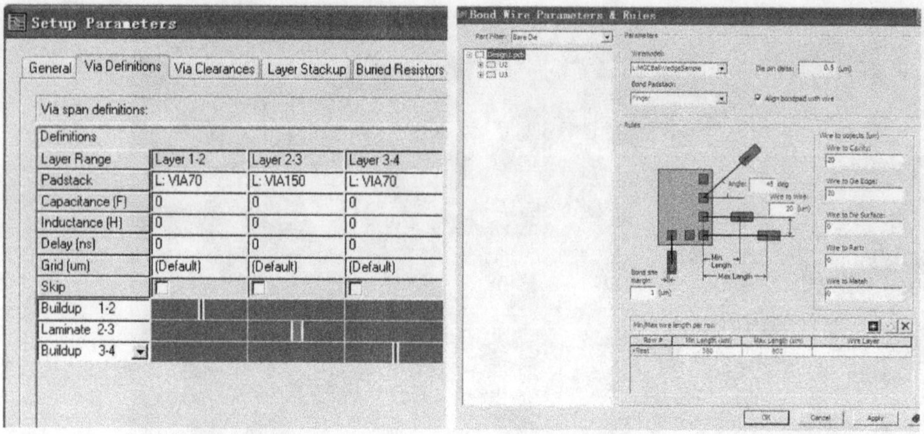

(a) Micro via definitions (b) Wire bonding parameters and rules

Figure 5.5 Blind and buried via setup and wire bonding setup.

The blind and buried vias in the buildup layers can be overlapped, but cannot overlap with the through-hole vias of the laminate layers, mainly due to limitations of the production process.

The wire bonding rule settings must be consistent with the corresponding reference data provided by the manufacturer, which describes the process capability.

The Package CDB function in DxDesigner can be used to transfer schematic data to Expedition, and call the corresponding component cells. The Place Parts and Cells commands can be used to place cells into the layout area, where BGA package must be put on the bottom side of the substrate.

The HDI plastic substrate used in SiP is commonly BT resin; the current processing capacity has reached line width/clearance 17/17μm for the buildup layer, line width/clearance 30/30μm for the laminate layer, and the buried and blind via diameter can reach 50μm.

Of course, for different manufacturers, the processing capability will be different; the designer needs to confirm with the specific manufacturer.

Compared to ordinary PCB FR4 substrate, BT resin substrate has higher line density, and can support more complicated design over the same area.

Expedition provides a net auto-optimizing function which can optimize the net allocation of die pins to BGA pads automatically, providing the reference for the actual distribution of BGA pads, as shown in Figure 5.6.

Figure 5.6 (a) shows a random net allocation of BGA pads by the designer; many nets cross and we can predict that the actual routing would be very difficult.

Figure 5.6 (b) shows the network allocation after auto-swap by Expedition; few netlines now cross, helping to improve design quality and efficiency. Use the automatic swap function to optimize net connection in layout, then back annotate to the schematic; the BGA symbol in the schematic will be updated accordingly.

In SiP layout design, we need to determine the BGA substrate layers according to the actual design requirements and the budget. At the same time, we also need to be aware, for M+N+M HDI substrates, that in general the electrical properties of the buildup

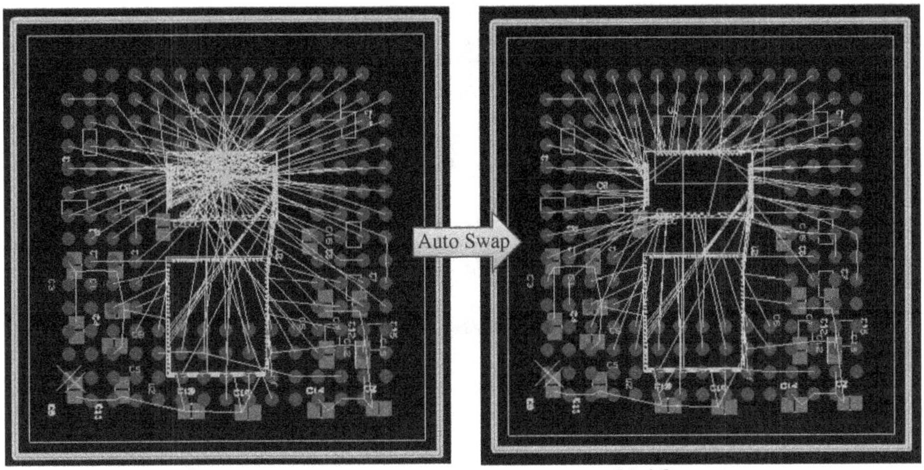

(a) Before auto swap (b) After auto swap

Figure 5.6 Expedition network automatic optimization.

layer are better than those of the laminate layer and the process is more accurate in the buildup layer.

For example, the dielectric constant is 3.4 in the buildup layer, and 4.7 in the laminate layer. In layout design, putting the critical signal or most of the signal in the buildup layer is beneficial to the whole SiP system electrical performance.

Expedition provides a dynamic plane shape function for use after the completion of routing. Even if after completion of plane shape modifications are made to routing or components are moved, the plane shape will be updated automatically, with no need to delete and regenerate, greatly improving design efficiency.

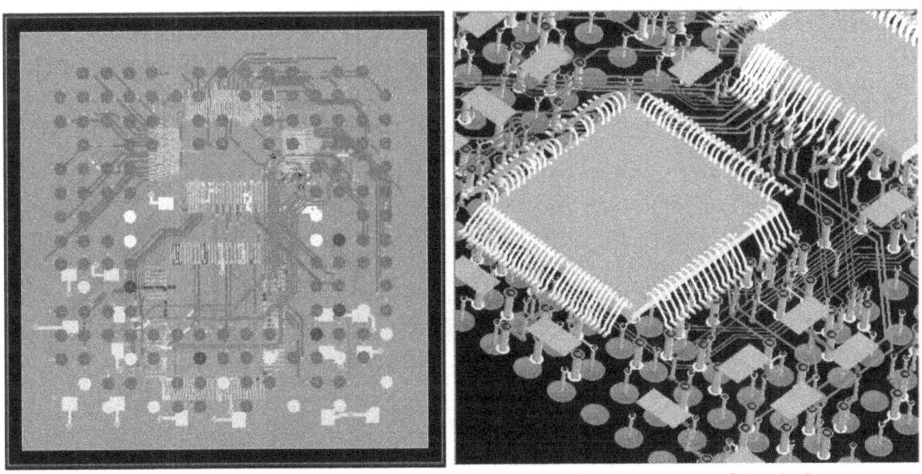

(a) 2D screenshot of the design (b) 3D screenshot of the design

Figure 5.7 Completed SiP design (2D figure and 3D figure).

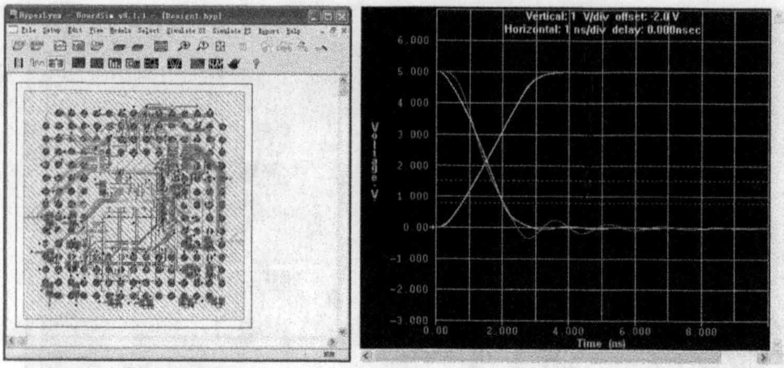

(a) Simulation UI (b) Simulation result

Figure 5.8 Export design to HyperLynx for simulation.

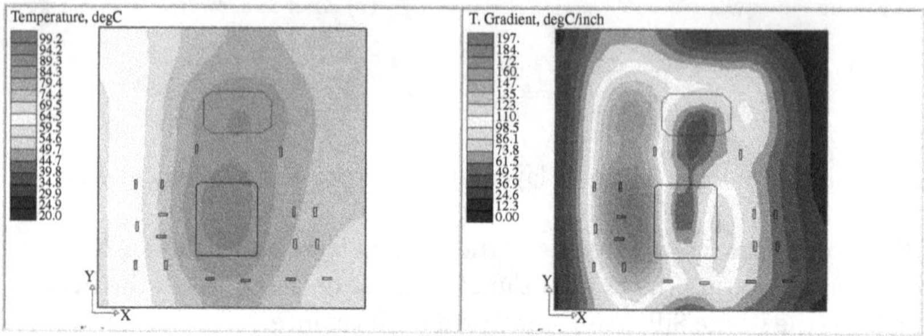

(a) Temperature (b) T. Gradient

Figure 5.9 The results of thermal analysis.

Usually there will be a three-dimensional structure in an SiP system, such as multiple rows of bonding wire, stacked chips, cavity, embedded chips or devices, etc.

Expedition allows for real-time 3D DRC checks so as to avoid errors in the design process, ensure design is correct, and improve design efficiency and accuracy.

Figure 5.7 shows a screenshot of an SiP layout design (in 2D and 3D).

5.2.4 Design Simulation

After the design is completed, it can be exported to HyperLynx SI and PI for signal integrity and power integrity analysis, to ensure that the results meet the design requirements, as shown in Figure 5.8.

Thermal analysis is very important in package or SiP design.

HyperLynx Thermal provides thermal analysis for the SiP system. In Expedition, we can export design data to HyperLynx Thermal via a menu key.

Figure 5.9 shows the results for thermal analysis; (a) is the temperature chart, (b) is the temperature gradient chart.

After all checks and analyses have passed, SiP design and simulation is complete.

At this time, we can export the Gerber data, drill data and other production files, and hand them over to the SiP manufacturers.

6

Central Library

6.1 The Structure of the Central Library

In the SiP design process, library creation is the first problem which the designer will encounter.

First we describe the structure of the Mentor central library.

In Mentor EE Flow, the design unit in the schematic is called symbol, while the design unit in layout is called cell. Depending on the different kinds of components, cell will refer to different types of padstacks.

For surface-mounted devices or bare chips, the padstack is usually composed of different layers of pads, such as the pad layer, soldermask layer, solderpaste layer, gold plated layer, etc.

For a through-hole device, the padstack is usually composed of different layers of pads and through-holes.

Symbol and cell combine via pin mapping to form a component called part. Figure 6.1 shows the structure of the Mentor central library.

The central library is not dependent on any project and exists independently; it can be referenced by many projects at the same time. The central library contains symbols, cells, parts and padstacks, reused blocks, and simulation models. We can create multiple central libraries, but each design can only be associated with one central library.

The part used in the SiP design schematic and the corresponding symbols, cells and padstacks, are all extracted from the central library.

The first task is to place symbols in the schematic and interconnect them with nets. Then, use the Package CDB function, and then, using Forward Annotation, extract the corresponding cells and padstacks automatically, and transfer to the layout environment. The relationship of the central library, schematic and layout are shown in Figure 6.2.

In each design there is a local library containing the parts, symbols, cells and padstacks only used in this design. Modifying the local library cell data will automatically update the cell in layout design, but does not change the central library data associated with the design.

SiP System-in-Package Design and Simulation: Mentor EE Flow Advanced Design Guide,
First Edition. Suny Li (Li Yang).

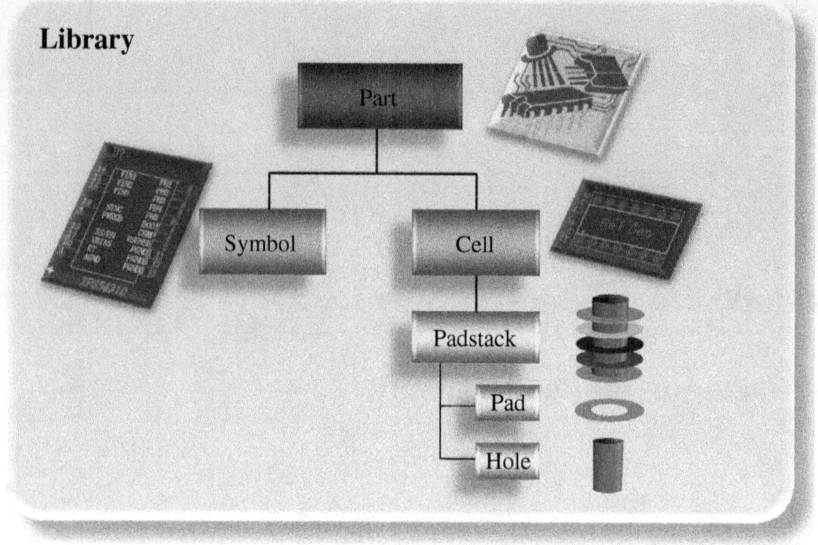

Figure 6.1 The structure of the Mentor central library.

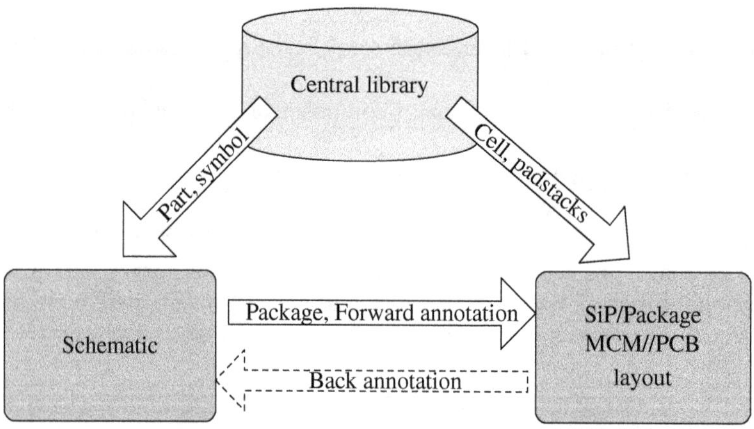

Figure 6.2 The relationship of the central library, schematic and layout.

6.2 Introduction to the Dashboard

Now we really come to the use of the Mentor tools.

After the installation of EE Flow in Windows, we see the icon on the desktop (different versions have different names). Double-click this icon to launch the Dashboard; the user interface of the Dashboard is shown in Figure 6.3.

The Dashboard contains a variety of tools used in SiP design; these tools can be started directly from the Dashboard.

Here we focus on the Folders bar and Shortcuts bar. There are projects, toolboxes, WDIR, etc. subdirectories.

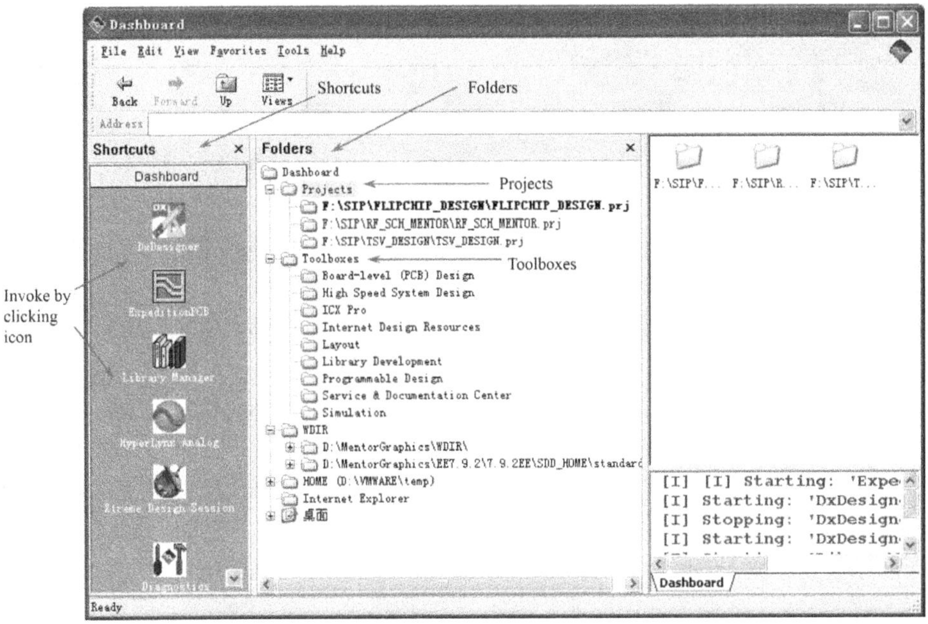

Figure 6.3 Dashboard interface.

The Projects subdirectory contains the current project under Dashboard; we can add a new project to Projects using the menu File→Add Project. The bold font stands for the active project; any project can be activated by a mouse click on its name, but there can be only one active project.

When the schematic tool DxDesigner is open, the active project will be opened automatically, as shown in Figure 6.4.

'Toolboxes' contains nine subdirectories, and includes various types of tools in EE Flow; they are arranged in different subdirectories according to the function of the tool.

Using the mouse to click each subdirectory, examine the design tools by selecting a tool icon and double-clicking the left mouse button to open the tool.

The Layout subdirectory, for example, contains Expedition, Job Wizard, PCB Planner, Xtreme Design Session, etc., as shown in Figure 6.5.

For frequently used tools, we can use the left mouse button to select the tool icon and drag it to the Shortcut bar.

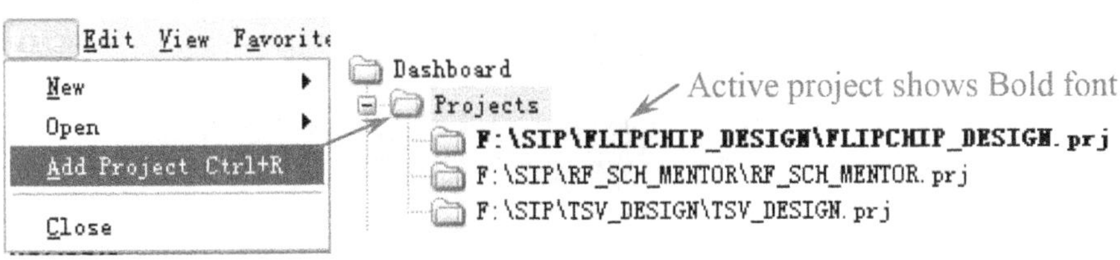

Figure 6.4 Active project.

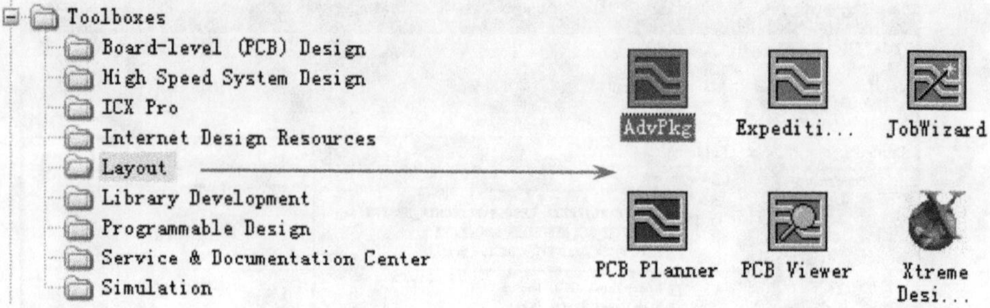

Figure 6.5 Toolboxes.

As shown in Figure 6.3, the Shortcuts bar shows: DxDesigner, Expedition, Library Manager, HyperLynx Analog, Xtreme Design Session, etc. In the Shortcuts bar a single click on the icon launches the corresponding tool.

Tools placed in the Shortcuts bar can be removed with the right-click menu item "Remove from Shortcuts".

6.3 Schematic Symbol Creation

Use the left mouse button to click the Library Manager Icon  in Shortcuts; this starts Library Manager. In Library Manager we can open the existing central library, as shown in Figure 6.6. Opening the library MENT_LIB, we can view the design elements, including symbol, cell, part, padstacks, etc. in this central library.

The designer can also choose File→New to create a new central library; the system will jump out to the "Select a new central library directory" window to set the path of

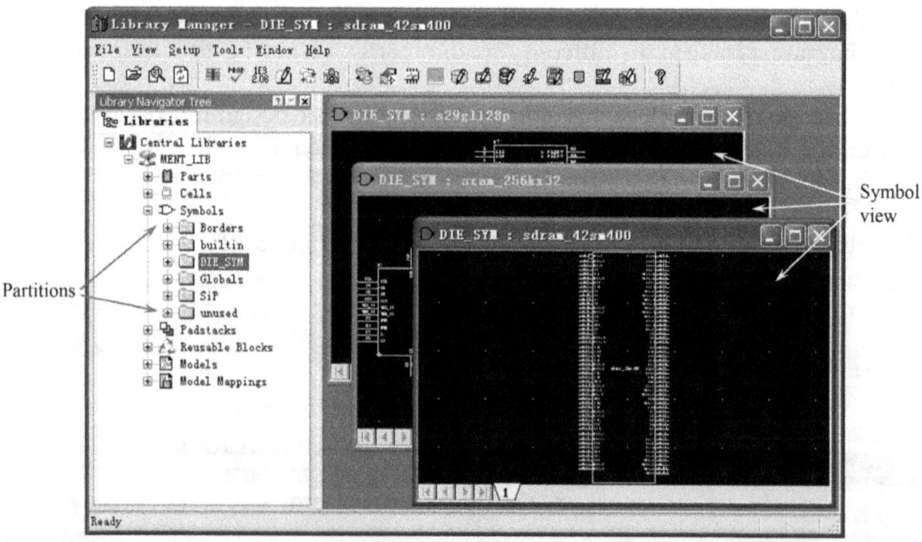

Figure 6.6 Library Manager user interface.

the new central library. Here we need to create a new folder, name it SiP_lib, and then double-click to enter this folder (this is important).

After entering the folder, the flow type by default is DxDesigner/Expedition; this remains selected, as shown in Figure 6.7. Click the OK button, wait a moment, and the system will automatically copy some basic elements to this newly created central library.

After central library creation, we can see that there are already three partitions under "Symbols". The three partitions "Boarders", "built-in" and "Globals" are generated automatically by the system, and include the schematic boarders, connectors, power supply and GND symbols, as shown in Figure 6.8.

Next, create a new "Symbol" in the central library; in order to distinguish this symbol, we need to create a new partition. Right-click on Symbols, select New Partition, input SiP_SYM in the pop-up window, as shown in Figure 6.9.

After clicking the OK button, we can see the SiP_SYM folder appearing in the list; right-click on SiP_SYM, and we can see that there are multiple options. Here we choose Symbol Wizard, and input "sym1" in the pop-up window, as shown in Figure 6.10.

Click the OK button; the Symbol Wizard Step 1 appears; according to the default option select "Block Type" for Module, and select Do Not Fracture Sym. Click the "Next" button to go to Step 2: Symbol Name and Path, and leave the default. Click the Next button to set some parameters in Step 3, such as the length of the pin, the pin number's visibility and position properties, as well as the symbol text size etc. Mostly these also keep the default values.

Click the Next button to set five basic properties of the symbols in Step 4:

1) Part Number: is unique in a central library.

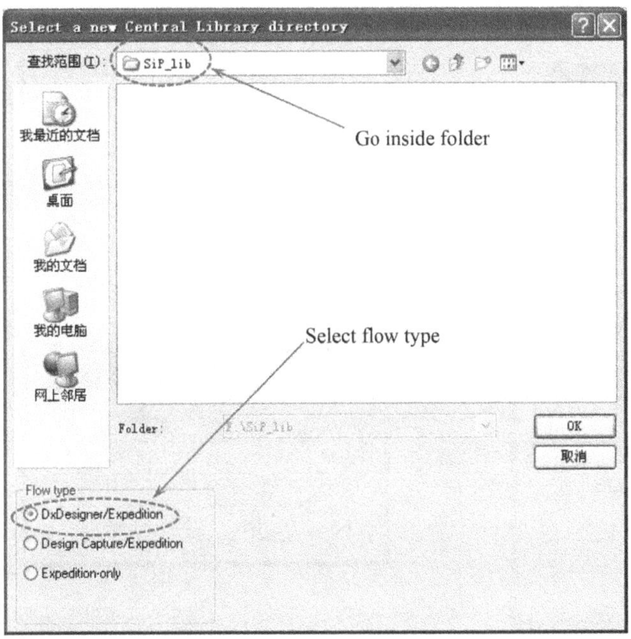

Figure 6.7 Create a new central library.

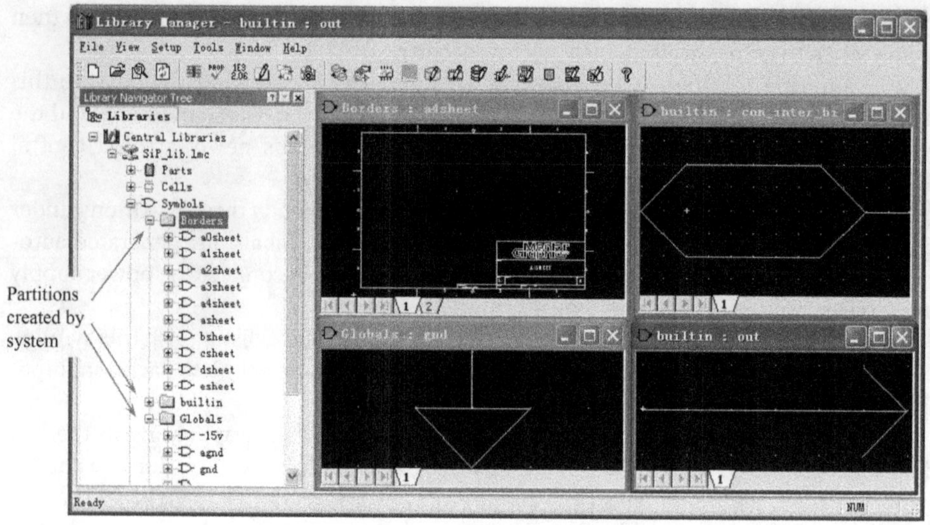

Figure 6.8 Boarders, built-in and Globals partitions created by the system.

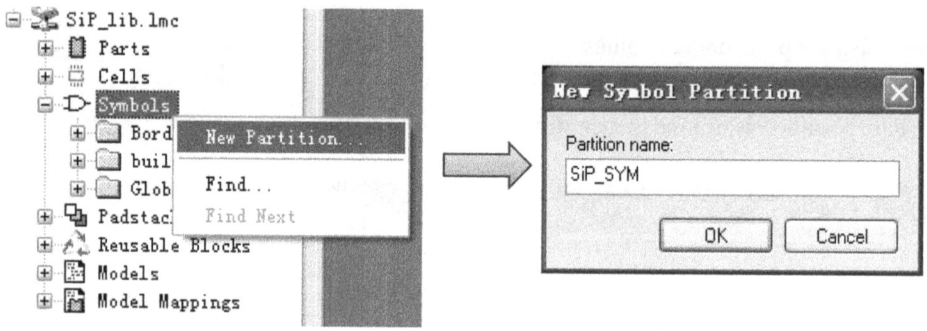

Figure 6.9 Create new partition.

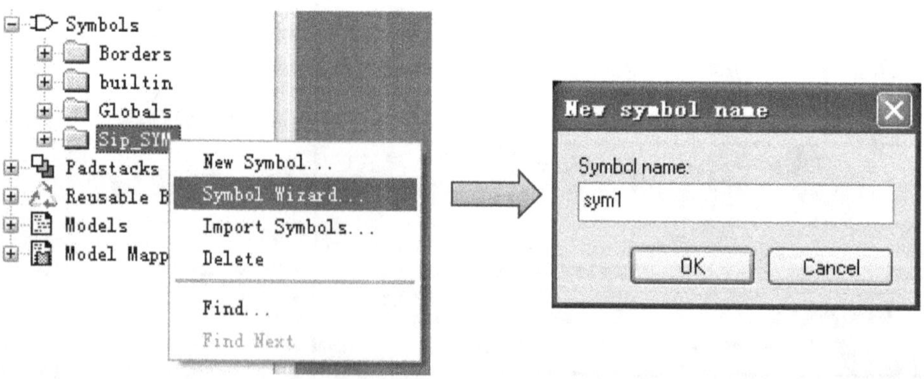

Figure 6.10 Create symbol via Symbol Wizard.

2) Ref Designator, U?: indicates the symbols in the schematic diagram with U as a prefix for sorting; the designer can modify these to N? or R?, etc.
3) PARTS: indicates how many of the same symbols are contained in a PART.
4) Level: the symbol level, usually keep as STD.
5) PKG_TYPE: assign the corresponding layout design unit cell.

Here, we define the Part Number property as PART1, the PKG_TYPE property as DIE_CELL1, with the others remaining as per the default, as shown in Figure 6.11.

Click the Next button; Step 5 is mainly used to set the symbol pins properties, a total of 11 items; the first four properties, pin name, pin number, type (input/output type) and symbol side, are the most important, as shown in Figure 6.12. We must fill in according to the relevant properties of the symbol pins.

Here we adopt a simpler method, namely copy-and-paste.

Before introducing the copy-and-paste method, we first introduce a management system for symbol and cell properties in an Excel spreadsheet.

	Property	Value	Visible	Color
	Part Number	PART1	Invisible	Automatic
	Ref Designator	U?	Invisible	Automatic
	PARTS	1	Invisible	Automatic
	Level	STD	Invisible	Automatic
	PKG_TYPE	DIE_CELL1	Invisible	Automatic
*			Invisible	Automatic

Figure 6.11 Properties definition for symbol.

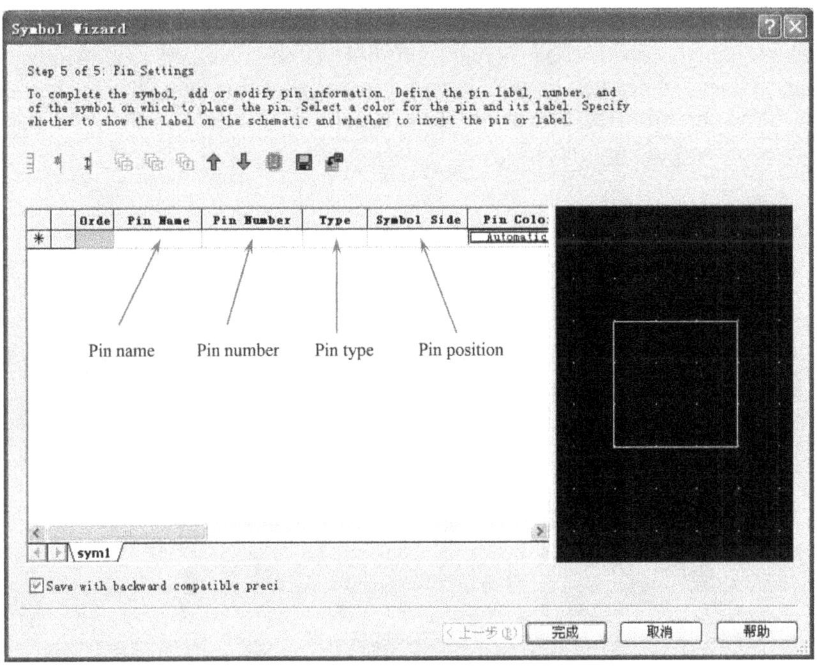

Figure 6.12 Set the pin properties of symbol.

In SiP design, properties information for all components can be managed using an Excel spreadsheet. The information on components used in a project can be managed as shown in Figure 6.13.

Cell properties are shown on the left; the right shows symbol properties. Different components have separate sheets, as shown in Figure 6.13 for Die1, Die2, Die3, Package, etc. In actual design, it is best to name sheets using the actual name of the component, making them easier to distinguish from each other.

Symbol and cell properties are based on the information from the IC chip supplier, this information may be in Excel, PDF or other formats.

No matter what kind of format is used, we first copy the useful information to Excel, and arrange as shown in Figure 6.13.

It is important to note that the pins of any symbol cannot be repeated, so repeated pins need to be renamed. To retain the original pin, we create two columns, respectively "pin (original)" and "pin (to remove repeated)", followed by pin number, input/output and pin location attributes.

Rename duplicate pins, for example, if one chip has many VSS pins, rename as VSS1, VSS2, VSS3, etc.

Cell properties are listed on the left of the Excel spreadsheet; this is the information needed to create a cell; the Die Wizard tool allows the importation of the information in order to create the cell automatically; see section 6.4 dealing with cell creation.

Edit the symbol and cell property information in Excel, and then select the corresponding area. To create a symbol, select the symbol properties area of the spreadsheet, ensuring it is surrounded by a dotted line, press < Ctrl + C > to copy, and then move the mouse to the first line of the "Pin Name" column in the Symbol Wizard window, press < Ctrl + V >, and the data is pasted into Symbol Wizard. Now in Symbol Wizard we can also see the symbols preview on the right side of the window, as shown in Figure 6.14.

Finally, click the Finish button; the schematic symbol has been created successfully.

At this point, in the Library Manager main window we can see that Sym1 has appeared under the SiP_SYM partition; the preview of the symbol is also shown on the right side of the window, as shown in Figure 6.15.

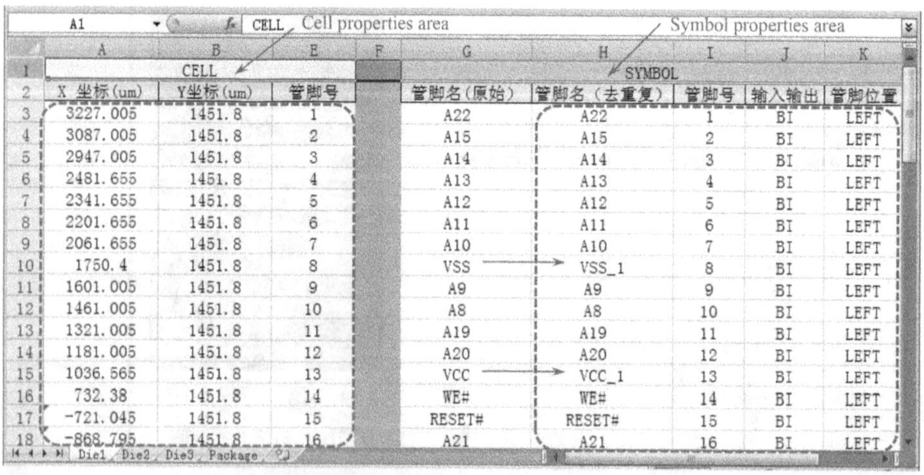

	A	B	E	F	G	H	I	J	K
1		CELL				SYMBOL			
2	X 坐标(um)	Y坐标(um)	管脚号		管脚名(原始)	管脚名（去重复）	管脚号	输入输出	管脚位置
3	3227.005	1451.8	1		A22	A22	1	BI	LEFT
4	3087.005	1451.8	2		A15	A15	2	BI	LEFT
5	2947.005	1451.8	3		A14	A14	3	BI	LEFT
6	2481.655	1451.8	4		A13	A13	4	BI	LEFT
7	2341.655	1451.8	5		A12	A12	5	BI	LEFT
8	2201.655	1451.8	6		A11	A11	6	BI	LEFT
9	2061.655	1451.8	7		A10	A10	7	BI	LEFT
10	1750.4	1451.8	8		VSS	VSS_1	8	BI	LEFT
11	1601.005	1451.8	9		A9	A9	9	BI	LEFT
12	1461.005	1451.8	10		A8	A8	10	BI	LEFT
13	1321.005	1451.8	11		A19	A19	11	BI	LEFT
14	1181.005	1451.8	12		A20	A20	12	BI	LEFT
15	1036.565	1451.8	13		VCC	VCC_1	13	BI	LEFT
16	732.38	1451.8	14		WE#	WE#	14	BI	LEFT
17	-721.045	1451.8	15		RESET#	RESET#	15	BI	LEFT
18	-868.795	1451.8	16		A21	A21	16	BI	LEFT

Figure 6.13 Edit symbol properties in Excel.

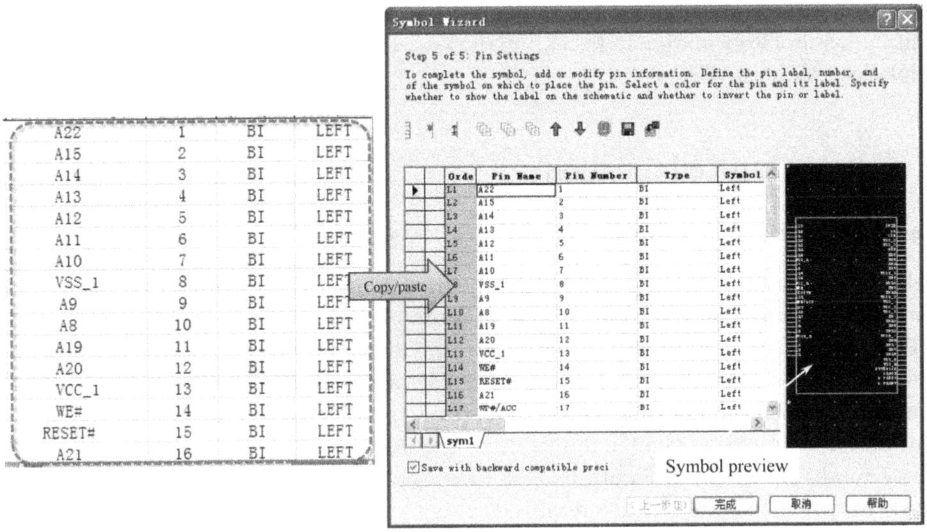

Figure 6.14 Copy-and-paste to create a symbol.

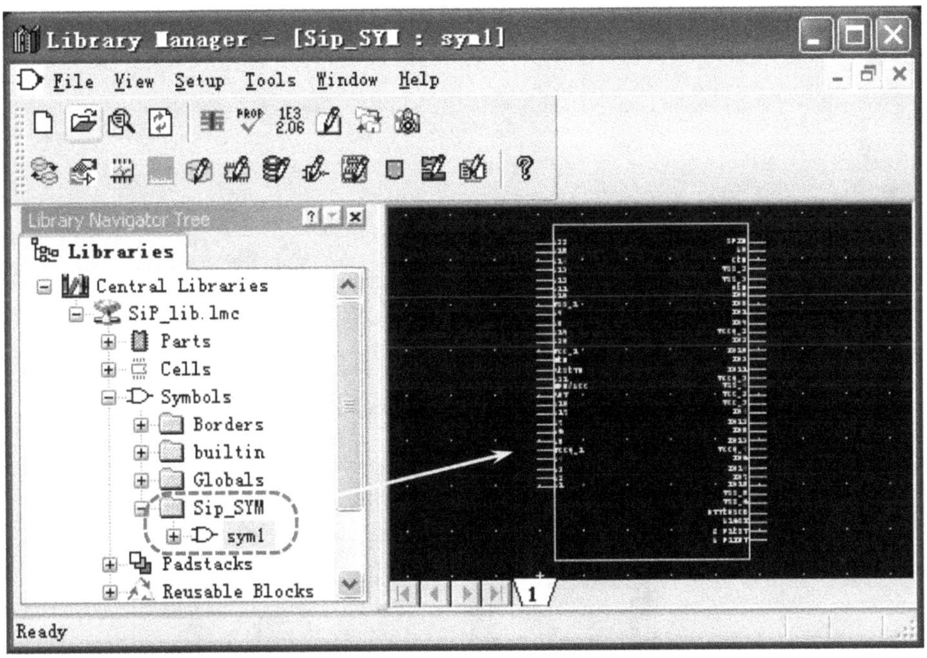

Figure 6.15 The completion of symbol Sym1.

Depending on the requirements of the design, for example, if the size of the symbol frame needs adjusting or the pin position needs rearranging, etc., we can use right button Edit command to open and edit the symbol in Symbol Editor.

Using the same method, we can create other symbols and the SiP package symbol. Edit the symbol information in Excel, and then use the copy-and-paste method to create symbols in Symbol Wizard.

6.4 Bare Chip Cell Creation

6.4.1 Create Bare Chip Padstack

Before building the bare chip cell library, the first thing to do is to create the padstack; click in the toolbar to invoke Padstack Editor.

First, switch to the Pads tab, click to create a new pad and set its size, select μm for units, select the shape "Square" on the right side of the window, and set the size to 62; the software will automatically name it Square 62. If it is necessary to name a pad manually, or, for example, to change the name of the 62μm square, we need to remove the check mark, as shown in Figure 6.16. If the designer directly changes the name, the check item will be automatically removed.

In the Pad tab, you can create a variety of shapes of pad; you can also create various types of thermal pad. If you need to create a specially shaped pad, switch to the Custom Pad & Drill Symbols tab; in this tab the designer can draw any shape of pad, as shown in Figure 6.17.

After the pad has been created, switch to the Padstacks tab, click to create a new padstack, and name it die_pad, as shown in Figure 6.18. Set the padstack type to Pin – Die,

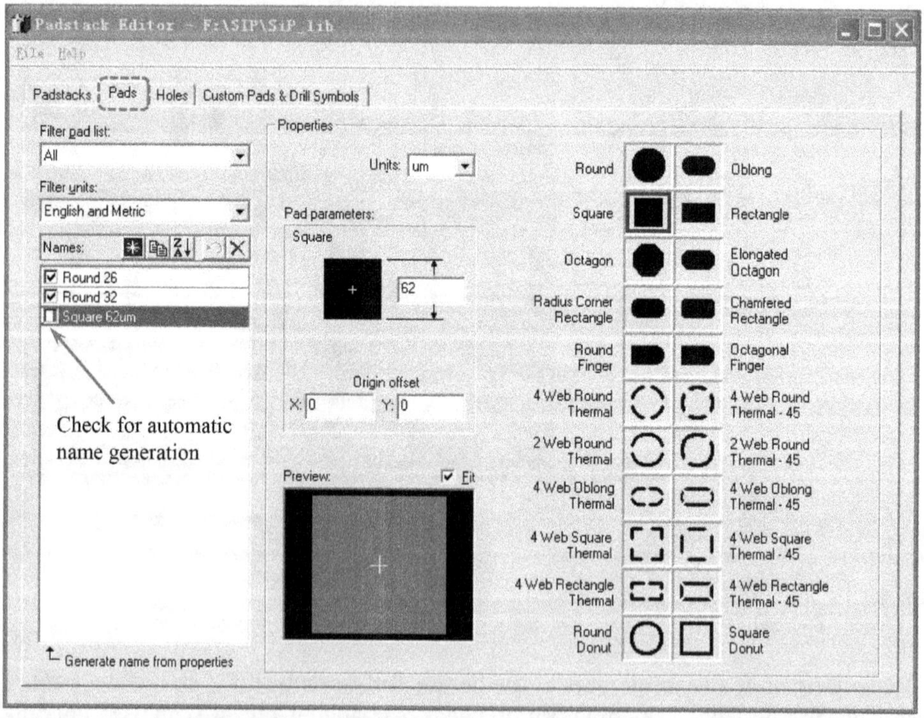

Figure 6.16 Create a new pad.

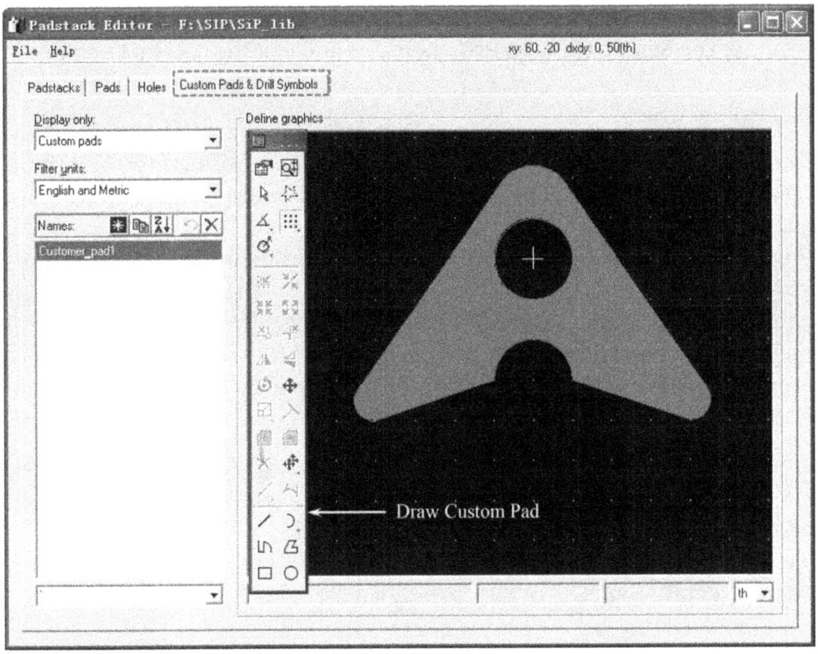

Figure 6.17 Create Custom Pads and Drill Symbols.

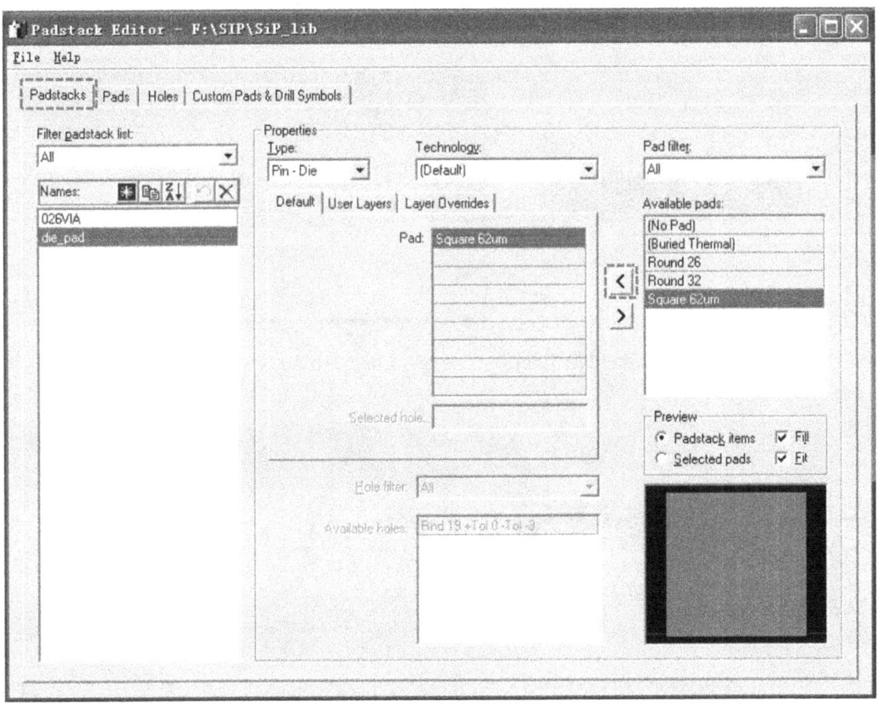

Figure 6.18 Create die pad.

because Pin – Die needs only one layer; select the Square 62μm pad and click the arrow to the left <, move the Square 62μm to Pad, then save and exit the padstacks editor.

6.4.2 Create Bare Chip Cell

Go back to the Library Manager main interface, right-click on Cells, select New Partition to create a New Cell Partition, and input SiP_CELL as the name in the pop-up window, as shown in Figure 6.19.

After clicking the OK button, we can see that the SiP_CELL folder appears in the cell list; right-click on the SiP_CELL folder, select New Cell, input DIE_CELL1 in the pop-up window, then click the OK button, as shown in Figure 6.20.

In the Create Package Cell window, as shown in Figure 6.21, the cell name is that input in the previous step. Here, the cell name can be changed; in this example, retain the name.

- Total number of pins: in this example, enter 62, the input number and the pin number defined in symbols should be consistent usually, otherwise an error will occur when mapping parts.
- Layers while editing cell: usually input 2.
- Package group: select IC-Bare Die in the drop-down list.
- Mount type: select Surface.
- Click the Cell Properties button; in the new pop-up window input unit type and information such as the height of the chip. First choose μm in Units (namely set unit to microns), and then enter height 200. Pay attention to the height depending on the

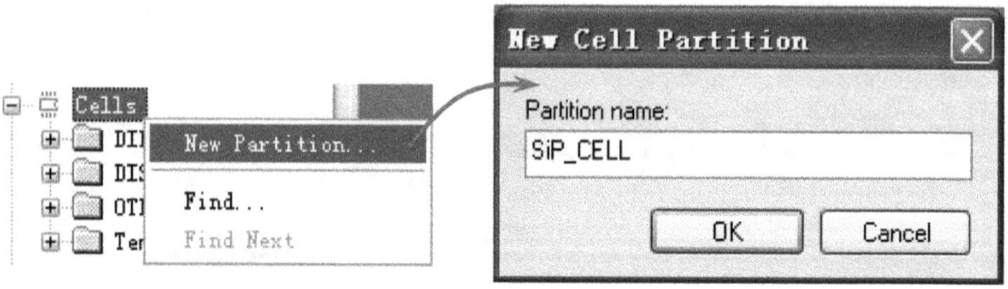

Figure 6.19 Create new cell partition SiP_CELL.

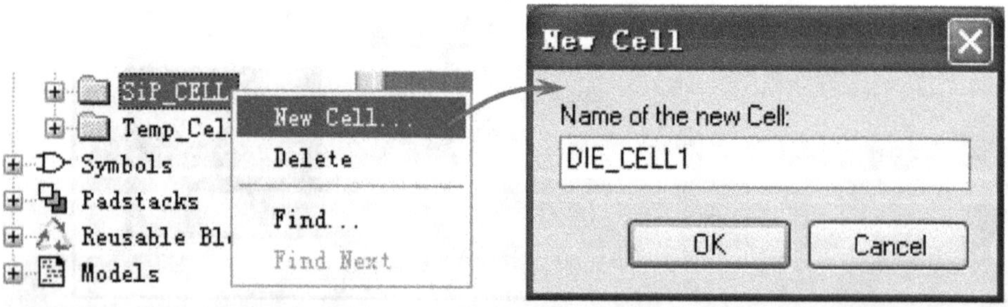

Figure 6.20 Create new cell DIE_CELL1.

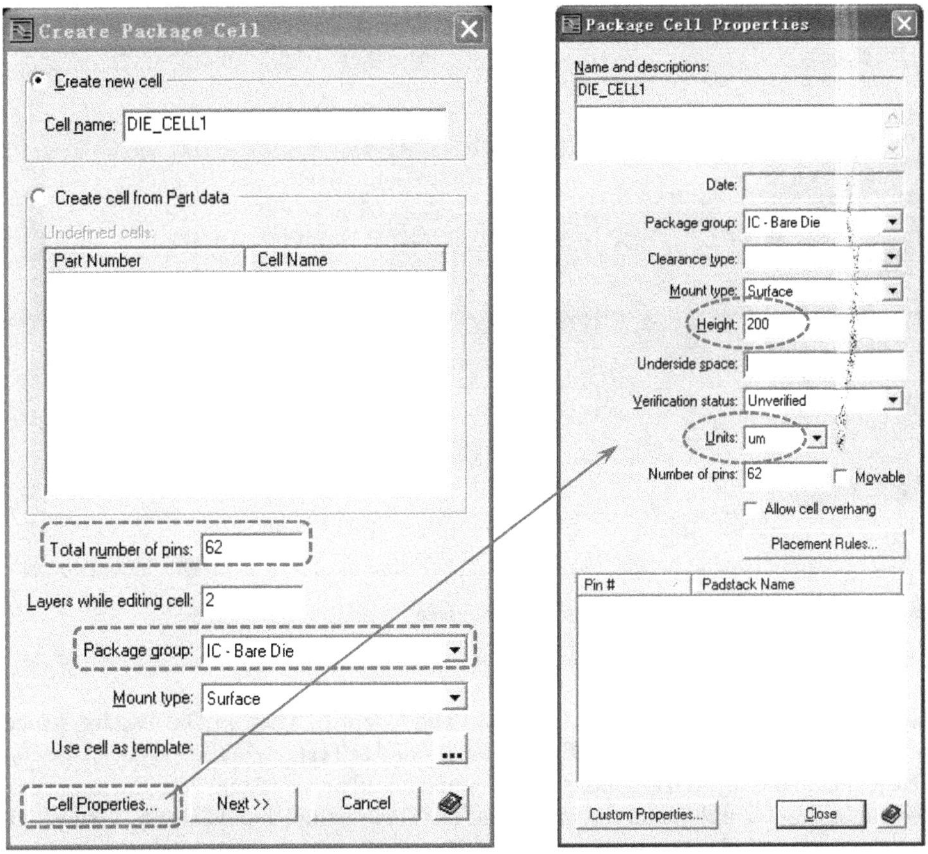

Figure 6.21 Create cell and edit cell properties.

actual chip thickness; usually this is the wafer thickness after thinning. It has a certain influence on the bond figure, the height of the bond wire starting point and the shape of the bond wire, thus affecting the consistency of the design and manufacture; this is a Design for Manufacture (DFM) consideration.

After setup is completed, click the Close button to exit the Cell Properties window, and return to the Create Package Cell window. Click the Next button to enter the Cell Editor window; the Place Pins interface will pop up automatically.

Select all of the pins and delete them (note: all pins must first be removed here, otherwise they will conflict with pins imported from Die Wizard), as shown in Figure 6.22, then click the Close button to close the Place Pins window.

In the Cell Editor window, click the Die/Wizard icon, and select the chip pins definition file to import. This file can usually be obtained from IC designers; the file content includes information such as bare IC chip X coordinates and Y coordinates of each pin, pin number, etc. This information is already in the Excel sheet described above (see section 6.3 on managing symbol and cell data in Excel).

In Excel, select the cell properties area, ensuring that the contents are surrounded by a dotted box; first copy the content into a text file and save to die_cell.txt (the data is

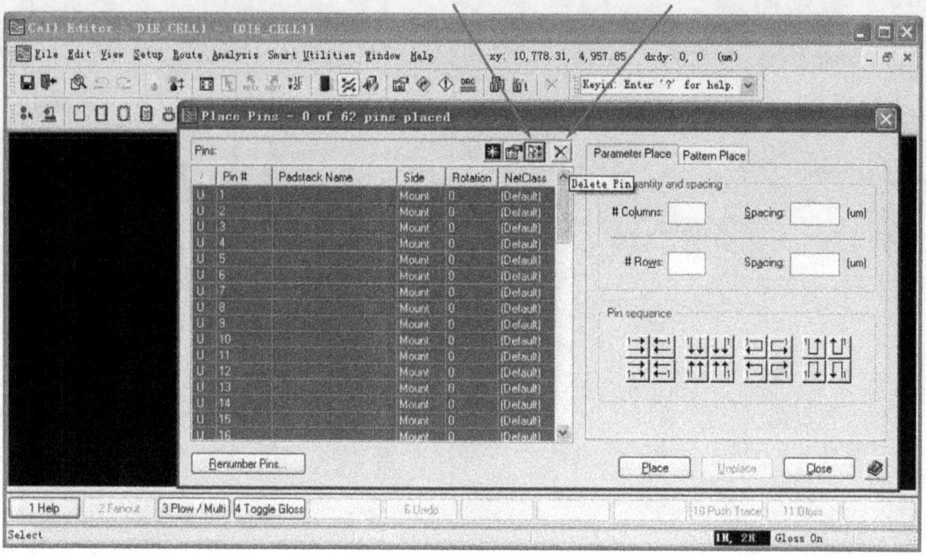

Figure 6.22 First delete all pins in the Place Pins interface.

copied into .txt format for easier import to Die Wizard). Then in Die Wizard select die_cell.txt file, in this example as F:\SIP\SiP_lib\die_cell.txt.

Note that in the Die Wizard interface, the unit should be consistent with the imported file (μm in this example), and the format set (X, Y, pin name) should also be consistent with the imported file, as shown in Figure 6.23.

Then choose the Separator according to the format of the imported file, such as tab in this example. Under Pad stacks, select User defined, and under Pad stack name select

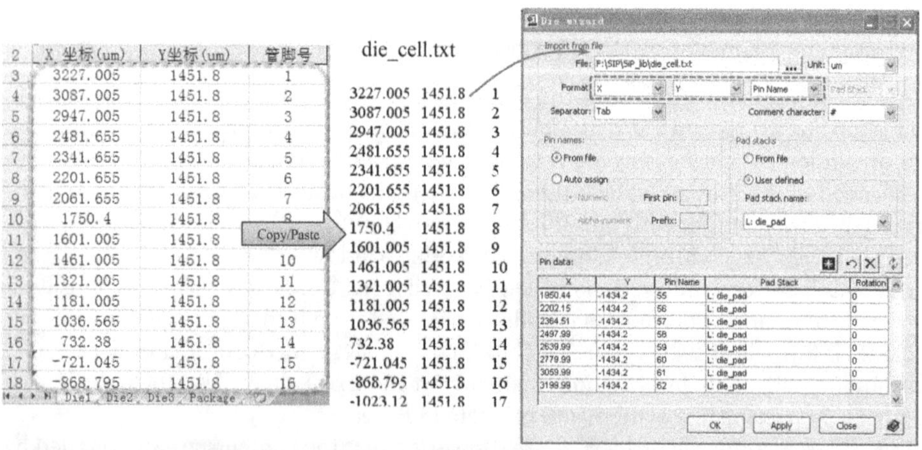

Figure 6.23 Import the die pin definition file via Die Wizard.

die_pad from the drop-down list. Then click the new and import buttons to complete the die_cell data import; see Figure 6.23.

After importation, click OK, and you can see the bare IC chip pins placed in Cell Editor Workspace, and arranged in accordance with the coordinates in the file. Depending on the requirements of the design, the designer can manually add Placement Outline, Assembly Outline, Silkscreen Outline, Assembly Reference Designator, and Silkscreen Reference Designator, etc. The Add Tools buttons are located in the window toolbar, as shown in Figure 6.24.

If the designer doesn't add any outline, when saved, the system will automatically add Placement Outline; the adding principle is that all feet are surrounded by the smallest rectangle. At the same time, assembly reference designator, silkscreen reference designator and assembly part number will also be added to this cell automatically.

It is important to note that automatically added text in a cell only plays the role of placeholder here; in design it will be replaced by the actual reference designator and part number.

Choose File→Save and exit Cell Editor. If you open the cell again, you can see the die cell as shown in Figure 6.25; the Placement Outline has been added by Expedition automatically.

By clicking the icon in the toolbar, you can enter Cell Editor, then click the New Cell button to create a new cell according to the above description.

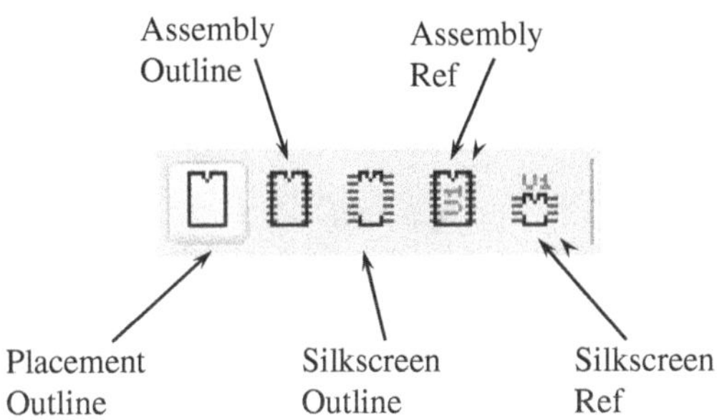

Figure 6.24 The outline and reference options of Add Tools.

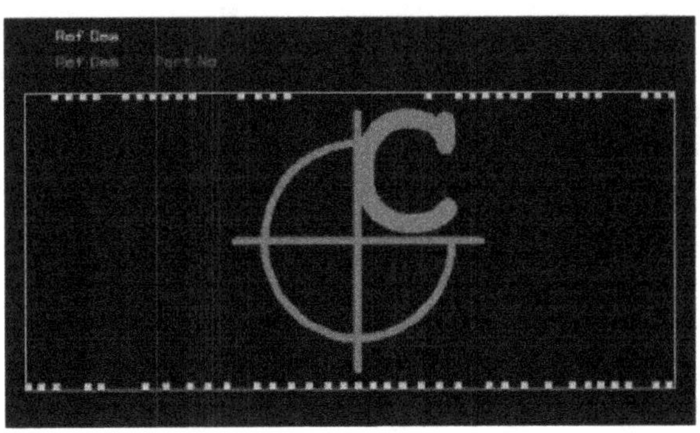

Figure 6.25 Bare die cell.

6.5 BGA Cell Creation

6.5.1 Create BGA Padstack

Before creating BGA cells, we need to create the padstack first; in the Library Manager toolbar, click to start Padstack Editor.

First, in the Pad tab set the size of the pad, select μm for the unit, in the right-side list choose round shape, set the size to 500, and the software will name it Round 500 automatically, as shown in Figure 6.26.

If you need to name the pad manually, remove the check in front of the name; if you don't want to rename, leave the check item. In the same way create a 450μm-diameter circular pad named Round 450.

Switch to the Padstacks tab, create a new padstack named BGA_pad_500um, and for Type select Pin-SMD. Because the Pin-SMD needs multilayer definition, hold the <Ctrl> key to select Top mount and bottom mount; in the right-hand column select Round 500, and then click the arrow to the left; Round 500 will be assigned to the top and bottom pad layers, as shown in Figure 6.27.

Then hold the <Ctrl> key to select Top mount soldermask, bottom mount soldermask, Top mount solderpaste and bottom mount solderpaste, in the right-hand column select Round 450, and then click the arrow to the left; Round 450 will be assigned to the corresponding layer. Finally, save, and exit Padstack Editor.

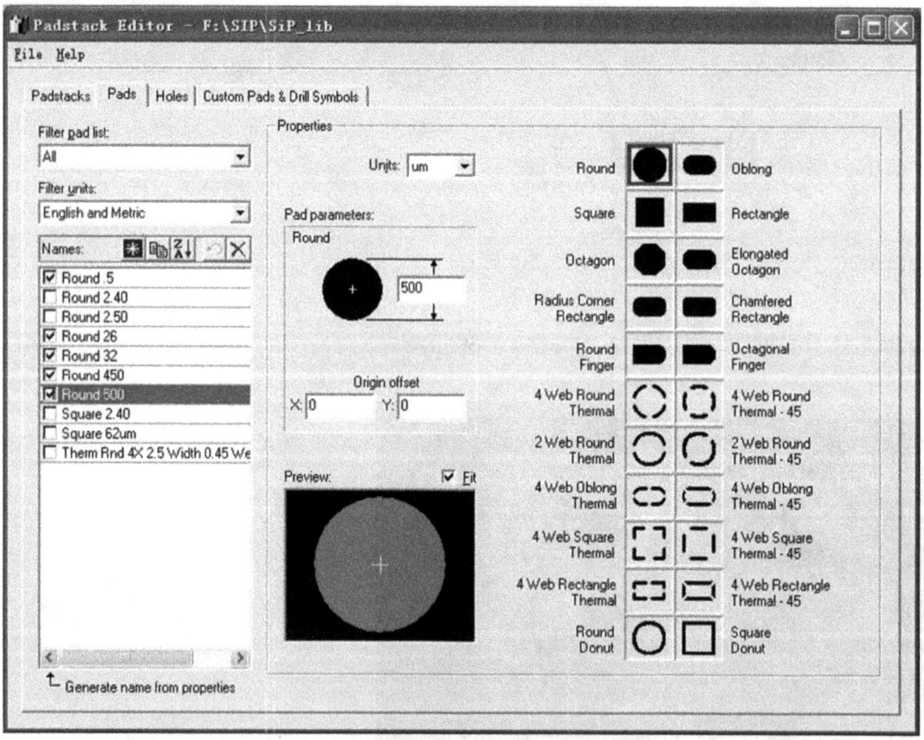

Figure 6.26 Create round pad in Pads tab of Padstack Editor.

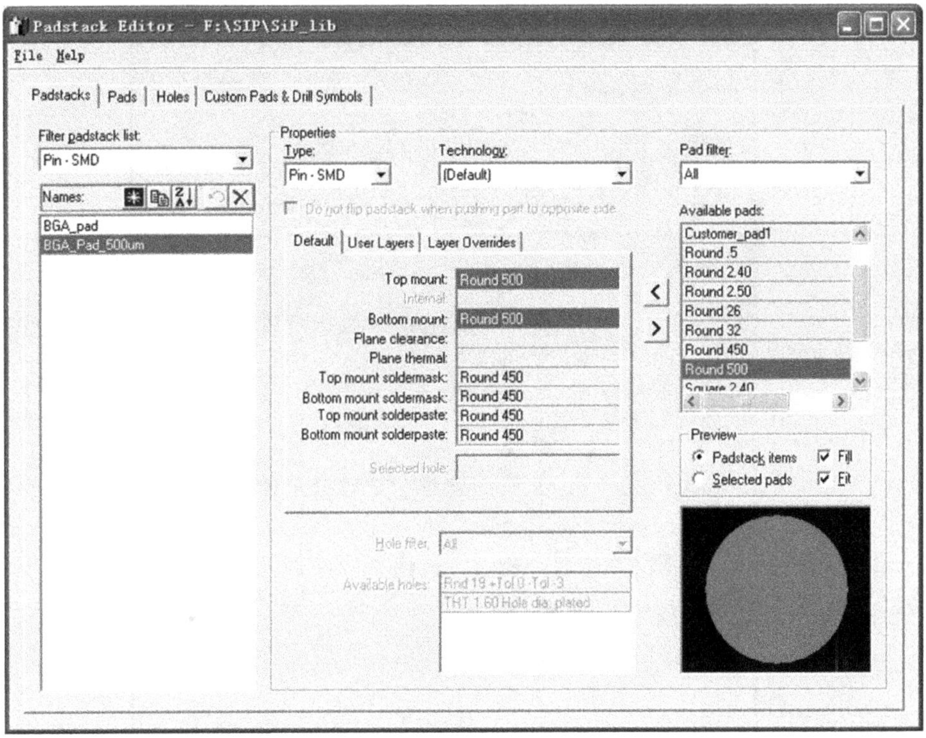

Figure 6.27 Create a BGA padstack in Padstack Editor.

In actual design, depending on the different processes, the padstack size of each layer is not necessarily exactly the same. For BGA padstacks for package or SiP substrate, soldermask is usually smaller than copper pad; for BGA padstacks in PCB design, soldermask is usually larger than copper pad; the dimension of soldermask general control is +/−10% relative to copper pad.

For the solderpaste layer, the size and thickness of steel mask controls the actual amount of solderpaste, which must be set according to the actual process, not only to ensure firm welding but also to avoid short-circuiting caused by too much solderpaste.

For the plane clearance layer, if it is not defined here, it will be defined as set in the layout design rules in CES definition. If its size is defined here, in negative plane generation, the clearance to copper will be according to the size of the plane clearance layer defined here; for the positive plane this has no effect.

For the plane thermal layer, if not defined here it will be defined at layout design according to the parameter settings in Plane Classes and Parameters. If its size is defined, and Use thermal definition is selected in the padstack options, the system will opt to select the plane thermal definition from padstacks, as shown in Figure 6.28. This is very useful for particular padstacks, if it is necessary to define different thermal pads from other padstacks.

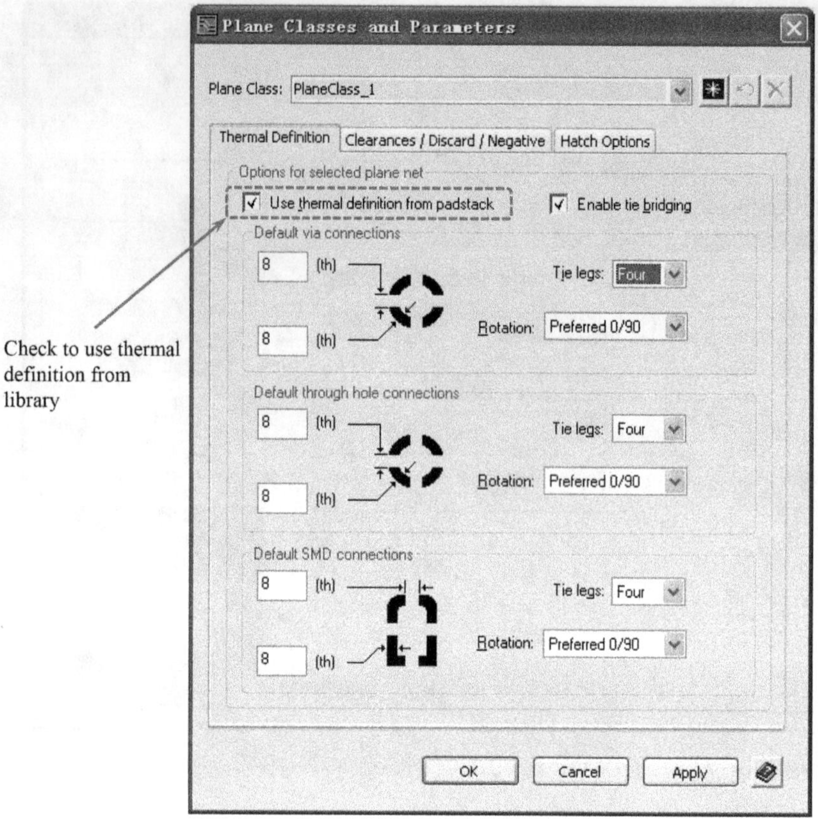

Figure 6.28 Use the thermal definition from padstack of central library.

6.5.2 Create BGA Cell Manually

After padstack creation is finished, in the Library Manager main window click [icon] to enter Cell Editor Interface. The cell DIE_CELL1 created above can be seen in the cell list; select the New button [icon] to create a new cell.

In the pop-up Create Package Cell interface input BGA_CELL in Cell Name, as shown in Figure 6.29.

Input BGA pin number in the Total number of pins bar; here it is necessary to consider whether the BGA pin arrangement is a full matrix or in cavity form. The full matrix is simpler; directly input the actual pin number. If it is the cavity form, then you need to use the actual pin number + pin number in cavity. For example, where the BGA actual pin number is 504, and the cavity is 20 × 20–4 (four cavity angle pins retained), you need to input the actual number (504 + 20 × 20–4) = 900.

In the Layer while editing cell bar, insert 2, as two layers need editing for this cell.

In the Package group bar, choose IC-BGA from the chip-type drop-down list.

In the Mount type bar, the system automatically selects Surface mount.

Click the Cell Properties button; in the new pop-up interface input information such as unit and height of the chip. Select mm for Units first, and then input chip height zero,

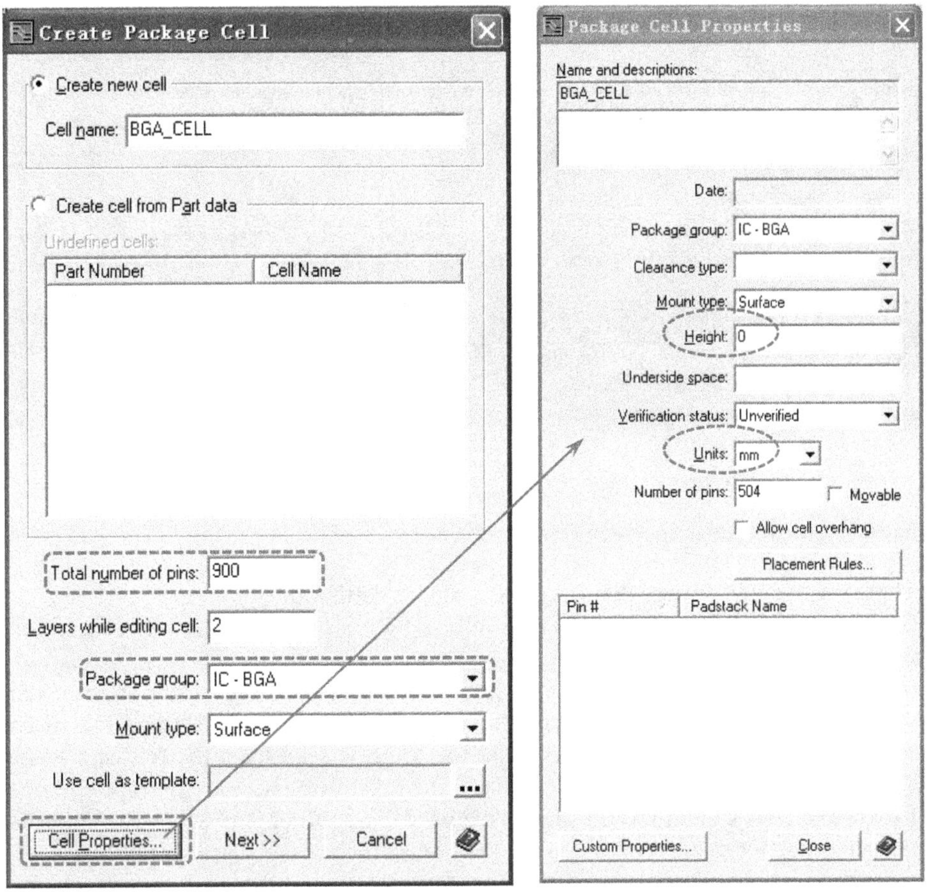

Figure 6.29 BGA_cell properties settings.

because the function of the BGA cell is as SiP pin fan-out – it is located in the bottom of the substrate, which itself has no thickness, so the height is set to 0.

After the completion of all settings, click the Next button to enter Cell Editor window.

In the pop-up Place Pins interface, first specify the BGA_pad_500um created previously for Padstack Name. Here's a tip: hold the <Shift> key, select the first pin, and then select the last pin, all pins are then selected. Keep the <Shift> key pressed, in the drop-down list select BGA_pad_500um; now you can see all of the pins are assigned with BGA_pad_500um.

In the Pin# column, the default pin numbers are 1, 2, 3, 4, 5, ...; these need to be renamed. As shown in Figure 6.30, first select pins 1 ~ 30, click the Renumber Pins button and set Prefix to "A" in the pop-up Auto Generate Numbers window, with Starting number 1 and Increment 1. Then click the OK button, and you can see the pin numbers have been changed to A1, A2, A3, ..., A30.

In the same way, select pins 31 ~ 60, rename to B1 ~ B30; select pins 61~90, rename to C1~C30, etc.

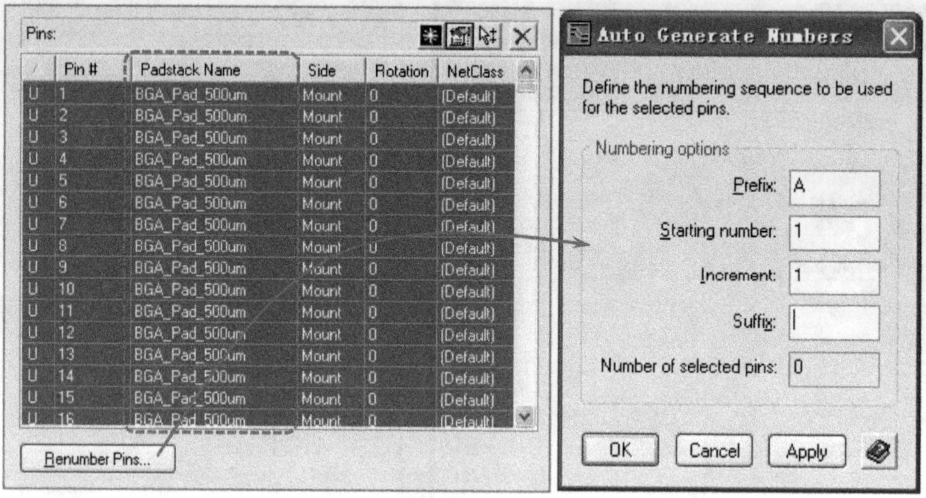

Figure 6.30 Specify Padstack Name and rename BGA pins.

There are two points here that need the designer's attention.

(1) Some letters are not allowed in the prefix of BGA, such as I, O, Q, S, X, etc., so you should skip these letters.

(2) Because BGA cells are placed in the bottom surface of the substrate, so the order of the pins must mirror this when creating cells, that is, for A1–A31, the actual order should be A31~A1, which is the pin order for A1–A31 seen from the top view when BGA cells are placed on the bottom surface of the substrate. There are two options, one is to set BGA pins side to mount (mirroring needed), another is to set to the opposite side (mirroring not needed).

After completion of Pin# rename, there are several ways to place pins; here we choose Pattern Place; for Pattern type select BGA, with other settings as shown in Figure 6.31.

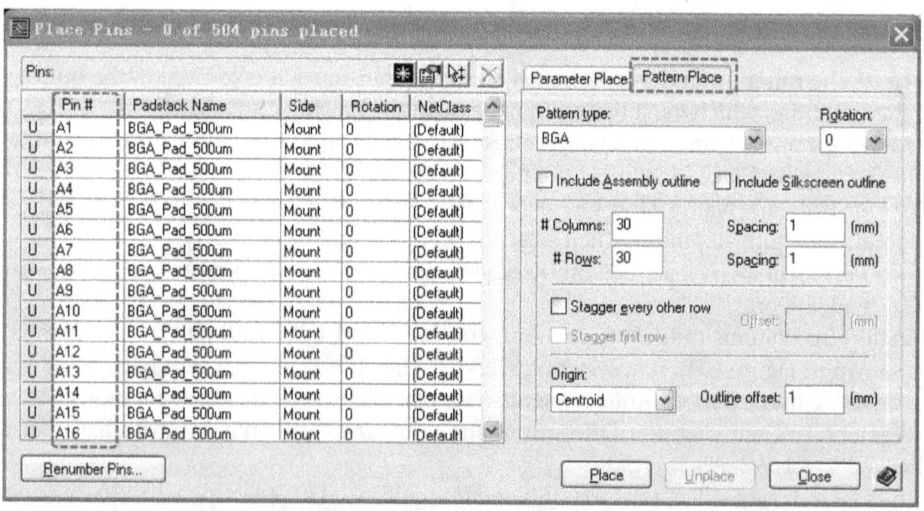

Figure 6.31 Rename the BGA Pin# and set placement parameters.

The placed pins are 30×30 full matrix, so we must delete the middle of the cavity manually. Select the pins to remove and then press the <Delete> key, as shown in Figure 6.32.

Finally, add Placement Outline, and adjust the size and location of letters. Then File→Save to save BGA_CELL; the dialog box pop-up as shown in Figure 6.33 appears, showing that only 504 pins were placed. To "Do you want to save this cell with the new pin count?", answer "yes".

6.5.2.1 Tips for Renaming the Pin Numbers

Click the Renumber Pins button, keep the Auto Generate Numbers window open, and define the renaming rules, then press the left mouse button and move across the BGA pins; depending on the direction of the mouse movement, the BGA pins are renamed according to the rules defined in Auto Generate Numbers window; see Figure 6.34.

6.5.2.2 View Layers Defined in Padstacks

In Padstacks, define the BGA pad as 500 μm, soldermask and solderpaste as 450 μm. In Cell Editor, select any pin, view its properties, and open the soldermask and solderpaste layer in display control to view, as shown in Figure 6.35.

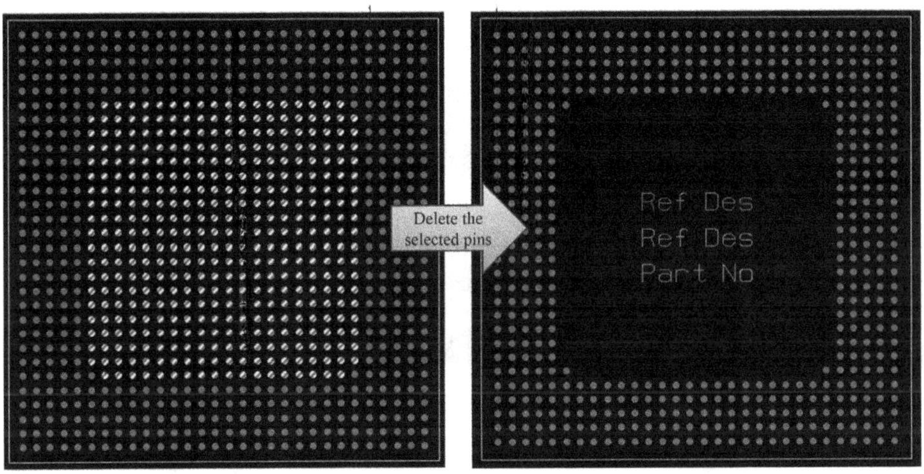

Figure 6.32 Delete the pins in the middle of cavity.

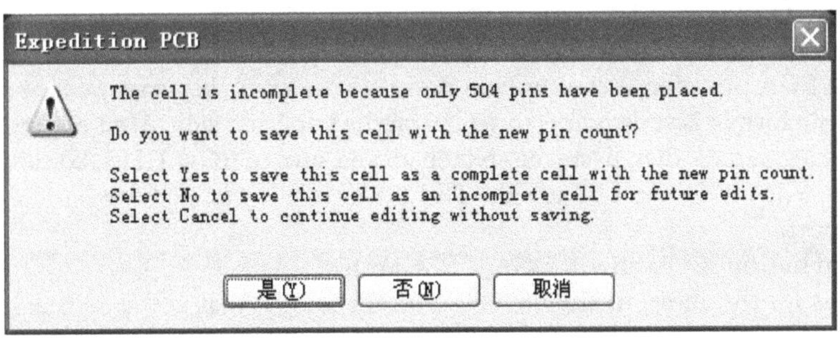

Figure 6.33 Prompt to update pin number information.

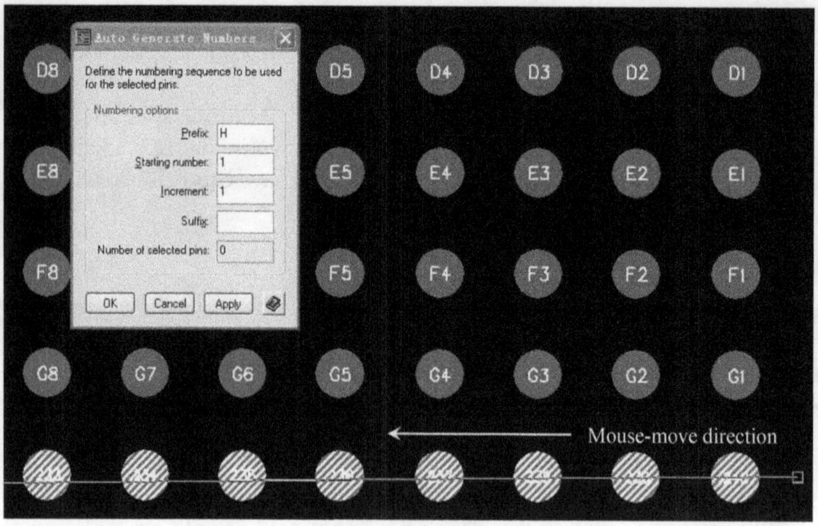

Figure 6.34 Renaming pin numbers using mouse movement.

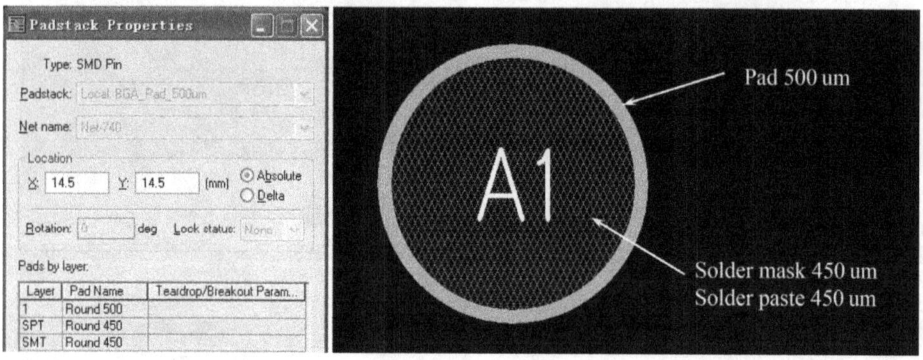

Figure 6.35 View the pin properties and different process layers.

6.5.3 Create BGA Cell with Die Wizard

We can also use Die Wizard to create BGA cells. First, turn off the Place Pins window, but make sure to delete all the predefined pins before turning off the Place Pins window, otherwise the pins will conflict with the pins imported by Die Wizard.

Second, edit the BGA pin coordinates and pin numbers. In Excel this is very easy to implement; the basic format is shown in Figure 6.36 on the left-hand side. After editing in Excel, copy the file content, then paste into Notepad, and save to BGA_CELL.txt file, as shown in Figure 6.36 on the right-hand side. The .txt format is more convenient for Die Wizard to import.

Click Die Wizard button to import BGA_CELL.txt file.

In Die Wizard, as for the above method, it is important to note that in the Padstack Name list we can only choose die_pad (because BGA pad is not in the list). After import is complete, replace die_pad with BGA pad in Cell Editor.

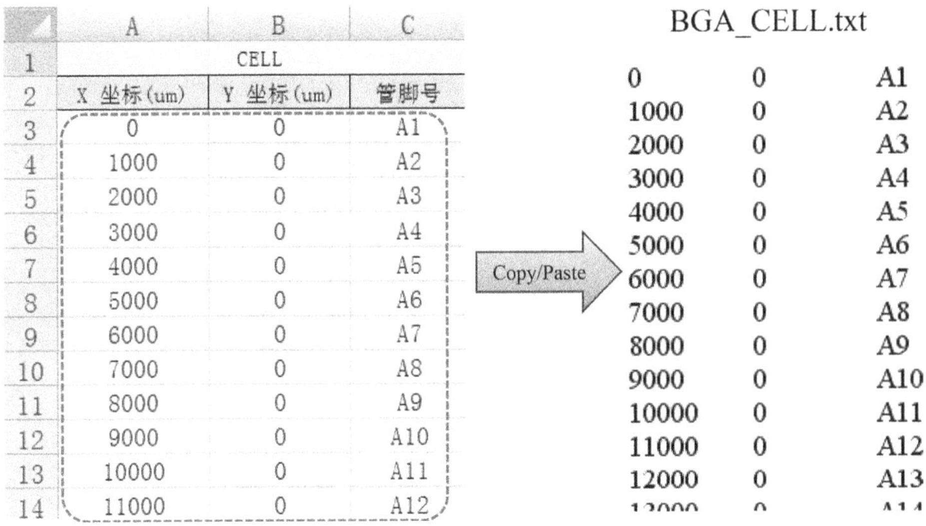

Figure 6.36 BGA pin coordinates and pin number definition.

Enter Cell Editor and select all pins in the Place Pins window. The selection method is as follows: first select the first row, hold the < Shift > key, and then select the last row, keeping the < Shift > key held down, then click Padstack Name in the last row, and in the drop-down list, select BGA_pad_500um (created above), complete the padstack replacement, then save and exit.

6.5.4 LP Wizard Professional Library Tool

If you need to quickly create all kinds of complicated BGA cells, you can use Mentor professional library tool LP Wizard, as shown in Figure 6.37. This tool is based on the IPC-7351 standard, and includes various process conditions, a variety of built-in BGA templates and a standard components library; you can quickly create high-quality component cells with this tool.

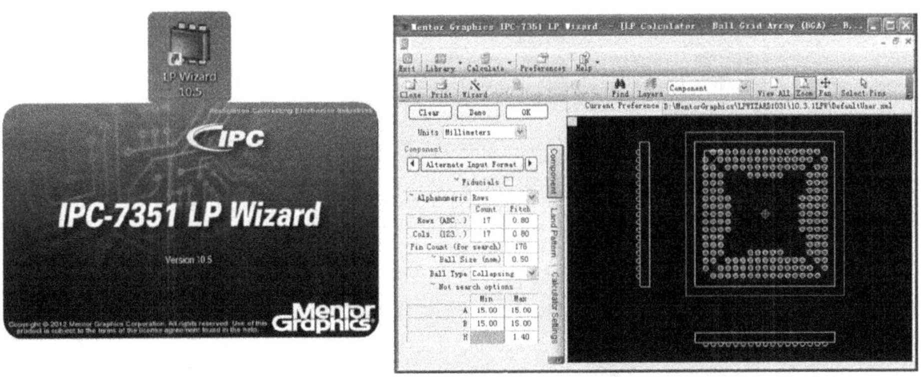

Figure 6.37 LP Wizard professional library tool.

6.6 Part Creation

After completing symbol and cell creation, we need to map the symbol and cell to a part and input component attributes such as power consumption, cost, simulation model, etc.

Part creation is the same as symbol and cell creation; the first thing to be done is to create a partition, such as SiP_PART. Then, on the Library Manager toolbar, click the Part Editor button, and then click the new button to create a new part, name DIE_PART1 as the part number.

Usually, the part number uniquely identifies a component; it cannot be the same as other components. Part Name and Part Label can be set by the designer according to the design need; the naming is relatively flexible, as these mainly help users to identify the component conveniently.

At the same time, according to the requirements of the design, the designer can input parameters such as power, process, model, etc. in Component properties, as shown in Figure 6.38.

Then click the Pin Mapping button at the bottom of Part Editor; the Pin Mapping window appears as shown in Figure 6.39. On the left side of the window, in the Assign symbol area, click to import symbols. In the pop-up window, select the required

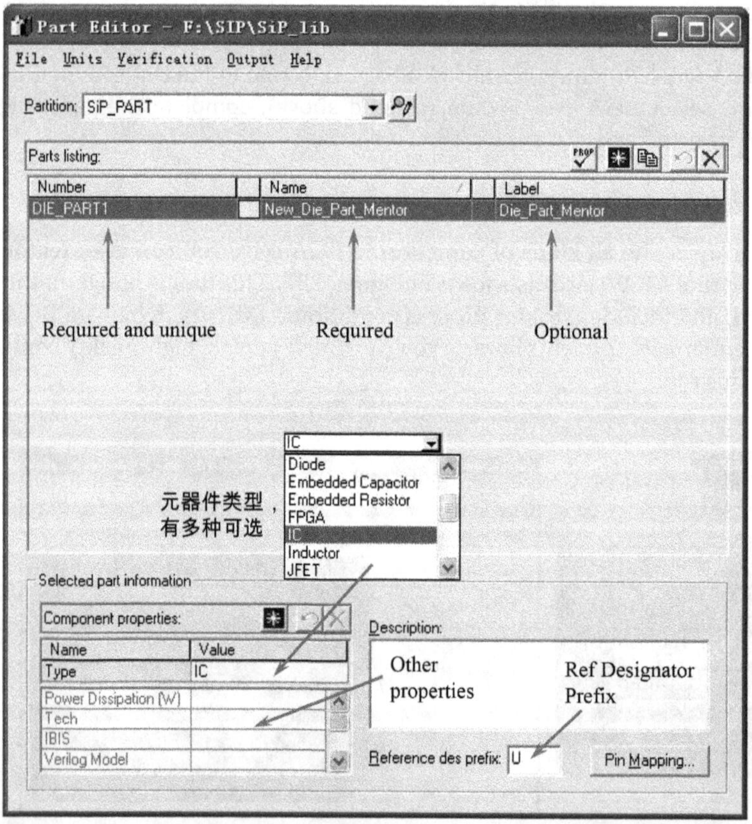

Figure 6.38 Create a new part in Part Editor.

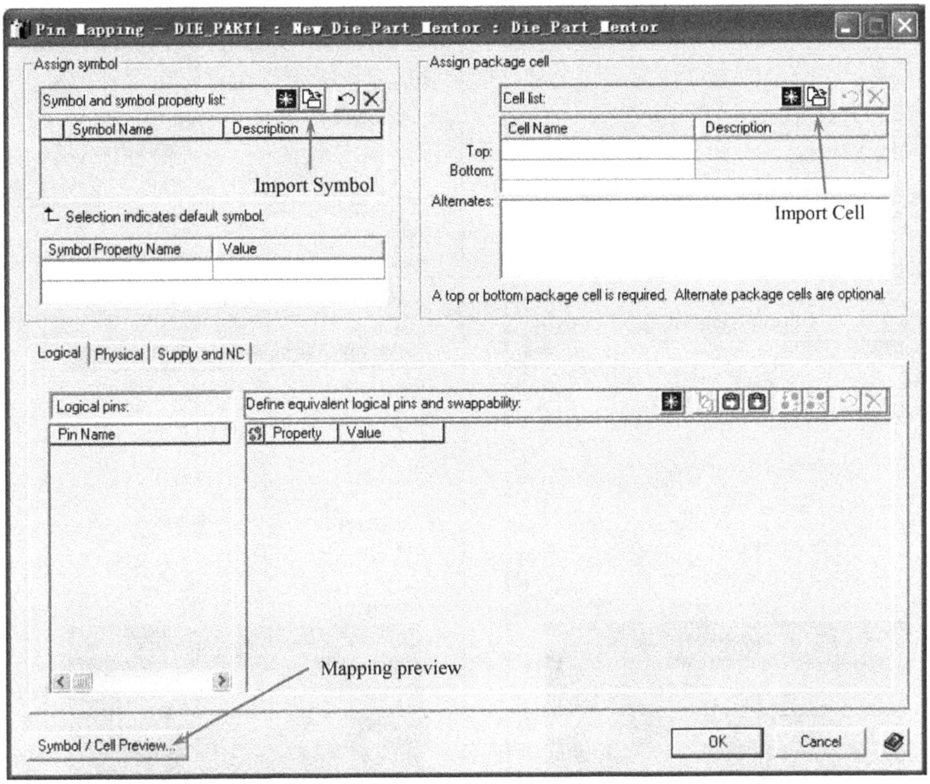

Figure 6.39 Pin Mapping window.

symbols, such as Sym1 under partition Sip_SYM, and input 1 in the number of slots in the component bar, indicating that the part contains only one such symbol. Then, as per Figure 6.40, check the corresponding option, and click the OK button.

Similarly, on the right side of Figure 6.39, in the Assign package cell area, click the import icon to import a cell. In the pop-up window, select the required cell, such as DIE_CELL1, then click OK.

A more efficient method is to switch directly to the Cells tab in the Import window, as shown in Figure 6.40 on the right side.

After import is complete, we can see the symbol and cell have been mapped, and the pin name and pin number have been mapped accordingly, as shown in Figure 6.41.

Click the Symbol/Cell Preview button in the lower right corner of the Pin Mapping window to preview symbols and cell graphically; visually check if the right objects have been mapped to each other, as shown in Figure 6.42.

If the symbol/cell mapping has problems, warning windows will pop up on exit or save, notifying you of the problem. If the symbol and cell have been mapped correctly, the part can be saved normally, thus effectively avoiding any problems when using this part.

Click the OK button to exit the Pin Mapping window, and save the part in Part Editor; DIE_PART1 has been created successfully.

Using the same method as above, create BGA part and map symbol/cell; on completion of mapping, the symbol and cell preview should be as shown in Figure 6.43.

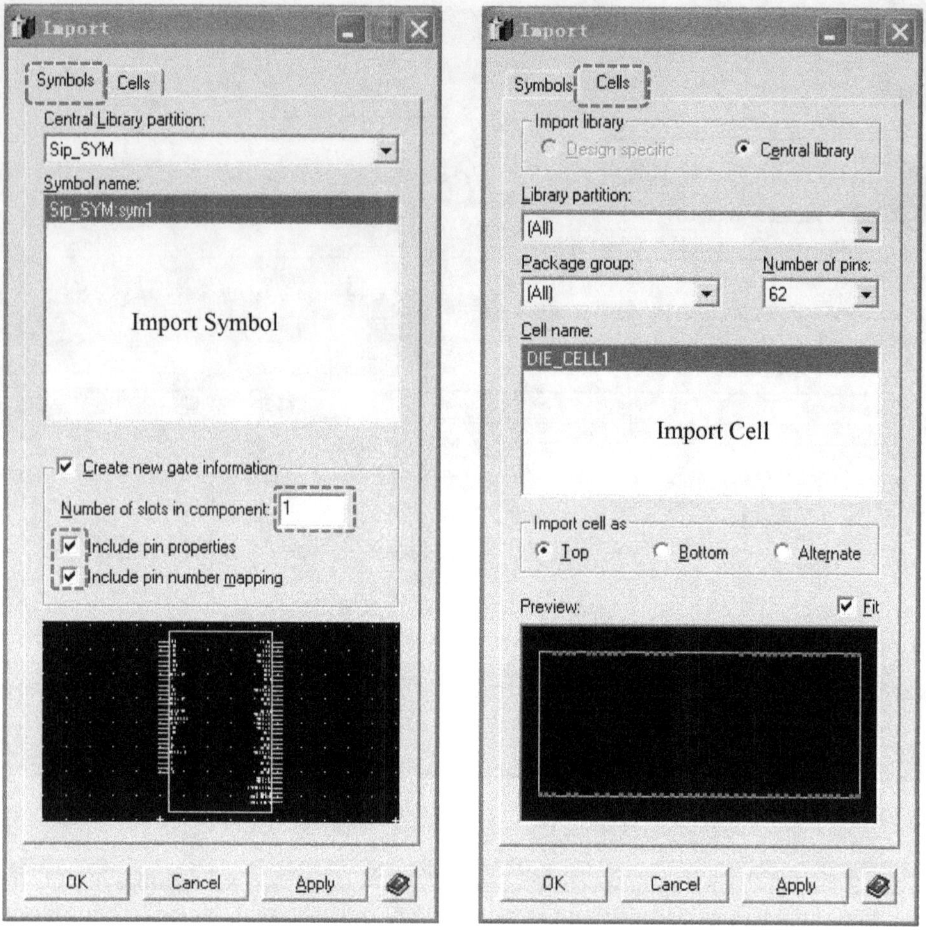

Figure 6.40 Import symbol and cell.

In the EE design process, all the components including bare chips and BGA package need to be mapped to parts; they can then be used in the design.

However, some special symbols such as power supply, GND, net connections, frame, etc. don't need a part because there is no corresponding cell to map them to; they can be added to the schematic directly from symbol partitions.

6.7 Create Cell Via Part

When you manually create a BGA cell, you need to rename pin numbers (Pin#), which requires a certain amount of effort. If you have created a symbol, the symbol has the pin number attribute set. To take advantage of this, we can create a part first, and then create the cell; the specific operation process is as follows:

1) After the symbol has been created, create a new part. Import the symbol in the Assign symbol area, paying attention to "Include pin properties" and "Include pin

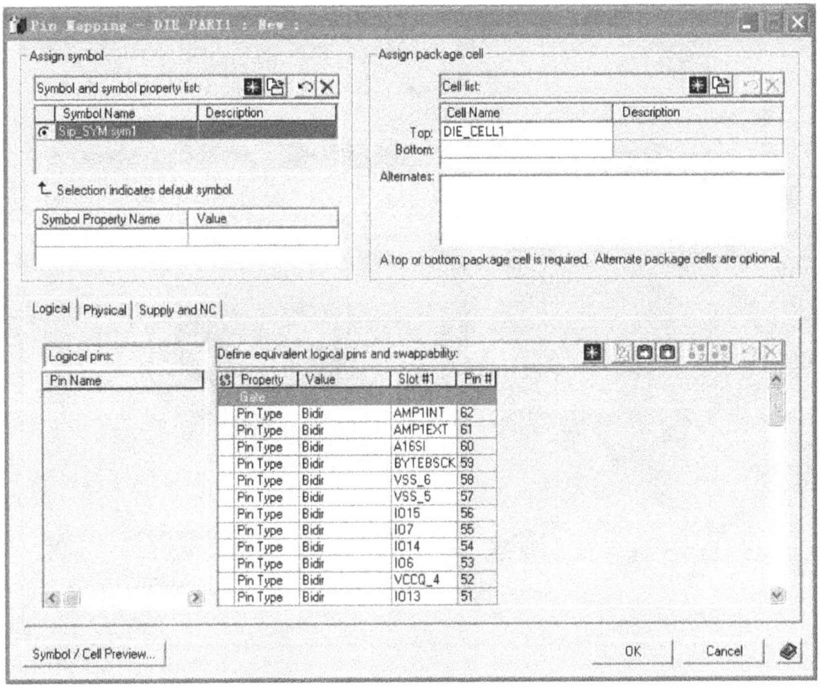

Figure 6.41 Symbol and cell have been mapped.

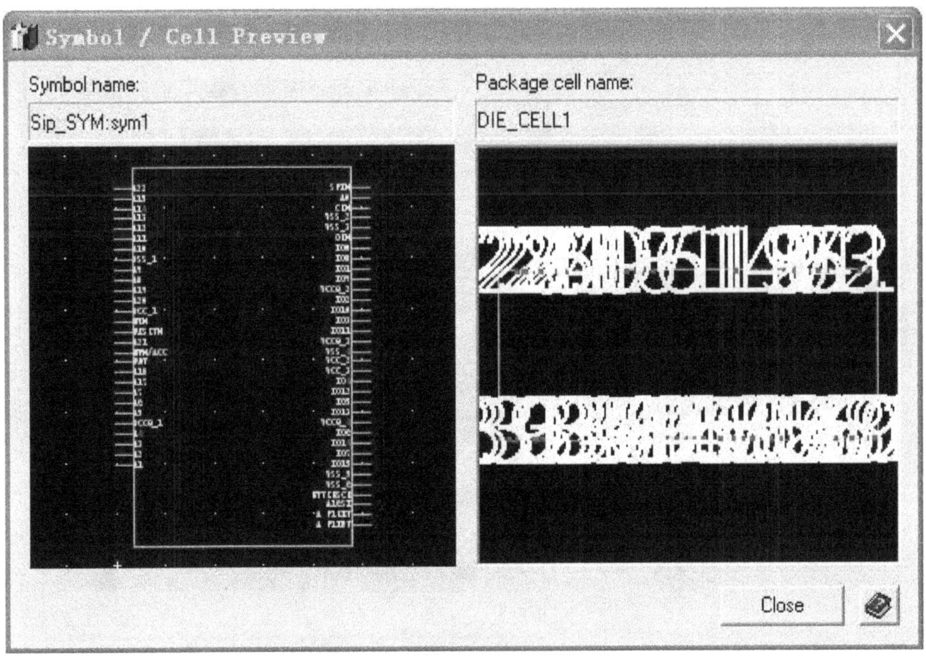

Figure 6.42 Symbol/Cell Preview window.

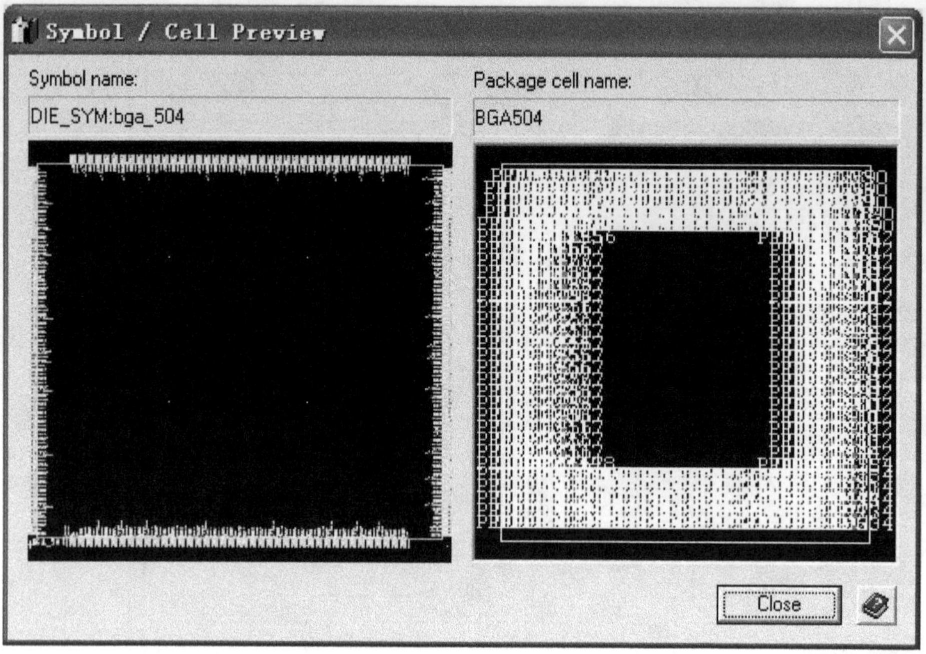

Figure 6.43 BGA Symbol/Cell Preview window.

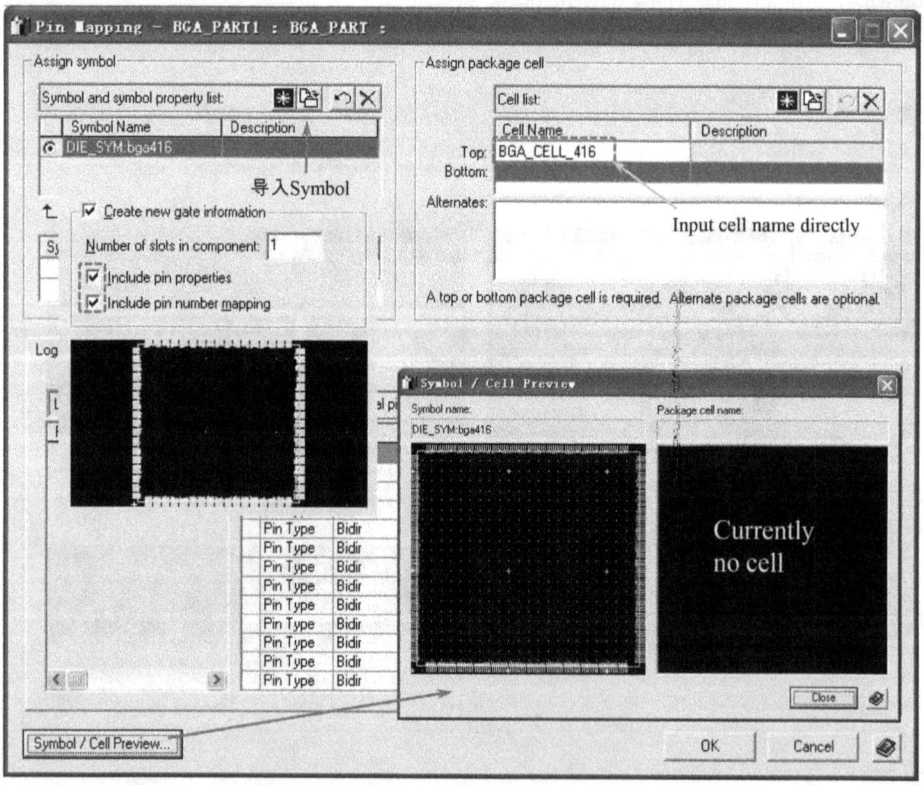

Figure 6.44 Create part with no corresponding cell.

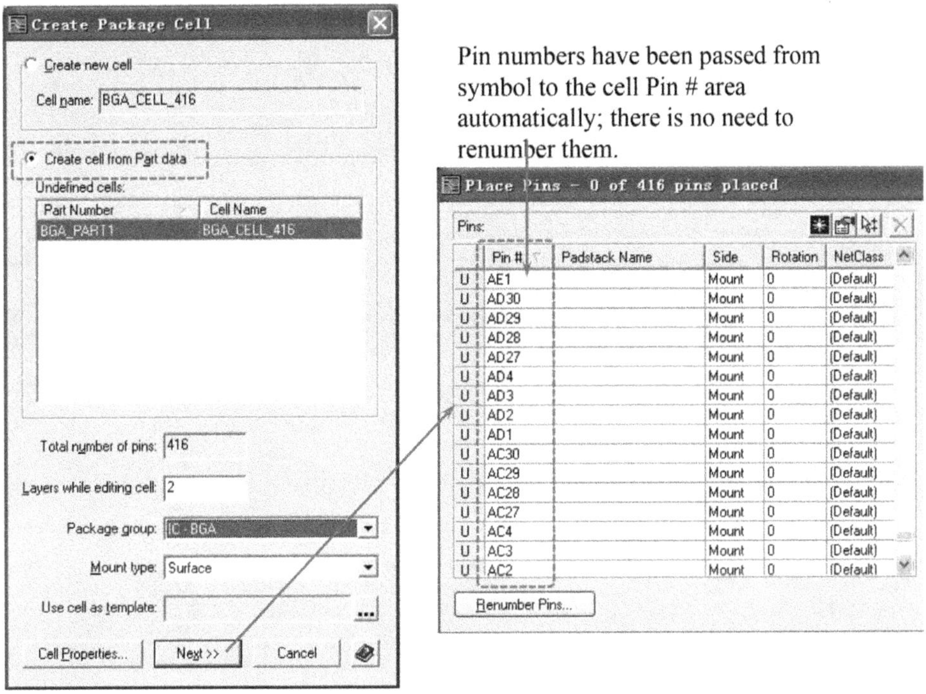

Figure 6.45 Cell inherits the pin numbers defined in symbol automatically.

number mapping"; both should be checked when importing. In the Assign cell area, directly input the name of the new cell; at this time no cell corresponds to the symbol, as shown in Figure 6.44.

2) On completion of part creation, create a new cell; select Create cell from part data, then we can see that part number and cell name as previously defined have already appeared in the list. After inputting the relevant properties, click the Next button to enter Cell Editor. In the Place Pins window we can see the Pin# area has inherited the pin numbers from the symbol definition, so there is no need to rename them; see Figure 6.45.

7

Schematic Input

7.1 Netlist Input

In system in package design, because there are connections between multiple chips, and chips also have interconnections with package, the connection relationship is complicated, so generally a schematic diagram must be drawn as standard input.

However, in the design of single-chip package, because the connection relationship is relatively simple, usually it is just necessary to connect bare chip pads to package pins according to certain mapping rules. So, many designers have used netlists as design input.

First of all, we discuss the netlist format of EE Flow.

EE Flow can support a variety of netlist formats; we introduce here the most commonly used: the keyin netlist. This netlist document is based on a .kyn suffix ASCII file, usually saved as *.kyn. Its main content is divided into two sections: the section "%net" for nets and the section "%Part" for components, as shown in Figure 7.1.

The net section begins from page%=NEWSCHEMATIC1; the first column is the network name, such as \AD0\. The pin name connected to this net is next, such as \C1\ - \1\; this indicates the first pin of C1, \U1\ - \11\ indicates pin 11 of U1.

The content after %Part is the list of components; the first column is the component part number, such as \BGA100\, the second column is the reference designator of the component, such as \U4\.

If it is a single-chip package, the written netlist will be very simple, according to the above file format: basically signal pins are assigned to package pins one to one; power supply pins and ground pins connect to outside package pins one to many, according to the design requirements. In this example, there are only two parts in the netlist, IC chip and package

Edit the netlist according to the standard netlist format, then the project can be created with that netlist. First, in the Dashboard Layout toolboxes find JobWizard tool 
and click to start Job Wizard, or use Start→All Programs→Mentor SDD→Job Manage Wizard to start the tool.

After startup, in the Job management tasks list select Create a new PCB job, click the Next button to set the project path, as shown in Figure 7.2, and then click the Next button.

In Project Editor, select the central library and the netlist.kyn previously edited; please note that in Netlist Type you should select Keyin Netlist, and then click the OK button. The central library and netlist are assigned to this new project, as shown in Figure 7.3.

SiP System-in-Package Design and Simulation: Mentor EE Flow Advanced Design Guide,
First Edition. Suny Li (Li Yang).

Netlist.kyn

```
;; V4.1.0
%net
%Prior=1

%page=NEWSCHEMATIC1
\AD0\   \C1\-\1\ \R1\-\1\ \U1\-\11\ \U2\-\34\ \U4\-\A2\
\AD1\   \C1\-\2\ \R1\-\2\ \U1\-\12\ \U2\-\35\ \U4\-\D1\
\AD2\   \C3\-\1\ \R3\-\1\ \U1\-\13\ \U2\-\36\ \U4\-\E1\
\AD3\   \C3\-\2\ \R3\-\2\ \U1\-\14\ \U2\-\23\ \U4\-\F1\
\GND\   \C2\-\2\ \R2\-\2\ \U1\-\10\ \U2\-\33\ \U4\-\A15\
\NET1\   \U1\-\15\ \U2\-\40\ \U3\-\15\ \U3\-\40\ \U4\-\G1\
\NET2\   \U1\-\16\ \U2\-\20\ \U3\-\16\ \U3\-\20\ \U4\-\H1\ \U9\-\1\
\NET3\   \U1\-\21\ \U3\-\1\
\NET4\   \U3\-\18\ \U3\-\19\
\VCC\   \C2\-\1\ \R2\-\1\ \U1\-\9\ \U2\-\32\ \U4\-\B2\

%Part
\BGA100P\   \U4\
\C0402\   \C1\ \C2\
\C0504\   \C3\
\R0402\   \R1\
\R0402_2\   \R2\
\R0603\   \R3\
\SIP_1\   \U1\
\SIP_2\   \U2\
\SIP_3\   \U3\
\SIP_4\   \U5\
\SIP_5\   \U6\
\SIP_6\   \U7\
\SIP_7\   \U8\ \U9\
```

Figure 7.1 Keyin netlist format.

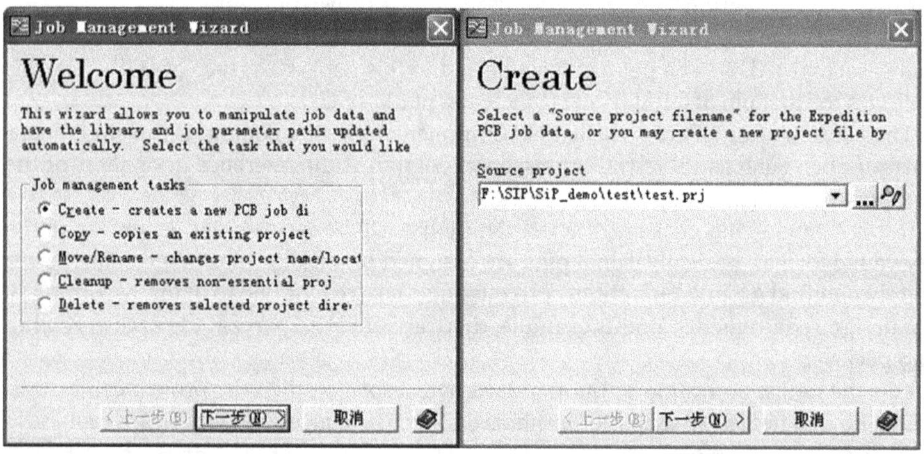

Figure 7.2 Create project with Job Management Wizard.

The next step is to select the layout design path and the layout template (the template is located in the central library), and then click the Next button; finally, click the Finish button, as shown in Figure 7.4.

The software automatically goes into the layout design tool, Expedition; execute Forward Annotation in Setup→Project integration; after forward annotation has run successfully, we can start layout design.

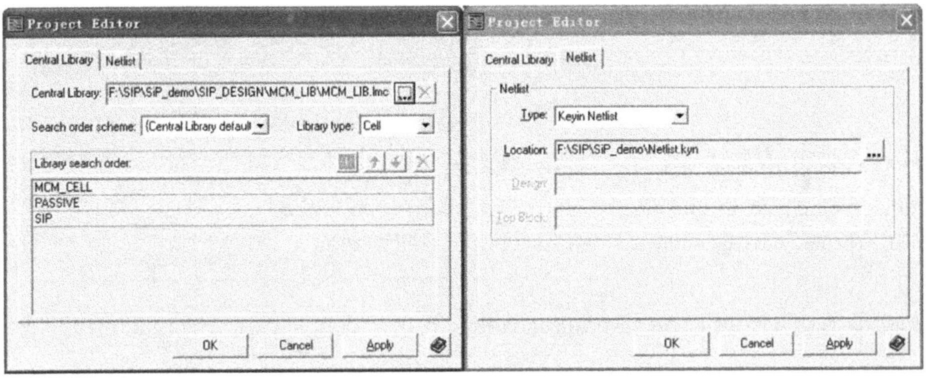

Figure 7.3 Assign central library and netlist for the new project.

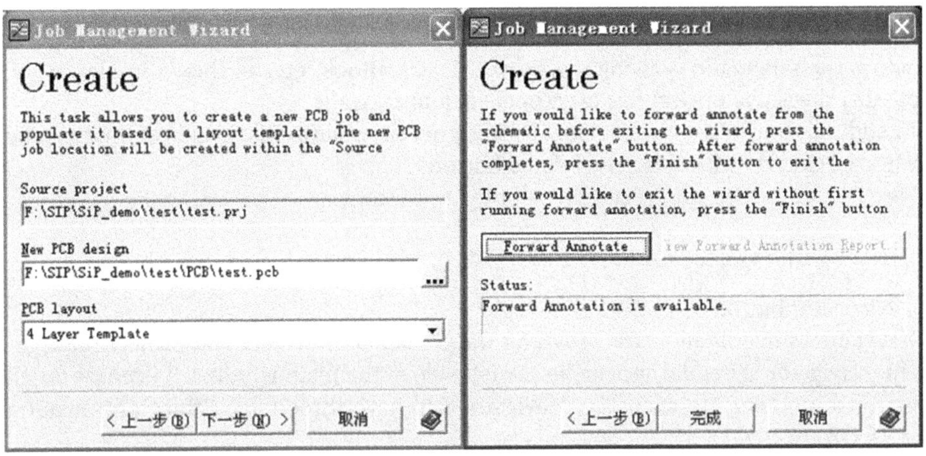

Figure 7.4 Specify layout design path and templates, complete project creation.

Expedition also supports other netlist formats, and the operating method is the same as above; however, due to space limitations of this book, we will not discuss them here.

7.2 Basic Schematic Input

7.2.1 Start DxDesigner

In SiP system-level package design, the schematic is usually the standard input method.

In the EE design process, DxDesigner is the standard schematic input tool. DxDesigner supports basic schematic input, schematic input based on DxDataBook component information, schematic simulation based on HyperLynx Analog, RF schematic input based on DxRFEngineer, etc. These modules require the relevant licenses; see Figure 7.5.

In the Dashboard Shortcuts quick launch bar, click the DxDesigner icon  to start DxDesigner.

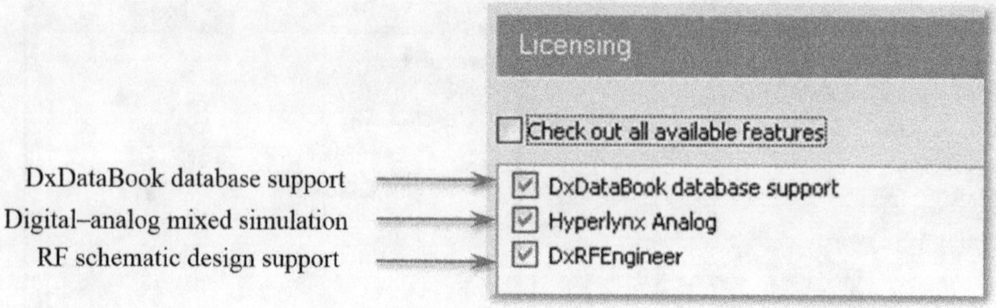

Figure 7.5 Extra licenses support extra functionality.

DxDesigner will automatically open the project in Dashboard which is set to active, as shown in Figure 7.6 (bold font display); we can switch the active project in the projects list by clicking on the required project name.

After startup, the DxDesigner window contains multiple sub-windows – the navigator window, the schematic drawing workspace, DxDataBook, etc. as shown in Figure 7.7. Note that the active project has been opened automatically.

In addition to the above interface, DxDesigner has many child windows that can be opened or closed by pressing the related button.

Next, we introduce some functions of the key buttons.

7.2.1.1 General Toolbar

Navigator button

Navigator is mainly used for browsing the elements in Project and Project management. Navigator generally appears in the left side of the main window. If licenses for all functional modules are available, Navigator will have three tabs: Project, Simulation and RF Groups.

The Project tab supports project management. DxDesigner supports multiple board management in one project, as shown in Figure 7.8, allowing management of the RDL board and the FC_PACKAGE board in project FLIPCHIP_DESIGN.

Each board contains its own schematics; each schematic can consist of one page or multiple pages.

Usually, each board only has one valid schematic; other schematics will be placed in blocks. The designer can use the right-mouse-button command Create Board to switch from block schematic to board schematic. He or she can also use the right-mouse-button Delete command on a board to delete the board; its schematic will automatically be

Figure 7.6 Bold font indicates that FLIPCHIP_DESIGN is the active project.

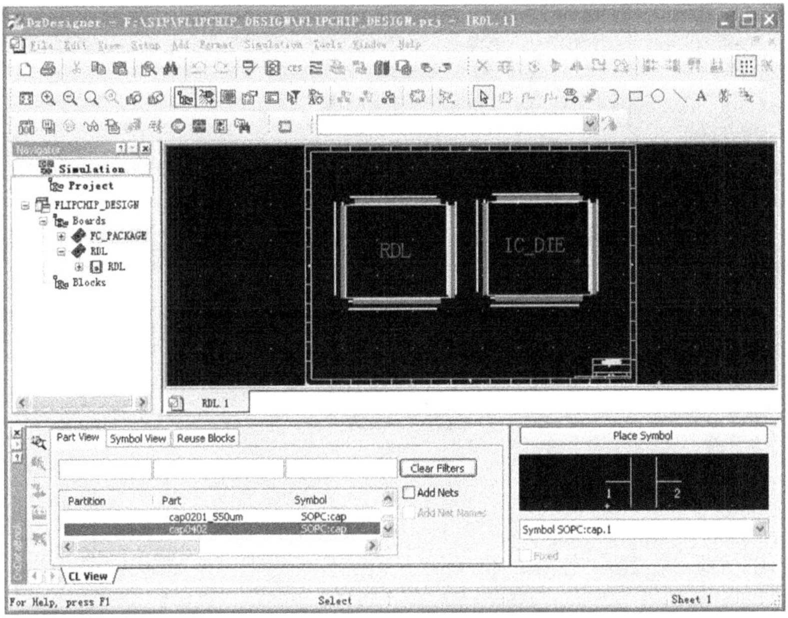

Figure 7.7 DxDesigner opens the active project automatically on startup.

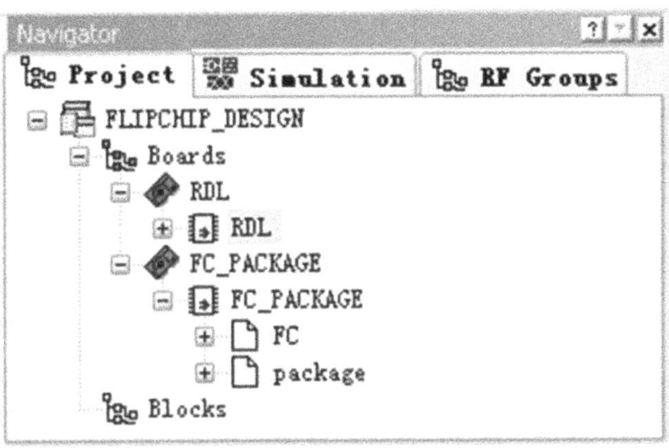

Figure 7.8 Project tab in Navigator.

converted to a block schematic. The schematics included in blocks cannot be package and generate PCB board, but can be referenced by other schematics.

The simulation tab is primarily used for function management of the simulation tool HyperLynx Analog, including the generation of test benches, and management of the simulation model library, such as import and management of model libraries, as shown in Figure 7.9.

The RF Groups tab is primarily used for RF design function management, including RF group creation and management, and other RF functions, as shown in Figure 7.10.

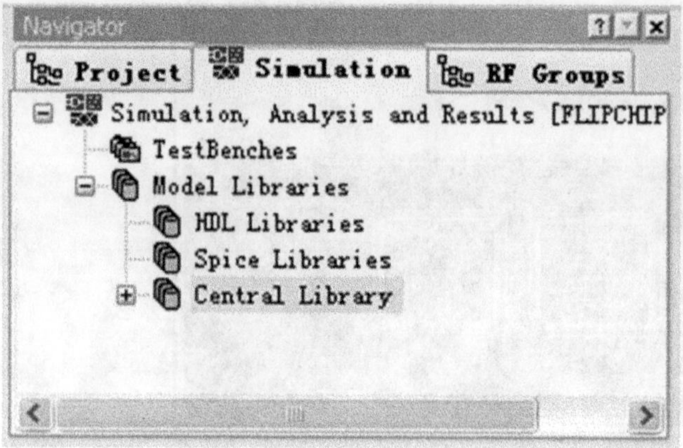

Figure 7.9 Simulation tab in Navigator.

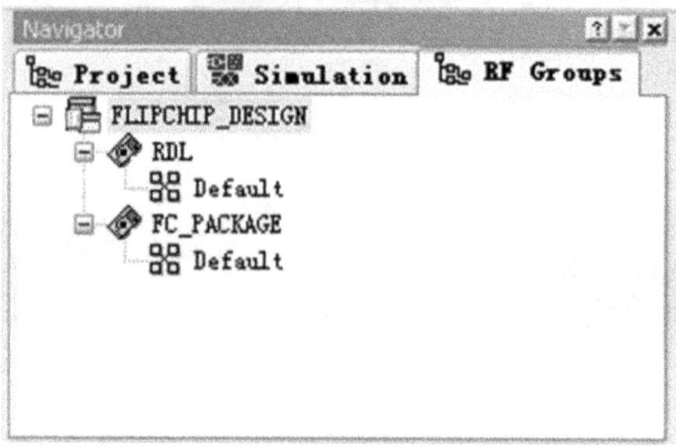

Figure 7.10 RF Groups tab in Navigator.

DxDataBook button

DxDataBook is mainly used to search for components and to place parts, and contains the Part View, Symbol View and Reuse Blocks tabs.

Components selected in Part View usually have physical cell units; the Symbol View is mainly used for such items as borders, power supply and connection symbols. Reuse Block is mainly concerned with the reuse of block modules, including logic reuse (schematic reuse) and physical reuse (schematic + layout reuse).

Select the needed component from the list, and then click the Place Symbol button or drag the symbol directly to the schematic window; the designer can use the Input filter box at the top of the list so that only components that meet the filter requirements will appear in the list, helping the designer to find components more easily. See Figure 7.11.

When placing components, checking the Add Nets and Add Net Names boxes will mean that the net and net name will be generated automatically in the schematic, as shown in Figure 7.12.

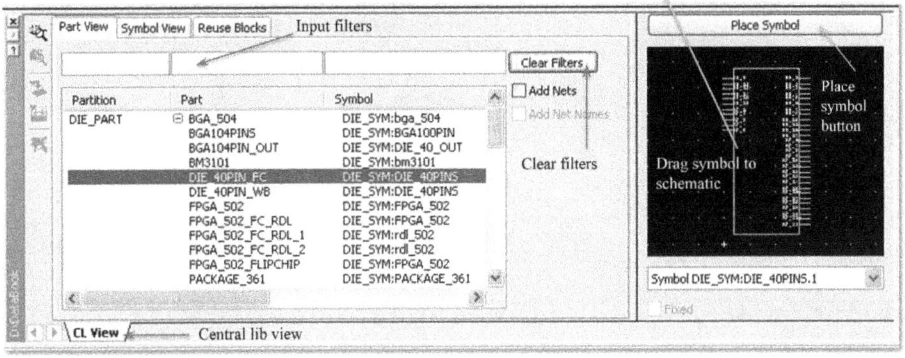

Figure 7.11 DxDataBook window.

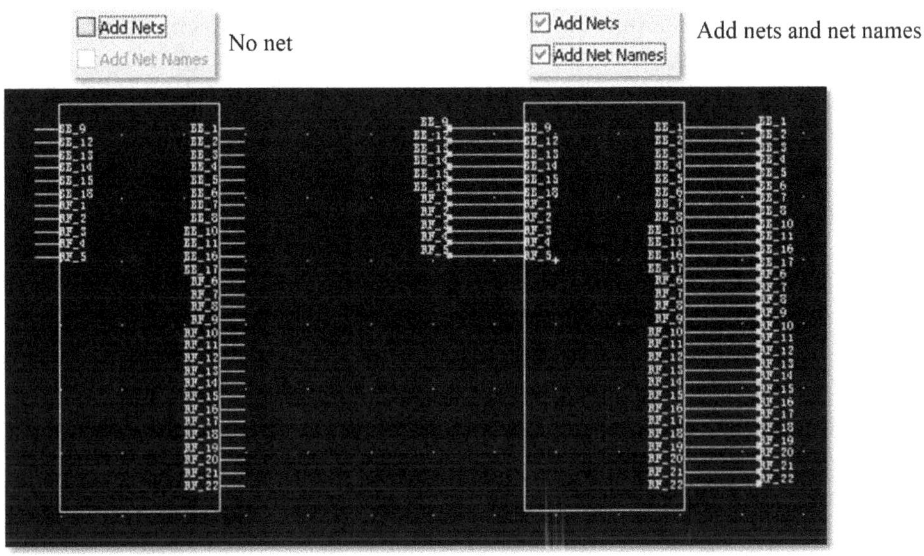

Figure 7.12 Add nets and net names in schematic.

The Search window is another powerful function of DxDataBook; click the New Search window button on the top-left corner of DxDataBook to start a new search window; a detailed description of this function will be found in Chapter 7, section 7.3.

ICT Viewer button

ICT (Inter Connectivity Table) Viewer contains three tabs: Hierarchy, Net Properties and Symbol Properties. The three tabs respectively allow viewing of net connection relationships, component attributes and net attributes. The tool can be used as a fast filter for design elements, and for sorting and searching nets and components, etc.; see Figure 7.13.

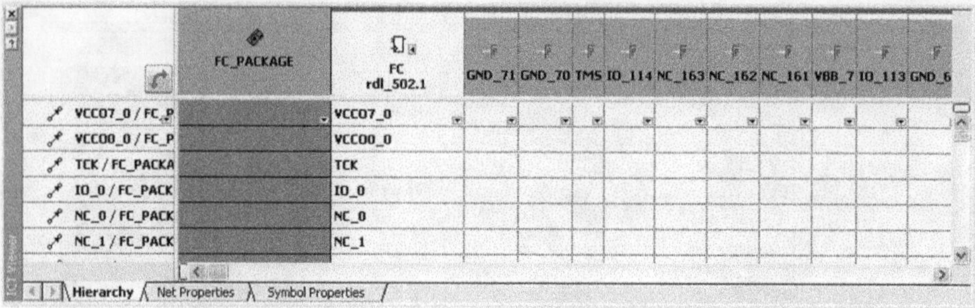

Figure 7.13 ICT Viewer.

eExp View button

The eExp View button allows viewing of Expedition layout design in the schematic environment. Using this button the designer can view the board design in DxDesigner. It is activated by checking the option "Create eExp View during Back Annotation" (this option is found in layout design; see Chapter 9) and successfully completing back annotation (i.e., the data are updated from layout to schematic).

With eExp View, the designer can easily see the layout design in the schematic environment and inspect and interact with it, without needing a license for layout design, as shown in Figure 7.14.

Property button

The Property button is used to display the properties of selected components or nets. The Property sub-window is usually placed on the right side of the schematic window.

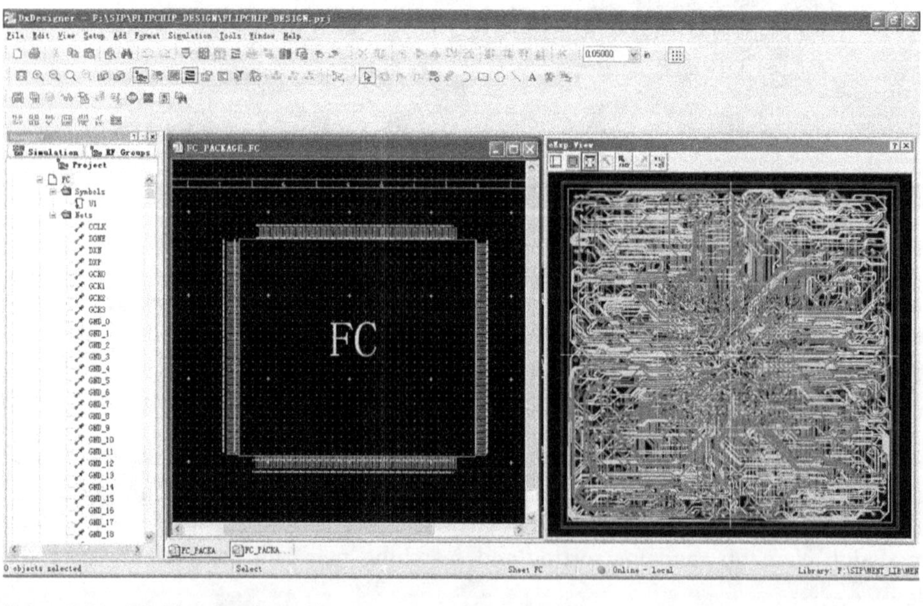

Figure 7.14 eExp View in schematic.

Output button

The Output button is mainly used to display compilation reports and other information; reports are usually placed below the schematic window.

Selection Filter button

The Selection Filter button allows designers to select different objects in the schematic; the 'All' option allows designers to quickly select or deselect all options, as shown in Figure 7.15.

Add Properties button

The Add Properties button allows designers to add attributes to a class of objects. Figure 7.16 shows how to add attributes to components and nets.

7.2.1.2 Digital/analog Simulation Toolbar

This toolbar is mainly used to support digital/analog simulation; however, because of space limitations, this function is not described here; the reader can expect the next book to deal with the simulation functions of Mentor SiP or PCB simulation.

7.2.1.3 RF Circuit Design Toolbar

This toolbar is mainly used for the design of RF circuits and data transfer with RF simulation tools (ADS/AWR); see Chapter 16 section 16.2 for RF schematic design operations.

7.2.2 Create New Project

7.2.2.1 How to Create a New Project

Start DxDesigner, choose File→New→Project, the New Project dialog pops up as shown in Figure 7.17; in Project Templates on the left, select Expedition→Default. If you need to do schematic simulation, select from HLA Eldo Library or HLA Library.

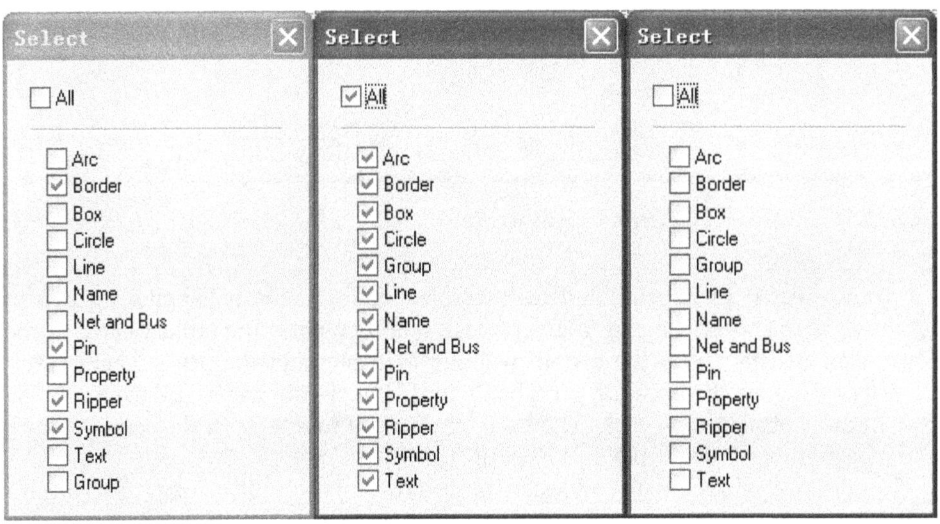

Figure 7.15 Selection Filter in DxDesigner.

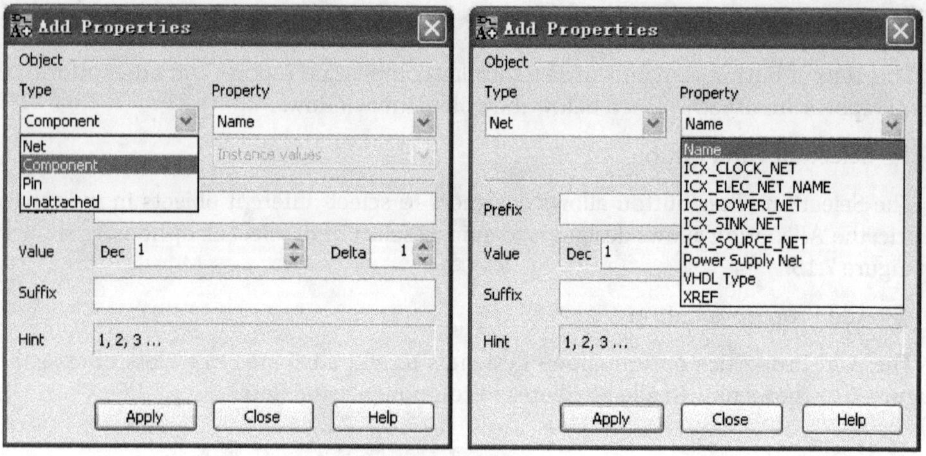

Figure 7.16 Adding properties to components and nets.

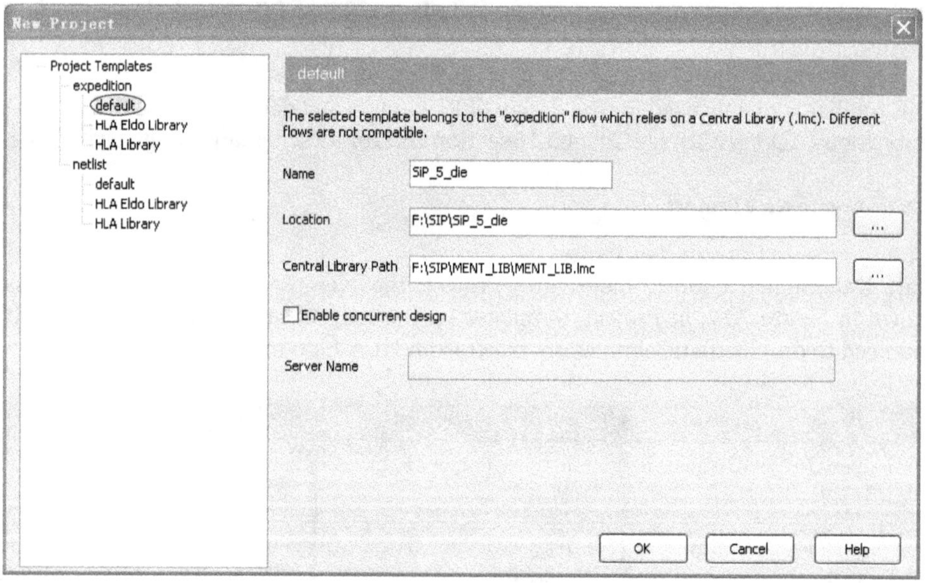

Figure 7.17 Create a new project in DxDesigner.

Having selected the Expedition default template, in the Location bar click the button on the right to choose the path of the project, and then enter the project name in the Name bar. At this point, the system will automatically add the project name after the selected path in the Location bar. The Central Library Path bar is used to choose the path of the central library; click the button on the right to search for the central library path, for example, F:\SIP\MENT_ LIB\MENT_LIB.lmc.

The option "Enable concurrent design" is used to support multi-person collaborative schematic design; usually you do not check this. If you do want to carry out concurrent schematic design (that is, many people designing one schematic at the same time), check this option and enter the server name. See Chapter 8 section 8.2 in relation to concurrent schematic design.

After completing all options, click the OK button; a new project is created successfully.

Click the New button in the new project, select new Board, and the system will automatically create a new board and corresponding schematic. Using the right-click menu select Rename, rename the board to SiP_Board, and rename the schematic to SiP_Sch, as shown in Figure 7.18.

The next step is to select components from the library and put them into the schematic. Usually in SiP design, there are two categories of components: bare chips and package.

Click the DxDataBook button ; in CL View mode select and place components. We can use the filter function to search for parts, allowing more rapid and accurate finding of components. For example, in the upper blank of Part, input "DIE", and the system will automatically display the components whose names begin with "DIE", as shown in Figure 7.19. Filtering can be done for all columns, such as partition, part and symbol.

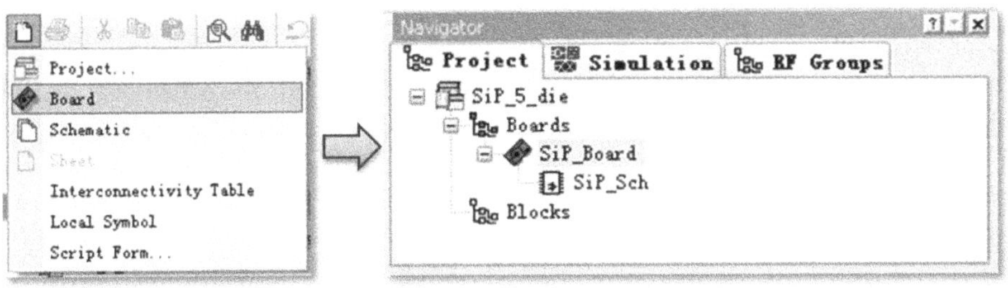

Figure 7.18 Create new board and corresponding schematic, then rename.

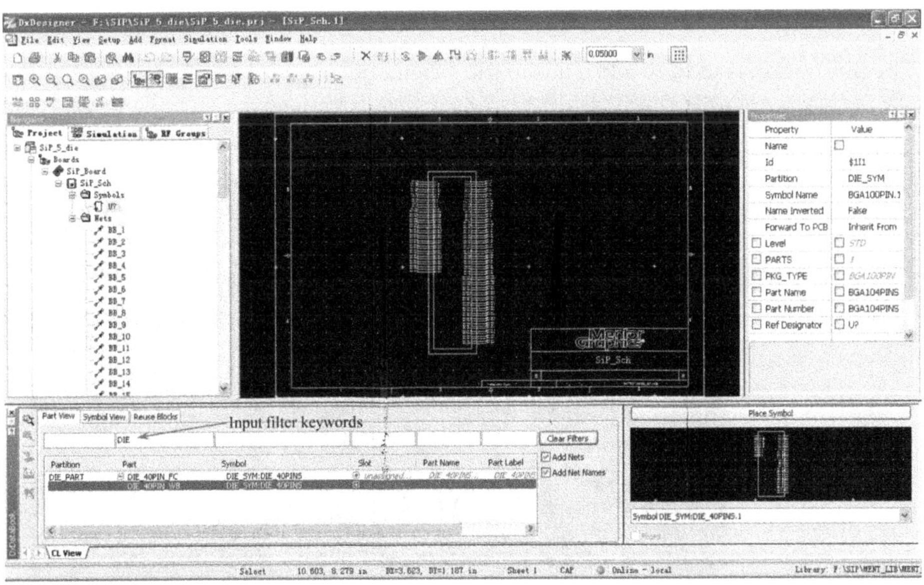

Figure 7.19 Search for and place components using filter function.

After selecting appropriate components from the central library and placing them into the schematic, we can close the DxDataBook window to avoid taking up workspace area.

The next step is to create interconnections between components; there are different ways to do this. If Add Nets and Add Net Names are checked during component placement, the system will automatically add nets and net names for the placed components; the net name is the same as the pin name, as shown in Figure 7.20.

At this point, nets with the same pin name will automatically connect in the schematic; this is very convenient for design in which the signal name is the same as the pin name.

7.2.2.2 Net Connection and Draw Toolbar

The net connection and draw toolbar is mainly used to add and link elements. Here we introduce six tools, the net connection tools.

Select button

The Select button is used to select elements in the schematic; it can be used together with the Selection Filter.

Block button

The Block button is used to draw function blocks directly in the schematic; as the top-level module of a hierarchical schematic design, a block can be introduced into the underlying schematic by the Push command. A block pin will be added automatically when connecting to a net, and the pin name is the same as the connected net name, as shown in Figure 7.21.

Net button

The Net button is used to add net connections in the schematic.

Bus button

The Bus button is used to add a bus in the schematic.

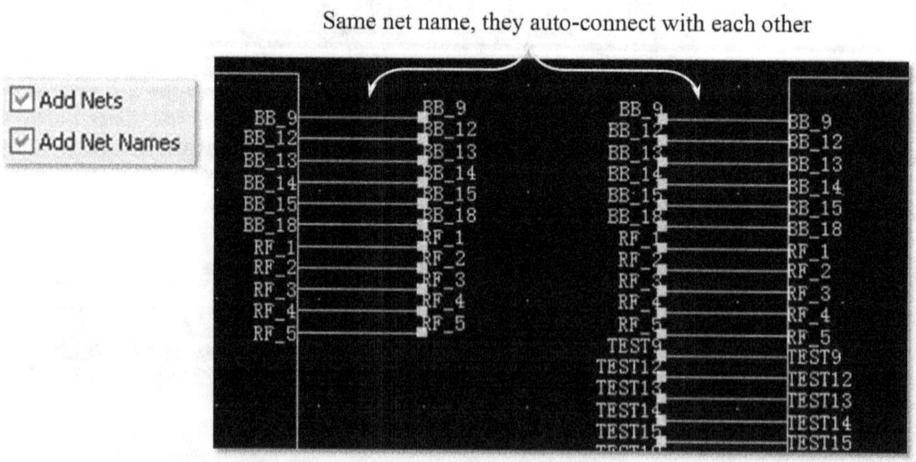

Figure 7.20 Automatically add nets and net names.

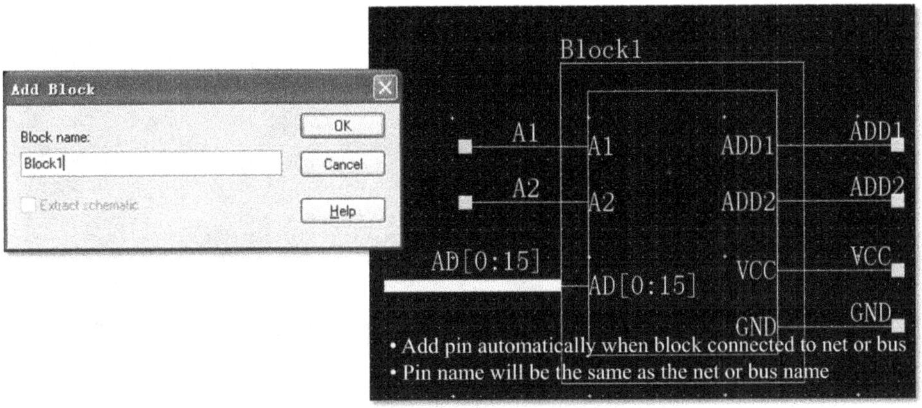

Figure 7.21 Draw function block and add net connections in schematic.

Special Components button

The Special Components button is mainly used to define input/output cross-page connection, power supply, ground, etc. When first used, it is necessary to define these via the Settings window; select the corresponding symbol in the central library to match. Special components can be used directly after they have been defined.

In the Special Components drop-down list, select symbols not defined before, such as Port OUT; the Settings window automatically pops up. Click the new button in the Settings window, select the corresponding symbols in the pop-up window, and then click OK. In the Special Components drop-down list we can see that the symbol has been defined; at this point, this special symbol can be selected and placed in the schematic directly, as shown in Figure 7.22.

Array copy circuit button

When using this feature, first select the circuit that needs to be replicated in the schematic, then click on the Array icon; the Array windows pop up. For (1) the Rectangular

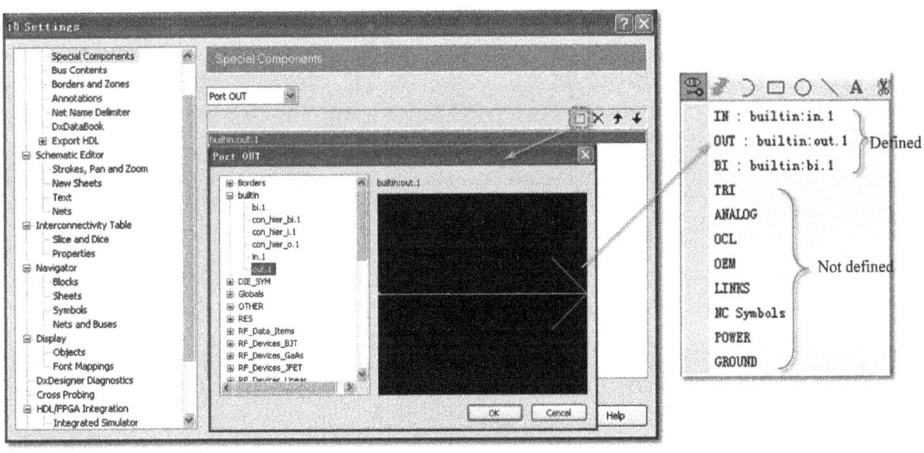

Figure 7.22 Special components definition and use.

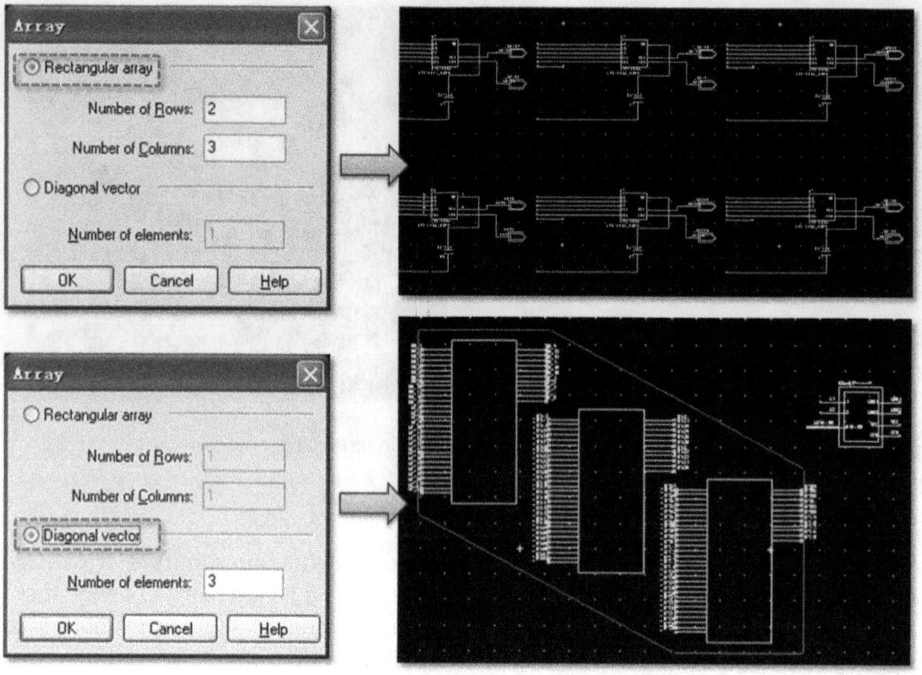

Figure 7.23 Array copy circuit diagram.

array, input rows and columns, moving the mouse to adjust the spacing; the replication result is shown in Figure 7.23 (upper graphic). For (2) the Diagonal vector, input the number of elements that need to be replicated, and move the mouse to the appropriate position; the result is shown in Figure 7.23 (lower graphic).

7.2.3 Schematic Design Check

After completion of schematic input, the designer needs to check the schematic design. Click the Verify button, select the Settings tab in the DRC (Schematic) pop-up window, and check items such as sheet, block, board, etc.; usually the default options can be kept, as shown in Figure 7.24.

Switch to the Rules tab and select the DRC Rules item. The designer can change the value of relevant items, as well as the severity of inspection reports, notes, warnings or error tips. In the absence of special inspection requirements, the default can be kept, as shown in Figure 7.25.

After DRC check is complete, we can check the results in the DRC report in the Output window, as shown in Figure 7.26. If there is no error or warning in the DRC results, or the designer can establish the cause of any such messages, the next step can be taken.

7.2.4 Design Rules Setup

In the DxDesigner main interface, click the Constraint Editor System (CES) button CES ; here the designer can enter the CES interface to set the relevant design rules, as shown in Figure 7.27.

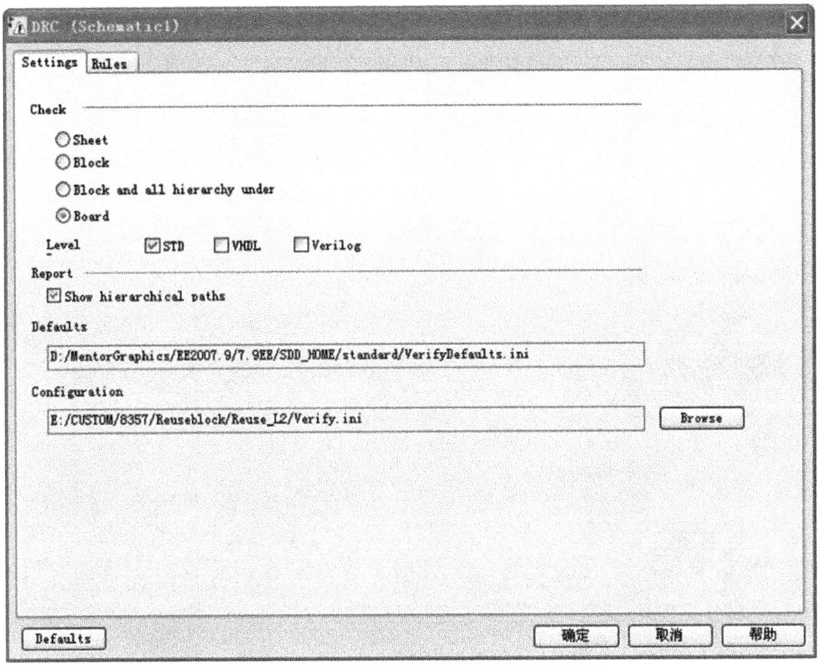

Figure 7.24 DRC (Schematic) Settings window.

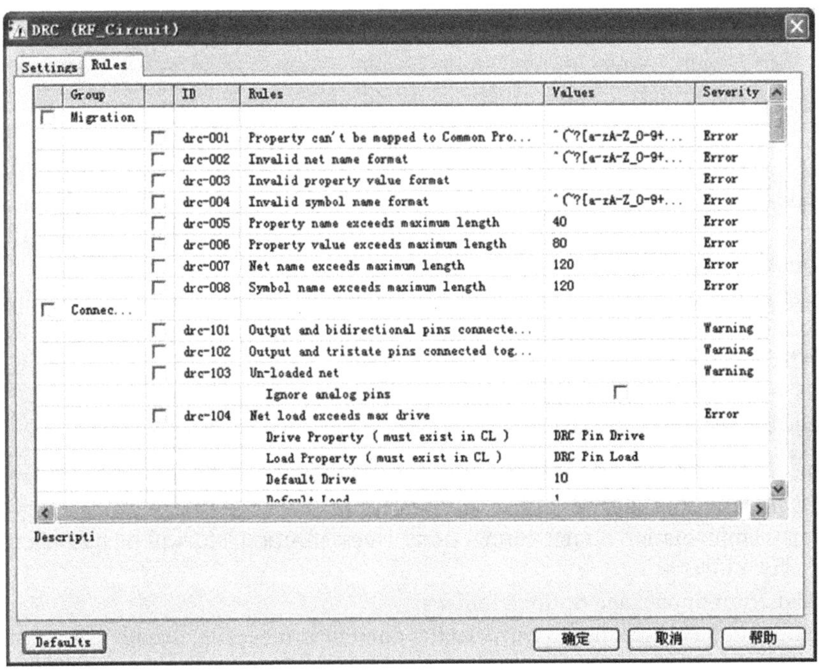

Figure 7.25 Schematic DRC Rules window.

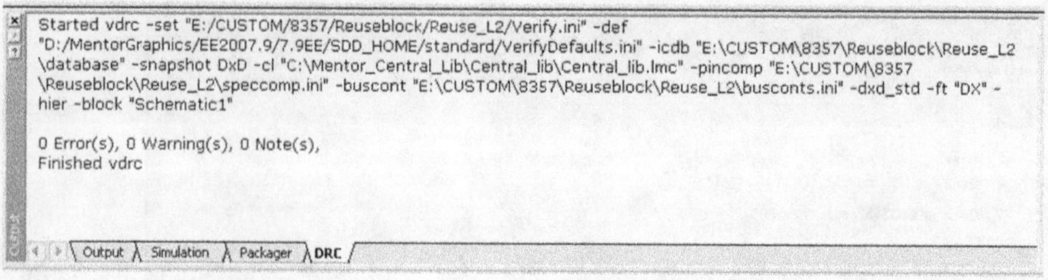

Figure 7.26 Schematic DRC report.

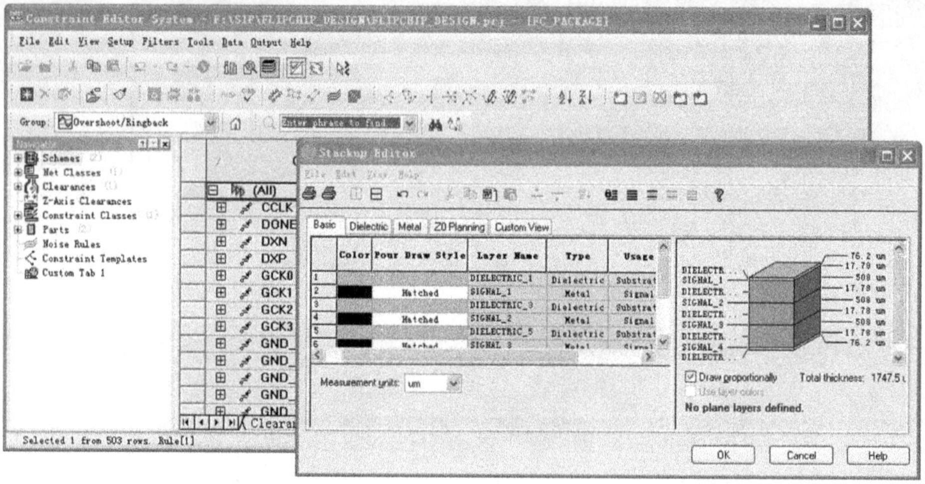

Figure 7.27 CES interface.

Because at this time we have not yet entered layout design, the relevant physical parameters cannot be set. Other rules, such as line width, clearance, differential pair, length match, net topology planning, etc. can be set at schematic stage.

The rules set in the schematic can be passed to the layout environment automatically using package and forward annotation, so as to guide layout design. Please refer to Chapter 10 for details of CES setup and constraint rules management.

7.2.5 Package Design

In the toolbar, click the Package button; the Packager window will pop up, as shown in Figure 7.28. Please note that "package" here refers to "package design"; in package design, schematic information on net connections, rule definition, etc. will be packaged and passed to layout design.

A brief explanation of package options follows.

The Project bar, which by default points to the current project file, usually does not need to be changed.

The control options include two parts: Packaging Options and PDB Extraction Options, as shown in Figure 7.28.

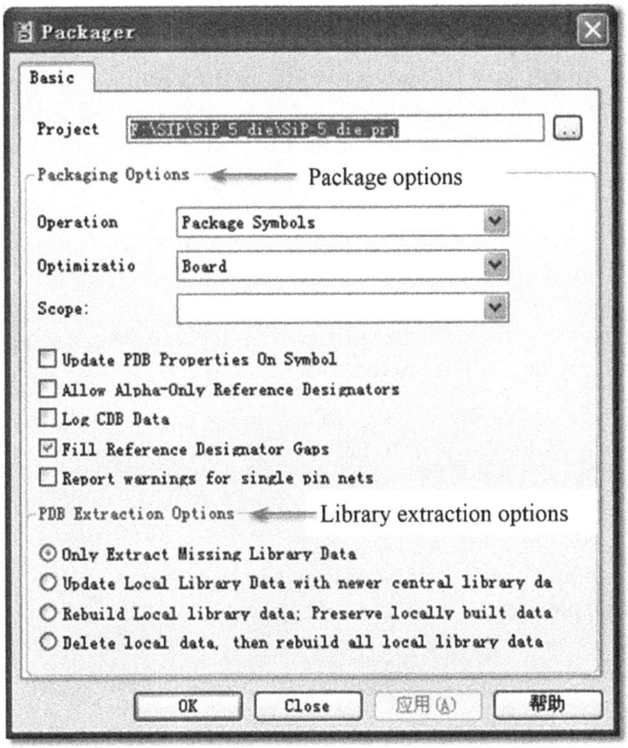

Figure 7.28 Packager window.

7.2.5.1 Packaging Options

3) Operation includes four items: Package Symbols, Repackage All Symbols, Repackage Unfixed Symbols, Verify Packaging, as shown in Figure 7.29.

   Package Symbols is the most commonly used item, and is used to package and assign a Ref designator to components which have not previously been packaged and had a Ref specified.

   Repackage All Symbols is used to package all components, include packaged components, and can cause the "Frozen Package" component property to be removed.

   Repackage Unfixed Symbols is used to package all components in addition to components with the "Frozen Package" property; the system will package unpackaged "Frozen Package" devices.

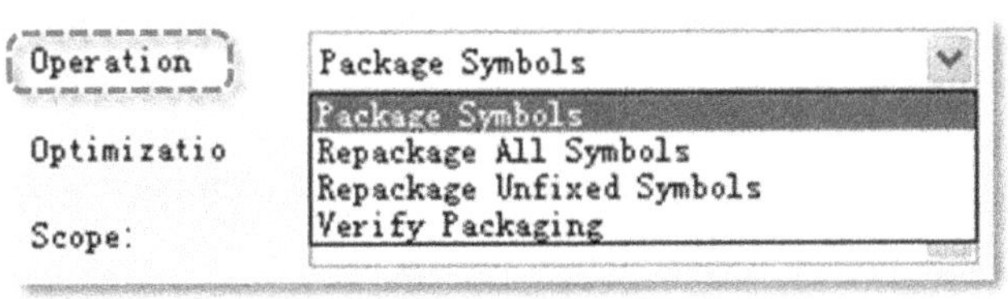

Figure 7.29 Operation items.

Verify Packaging is used to check the correctness of package; if there are package errors, they will be written to the Partpkg.log file.

If the component property “Frozen Package=Fix” is set, the options Package Symbols and Repackage Unfixed Symbols cannot change the existing reference designator and pin mapping, but the Repackage All Symbols option will remove this property and repackage the reference designator and pin mapping of this component.

4) Optimization options include the three options, board, block and page, applying to the entire board, or a block or page; this allows the same kind of parts to be merged into the same components for packaging (part can include multiple symbols), as shown in Figure 7.30.
5) Scope allows the designer to set the range of packaging; options include design, schematic, page, etc. If left blank, package applies to the whole design.
6) Other items, as shown in Figure 7.31:

   a) Update PDB Properties on Symbol – update part database properties to schematic symbol;
   b) Allow Alpha-Only Reference Designators;
   c) Log CDB Data – write packaging information to log file;
   d) Fill Reference Designator Gaps – make reference designators consecutive;
   e) Report warnings for single pin nets.

7.2.5.2 PDB Extraction Options

PDB Extraction Options mainly contains the following four options, as shown in Figure 7.32.

1) Only Extract Missing Library Data. Extracts only those parts that do not already exist in the local copy. Does not check for newer versions of local parts.
2) Update Local Library Data with newer central library data. Extracts only those parts that do not already exist in the local copy. Overwrites existing local parts with newer versions if they exist in the central library.

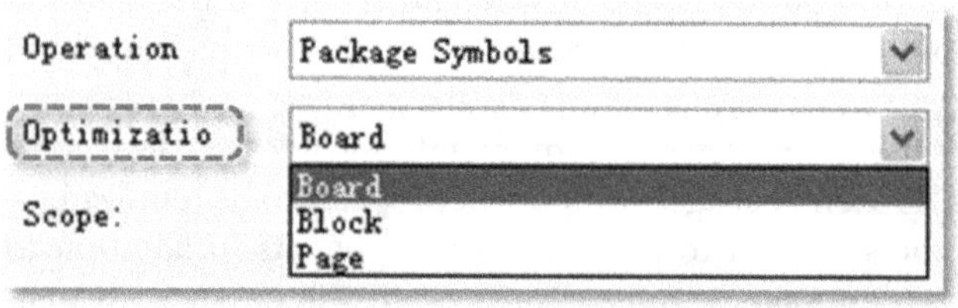

Figure 7.30 Optimization options.

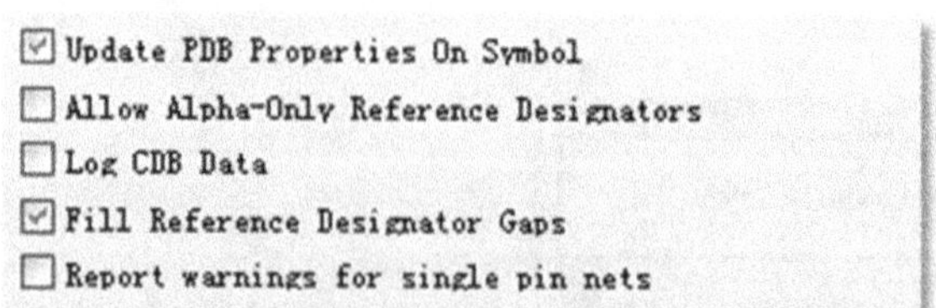

Figure 7.31 Other options.

Figure 7.32 PDB Extraction Options.

3) Rebuild Local library data. Preserves locally built data. Deletes local data and re-extracts. Exception: If you have imported any data directly into the local copy (for example, an FPGA Pin File), that data is preserved.
4) Delete local data, then rebuild all local library data. Deletes all local data and re-extracts.

In actual design, when packaging we can select options according to the design needs. Usually in a new design package, select option 1) Only Extract Missing Library Data.

If the central library data is updated during the design process, and the central library data update includes symbol, cell or part data, depending on the design needs, we need to select items 2), 3) or 4) to update the local data from the central library.

Designers should be aware that, as well as choosing a specific option here, in the Expedition layout design tool's Project integration window, we need to select a corresponding option, and the two options must remain consistent. For example, if item 2) is selected in PDB Extraction Options, it must also be selected in the Project integration window of Expedition, as shown in Figure 7.33, to achieve the desired effect.

After the completion of packaging, if a report such as shown in Figure 7.34 appears, packaging has been successful. If packaging is not successful, an error message will appear and the packager window will automatically pop up again. The designer must check the error message, and modify the components until packaging is successful; he or she can then proceed to layout design.

7.2.6 Partlist Output

In the DxDesigner main interface, selecting Tool→Part Lister allows output of the partlist for production or procurement. The Part Lister window is divided into three tabs, General, Columns and Advanced.

The General tab is mainly used to control general options, such as the partlist scope definition of a project, board or block. If there are multiple boards or blocks in this project, selection can be made from the drop-down list.

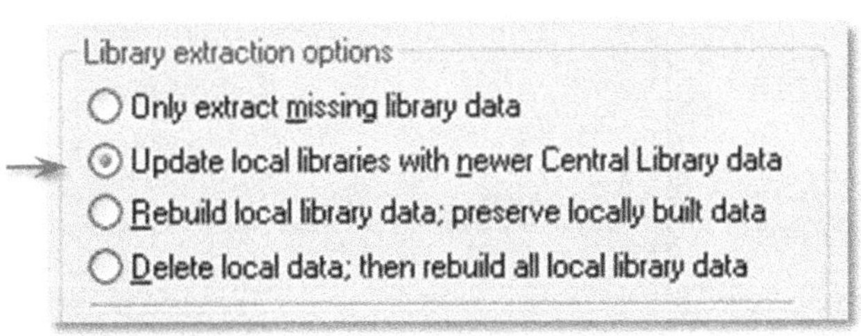

Figure 7.33 Library extraction options in Expedition.

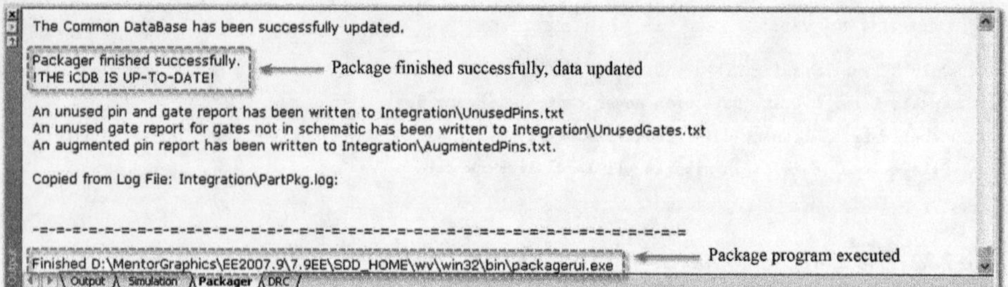

Figure 7.34 The report information after packaging has finished successfully.

The Output format bar is used to select the Output file format, including Text File, HTML and EXCEL. The Output file bar is used to set the name of the output file.

When "Open the generated file" is checked, the generated partlist file will automatically open, as shown in Figure 7.35.

As shown in Figure 7.36, the Columns tab is mainly used to set up the content of the partlist output file. The list on the left shows the output partlist entries; we can use the

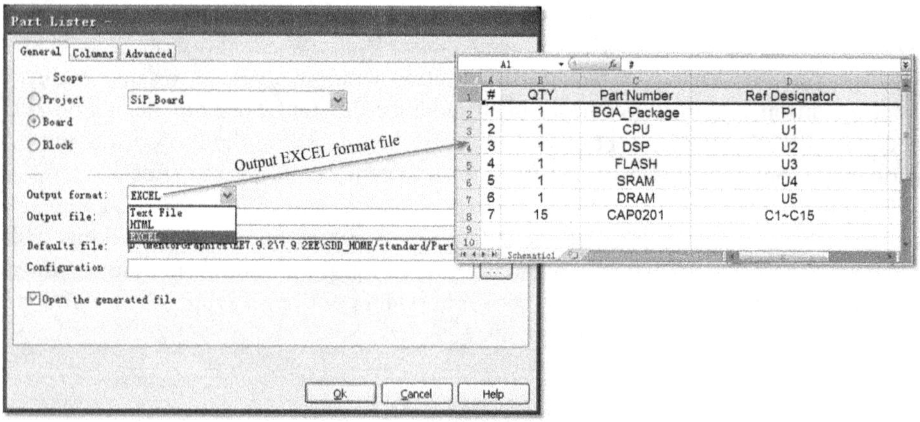

Figure 7.35 Output Excel format partlist.

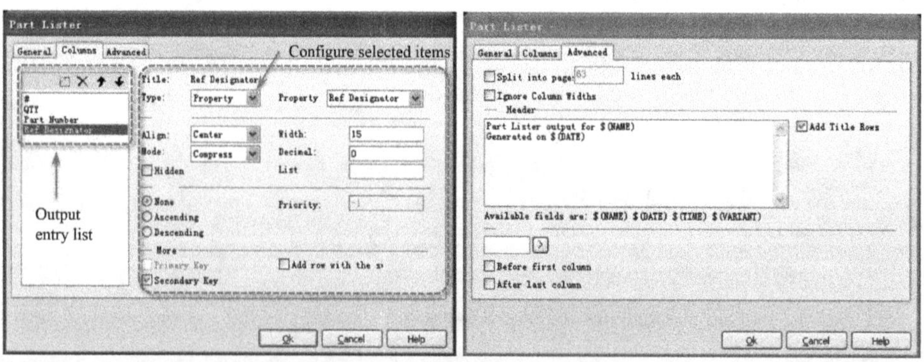

Figure 7.36 Columns and Advanced tabs of Part Lister.

button at the top to create a new entry, delete an entry or arrange the order of the entries. The right side is used to configure the selected item.

The Advanced tab is mainly used for Text File format configuration; if HTML or Excel format is selected, the Advanced tab is grayed out and not available.

After completion of the settings, click the OK button; you can now output the part-list file.

7.2.7 Chinese Input in Schematic

For Chinese designers, English cannot replace Chinese as the mother language. In the process of schematic input, sometimes designers need to write Chinese characters in the schematic diagram, as design specification and description.

In DxDesigner, inputting Chinese is very easy. First, select the Chinese character configuration. Select from the menu→Settings→Display→Font Mappings, in the pop-up window choose one font, such as Kanji, and then click the Edit mapping button; in the new pop-up window specify mapping font, such as "宋体". After clicking the OK button, you can see "宋体" is mapped to the Kanji font. Similarly we can map Fixed as "宋体", Sans Serif as "宋体", Plot as "华文隶书", etc., as shown in Figure 7.37.

After font mapping, click Apply or OK; now you can directly type Chinese words. Click the Text button A, then click the left mouse button in the DxDesigner workspace, and type Chinese words directly into the pop-up Text Properties window.

Kanji is the default font for Chinese font mapping, because Kanji has been mapped to "宋体" above, so it is shown as song font. If you want to show other fonts, such as "华文隶书", change the font to Plot, which has been mapped to "华文隶书" font.

In Figure 7.38 we can see that, on the same page of the schematic, the designer can input a variety of Chinese fonts.

In addition to the need to map Chinese words, sometimes, in order to make the design clean and trim, we also need to map the font for English letters. In DxDesigner the

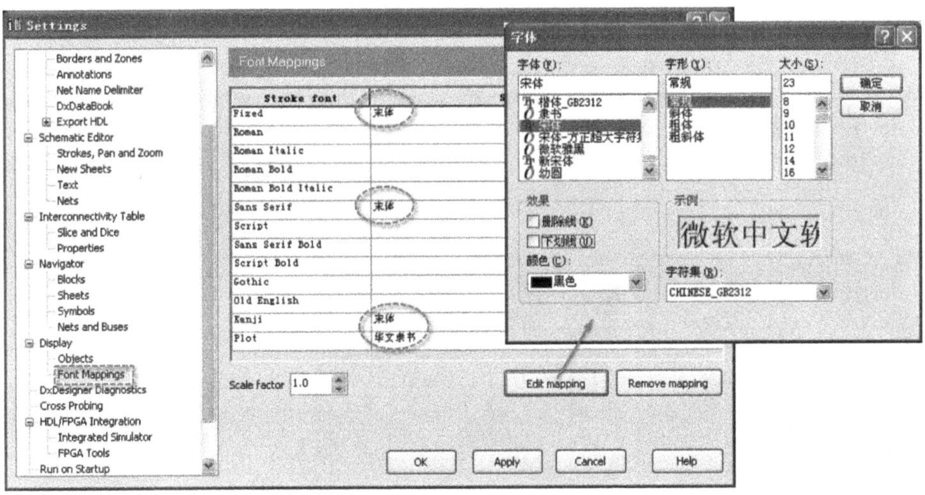

Figure 7.37 Chinese font mapping.

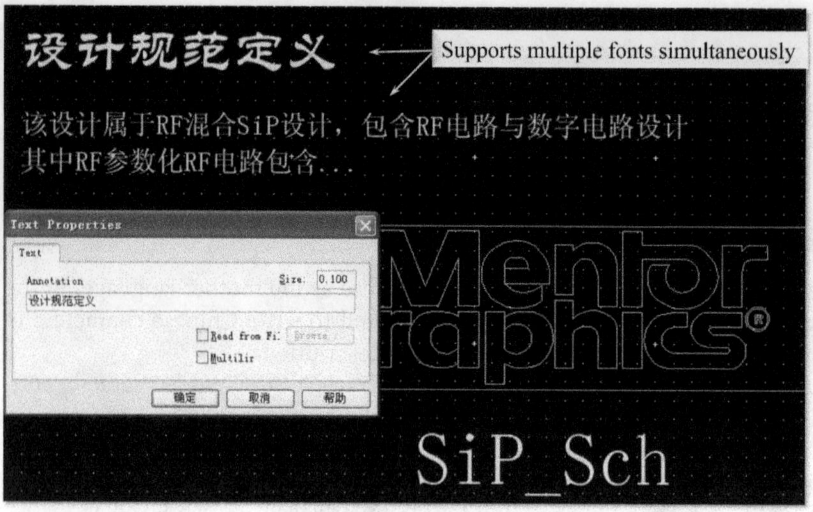

Figure 7.38 Input a variety of Chinese fonts in schematic.

default font is fixed font, its character spacing is loose, and its stroke is thin; the display effect will be significantly improved after font mapping.

Figure 7.39 shows a comparison of the display effect of Fixed font and "宋体" font, and it can be seen that there is a very big difference in the display effect after fixed font mapping to song font.

In a single-computer environment, fonts need to be mapped only once. Later, no matter whether DxDesigner Schematic Editor or the Edit symbol in Symbol Editor of Library Manager is opened, the font mappings are valid. But if EE Flow software is reinstalled, font mapping must be redone.

7.2.8 Enter Layout Environment

In the DxDesigner main interface, click the button, and on the DxDesigner to Expedition PCB pop-up, select layout template and PCB directory, as shown in Figure 7.40.

In MENT_LIB, the default template has two options: the four-layer and the eight-layer template. The designer can also customize the layout template; in the template

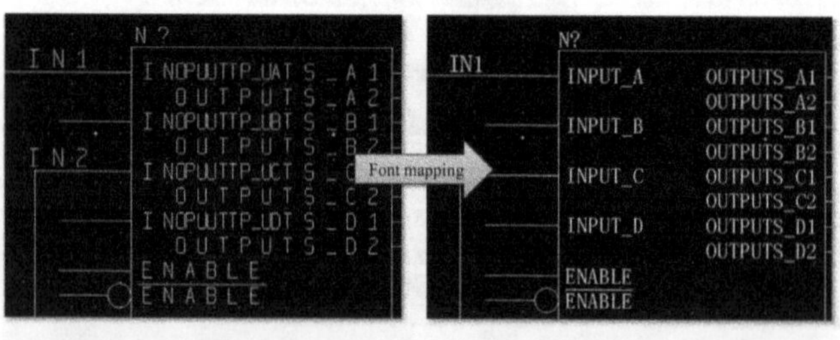

Figure 7.39 Display effect comparison before and after font mapping.

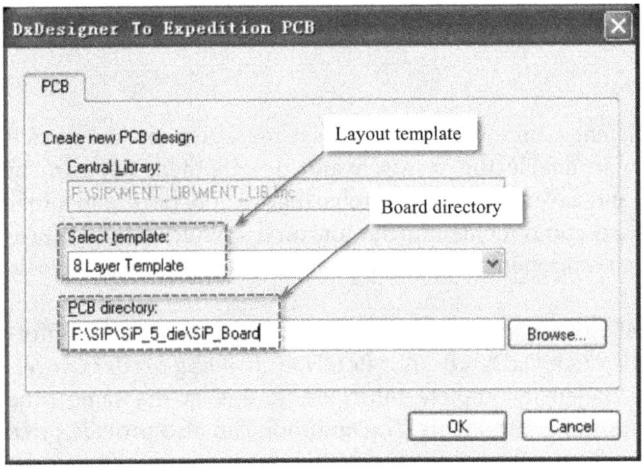

Figure 7.40 Select layout template and PCB board directory.

we can set substrate structure, material parameters, substrate shape and size, etc. For more about the definition of templates, see Chapter 9 on layout board creation and setup.

Click OK button after selection, and the DX2EXP window automatically pops up, as shown in Figure 7.41, with the prompt "PCB directory does not exist, create?" Choose yes, and the system will enter Expedition and open the layout design environment.

The first time you enter Expedition from the schematic, the system will automatically pop up the dialog box shown in Figure 7.42, stating "Back annotation has been disabled for current project".

This is a warning message; just click the OK button, and go into the Expedition layout design environment. For a detailed introduction to the layout design environment, please refer to Chapter 9 on layout creation and setup.

Figure 7.41 Create PCB directory.

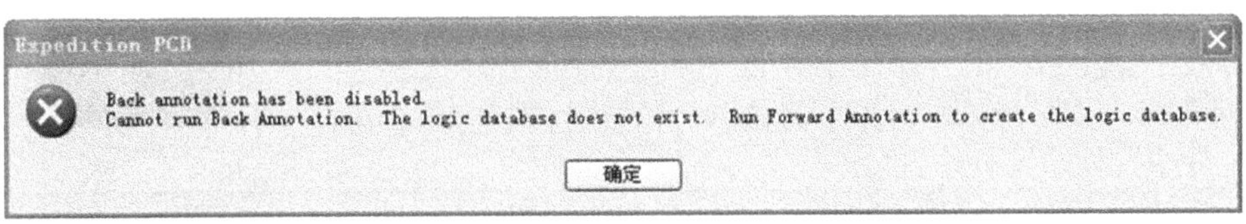

Figure 7.42 Back annotation has been disabled for current project.

7.3 Schematic Input Based on DxDataBook

7.3.1 DxDataBook Introduction

DxDataBook provides an efficient component information management and search function for design work; it can enable the research and development department (including schematic design and layout design), purchasing, production, and other functional departments to share component information on a unified data platform and to work together to ensure components in design are in accord with the enterprise standard.

The latest research and development, component purchasing and production information can be quickly linked through DxDataBook, effectively avoiding product development or production delay due to incomplete information, and at the same time helping to optimize the selection of components. DxDataBook can also provide price information on the component material in the system; this can be carried out in the schematic design phase for cost accounting purposes. This is shown in Figure 7.43 for the DxDataBook interface.

DxDataBook is a bridge between the symbol library and the component properties database; the designer only needs to create one symbol for each kind of component. Properties are not attached to the symbol itself; all properties are available from DxDataBook at design time.

DxDataBook has a property verification function which checks property information for components in design according to the material standard components in the system, finds non-standard components in design and gives an appropriate warning, thus providing an effective guarantee for product preparation and smooth production process implementation.

DxDataBook can connect the component information databases of a company; when placing components from DxDataBook, information and component are put together in the schematic, such as the parameters, price, inventory and manufacturer of the component, etc.

DxDataBook can effectively reduce the library build work. Looking, for example, at resistors, there are many types, resistances, powers, precisions, etc. While the parameters

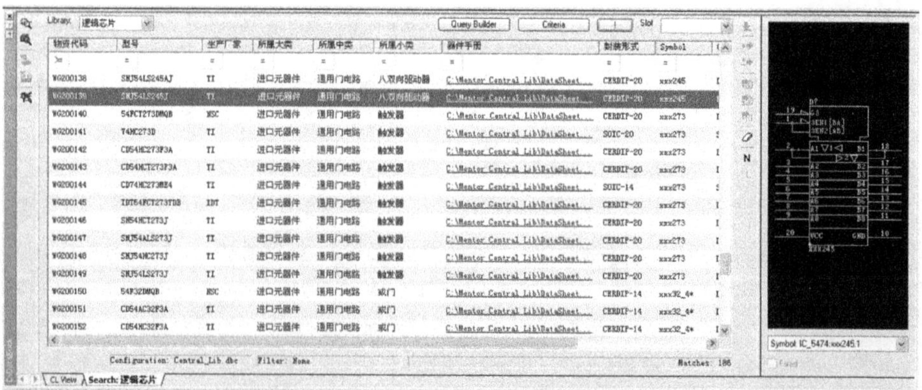

Figure 7.43 DxDataBook interface.

are different, the symbol in the schematic is the same. Without DxDataBook it would be necessary to map many parts in the library, a laborious task. With DxDataBook, the designer can create just one part, and then input a variety of property information to the database using DxDataBook.

DxDataBook first requires configuration and it can then be used normally. Due to space limitations, we simply describe its usage here; we do not deal with the detailed configuration of DxDataBook.

7.3.2 DxDataBook Usage

Click the New Search window button in the upper left corner of the DxDataBook window, open a new search window, in this window, click the right mouse button and select the Configure→Open, Open Central_Lib.dbc file, a file which has been configured in advance in the central library, as shown in Figure 7.44.

Then click the drop-down arrow to the right of the Library bar; there will be a list similar to that shown on the left of Figure 7.45. Select any of these items, such as common chips; the Select database window pops up automatically; then select the CIS.mdb database file, as shown on the right of Figure 7.45.

After clicking the OK button, the component list appears, as shown in Figure 7.46.

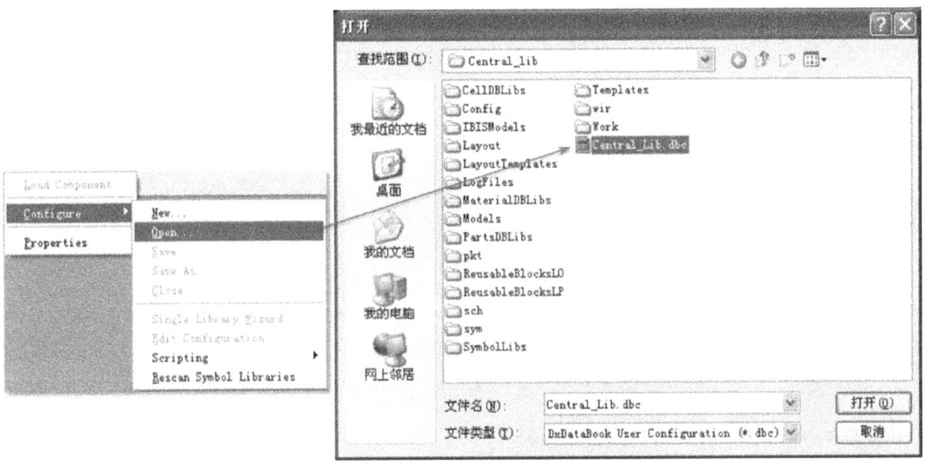

Figure 7.44 Open the Central_Lib.dbc file.

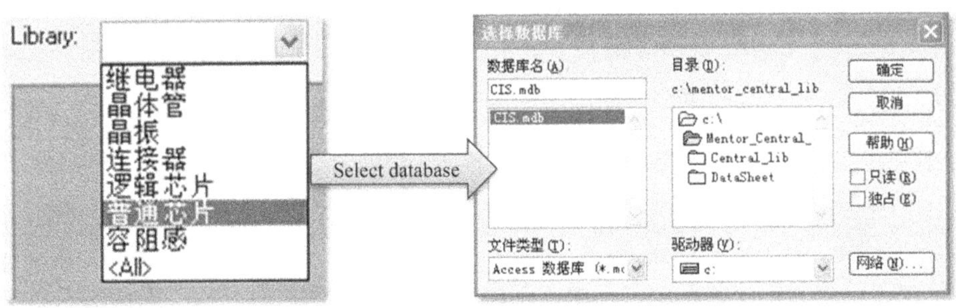

Figure 7.45 Select CIS.mdb database file.

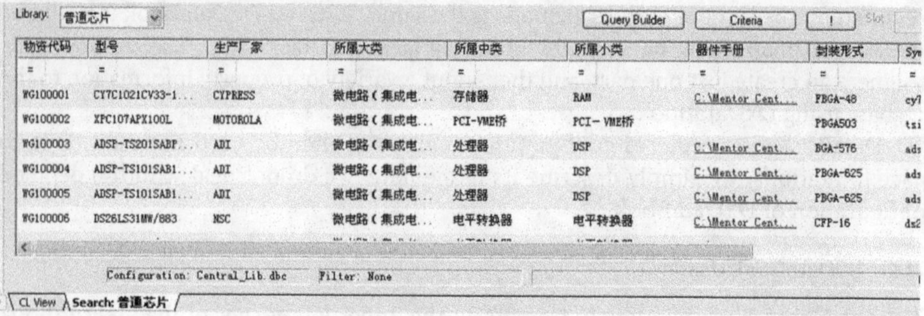

Figure 7.46 DxDataBook component list.

In this list, we can see the material code, type, manufacturer, whether the component belongs to the big, mid-size or small category, device manual and packaging type, etc. All these properties are not only convenient for designers in selecting components, but are also automatically loaded into the component properties in the schematic when components are placed in the schematic, as shown in Figure 7.47.

After the completion of design, in the process of standard file archiving, these properties can be extracted automatically by archiving software to a standardized document, without the need for the designer to fill it in manually, saving time, and at the same time effectively avoiding error-prone manual tasks, guaranteeing the consistency of design and documents.

In each column below the property name there is a corresponding filter to help the designer to filter components. After setting up the search criteria, such as "manufacturer is TI", click the exclamation mark button [!]; the list will contain only components of manufacturer TI, as shown in Figure 7.48.

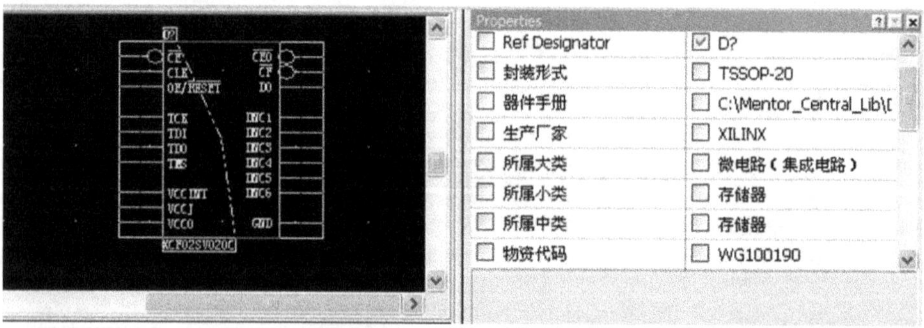

Figure 7.47 A component in the schematic and its properties.

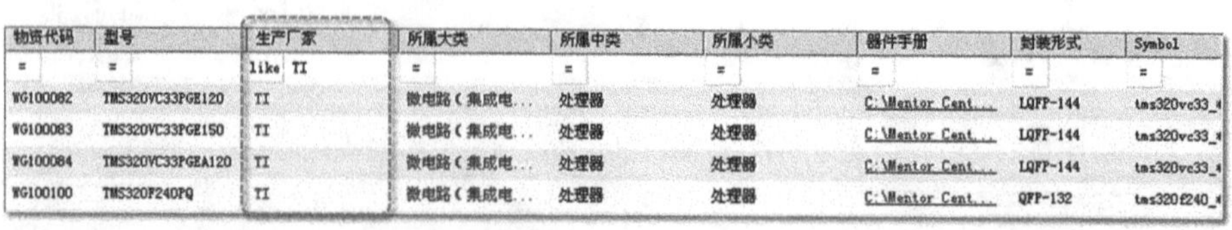

Figure 7.48 List components of manufacturer TI.

Clicking on the dynamic link on the Device manual bar, the component datasheet associated with the device can be opened directly, as shown in Figure 7.49; this is very convenient for the designer.

If you want to search components in another partition, just select the other partition in the drop-down list. Figure 7.50 shows the transistor partition screenshot of DxDataBook. Figure 7.51 shows the connector partition screenshot of DxDataBook.

7.3.3 Check and Update Component Properties

One of the important functions of DxDataBook is to check and update component properties. For example, if some components have been placed in the design schematic, and for some reason their properties are updated in the central library later, DxDataBook can detect the inconsistency between design and the central library, and update the component properties in the schematic.

We can simulate this process. First, in the schematic, delete or change the property value of a certain component, as shown in Figure 7.52.

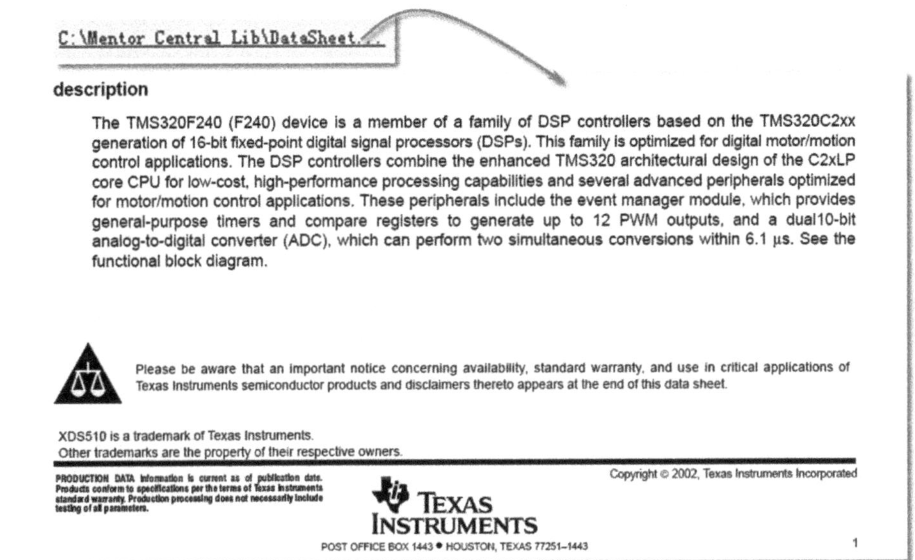

Figure 7.49 Open the device manual via the dynamic link.

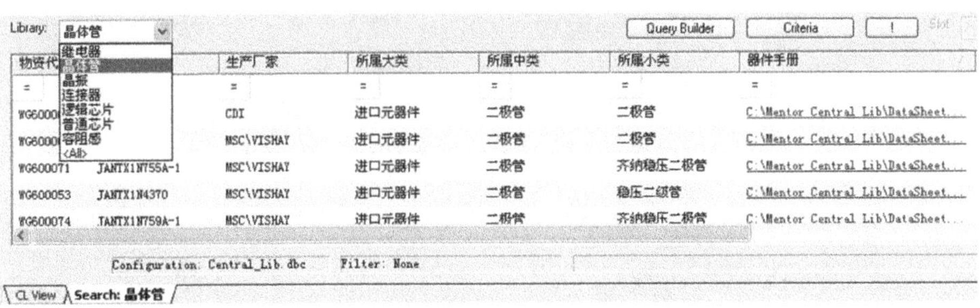

Figure 7.50 Transistor partition of DxDataBook.

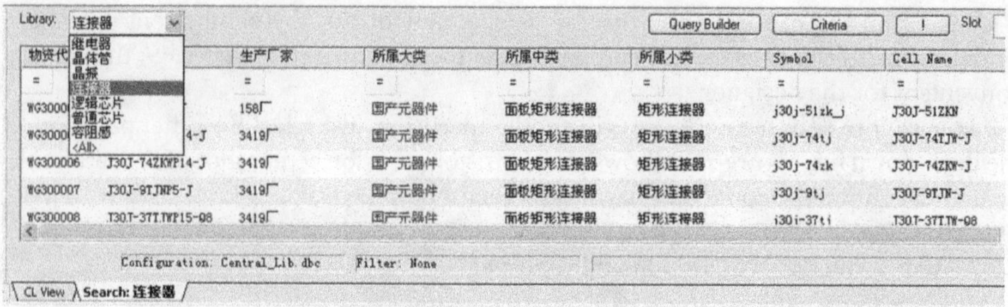

Figure 7.51 Connector partition of DxDataBook.

Figure 7.52 Delete the property value artificially.

Click the New Live Verification window button on the left side of the DxDataBook toolbar, and open the System Verify window as shown in Figure 7.53.

On the left we can see that the component's Status lamp is yellow, indicating that the property of the component in the schematic is inconsistent with DxDataBook, and in the Status bar the display indicates that properties are missing in the schematic and do not match the properties in DxDataBook.

Click the Update All Unique Match button at the top right of the window; we can see that in the component list on the right side of the window, the property value has been updated, and at the same time the status light changes from yellow to green and

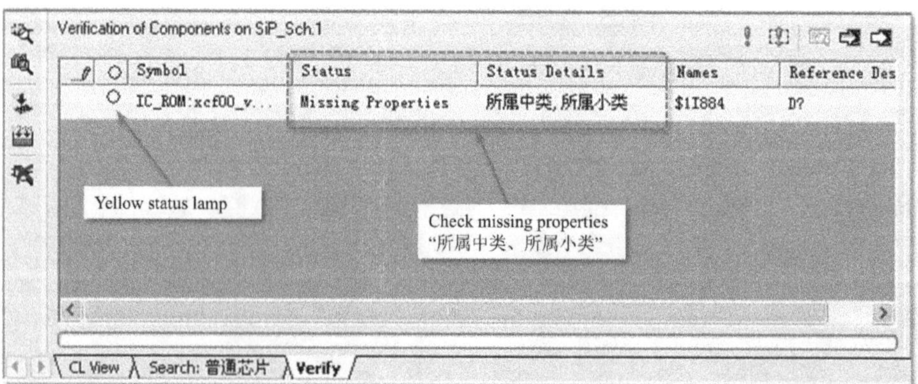

Figure 7.53 Component attribute status display.

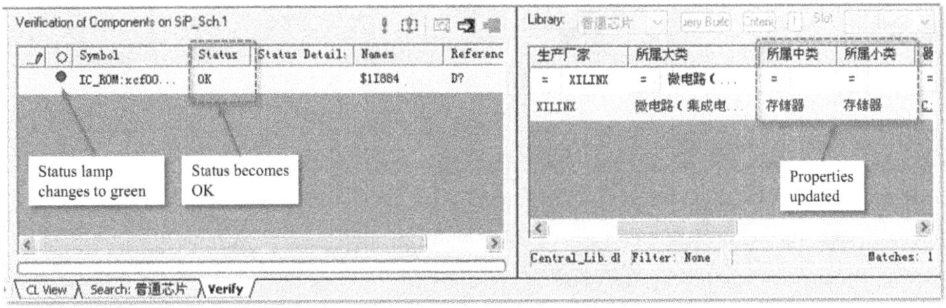

Figure 7.54 Update the component properties with DxDataBook.

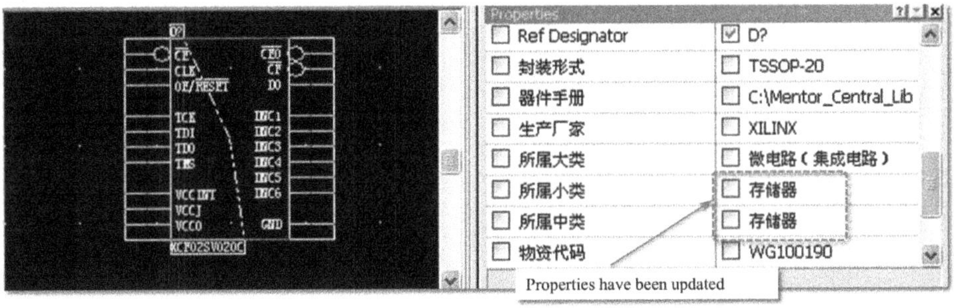

Figure 7.55 Component properties have been updated in the schematic.

the status bar shows OK, indicating that the component properties have been updated by DxDataBook, as shown in Figure 7.54.

Going back to the schematic, check the selected component, and we can see that the properties have been annotated to the relevant bar automatically, as shown in Figure 7.55.

8

Multi-board Project Management and Concurrent Schematic Design

8.1 Multi-Board Project Management

8.1.1 SiP and PCB Collaborative Design

In recent years, SiP technology has developed rapidly and brought about a new trend – the chip package. Gradually, system users have begun to handle package design and production traditionally undertaken by chip vendors. Of course such package is essentially composed of multiple chips, forming a complete system, namely system-level package SiP.

In the past, chip manufacturers usually sold packaged chip to users; users got the packaged chip, then directly applied them in PCB design; in this design mode the majority of package is single-chip package.

Now, with the development of SiP technology, more and more users want to get bare chip, in order to design systems and package on the basis of bare chip. The demand for bare chip will greatly increase, followed by the market, and some traditional chip agents will expand their bare chip business gradually, in order to meet the growing demands of the bare chip market.

At present, as the demand for system miniaturization and low power consumption grows, many system users will turn to SiP; these users include some large international companies and research institutions, and also many domestic research institutes such as aerospace, electronics, weapons, shipping, etc.

Because the design of SiP or package was transferred from IC vendors to system users gradually, and system users focus most attention on system function design, so the unified management of SiP design and PCB design is becoming more and more important.

SiP itself has become a key point in whole-system design; it needs a unified platform for function realization.

This is illustrated by a concrete example.

One designer needs to design a system; this system contains two SiPs, and the two SiPs will be placed on the same PCB, forming a complete system. The system structure and signal pathway are as shown in Figure 8.1. Here, we will introduce how to implement this system in Mentor EE Flow.

SiP System-in-Package Design and Simulation: Mentor EE Flow Advanced Design Guide,
First Edition. Suny Li (Li Yang).

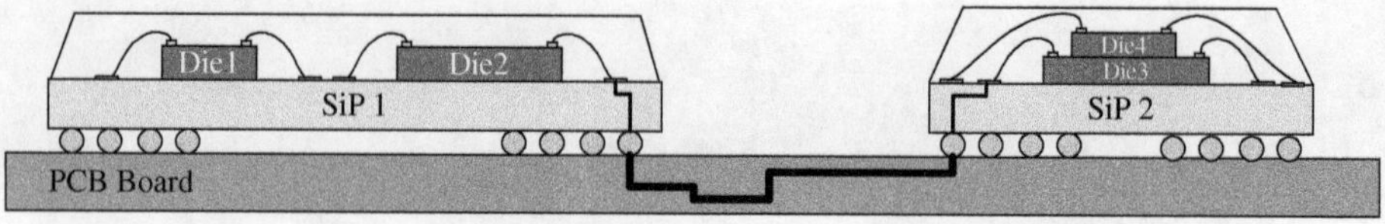

Figure 8.1 Chip, SiP and PCB relationship and the signal pathway.

8.1.2 Multi-Board Project Design Flow

In DxDesigner, we create the following project: a project named SiP_PCB containing both SiP and PCB. Then, we create three new boards in turn; at this time there are three boards in the project, named SiP1, SiP2, and PCB_Board, as shown in Figure 8.2.

The three boards each has its own corresponding schematic; they can be designed independently, and can also share data. For example, if SiP1 and SiP2 both use the same schematic design module, it can be designed to a block, and then the SiP1 and SiP2 schematics can both refer to this block, which is very convenient for real projects.

Figure 8.3 shows three schematics corresponding to the three boards.

After design is complete, the designer carries out DRC inspections on each schematic and package one by one, then starts the layout design tool Expedition, assigns different folders to each layout, and selects the corresponding template according to the actual demands of the design, as shown in Figure 8.4.

After the completion of each board's settings, click the OK button and enter the layout design environment; designers can do each layout design separately, without causing interference to each other, as shown in Figure 8.5. In an actual project, the designer should arrange the order of the three layout designs according to the project context.

Using one project to manage multiple boards is a design method that is more convenient for project management, because all of the boards belong to a larger project, and can exchange data easily.

At the same time, the schematics of the boards are interconnected and can share resources. For example, in relation to SiP package shell, SiP itself belongs to the external interface, but in PCB design, package is a component cell in layout, and makes a connection between internal bare chips and external PCB.

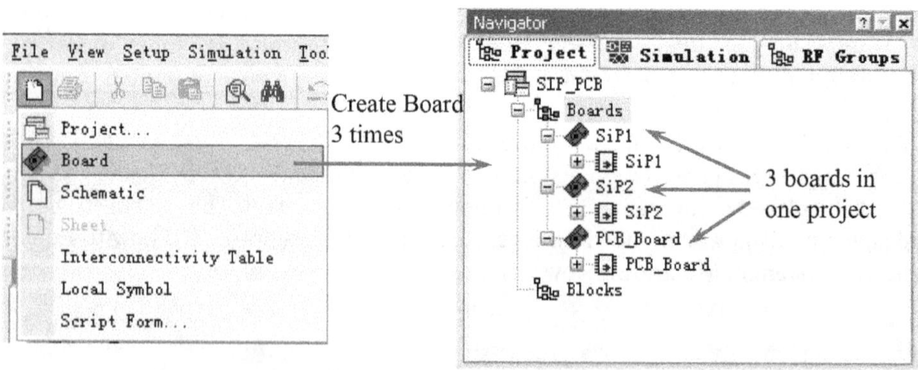

Figure 8.2 One project manages multiple boards.

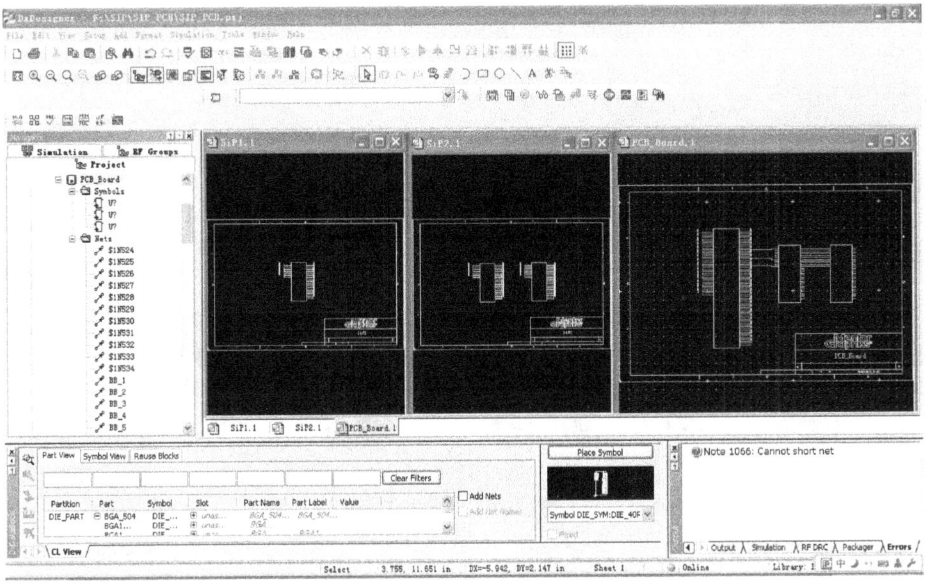

Figure 8.3 Multi-board project schematic design.

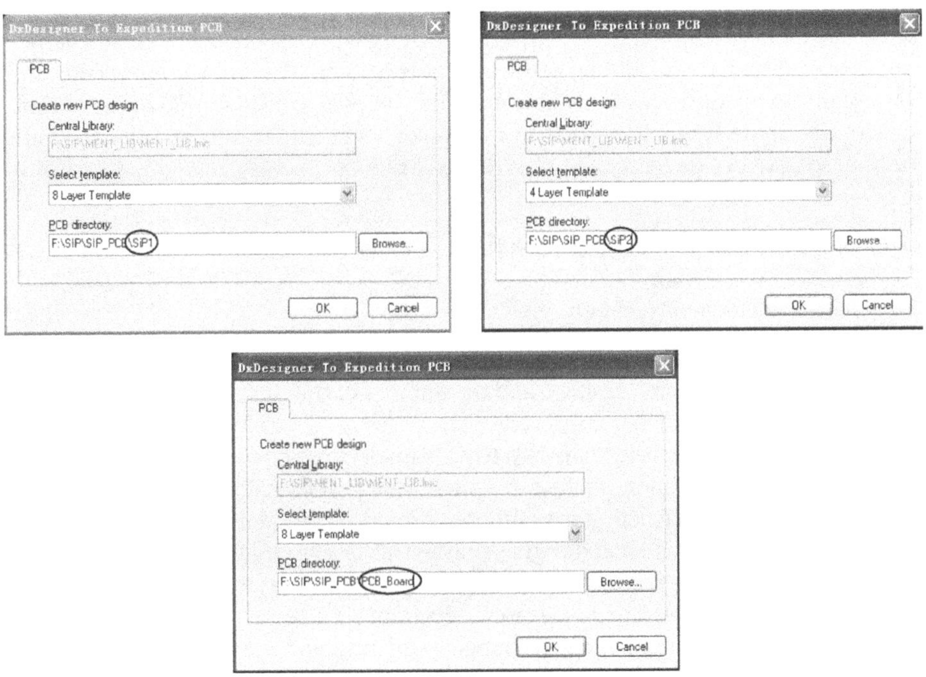

Figure 8.4 Three boards are placed under different paths.

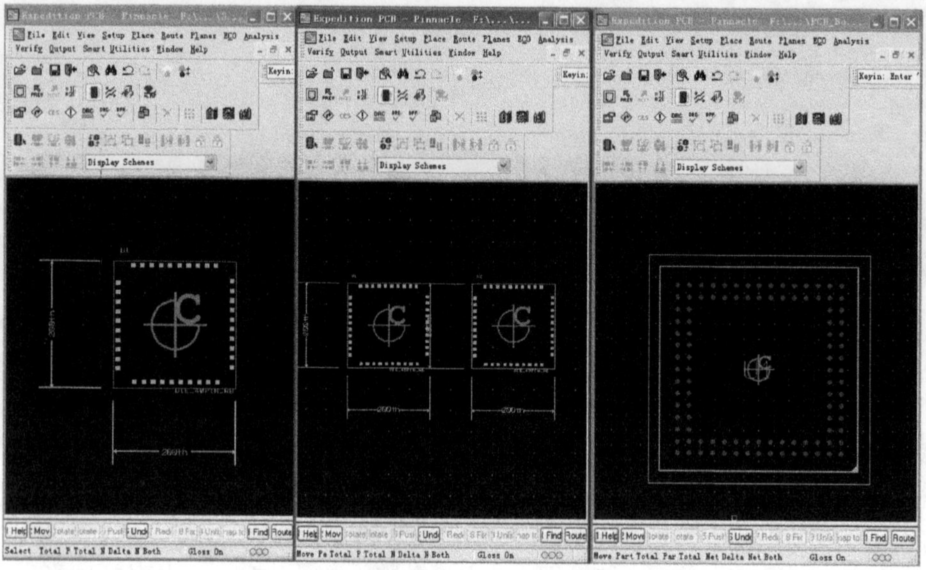

Figure 8.5 SiP1, SiP2 and PCB_board; each has its own independent design environment.

Through collaborative design between several boards in one project, in adjusting SiP package shell pin assignment, not only can the designer consider SiP internal connections and net optimization, but he or she can also consider the connection relations of PCB design and net optimization, allowing more complete consideration of the optimization of the whole system, rather than making one-sided adjustments and local optimization, so as to design better, more advanced products.

8.2 Concurrent Schematic Design

8.2.1 Concurrent Design Thinking

With the development of technology and the increase in system function, electronic systems are becoming more complex, and the number of staff involved in system design will also increase accordingly.

In the design of complex electronic systems, some system circuit schematics contain different functional modules, such as the digital part, the analog part and the radio frequency part. Usually, the circuits of different functional modules are handled by different schematic designers; each designer finishes his or her circuit and then copies the schematic to form a system schematic diagram, as shown in Figure 8.6. This traditional method is still feasible for the mature circuit diagram.

But, in a new project the design will change frequently, and the problem is that copying will be required too often. Repeated copying will cause a lot of problems. For example, designer A makes two changes after finalizing, so he will copy three times to the main schematic; designer B makes three changes after finalizing, so he will copy four times to the main schematic.

If designer A's final version and designer B's final version are not copied to the unified schematic, the design may appear inconsistent. At the same time, the replication can

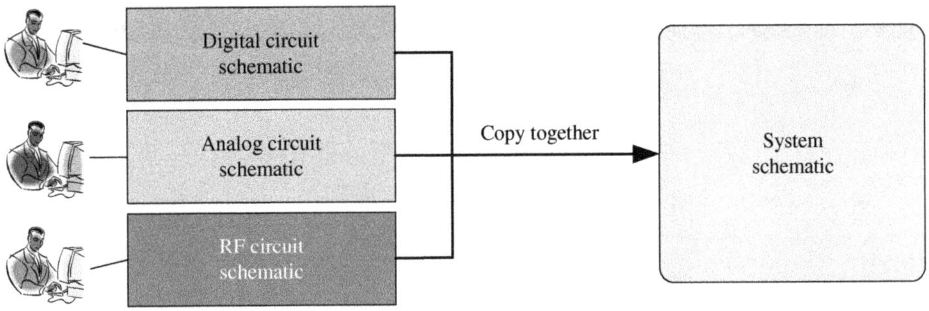

Figure 8.6 Traditional circuit design separately and then copy together.

produce a variety of versions of the schematic, making the versioning chaotic and difficult to control.

Concurrent schematic design provides new thinking to solve the problems that traditional methods couldn't solve, and provides a concrete realization method.

The schematic of a complex system is divided into different parts, usually in different sheets to distinguish the parts; in concurrent design, under unified management, different designers can enter and work on the required sheets of the schematic.

When designer A is editing sheet 1 of the schematic, this sheet is read-only for other designers, thus effectively avoiding conflicts. At this point, if one of the other designers, say designer B, also wants to edit sheet 1, he or she simply needs to ask designer A's permission; if designer A agrees, he or she stops editing and closes the sheet, and the sheet is automatically protected. While B is editing this sheet, the sheet becomes read-only for other designers, including designer A.

The train of thought in relation to avoiding conflict in concurrent schematic design can be summarized as follows: the first designer to open and edit the sheet shall enjoy the right to operate on this sheet, which can simply be understood as the principle of "first come, first served". As shown in Figure 8.7 the concurrent schematic design method is based on sheet operation.

8.2.2 Operating Method for Concurrent Schematic Design

During concurrent schematic design, the hardware environment is prepared first. A computer acts as file server; the project folder is put on the server, and all personnel involved in the project have read-write permission.

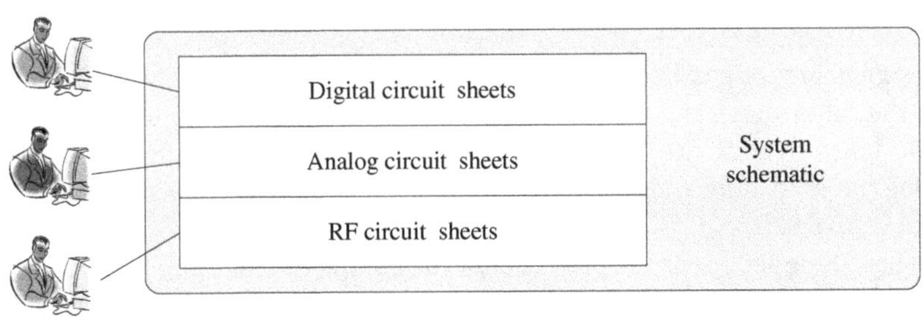

Figure 8.7 The operation of concurrent schematic design.

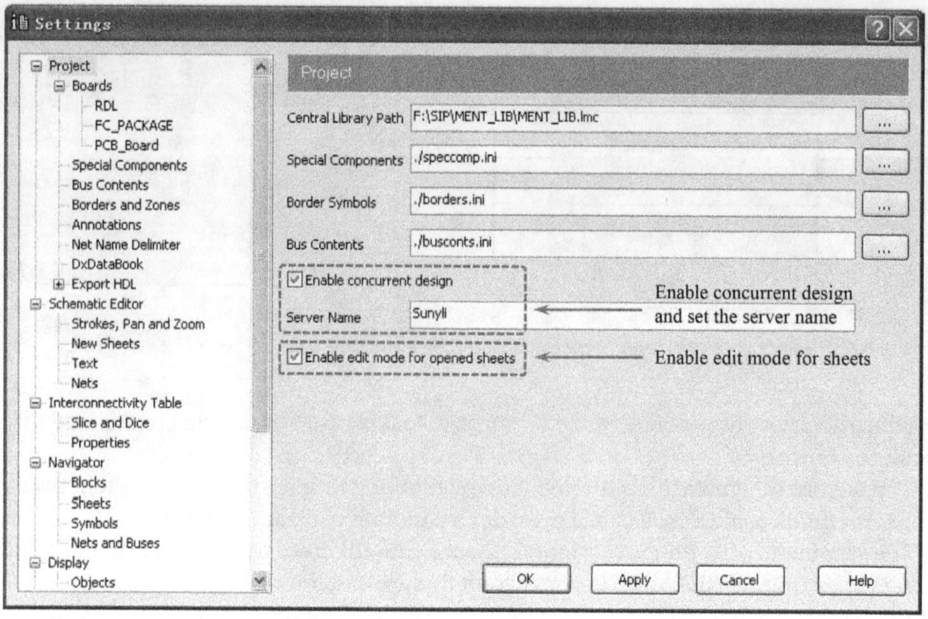

Figure 8.8 Enable concurrent schematic design and set the server name.

Of course, the file server may be a particular designer's personal computer; he or she just needs to put the project file in a folder on the computer and give read/write access to the other designers involved in the project.

After the completion of hardware environment configuration, any of the designers can open the collaborative design project; in Setup→Settings, check the Enable Concurrent Design option, and write the name of the server in the Server Name bar, for example, Sunyli, as shown in Figure 8.8.

Enable the Edit mode for opened sheets option; the system will enter sheet edit mode automatically.

If this option is checked, the first time a designer opens a sheet, he or she will enter edit mode automatically; then, if other designers open this sheet, the read-only prompt will appear at the top of the edit window, stating that the sheet has been locked by the first designer, as shown in Figure 8.9.

If this option is not checked, when the first designer opens a sheet, edit mode will not be entered automatically; the system will state that the mode is read-only at the top of the edit window, and the "Click for Edit button" needs to be manually clicked to enter edit mode.

If the first designer just opens the sheet but does not enter edit mode, when other designers open the same sheet, the window above will show the same content. At this time, the first designer to click the "Click for Edit" button will enter edit mode; other designers will see the read-only and locked by the first designer message, as shown in the lower part of Figure 8.9.

Currently, the first designer opens the project after the completion of setup for concurrent design, and the other designers involved in the project can open the project file directly. Usually he or she should start DxDesigner first, then Open Project via File→Open.

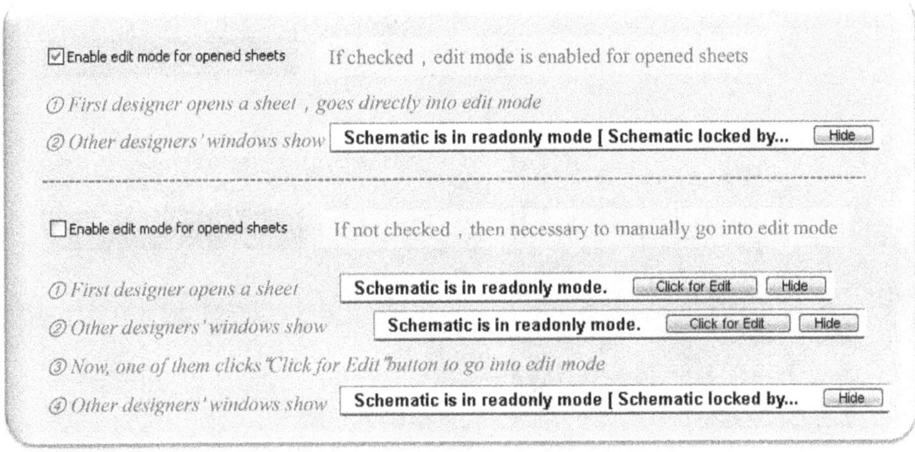

Figure 8.9 Enable edit mode for opened sheets option checked and unchecked comparison.

According to whether the "Enable edit mode for opened sheets" option is checked or not, the status bar at the top of the window will display different information for each designer, as shown in Figure 8.9.

When a sheet is in use, other designers can only operate on different sheets; the write permission is based on above-mentioned principle "first come, first served". When a designer edits a sheet, other designers can't edit this sheet, but can see changes to this sheet in real-time, as shown in Figure 8.10.

At this point, if other designers attempt to edit this sheet, the system will automatically pop up the "Sheet is read-only" window, warning the designer that this sheet is

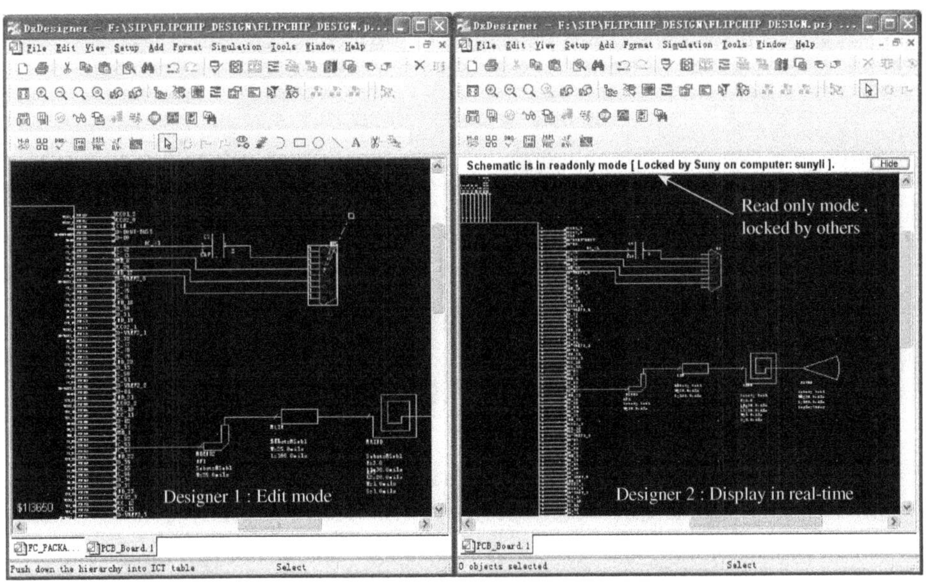

Figure 8.10 When a designer edits the sheet, other participants can see in real-time.

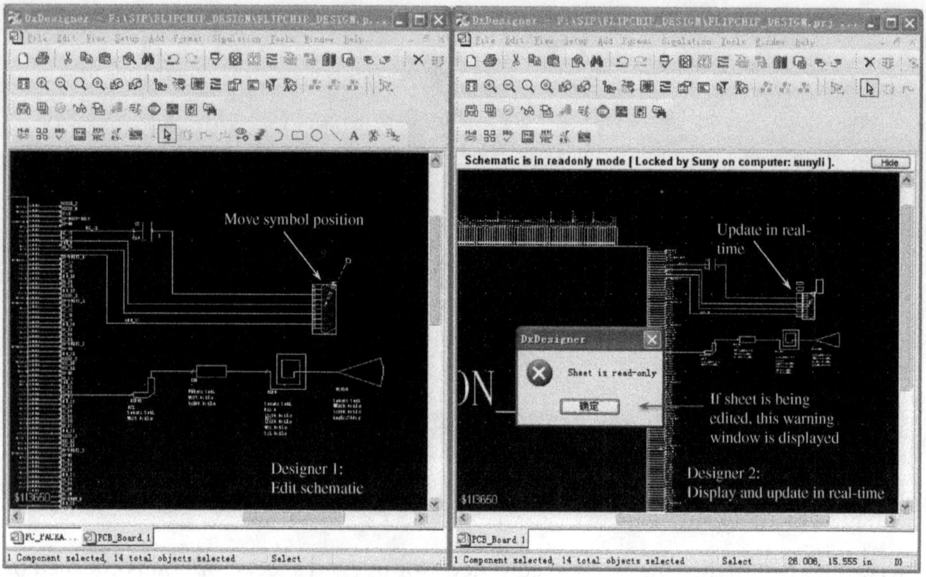

Figure 8.11 For other designers, the sheet is read-only.

read-only and is being edited by the first designer to open and edit the sheet. If one designer keeps opening the sheet and clicking the edit button [Click to Edit] (even if he or she does not edit at that time), for other designers this sheet is read-only, as shown in Figure 8.11.

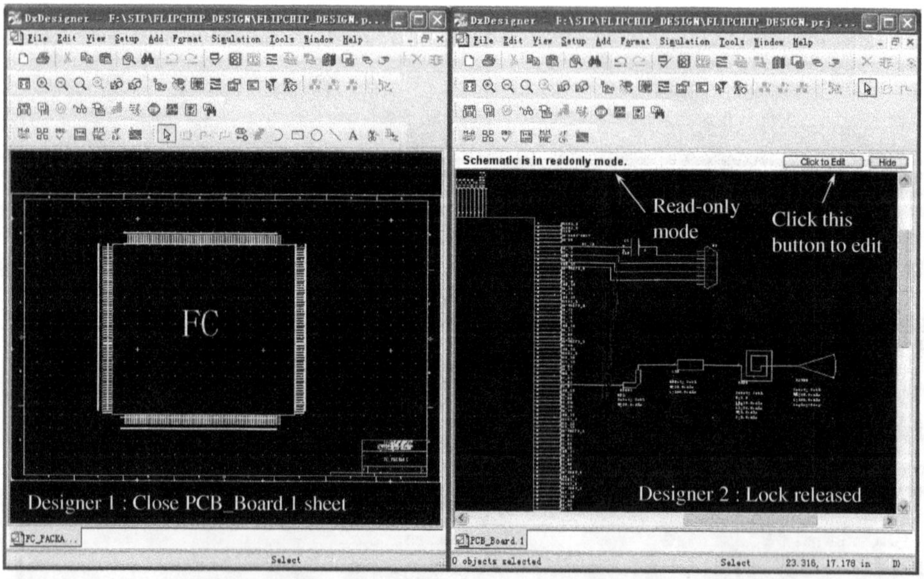

Figure 8.12 The designer who first clicks the "Click to Edit" button has write permission to the sheet.

Only when the designer is no longer editing and closes the sheet will write permission for this sheet will be released to other designers. For other designers, the status bar above the edit window will display the “Sheet lock released” message, but the sheet remains in read-only mode; the designer can click the button Click to Edit to enter edit mode.

Thus, based on the principle of “first come, first served”, the designer who first opens the sheet and clicks the button Click to Edit will have write permission to the sheet, as shown in Figure 8.12.

The status bar will update at the same time, showing that the sheet has been locked by that designer.

Concurrent schematic design based on mutual cooperation can greatly improve the efficiency of design, and effectively maintain the consistency of design data.

The concept of concurrent design has been accepted by more and more designers, and for large and complex projects, concurrent design has become an inevitable trend.

It can effectively accelerate project progress, improving the competitiveness of the product.

9

Layout Creation and Setting

9.1 Create Layout Template

9.1.1 Layout Template Definition

The layout template is defined in the central library, and can be referenced by layout design.

The information contained in the layout template includes layer stackup, substrate outline, mounting holes, fixed position connectors, etc. Frequently used layer stackup or substrate outlines can be defined in the central library as layout templates, and when doing layout design, such templates can be referred to directly, saving much design time. This is also advantageous for the standardization and normalization of layout design.

In the layout design of SiP, HDI (High Density Interconnection) substrate was usually used. Now, however, an m+N+m-type layer stackup substrate is more commonly used; "m" refers to the micro-via layer, usually called the buildup layer, "N" refers to the mechanical drilling layer, usually called the laminate layer. Figure 9.1 shows a typical 2+4+2 HDI layer stackup.

In EE Flow, the layer stackup can be set in the process of layout design, and can also be defined in advance in a central library layout template, then referred to in the design process. This is suitable for more mature and commonly used layer stackups.

9.1.2 Create SiP Layout Template

Open the central library with Library Manager (for example, the MENT_LIB.lmc file), select Tools→Layout Template Editor, open the window as shown in Figure 9.2. The system default is a four-layer template or an eight-layer template.

In the Template Editor interface, click the new button, in the Source design file bar of the pop-up window New Layout Template, input the name of the source design which the layout template needs to reference, and input the name of the template in the Template name bar, as shown in Figure 9.3.

After clicking the OK button, the system will automatically copy the source design to the new template and the source design elements including layer stackup, substrate outline, mounting holes, etc. are inherited by the new template and can be used in a new design referring to that template.

SiP System-in-Package Design and Simulation: Mentor EE Flow Advanced Design Guide,
First Edition. Suny Li (Li Yang).

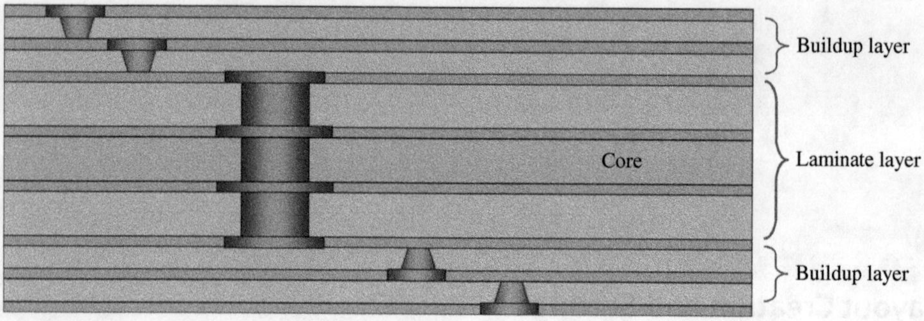

Figure 9.1 2+4+2 HDI layer stackup.

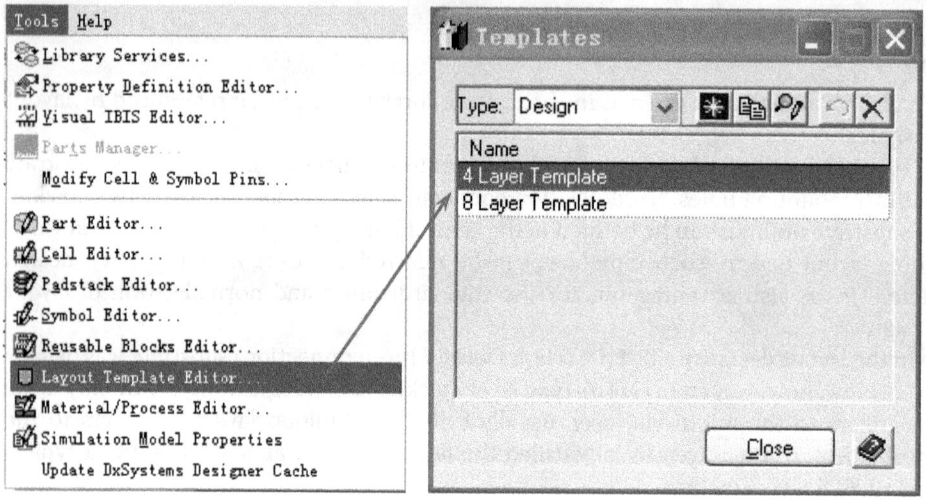

Figure 9.2 Start Layout Template Editor.

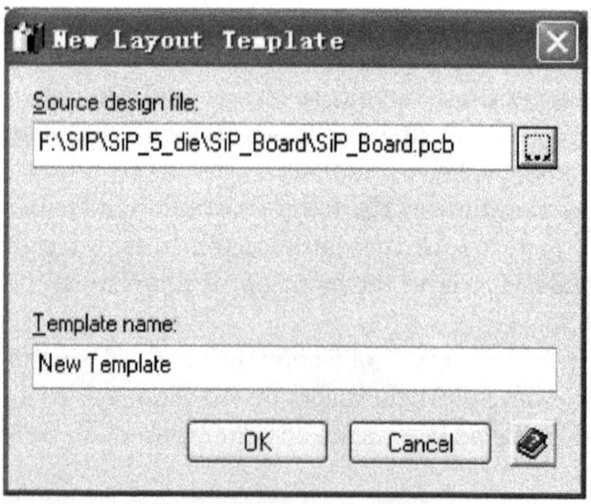

Figure 9.3 New Layout Template.

In addition, the designer can copy any existing template and edit it to create the new one. For example, select "4 Layer Template", then click the Copy Template button , and then rename the template to 2+4+2 Template, as shown in Figure 9.4.

Then click the Edit template button ; the system will open the Expedition window and go into an empty layout design; in the menu bar select Setup→Setup Parameters and go into the template Setup Parameters window as shown in Figure 9.5.

First, in the Display units area, set Design units to Microns (μm) and Velocity of propagation units to Meters/s. Then, input 8 instead of the original 4 in the Number of physical layers bar, as shown in Figure 9.5.

Then click the Remap Layers button, as shown on the left of Figure 9.6; the current layers are shown as 1, 2, 3, 4 and the new layers as 1, 2, 3, 8. For layers 1 and 8, the mapping cannot be changed, because devices may be placed on surface layers 1 and 8, so we can only map them as surface layers, that is, the current layer 1 will be mapped to new layer 1 and layer 4 will be mapped to new layer 8.

But for layers 2 and 3, designers can change the mapping, they can be mapped to the new layers. If you don't need to change the mapping relationship, just click the OK button.

Re-click the Remap Layers button; we can see on the right of Figure 9.6 that the template layers have successfully changed to eight layers. Layer change is complete.

Switch to the Via Definitions tab in the Setup Parameters window; we can see the default via is 026VIA, as shown in Figure 9.7.

At this point, if the central library hasn't defined micro-via, we need first to define micro-via in the central library padstack definition.

Select Setup→Padstack Editor and enter the padstack editor interface.

First, create two kinds of holes with diameters 75 μm and 150 μm, set tolerance to 2; if the "Generate name from property" option is checked, we can see the hole is

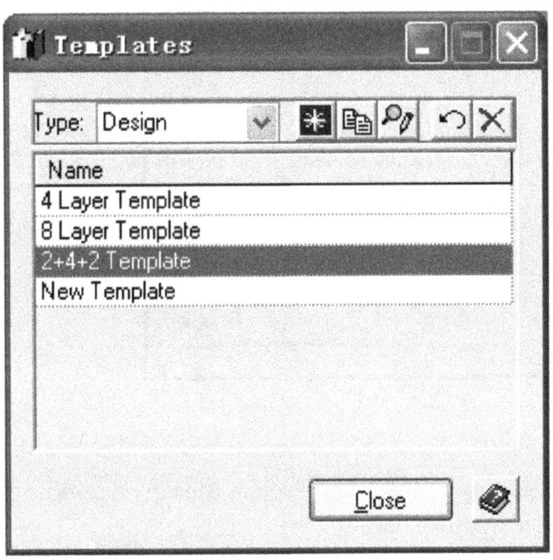

Figure 9.4 Copy template and rename.

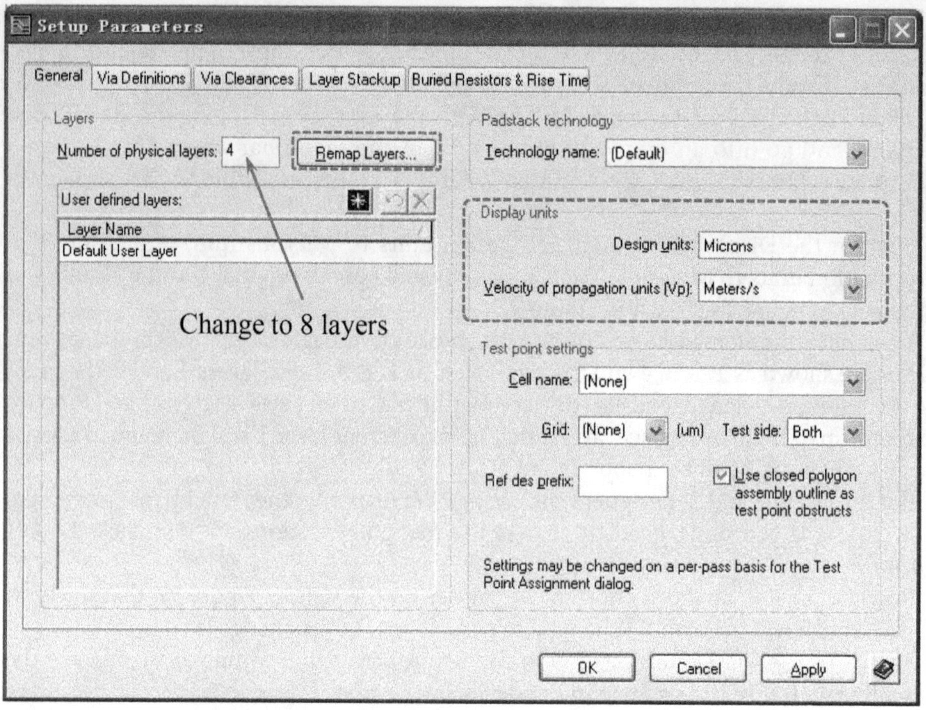

Figure 9.5 Start the template Setup Parameters window.

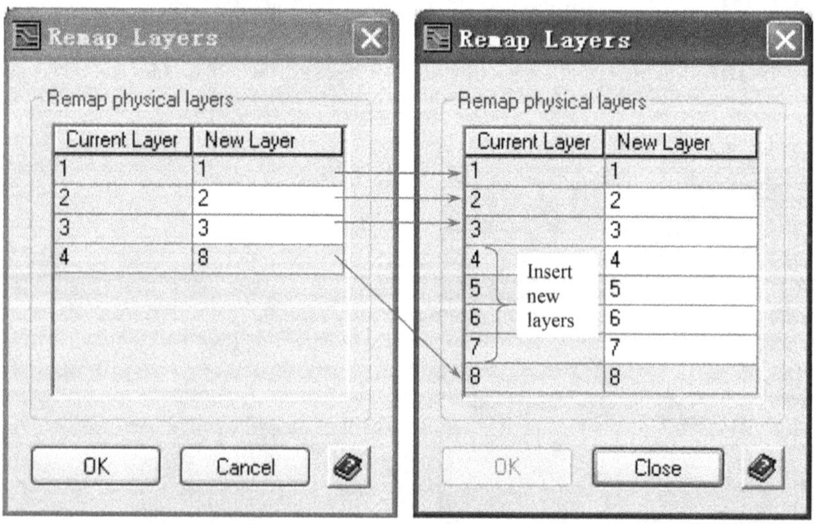

Figure 9.6 Remap layers.

automatically named Rnd75+/-Tol2 and Rnd150+/-Tol2. Then select the corresponding drilling symbol for the two holes in the Drill symbol list.

The drill symbol is designed to represent the hole in the drill drawing output after completion of design for drilling personnel to view and check.

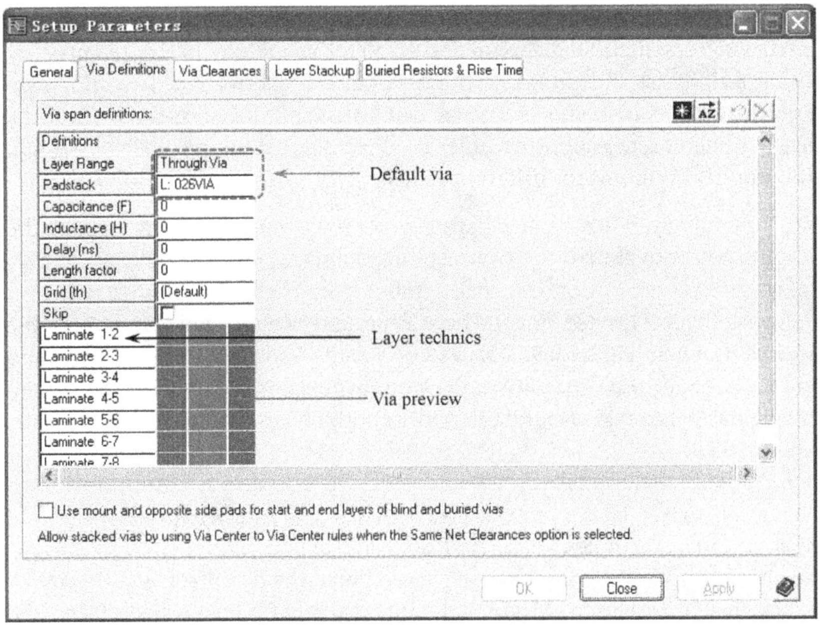

Figure 9.7 Via definition of layout template.

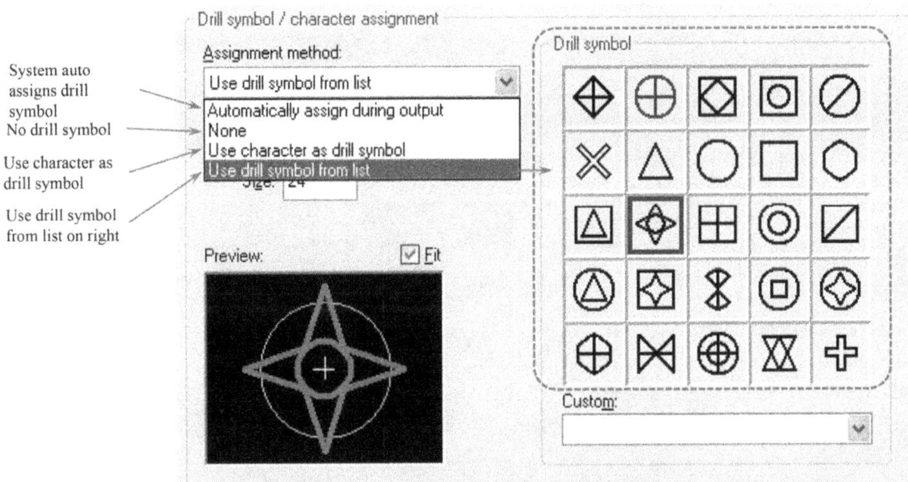

Figure 9.8 Four types of drill symbol.

There are four options for the drill symbol, as shown in Figure 9.8, which are explained below.

1) Automatically assign during output, indicates that the drill symbol is assigned automatically by the system during output.
2) None, indicates that no drill symbol is set for the hole.

3) Use character as drill symbol, indicates that the designer can select a drill symbol from 26 English letters, including capital and small letters, a total of 52 options.
4) Use drill symbol from list, indicates that the designer can choose a drill symbol from the panel on the right, which shows 25 drill symbols; the blue color shows that the drill symbol has been used by another hole. To avoid confusion, the designer should choose different drill symbols for different holes.

The Size bar is mainly used for the size definition of the drilling symbols; usually the symbol size should be set to almost the same as the hole size.

In this case, we choose the option Use drill symbol from the list, and select one drill symbol from the list. Select "Drilled" for the hole Type, and "Round" hole, and check the "Plated" option, as shown in Figure 9.9. Then, switch to the Pads tab, create 175 μm and 250 μm Round pads, check the Generate name from properties option, and the system will automatically name the pads Round175 and Round250 according to the attribute, as shown in Figure 9.10.

Then, switch to the Padstacks tab, create padstack Via_175 um, select Round175 in the pad list, using the arrow to the left < assign to Via_175 um Mount side, Internal and Opposite side, and in the holes list select Rnd75+/-Tol2.

Then create padstacks Via_250 um, choose Round250 in the pad list, using the arrow to the left < assign Via_250 um to Mount side, Internal and Opposite side of Via_250 um, and in the holes list select Rnd150+/-Tol2, as shown in Figure 9.11.

After the two vias have been created, select File→Save, save the via, and then exit.

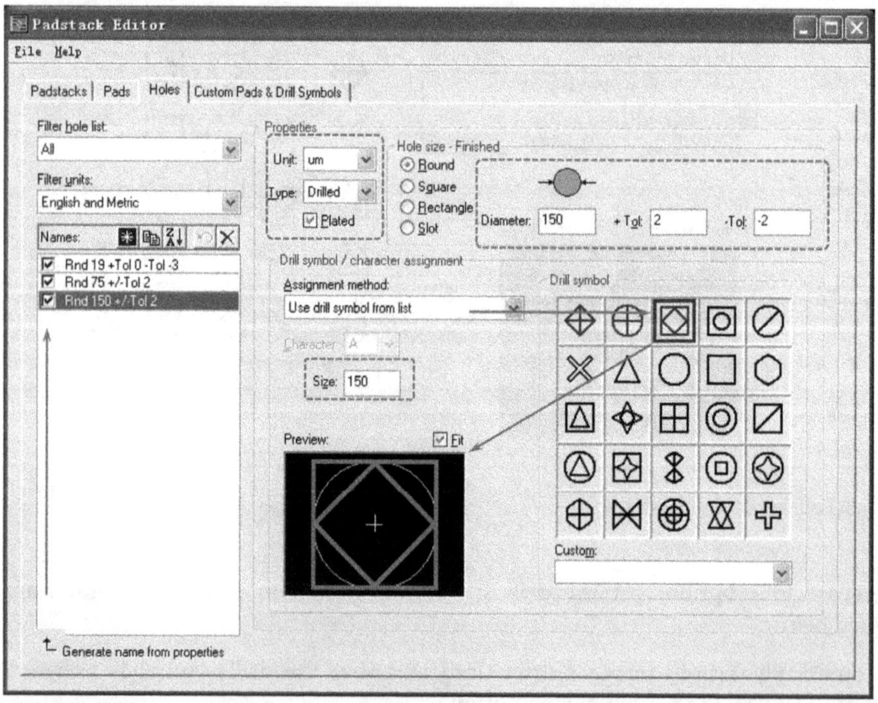

Figure 9.9 Create drill hole and specify the corresponding drill symbol.

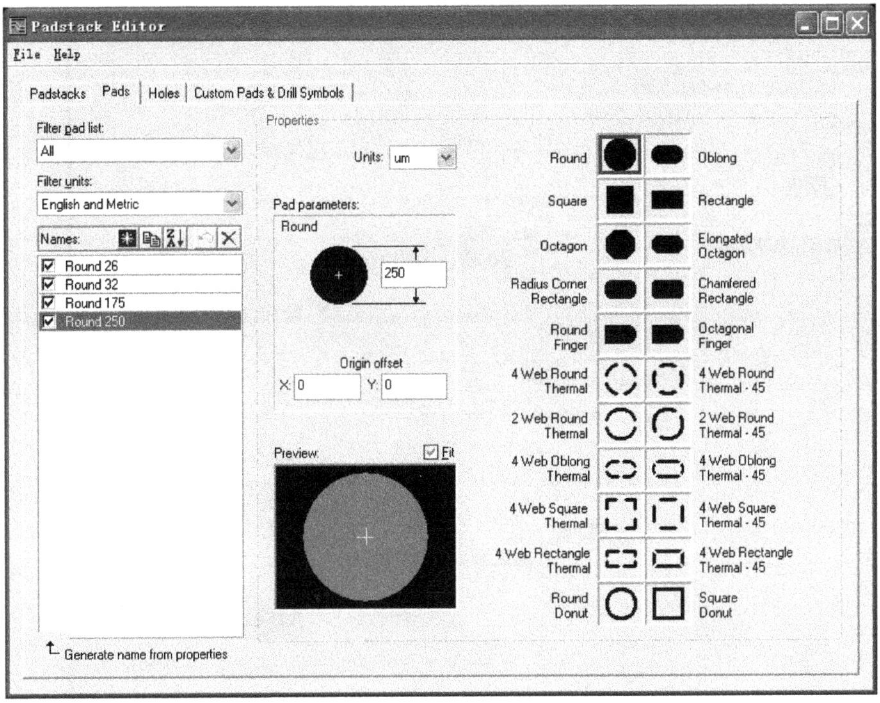

Figure 9.10 Create pads.

Reselect Setup→Setup Parameters and switch to the Via Definitions tab, as shown in Figure 9.12, to assign the via.

From Layer 1-2, Layer 2-3, Layer 6-7 and Layer 7-8 chose Via_175 um, for process type select buildup; for Layer 3-6 select Via_250 um, for process type select laminate.

Buildup and laminate are two process methods for substrate production; again, because of space limitations, it will not be discussed further here; readers can consult relevant materials.

Switch to the Layer Stackup tab, set up the Signal Thickness, Dielectric Thickness, Dielectric Constant, Solder_Mask Thickness and Dielectric Constant for each layer, as shown in Figure 9.13.

This should be set up according to the actual context of the project and the capability of the supplier, since the setup results have a certain influence on the simulation and transmission line impedance calculation at a later stage.

As an example here we are only using general settings; for a specific project, modify according to the specific context of the project.

After completion of the settings, click OK, and exit the Setup Parameters window.

Click the Draw Mode button in the Expedition toolbar, and enter draw mode. Then click the Dimension button in the Draw toolbar to create dimensioning on the board. In the pop-up dimension toolbar , click on the first tool, the Dimension Option button ; the Dimension Parameters window pops up as shown in Figure 9.14. It includes various parameters for dimension settings. Here we simply change units to mm, then click OK.

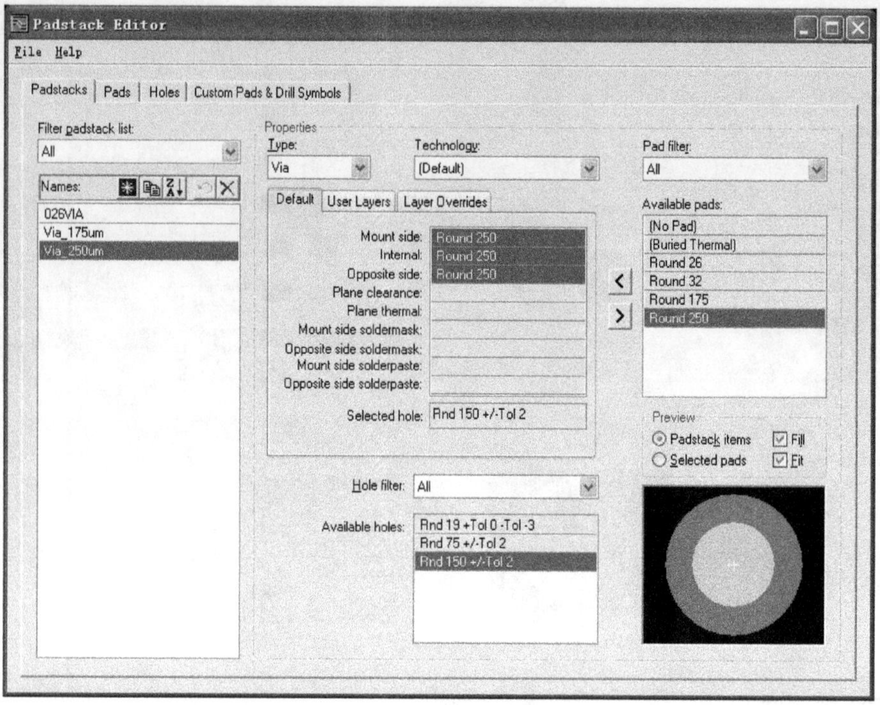

Figure 9.11 Create padstack for via.

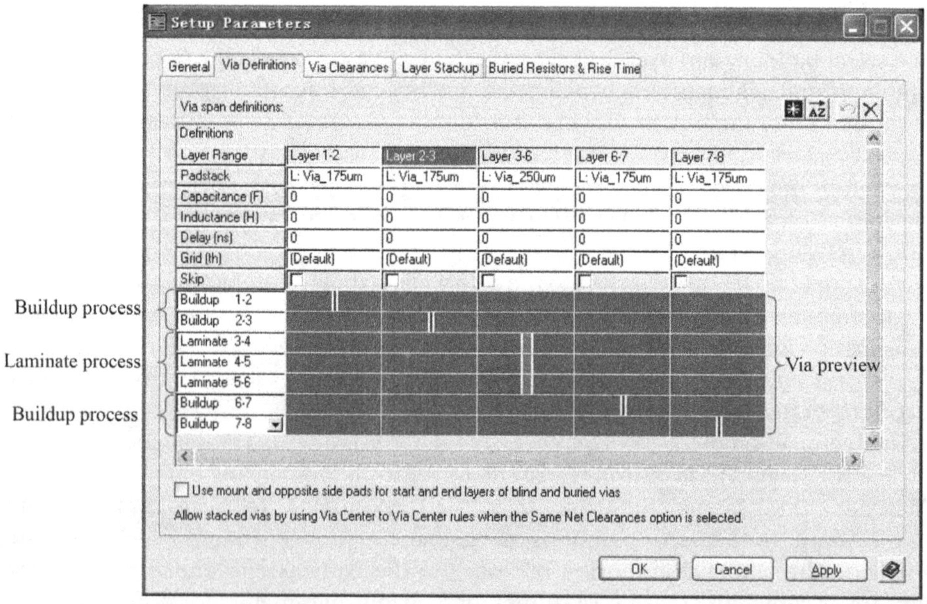

Figure 9.12 The definition of layer stackup 2+4+2.

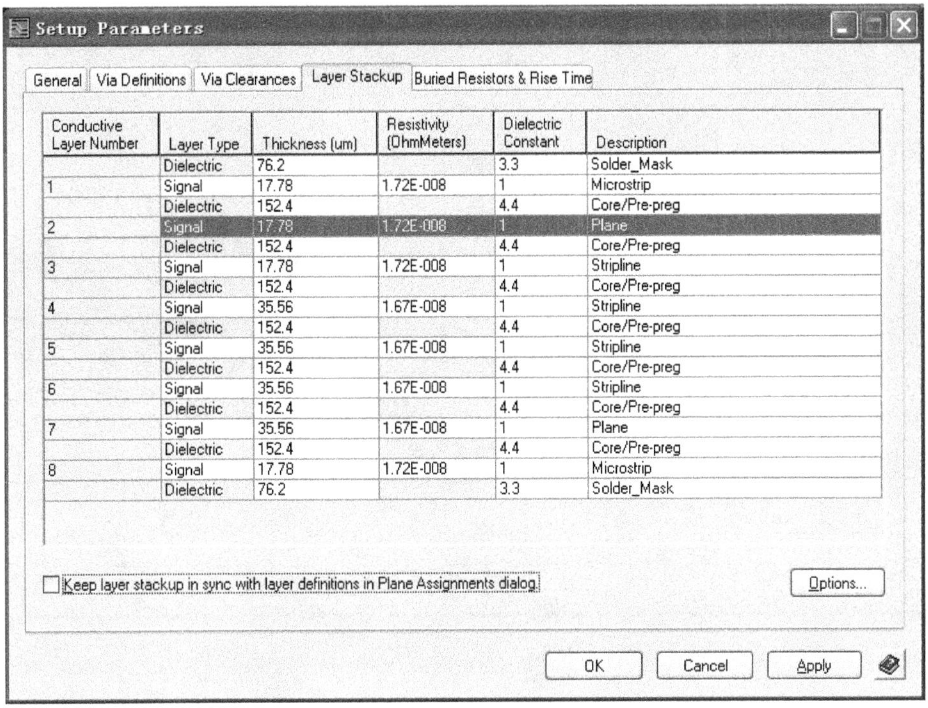

Conductive Layer Number	Layer Type	Thickness (um)	Resistivity (OhmMeters)	Dielectric Constant	Description
	Dielectric	76.2		3.3	Solder_Mask
1	Signal	17.78	1.72E-008	1	Microstrip
	Dielectric	152.4		4.4	Core/Pre-preg
2	Signal	17.78	1.72E-008	1	Plane
	Dielectric	152.4		4.4	Core/Pre-preg
3	Signal	17.78	1.72E-008	1	Stripline
	Dielectric	152.4		4.4	Core/Pre-preg
4	Signal	35.56	1.67E-008	1	Stripline
	Dielectric	152.4		4.4	Core/Pre-preg
5	Signal	35.56	1.67E-008	1	Stripline
	Dielectric	152.4		4.4	Core/Pre-preg
6	Signal	35.56	1.67E-008	1	Stripline
	Dielectric	152.4		4.4	Core/Pre-preg
7	Signal	35.56	1.67E-008	1	Plane
	Dielectric	152.4		4.4	Core/Pre-preg
8	Signal	17.78	1.72E-008	1	Microstrip
	Dielectric	76.2		3.3	Solder_Mask

Figure 9.13 Layer Stackup setup.

Next, click the Place Dimension along a Linear Element button, choose Mark location of board outline, click the left mouse button, and the dimension annotation will be placed outside the board outline automatically; the default dimension annotation will be placed in the Assembly Top layer. If we need to place other layers, such as the customer layer, we can define these in the Layer bar of the Dimension Parameters window in advance, as shown in Figure 9.14.

After placing dimension annotation, if the size of the board changes, the dimension will automatically update to follow the change of board size.

In this case, we set the board outline size to 30 mm × 30 mm, and on the basis of the board outline we set the route border shrink value to 0.5 mm (500 μm).

The route border size-setting method is as follows: (1) hold the < Ctrl > key, double-click the left mouse button on the board outline to copy out Draw Object type border; (2) change Draw Object type to route border; input −500 μm in Grow/Shrink bar and press < Enter >; (3) check the route border size in Properties and check that the route border has been updated in layout, as shown in Figure 9.15. This kind of method is very effective for complex board shapes.

At this point, the template has been created.

The three key elements of the layout template are as follows: (1) substrate shape and size (2) via definition (3) layer stackup and parameters definition, as shown in Figure 9.16.

Click File→Save and exit the template editing environment, then exit Library Manager.

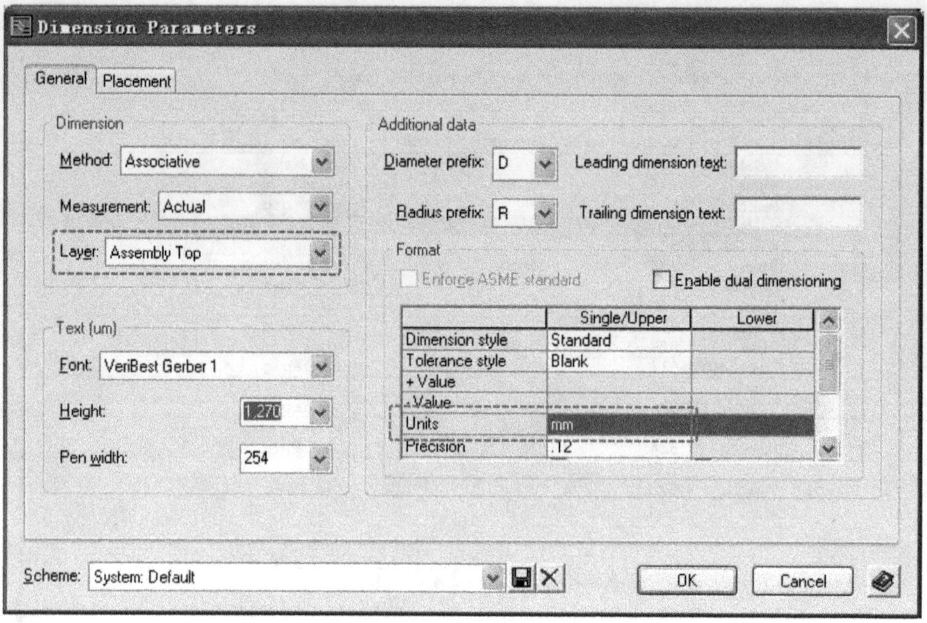

Figure 9.14 Dimension Parameters window.

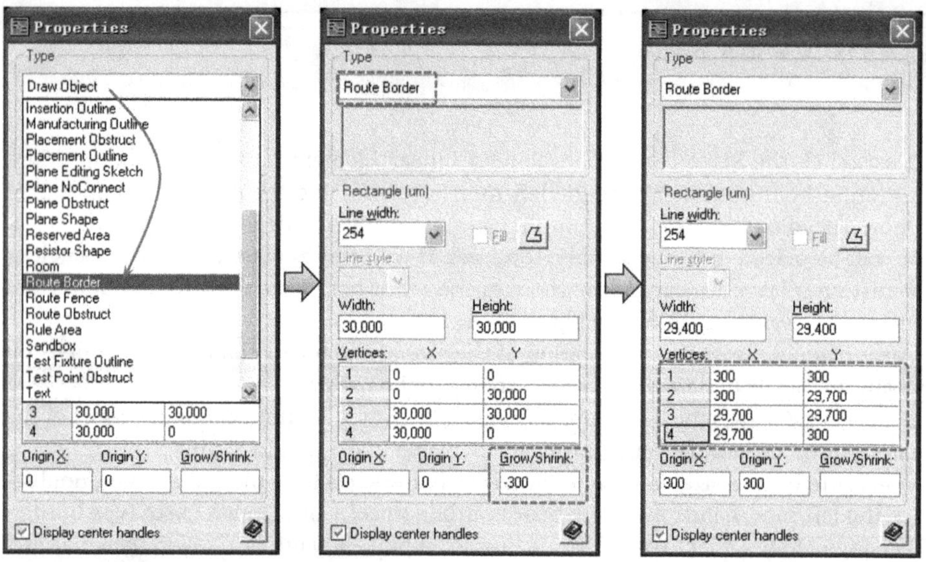

Figure 9.15 Copy route border and change the size.

9.2 Create Layout Project

9.2.1 Create New SiP Project

To create a new project in DxDesigner, select File→New→Project, choose expedition – default in Project Templates, set the project name, location and the corresponding

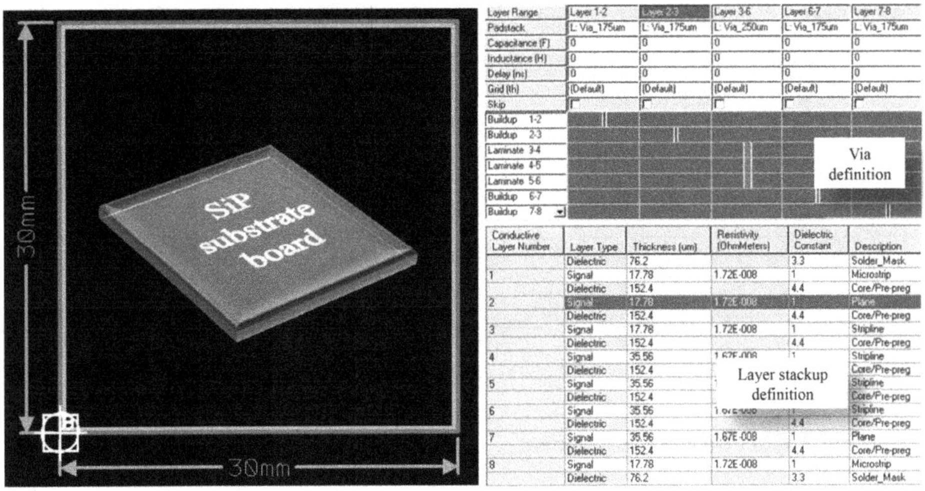

Figure 9.16 The three key elements of the layout template.

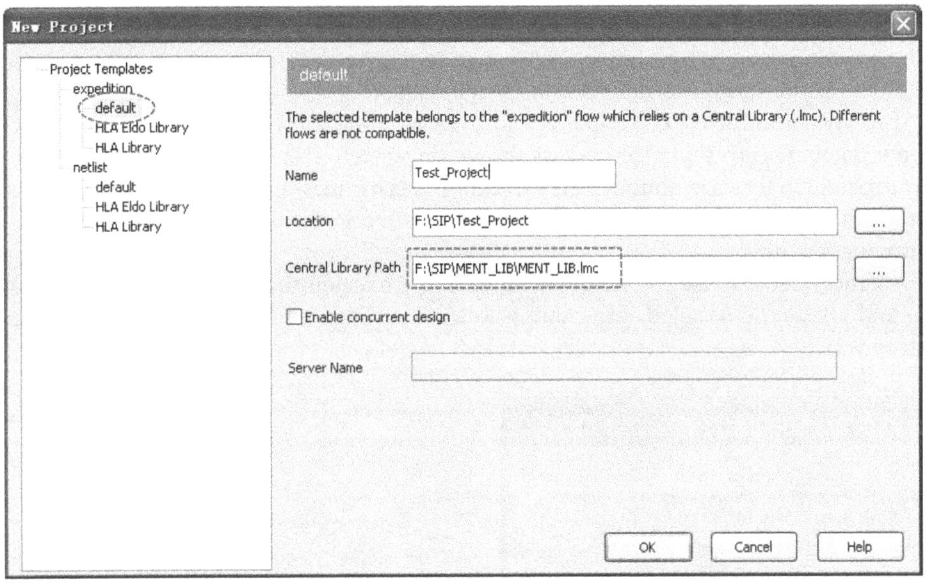

Figure 9.17 Create new project.

central library path. In this case, we select MENT_LIB.lmc, created previously, as the central library, as shown in Figure 9.17.

Then create a new schematic and draw the schematic; see section 7.2 for a detailed description.

After schematic drawing has been completed, verify the schematic, and then package the design. After successful packaging, click the Expedition button, in the pop-up window choose the 2+4+2 template in the drop-down list, and set the layout directory to ...\ SiP_Board, as shown in Figure 9.18.

Then click OK and enter the Expedition layout design environment.

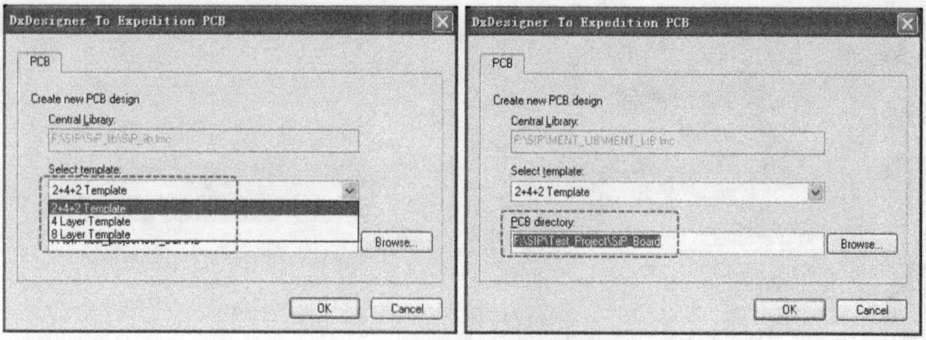

Figure 9.18 Layout template selection and directory settings.

9.2.2 Enter Layout Design Environment

After entering the Expedition environment, the system will automatically prompt a schematic update; click the Yes button on the prompt dialog box, the system opens the Project Integration window automatically, click the orange button in front of Forward Annotation (orange indicates new update required), as shown in Figure 9.19.

A new window pops up, stating that system synchronization is ongoing; the synchronization includes package, database load, net load, etc.

After synchronization is complete, a prompt window shows that the schematic and layout databases are synchronized, as shown on the left side of Figure 9.20. If there is any problem in forward annotation, a warning or error message window will appear, as shown in Figure 9.20 right side, and detailed information will be written to the corresponding text file.

Designers need to use the File View button  to open the ForwardAnnotation.txt file and check the detailed information in the warning or error report, as shown in Figure 9.21.

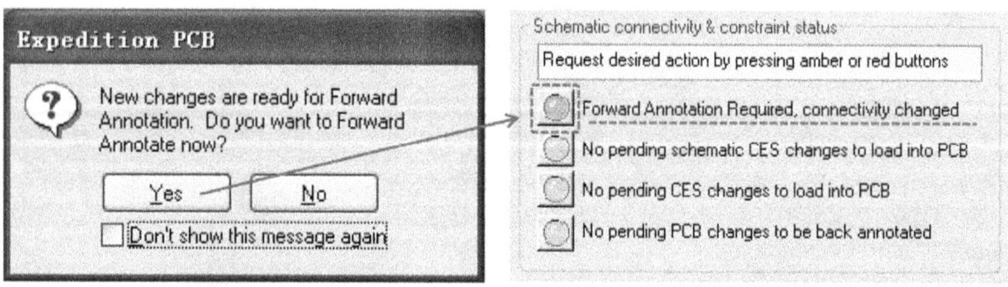

Figure 9.19 Forward Annotation prompt and indication.

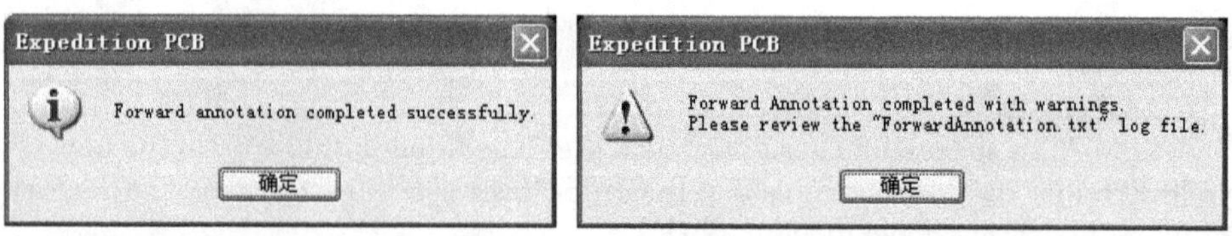

Figure 9.20 Prompt window indicates forward annotation successful or shows warning.

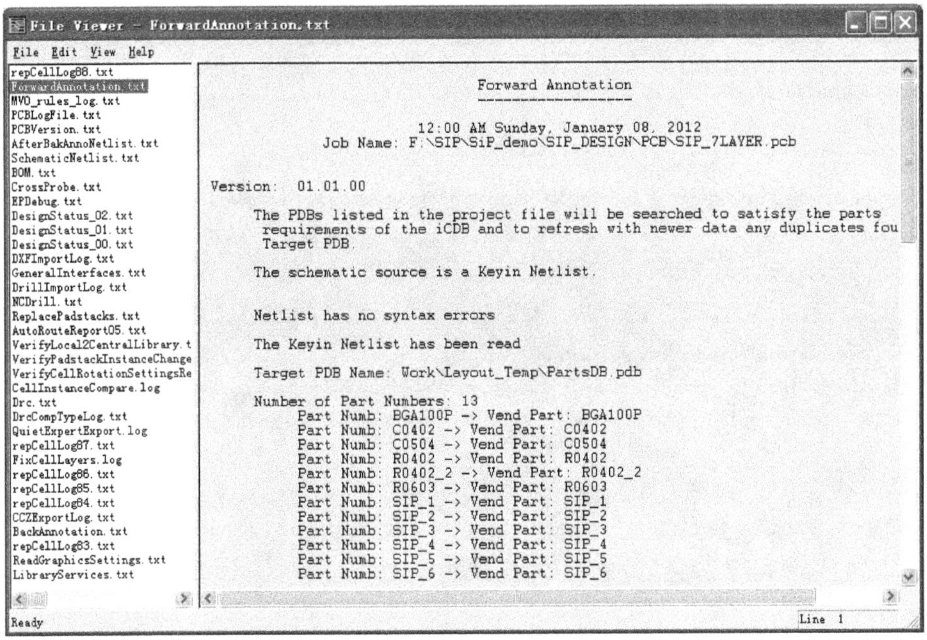

Figure 9.21 Check ForwardAnnotation.txt file.

Using this information we can confirm whether the error or warning will affect the design, in which case appropriate changes may be needed, and then re-package and forward annotation must be redone until the errors no longer appear.

Some warning messages do not have a substantial influence on the design and can be ignored.

Click the confirm button in the prompt window and enter the Expedition design interface, as shown in Figure 9.22.

9.3 Layout-Related Setup and Operation

9.3.1 Introduction to Layout License Control

First we discuss Expedition license control.

From the menu select Setup→License Modules; we can see the license control menu as shown in Figure 9.23; we will discuss the entries one by one as follows:

1) ICX Pro Verify: High-speed circuit design module, used to support high-speed circuit design capabilities.
2) Advanced Technology Pro (AdvPkg): Advanced package and SiP design module, supporting wire bonding, cavity, die stacking, real-time 3D DRC, etc.
3) Advanced Technology Pro (EP): Embedded resistor and capacitor generation module; discrete resistors or capacitors can be directly integrated into embedded planar-type resistors or capacitors via a menu command.
4) Advanced Technology Pro (RF): RF circuit design module, specifically for Expedition environment RF circuit design.

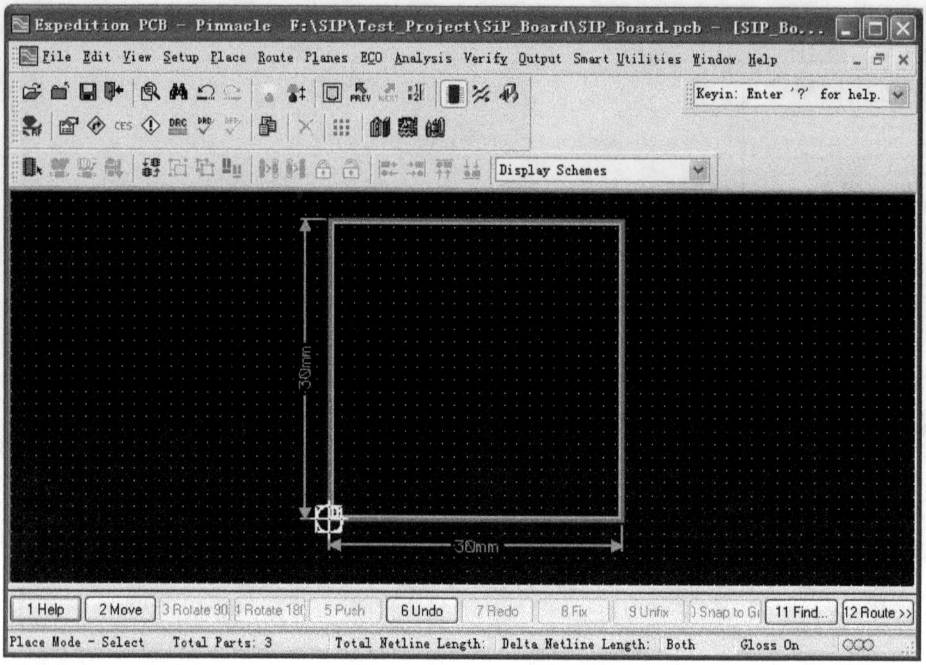

Figure 9.22 Expedition interface.

5) Advanced Technology Pro (Flex): Flexible circuit design module, specialized for flexible circuit design in the Expedition design environment.
6) Advanced Interconnect: Advanced interconnection module, supporting auto-routing in HDI complex layer stackup, such as blind and buried hole auto-routing.
7) IFF Interface: IFF interface module, used to support IFF format data input and output.
8) Reusable Block: Design reuse modules, used to support mature circuit design reuse; circuits from the central library can be directly placed in the Expedition layout design.
9) Team PCB: Team layout design module, supports multiplayer layout design, non real-time, need to design split and merge operations.
10) Design for Fabrication (DFF): DFF module, supports DFF process rule checks.
11) Design for Testing (DFT): DFT module, supports DFT analysis, which includes setting test probe placing and test probe simulation.
12) FabLink XE Pro: Manufacturing-related module, mainly for supporting production-related data generation and inspection. FabLink XE Pro is the professional version, FabLink XE is the simpler version.
13) Xtreme Design Session: Real-time collaborative design module; unlike TeamPCB, Xtreme does not need any design split, many people can join in the same design in real-time, each designer's editing is visible to other designers in real-time.
14) Topology Planner: Topology planning module, used for bus topology planning.
15) Topology Router: Topology routing module, used for bus topology routing.

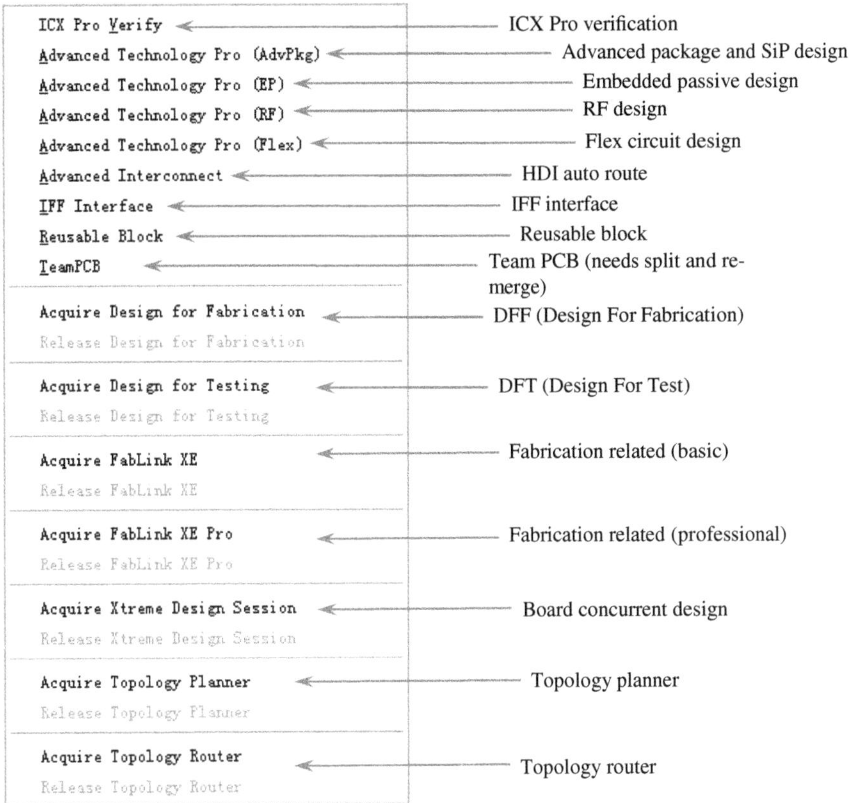

Figure 9.23 Expedition license control.

The licenses described above are additional functions beyond the basic functions in Expedition, and can be selected according to user requirements. If we do not enter Expedition from the schematic environment, but start Expedition on its own, before entering the Expedition interface a window will automatically pop up from which the user can choose a license/s, as shown in Figure 9.24.

In addition to the basic license selection items, additional license options are almost consistent with the design menu license options in Expedition.

There are four basic license selection items:

- Pinnacle has the most comprehensive functionality apart from the additional license options; it supports various functions, including high-speed routing;
- Ascent LX does not support the high-speed routing function;
- Ascent does not support high-speed routing or automatic routing;
- Xtreme Design Client is special; it is used for real-time collaborative design.

In addition, because designers use license functions differently, the license options that they see may be different from those shown in Figure 9.24. License functions not contained in the license file are not shown in Figure 9.24.

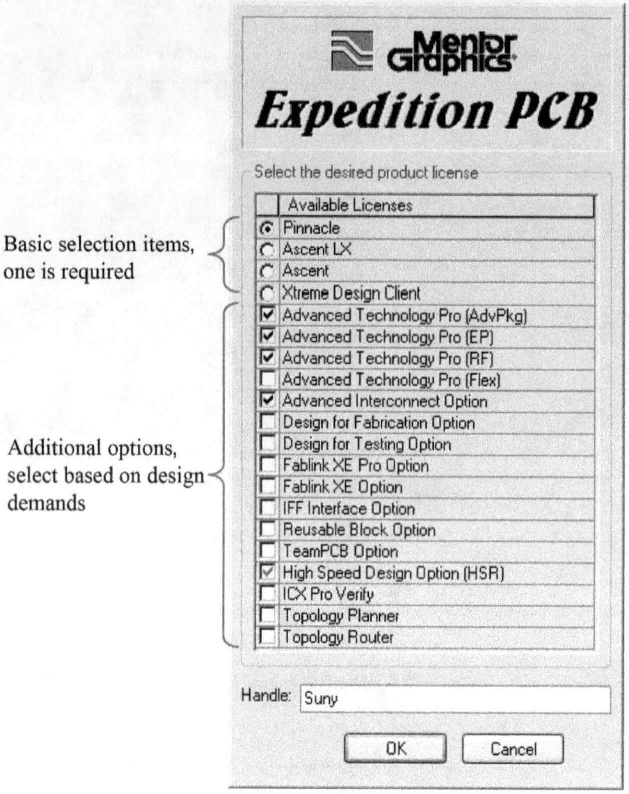

Figure 9.24 License control window before entering Expedition.

9.3.2 Mouse Handling

Mouse-handling methods in the Expedition environment are shown below.

- Selection: using left mouse button, click object for selection.
- Zoom: the middle wheel of the mouse is used for image zooming and panning. Roll the wheel forward, zoom in, roll the wheel backward, zoom out.
- Pan: push down the middle wheel and move the mouse, screen pan.
- Stroke: the right mouse button is used as the default stroke key; if the right button is pressed and the mouse moved simultaneously, the stroke function will be invoked.
- Right-click menu: right-click to view pop-up menu. If the right button is pressed and the mouse is not moved, the right-click menu will pop up.

The functions and operation methods of the mouse are shown in Figure 9.25.

"Stroke" (also called hand-drawn figure) is a quick way of calling commands using mouse operations. Press the right mouse button and move the mouse; draw a 3×3 matrix path as a foundation.

In the matrix, drawing a slide path with the right mouse button pushed down will generate a sequence of numbers; the digital sequence corresponds to an operation in the system commands.

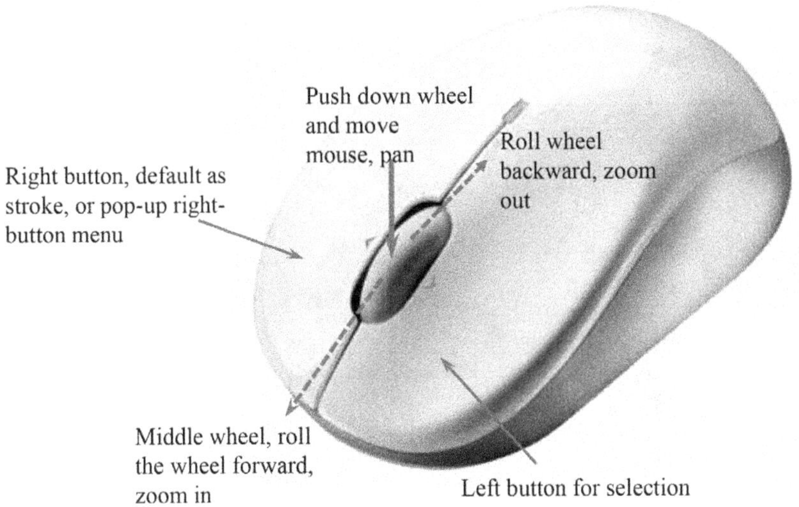

Figure 9.25 The functions of the mouse in Expedition.

For example, press the right button, starting from the upper-left corner of the matrix, slip the mouse along the diagonal to the lower-right corner; this generates a 159 sequence. It will call the View Area command. Starting from the lower-right corner of the matrix, slip the mouse along the diagonal to the upper-left corner; this generates a 951 sequence. It will call the View All command, as shown in Figure 9.26.

The stroke can also be realized using the middle mouse button; select the View menu→Mouse Mapping→Middle Button Strokes; here we can change the stroke button from right mouse button to middle mouse button, as shown in Figure 9.27.

After toggling to middle-mouse-button stroke, the result is the same as the traditional BoardStation user operations (BoardStation, also known as EN, is a layout design tool of Mentor based on UNIX; there is also a virtual Windows version). The function of the right button is swapped with the middle button, translate to panning.

Table 9.1 shows common stroke commands; it can be seen that can see the key stroke commands are very rich in Expedition, and the mastering of stroke commands can effectively improve design efficiency.

9.3.3 Three Kinds of Commonly used Operating Modes

In Expedition, there are three kinds of commonly used operation modes: place mode, route mode and draw mode. Different modes represent different functions; in each mode, the content and shortcut menu in the toolbar will be different.

To access place mode, click the Place Mode button and enter place mode. This mode is mainly used for component placement; we can move, push, align and lock components, etc.

To access route mode, click the Route Mode button and enter route mode. This mode is mainly used for routing, including manual routing, semi-automatic routing, automatic routing, etc.; it can also be used to adjust net connections. Traces, pins and routing-related elements can only be selected in route mode.

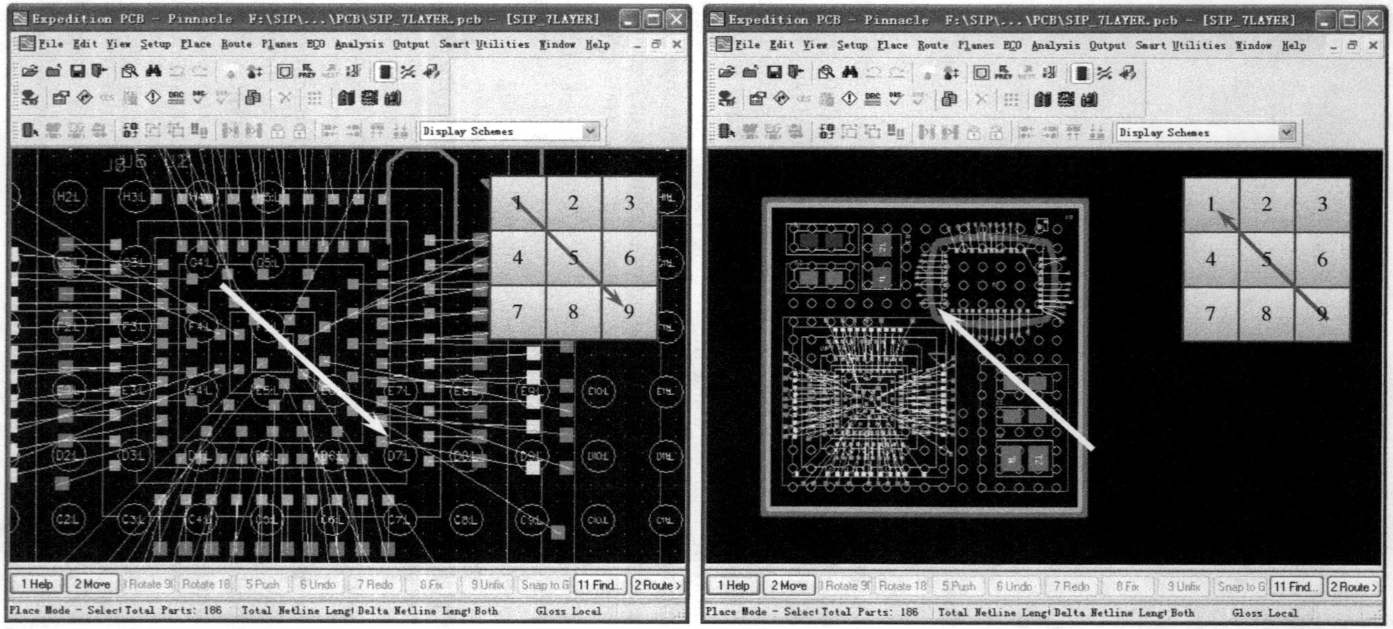

(a) 159 Zoom in to area

(b) 951 View all

Figure 9.26 Right mouse button strokes.

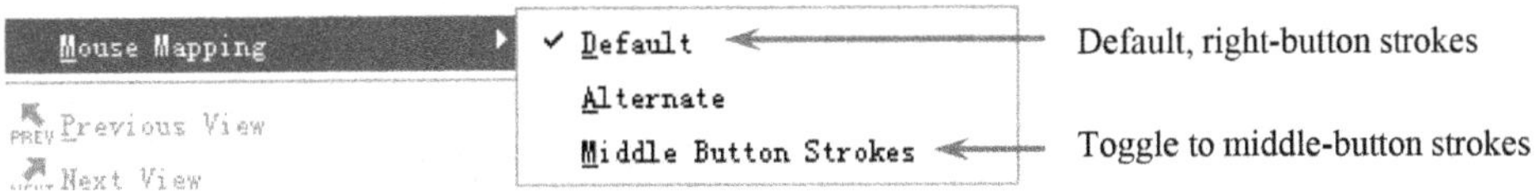

Figure 9.27 Switch from right-mouse-button stroke to middle-mouse-button stroke.

Table 9.1 Stroke actions and corresponding operands.

Stroke	Action	Operand
	Displays help on strokes	78952
	Undo	7412369
	Report Selected Item	1474123
	View Area	159
	View All	951
	Zoom In	357
	Zoom Out	753
	Delete	74123698
	Copy Trace / Segment	3214789
	Turns On / Off Netlines	321478965
	Tentative Snap	729
	Static Snap	927
ALT key +	Route Mode	123
ALT key +	Place Mode	321
ALT key +	Draw Mode	147
	Route Interactive	852
	Route Switch Ends	9632147
	Display Control	1478
	Toggle Display Active Layer Only	96541
	Editor Control	14569
	Auto-Finish (Plow)	258
	Unselect All	1478963
	Move (Place Mode)	74159
	Rotate	3698741
	OK	654
	Cancel	456

To access drawing mode, click the Draw Mode button and enter draw mode. This mode is mainly used for drawing graphics, for example, board outline, route border, plane shape, etc. Text can be added, and various dimensions, etc.

According to the current mode of operation (place mode, route mode or draw mode), the user can choose certain related objects. In place mode we can select components, but traces, pins and netlines can only be chosen in route mode, and drawing-type objects can only be chosen in draw mode.

9.3.3.1 Select Parts

"Select Parts" must be done in place mode; if a component is selected, the placement outline of the component will be highlighted. We can determine whether this component is selected or not by the color of the component outline.

In the Display Control dialog, enable the Fill Outlines on Selection option under the Part tab; when a component is selected, the outline will be filled automatically, as shown in Figure 9.28.

There are five methods for selecting components, listed below.

1) Single Select: Left-click anywhere inside the component placement outline; the component can be selected. The previously selected components will no longer be selected. If we want to select more components, press < Ctrl > and select other components. The < Ctrl > key can be used to select multiple consecutive components or to unselect one of these components. To unselect all components, simply click any blank space with the mouse button.
2) Group Select: To select a group of components, draw a box using the left mouse button; the components wholly or partly within the box will be selected.
3) Select All: In place mode, use the menu command Edit→Select All; all components can be selected.
4) Edit→Add to Select Set menu command: Contains options for fixed, unfixed and locked component selection.

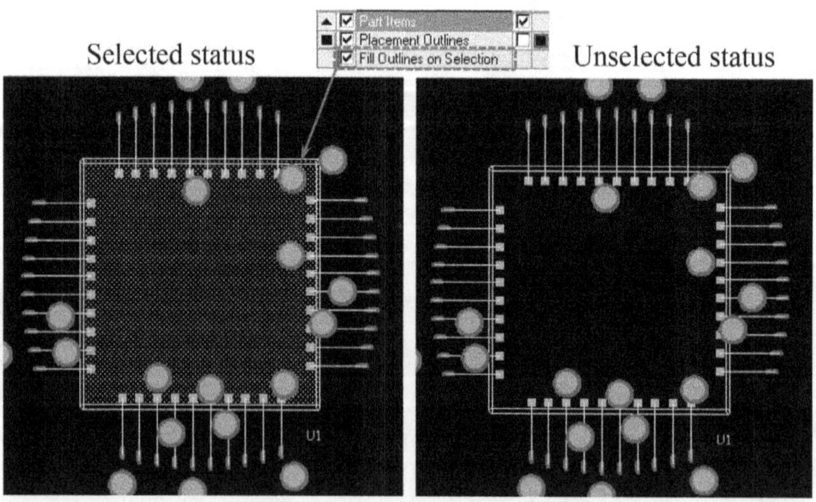

Figure 9.28 Comparison of components in selected state and unselected state.

5) Edit→Find Command: This can also be done with the < Ctrl > + < F > key combination; this function can be used to find, select, highlight or "Fit view" components, as shown in Figure 9.29.

9.3.3.2 Select Nets

Nets or traces can be selected in route mode; when a net or trace is selected, the color of it will be shown as highlighted stripes; we can judge whether the net or trace is selected by its color, as shown in Figure 9.30.

In route mode, we can select many pins or traces at the same time. After selection, commands such as fan out, route, re-route, smooth, adjust length, delete, etc. can be used.

1) Use the left mouse button to click on a pin; we can select the pin and all the nets connected to the pin will be selected at the same time. Click on a net line, the net will be selected. Click on a trace, the straight-line part of the trace will be selected.

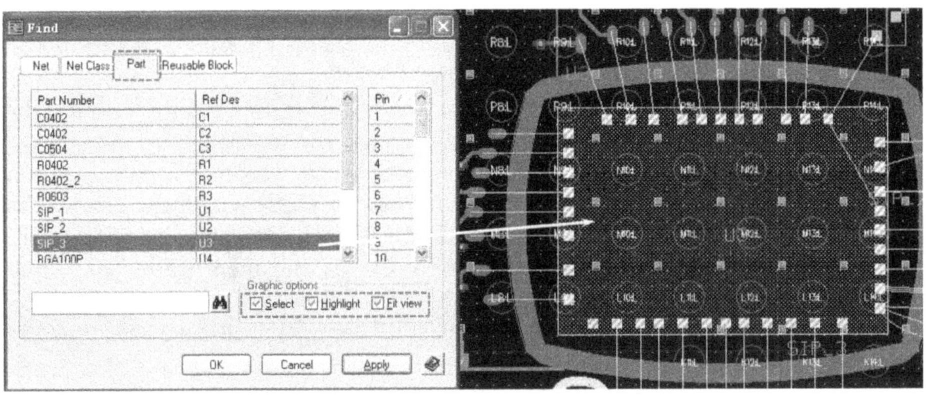

Figure 9.29 Find and select components with the Find command.

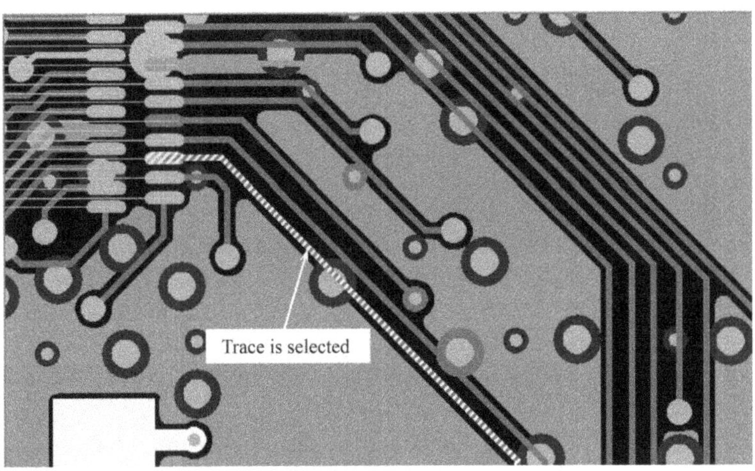

Figure 9.30 Select trace or net.

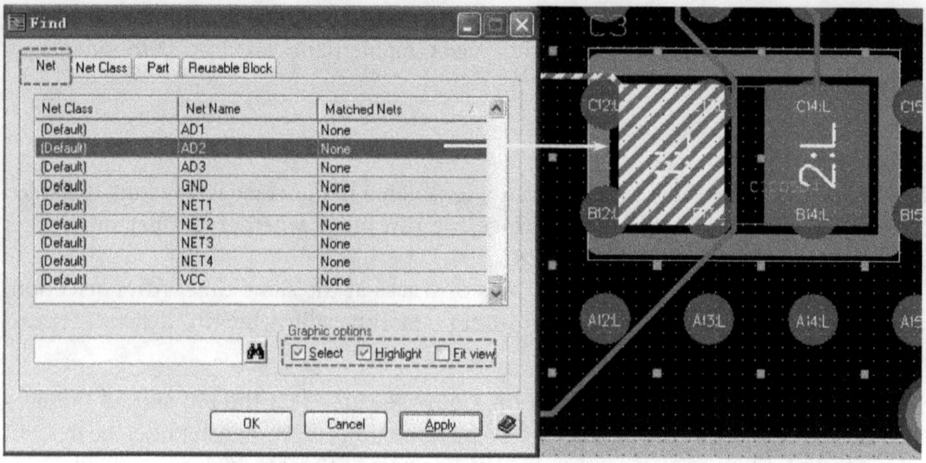

Figure 9.31 Find and select net with Find command.

2) Double-click select: Use the left mouse button to double-click on a trace; this will select the part of the trace between two pins.
3) Triple-click select: Triple-click with the left mouse button on a pin, net or trace; the whole net will be selected including all the pins, nets and traces.
4) Group select: To select a group of related objects, we can use the left mouse button to draw a box. Those nets, pins and traces wholly or partly within the box will be selected. Previously selected objects outside the box will no longer be selected. To keep the previously selected objects, press < Ctrl >, and then use the mouse to draw a box.
5) Select All: In route mode, run the Edit→Select All command; all the traces, pins and nets will be selected.
6) The Edit→Add to Select Set menu contains commands to select fixed traces, unfixed traces, holes, and other special objects.
7) The Edit→Find command (or the < Ctrl > + < F > key combination) can be used to find a net and to select, highlight or choose Fit view for that net, as shown in Figure 9.31.

9.3.3.3 Select Draw Objects

In draw mode, the designer can select multiple graphic objects at the same time. The selected objects can be modified or deleted.

Warning: Designers must note that in draw mode trace and component outlines can be selected and deleted. When editing or deleting objects in draw mode, we need to be careful not to edit or delete trace and component outlines by accident.

As in route mode, in draw mode the selected element is also highlighted by stripes overlapping the original color, making it convenient for the user to identify the chosen object. As shown in Figure 9.32, plane shape and component outline are selected and edited in draw mode.

In draw mode, the designer can draw as many as 32 kinds of graphics, as shown in Figure 9.33. According to the licenses, which support different functions, the number of graphics will also be different. After completion of graphics, the type can be changed,

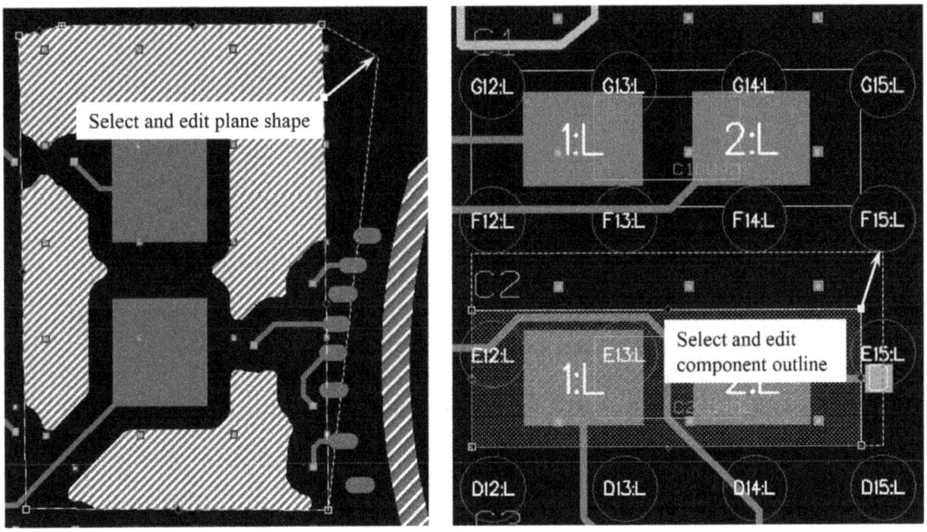

Figure 9.32 Select and edit plane shape and component outline in draw mode.

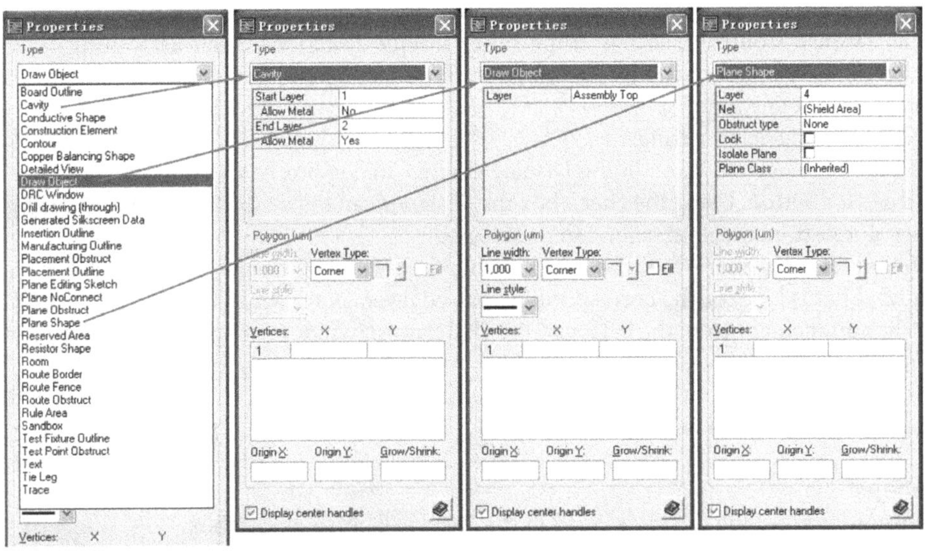

Figure 9.33 Draw various kinds of graphics in draw mode.

but not all types can be changed to other types. In addition, text can be added to the substrate, such as labels, annotations, etc., in draw mode.

In draw mode, we have two select functions, single select and group select:

1) Single select: Click on a drawing object to select it. To select more drawing objects or to cancel selected objects, press the < Ctrl > key and click objects at the same time.
2) Group select: To select a group of drawing objects, draw a box around them with the left mouse button.

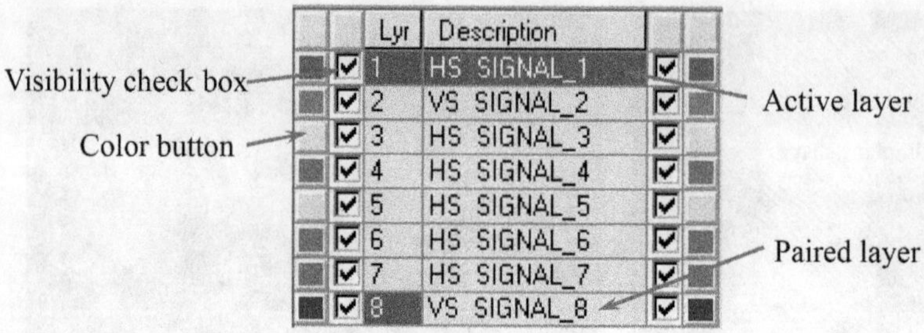

Figure 9.34 Layer visibility and color setting.

Below, we discuss Display Control, Edit control and the Parameter Settings dialog.

- Display Control – controls color, transparency and other display settings.
- Editor Control – place-, route- and grid-related settings.
- Setup parameters – basic parameters of layout settings.

9.3.4 Display Control

The Display Control dialog box is mainly used for graphic display settings. At the same time, Display Control also has some other unique functions, such as routing layer switch, etc. When carrying out routing, open the dialog and place it beside the work area, to use at any time.

Visibility and color settings

Each graphics item listed in the Display Control dialog box has a visibility check box and a color button. Using the check box the designer can toggle the item visibility; using the color button, the item color can be selected.

In the Layer section, the whole bar is blue for the active layer, while only the layer number bar is blue for the corresponding paired layer, as shown in Figure 9.34.

Click any color button; the Select a Color Fill Pattern window will pop up; the designer can select color and pattern in this window, which has a variety of options.

The designer can set the color and pattern according to his or her design habits and preferences, as shown in Figure 9.35.

Groups

Settings for grouping graphics listed in the Display Control dialog box.

The Group title bar is a color. Using the group check box, the designer can display or shut down all the graphics of a group at one time. In order to simplify the dialog, grouped items can be folded together, so that only the group theme shows. Using grouping, the designer can more easily find the options needed for set up, as shown in Figure 9.36.

Display Control Schemes

The designer can set the object's color display and then save this to the Display Control scheme; the scheme can then be called at any time. For certain graphics, such as placement display, route display, silkscreen display, drilling display, etc., the displays can be stored in multiple schemes.

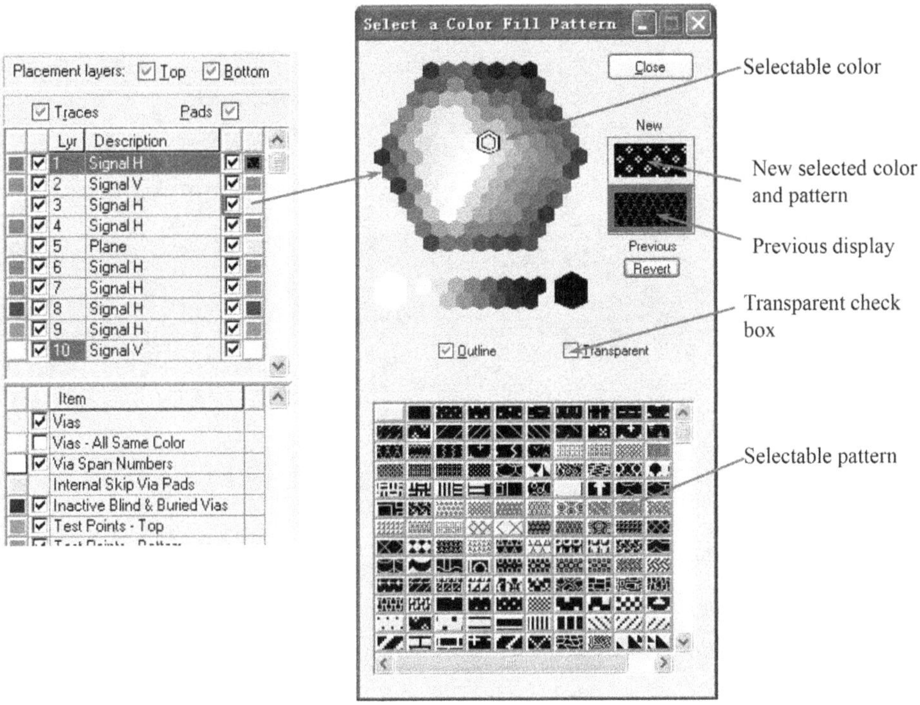

Figure 9.35 Display color and pattern setting window.

These display schemes can be stored in the same location as design (limited to the current layout design), and can also be saved with product system files (can be called in all the layout designs on the computer); the designer can also choose both options.

To call an existing display scheme, choose from the display scheme drop-down list.

To delete a saved display scheme, select this scheme from the drop-down list, then click the delete button X, as shown in Figure 9.37.

9.3.4.1 Layer Tab, General Tab, Part Tab

Figure 9.38 shows the Layer tab, General tab and Part tab of Display Control.

a) Layer tab

The Layer tab is used to set up the trace and pad layer display, the current routing layer, netline, grid, copper and many other display items.

The layer Display Control area is at the top of the Layer tab; we can open or close all trace or pad displays using the Traces and Pads check item. At the same time, we can open or close each layer separately. Here, it is important to note that the active layer cannot be closed, so if you want to close it, switch the active layer to another layer, and then close the required layer.

1) Netlines & Classlines Group: used to toggle the netlines and classlines visibility, and set the display color and display mode of netlines, such as netlines from the traces or netlines from route targets, etc.

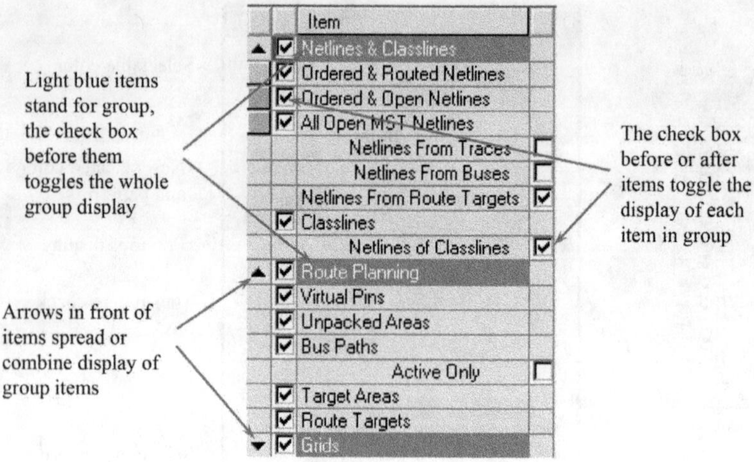

Figure 9.36 Display Control item grouping.

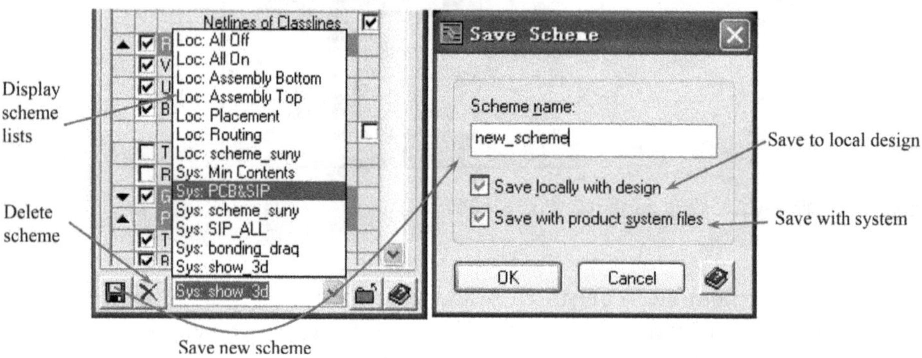

Figure 9.37 Save and delete display scheme.

2) Route Planning Group: used for route planning, including the bus routing channel and virtual nodes display.
3) Grids Group: used for grid display, including place grid, route grid, draw grid, jumper and test point grid.
4) Pads Group: used for pad display, including through-hole pad, bond pad, via pad, test point pad, etc.
5) Planes Group: used for copper plane display, including plane shape and obstruct, plane data and fill/hatch option.
6) Copper Balancing Group: used for copper balancing data and shape display.
7) Materials Group: used for all kinds of materials including resistance, capacitance and conductance materials, mainly used in the embedded technology of resistors and capacitors.
8) Spacers Group: used to display spacers; when this option is open, in addition to spacers, others display in shadow mode.
9) Selection & Highlight Group: used for display or highlight of the selected net. The "Selected 'only'" option in cooperation with shadow mode can be a good method for net topology view, as shown in Figure 9.39.

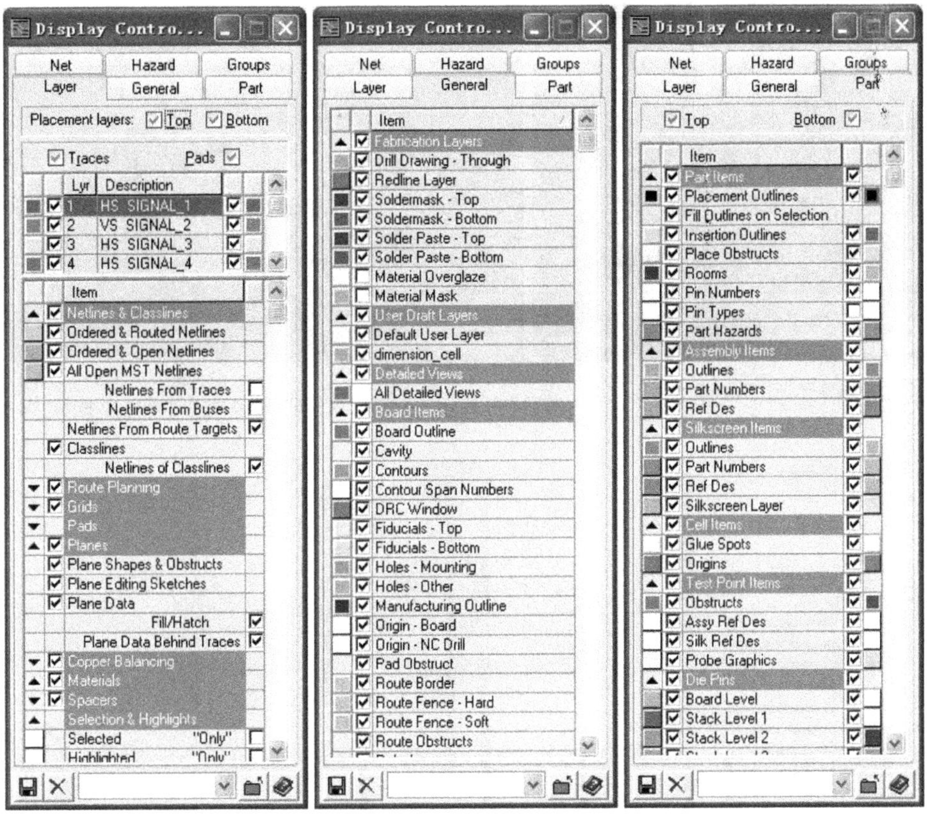

Figure 9.38 Layer tab, General tab and Part tab of Display Control.

b) General tab
   The General tab contains eight groups, which are mainly used to set display options related to production, user-defined layers and board-related items, as well as some other display items.
   1) Fabrication Layers Group: used to display soldermask layer, solderpaste layer, drill drawing layer, material mask layer, etc.
   2) User Draft Layers Group: used to display the user-defined custom layer in Setup Parameters, and also includes the Gerber or NC Drill data layer when importing these data for checking.
   3) Detailed Views Group: mainly used for toggling Detailed Views visibility.
   4) Board Items Group: used for enabling board-related graphics visibility. For example, the board outline is used to display board outline, cavity is used to display substrate cavity, origin board is used to display the origin of substrate, Origin-NC Drill is used to display the origin of drill, rule area is used to display the rule area border, etc.
   5) Text Items Group: mainly used for pin and cell attribute text display.
   6) Multiple Designers Group: mainly used for the display of protection area, etc. in multiple designer collaborative design.

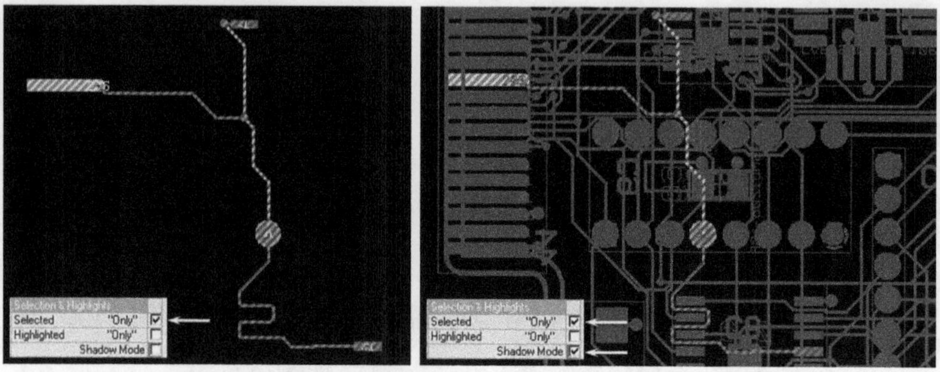

Figure 9.39 Use Selected only option to view net topology (with shadow mode shown on the right).

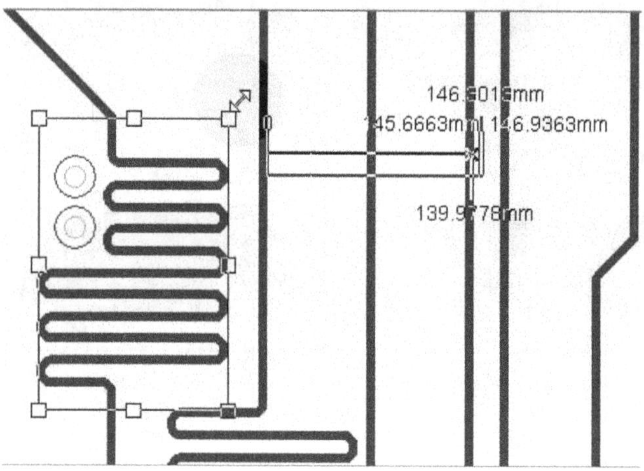

Figure 9.40 Display Tuning Meter.

7) Display Patterns Group: mainly used for the display pattern effect. When this option is unchecked, the pad, routing, copper, etc. will be displayed with hollow effect.
8) Options Group: includes many display options. For example, if the "Auto-Pan" option is enabled, then when the designer drags a trace or component along the edge of the working area, the software will automatically adjust the view area to fit the operation. If the "Full Cursor-move part" option is enabled, when the designer moves a component, the cursor will become a full screen cross. If "Display Tuning Meter" is enabled, if defining length or timing constraints in CES, when routing or moving trace for the constraint nets, the tuning meter will display the current length or delay, as shown in Figure 9.40. If the "Update Only On Apply" option is enabled, the Apply button will appear at the bottom of the Display Control window, and only when the designer clicks on the Apply button will the selected graphical display effect be updated. If the "Display Active Layer Only" option is enabled, the system will show the active layer

items only; this option is often used in interactive routing. The "Pan Sensitivity" option is used to adjust the sensitivity of mouse movements; by increasing or decreasing the value of this option, the graphic movement velocity will speed up or slow down.

c) Part tab
The Part tab is used to control component-related display; using the check items ☑ Top Bottom ☑ at the top of the Display Control window, the designer can switch on or off all the top-side or bottom-side parts display.
   1) Part Items: mainly used to display component placement outline, pin number, pin type, etc.
   2) Assembly Items: assembly-related display items, mainly used to display assembly outline, assembly Ref Des, assembly part numbers, etc.
   3) Silkscreen Items: silkscreen-related display item, mainly used to display silkscreen outline, silkscreen Ref Des, silkscreen part numbers, etc.
   4) Cell Items: mainly used to display the cell origin and glue spot of SMT.
   5) Test Point Items: mainly used to display test point-related items.
   6) Die Pins: mainly used to show bare chip pins; the pins can be set with different colors for multilayer chip stacks.
   7) Bond Wires: mainly used to display bond wire; the bond wire can be set with different colors for multilayer chip stacks.

9.3.4.2 Net Tab, Hazard Tab, Groups Tab

Figure 9.41 shows the Net tab, Hazard tab and Groups tab of Display Control.

1) Net tab
The Net tab is mainly used to display Net, Net Class and Constraint Class with user-defined colors, and to select whether to enable the selected colors using the check items. The option "Apply color to Netlines, Traces, Pads, Planes and Vias" enables the color settings for each item to be selected.
   a) Net Class Names Group: mainly used for netclass color setting; check to enable the color setting.
   b) Constraint Class Name Group: mainly used for constraint class color setting; check to enable the color setting.
   c) Net Names Group: mainly used for all net color setting; check to enable the color setting.

2) Hazards tab
The Hazard tab is used to set color for Online DRC and Batch DRC reported violations; it is convenient for error position view. If shadow mode is checked in addition to the enabled violations display, the other elements will show gray.
   a) Online Group: used to set colors for Online DRC violation report; check to enable color setting.
   b) Batch Group: used to set colors for Batch DRC violation report; check to enable color setting.

3) Groups tab
The Groups tab is mainly used for RF Group color setting and display in RF circuit design; if no RF Group has been defined, the column will be empty.

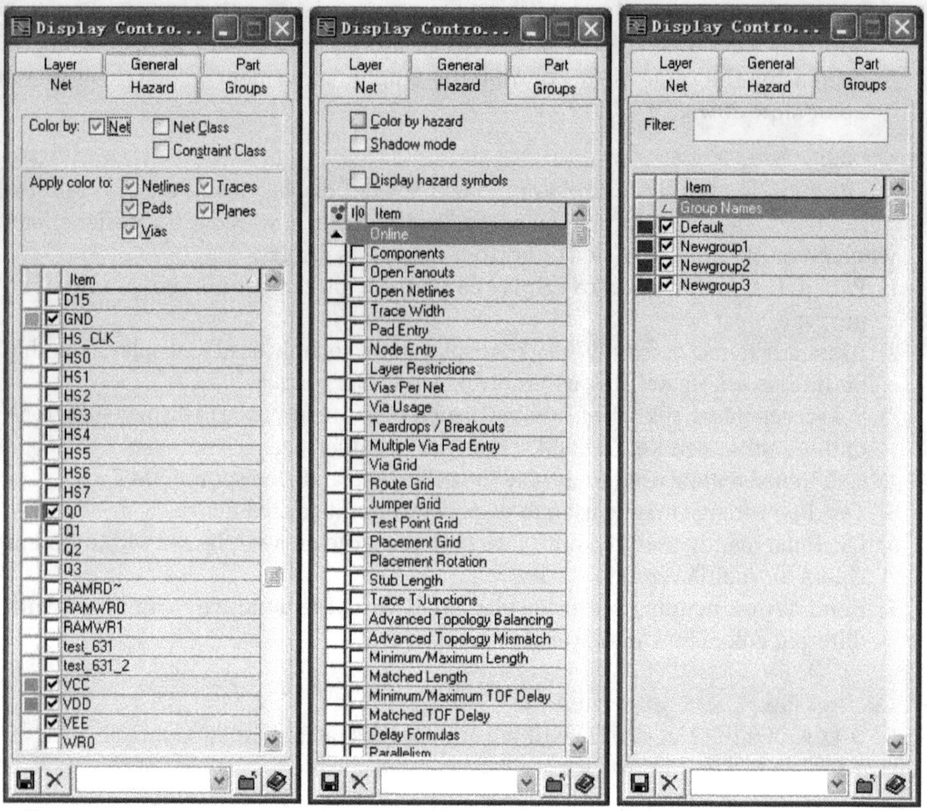

Figure 9.41 Net tab, Hazard tab and Groups tab of Display Control.

9.3.5 Editor Control

The Editor Control toolbar is divided into Common Settings and three tabs: Place tab, Route tab and Grids tab. Each is introduced below.

Figure 9.42 is the interface screenshot of Place tab, Route tab, Grids tab and Common Settings.

9.3.5.1 Common Settings

When the Editor Control window is switched between several tabs, Common Settings (at the bottom of the window), which contains the check option for Interactive DRC and the auto-save interval time setting, remains unchanged.

1) DRC setup

   By default, interactive DRC is always open, that is, the check item in front of Interactive Place/Route DRC is always selected, to ensure that in layout design operation errors will not occur. If it is necessary to close interactive DRC in the design process, deselect the check item before Interactive Place/Route DRC. At this point, the system will pop up a prompt window asking whether the DRC function is to be closed; click "Yes" to confirm and the system will automatically pop up a "DRC

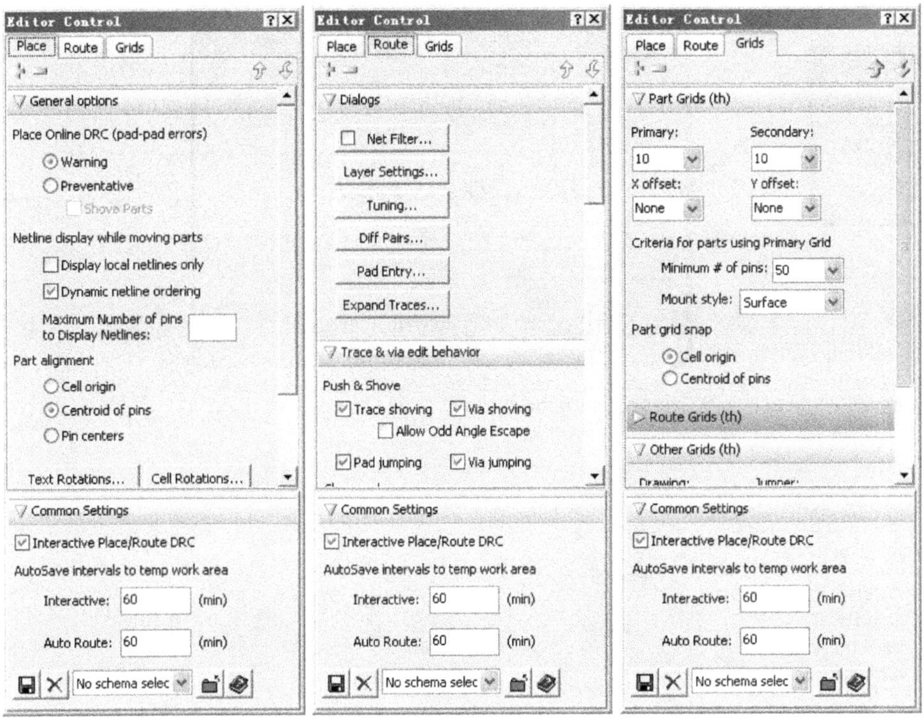

Figure 9.42 Place tab, Route tab, Grids tab and Common Settings.

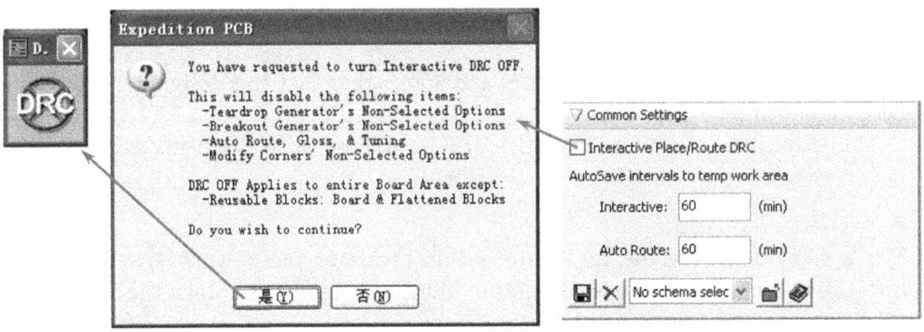

Figure 9.43 When interactive DRC is closed, the system shows a "DRC Off" warning.

closed" alert window, notifying the designer that interactive DRC was closed, as shown in Figure 9.43.

In interactive DRC closed state, the designer can carry out operations in violation of DRC rules, such as placing two component pads overlapping; this is often used in RF circuit design. Here we need to remind designers not to forget to re-open interactive DRC on completion of such operations.

The method is very simple; we just need to close the DRC Off prompt window. The system will automatically pop up the window shown below, stating that

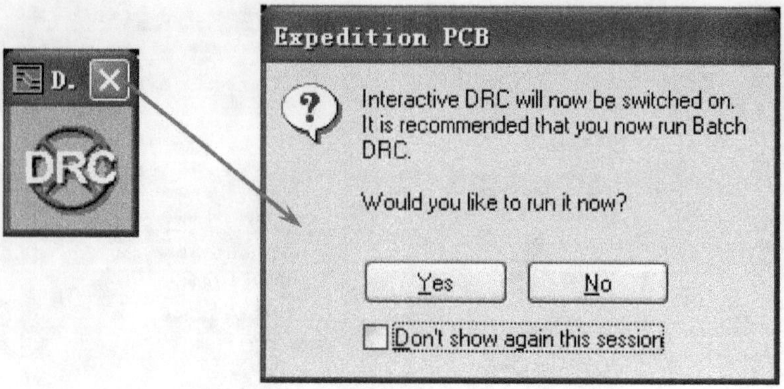

Figure 9.44 Interactive DRC re-open prompt window.

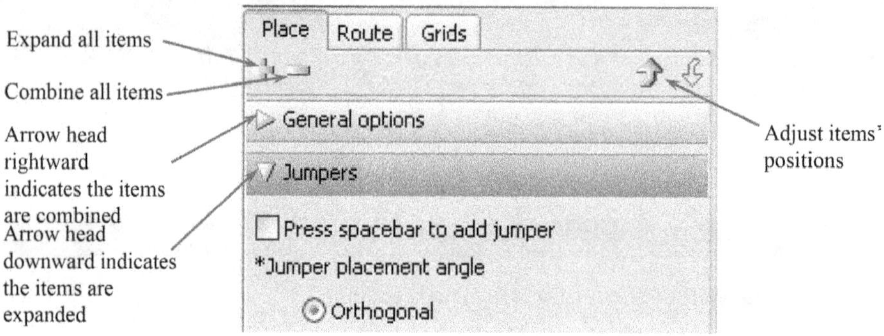

Figure 9.45 Edit control window operation introduction.

interactive DRC will be switched on, and suggesting that the designer runs Batch DRC, as shown in Figure 9.44.

Under normal conditions, click the "No" button, because after the design is completed, the designer will always run Batch DRC.

2) AutoSave interval setting
The system default value for "AutoSave intervals to temp work area" for both Interactive and Auto Route is 60 minutes but the designer can adjust the time interval according to the design reality.

Next, we introduce the Place tab, Route tab and Grids tab, but before introducing each tab, we need to understand their operation methods; see Figure 9.45.

Use the left mouse button to click on the plus sign to expand all the items, click the minus sign to combine all the items. Clicking the triangular arrow symbol to the left of each item expands or combines that item alone; when combined, the left arrow points rightward, when expanded, the arrow head points downward.

The right side up and down arrow is mainly used to adjust the relative position between items; using the up arrow the designer can make the selected item move up, so that designers can place commonly used items at the top of the window, convenient for selection.

9.3.5.2 Place Tab

The Place tab includes General options and Jumper options.

1) General options
   The general options are mainly associated with the placement rules settings, such as placement online DRC rules. If set to Warning, when components conflict the system will show a warning; if set to Preventative, the system will prevent component conflict happening; if the "Shove Parts" option is set, when components are in conflict, the system will automatically push other parts.

   In real design, choose which option accords with the actual conditions. For example, if we need to place part of some components outside the board outline, we should choose the warning option, so that even if a warning is given, the system allows for such component placement. If this is set to Preventative, components cannot be placed across the board outline at all because of conflict between components and board outline, and the system is set to prevent conflict.

   Due to space limitations, other options such as netline display while moving, part alignment, text rotation option, cell rotation option, clusters and rooms will not be described in detail here.
2) Jumper options
   The Jumper options are mainly used to set up the rules about jumper and are not discussed here.

9.3.5.3 Route Tab

The Route tab contains five main options: Dialogs, Trace & via edit behavior, General options, Plow and Via & fanouts. The location of each option can be adjusted using the right upper and lower arrows; the designer can place the most commonly used option to the top of the window, convenient for selection, as shown in Figure 9.46.

1) Dialogs
   1) Net Filter option, used to filter nets; only the nets in the list can be selected and operated. Here we need to enable Net Filter by checking the check box on the left side, otherwise the Net Filter is not effective; the method is shown in Figure 9.47.
   2) Layer Settings option, mainly used to set allowed routing layers, the default routing direction of each layer and layer pair (automatic switched layers in routing).
   3) Tuning option, mainly used to set tuning rules, including tuning shape, method, physical parameters, etc.

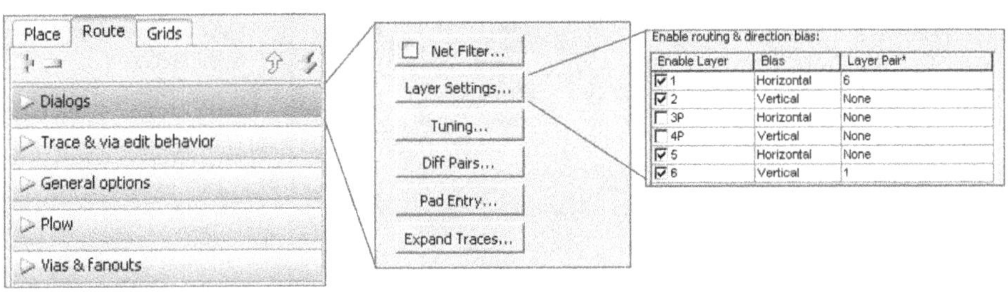

Figure 9.46 Route tab Dialogs option expanded step by step.

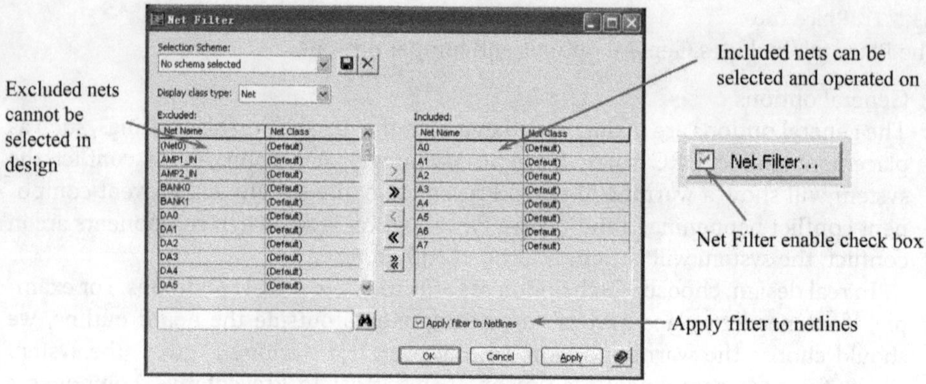

Figure 9.47 Net Filter method.

4) Diff Pairs option, mainly used to set the differential pairs, including same layer pairs or adjacent layer pairs, and the Diff Pairs physical parameter settings, etc.
5) Pad Entry option, mainly used to set pad entry and exit direction and allow via under pad. In SiP or package design, we often need to place via under pad, such as via on BGA pad. For example, in a four-layer SiP substrate design, if we need to place via under BGA pad, select BGA pad Round500, check the Allow via under pad option, and then select Layer Range 3-4, as shown in Figure 9.48. This indicates that via under pad Round500 is allowed as is via across from layer 3 to layer 4. Don't check the Allow off pad origin option, indicating that the via cannot be deviated from the pad origin.
6) ⑥ Expand Trace option, mainly used to set trace width expansion; with routing region change, the trace width can be adjusted automatically according to the setting.

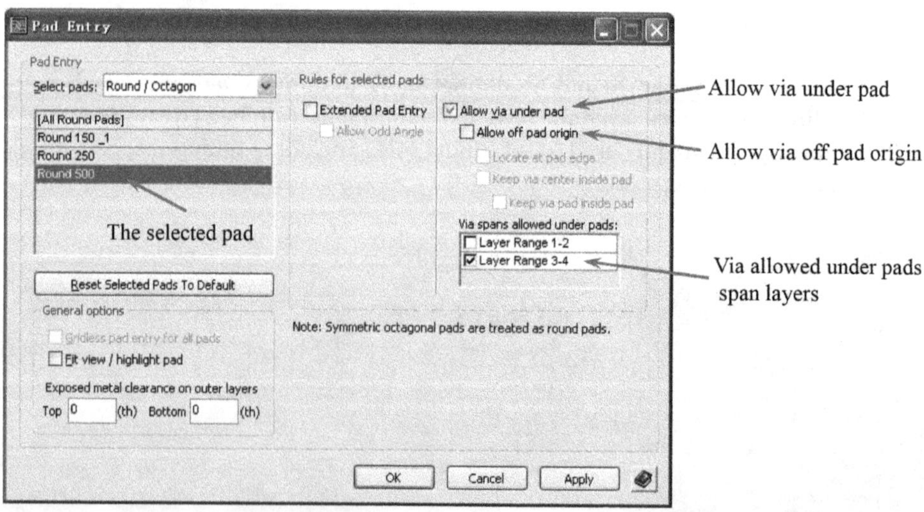

Figure 9.48 Allow via under pad settings.

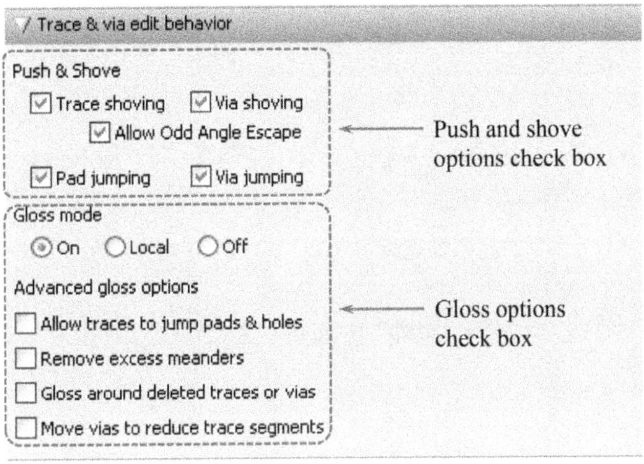

Figure 9.49 Trace & via edit behavior option.

2) Trace & via edit behavior
   The Trace & via edit behavior option is mainly used for the shoving and gloss mode setting in the routing process, as shown in Figure 9.49.
   1) Push & Shove options
   Trace shoving controls whether to push trace when routing, via shoving controls whether to push via when routing, pad jumping controls whether to jump over pad when routing, via jumping controls whether to jump over via when routing.
   2) Gloss mode options
   Gloss mode includes gloss on, gloss local and gloss off; in the process of routing we can also use the < F4 > key to toggle the three gloss modes.
   3) Advanced gloss options
   Advanced gloss options includes four items: Allow traces to jump pads & holes; Remove excess meanders; Gloss around deleted traces or vias; Move via to reduce trace segments. All are used to improve the quality of routing.

3) General options
   The General options are mainly used to open or close a specific rule check in the routing process, as shown in Figure 9.50.
   1) Turn on/off net rules include: stub lengths, used to check the length of stub; layer restrictions, used to check the layer rules; via restrictions, used to check the via rules; max delays and lengths, used to check delays and lengths rules. These options are only valid for interactive route, not auto route.
   2) Curved Traces controls the radius of circular arc routing, including the Radius and Min. radius; the designer can set the value according to the design requirements. If set to Variable, the routing radius will adapt according to the specific design situation.

4) Plow
   The Plow option is mainly used to control specific route operations, as shown in Figure 9.51:
   1) Modes, used to select plow (route) mode, including forced plow, angle plow and route plow; the designer can check multiple options at the same time, and toggle between them in the process of plow using the < F3 > key.

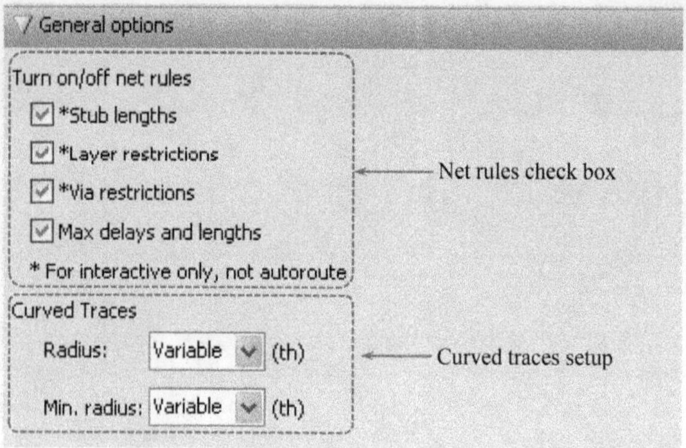

Figure 9.50 General options.

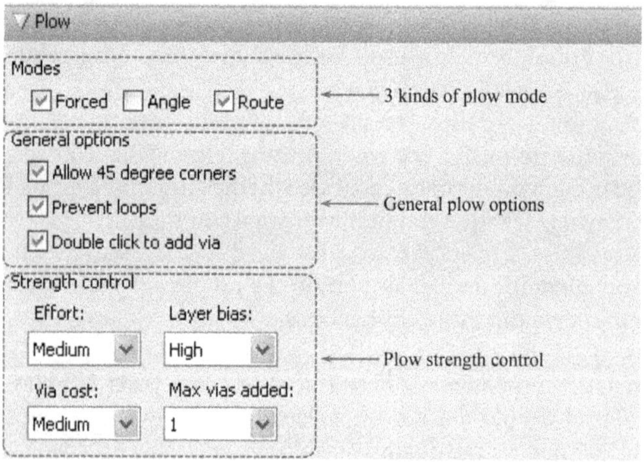

Figure 9.51 Plow options.

2) General options, mainly used to control whether to allow 45° routing in design, whether to prevent loops, and whether to add via by double-click.
3) Strength control, mainly used to control push ability, via cost, etc. for the algorithm used in the routing process.

5) Via & fanouts
This item is mainly used to set up via holes and fan out rules in the process of routing, as shown in Figure 9.52.
   1) Auto trim through vias option, used to cut off via to blind or buried via automatically according to the routing layer this via connects. Of course, the premise is that the design allows the use of blind or buried via.
   2) Allow one additional via per SMD pin option, used to control whether multiple vias can be placed on one pad. In real design, some component cooling pads need multiple vias.

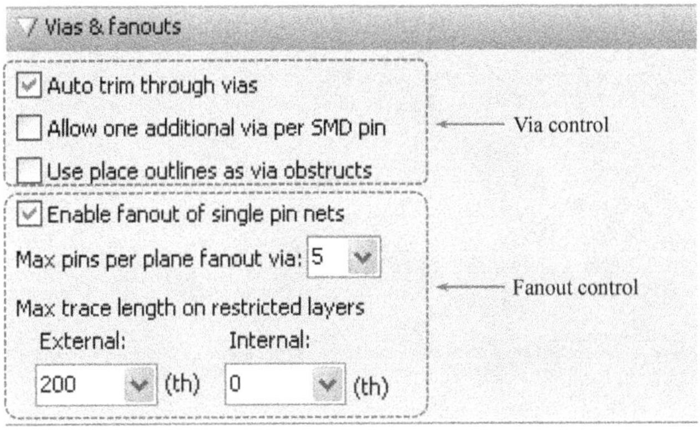

Figure 9.52 Vias & fanouts.

3) Use place outlines as via obstructs option, used to control whether to allow via placement inside the placement outline of components; usually this is allowed, so do not select this option.
4) Enable fanout of single pin nets option, used to control whether to allow single pin net fanout. If unselected, single pin nets will not fanout and route.
5) Max pins per plane fanout via, defines the maximum number of pins sharing the same via for fanout. If not defined, there is no restriction.
6) Max trace length on restricted layers, defines the maximum length of route in restricted layer, including external and internal layers.

9.3.5.4 Grids Tab

The Grids tab is mainly used for grid settings, including Part Grids, Route Grids and Other Grids, as shown in Figure 9.53.

1) Part Grids
   The designer can set up two kinds of part grids, Primary grid and Secondary grid; the "Criteria for parts" by pin number is used to define different grids for different components. In this way, a big device such as the IC with more pin numbers can be moved quickly with the primary grid, and small devices such as resistors and capacitors with less pin numbers can be moved slowly with the secondary grid. This is convenient for placing small devices around a big device in component placement. Offset is usually set to none.

2) Route Grids
   This is mainly used to set the route grid and via grid. Because Expedition supports "no grid" routing, you can set the route and via grid to none, making it easier to push trace or via when routing. Sometimes, if we need an accurate position to place via or trace, such as routing in a BGA pin array, we can set an appropriate grid, so as to make sure the trace or via is accurately located in the middle of two pads.

3) Other Grids
   Mainly used to set drawing grid, jumper and test point grid.

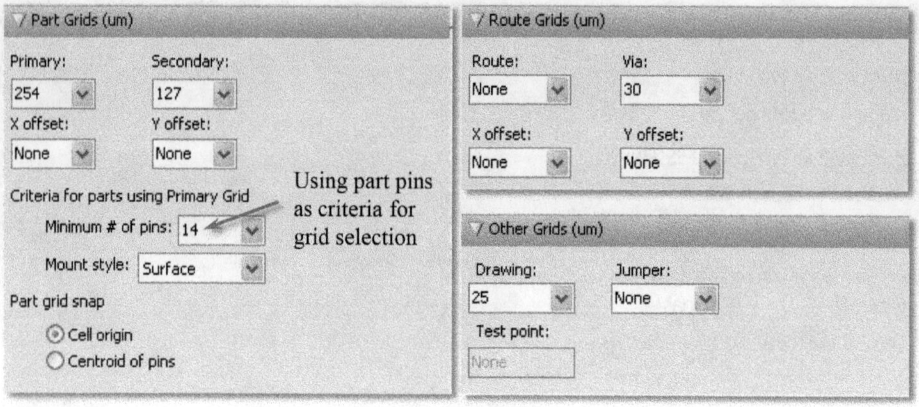

Figure 9.53 Grids tab.

9.3.6 Setup Parameters

In the process of layout template creation, we have set up many parameters related to layout; if a layout design refers to such template, all the parameter settings in the template will be automatically inherited. For example, the board outline, the blind and buried via definition, the layer stackup definition, etc. are all inherited from the layout template to layout design.

Thus, in layout design we need only adjust the size and origin of the board and so on according to the actual design situation. In this example, adjust the board size from the original 30 mm×30 mm to 31.5 mm×31.5 mm, add 1.5 mm in length and width, and adjust the board origin to the center of the board; the blind and buried via definition, layer stackup definition and other settings remain the same as in the layout template.

In real design, if the layout board often uses the same parameters as the template, we do not need any adjustment, which not only improves design efficiency, but also helps design standardization.

9.4 Substrate Layout

According to the specific design situation, the board size was expanded from 30 mm×30 mm to 31.5 mm×31.5 mm, and the board origin moved to (15750 μm, 15750 μm) the center of the SiP substrate.

9.4.1 Component Placement

Click the Place Mode button, switch to place mode, and then select the Place Parts and Cells button; the Place Parts and Cells window pops up, as shown in Figure 9.54.

Check the unplaced option, and the components not placed to the board are listed; select some components and then click the downward arrow to move the components to active lists, or click the downward double arrow to move all components to active lists, then select one of the components and click the Apply button to place.

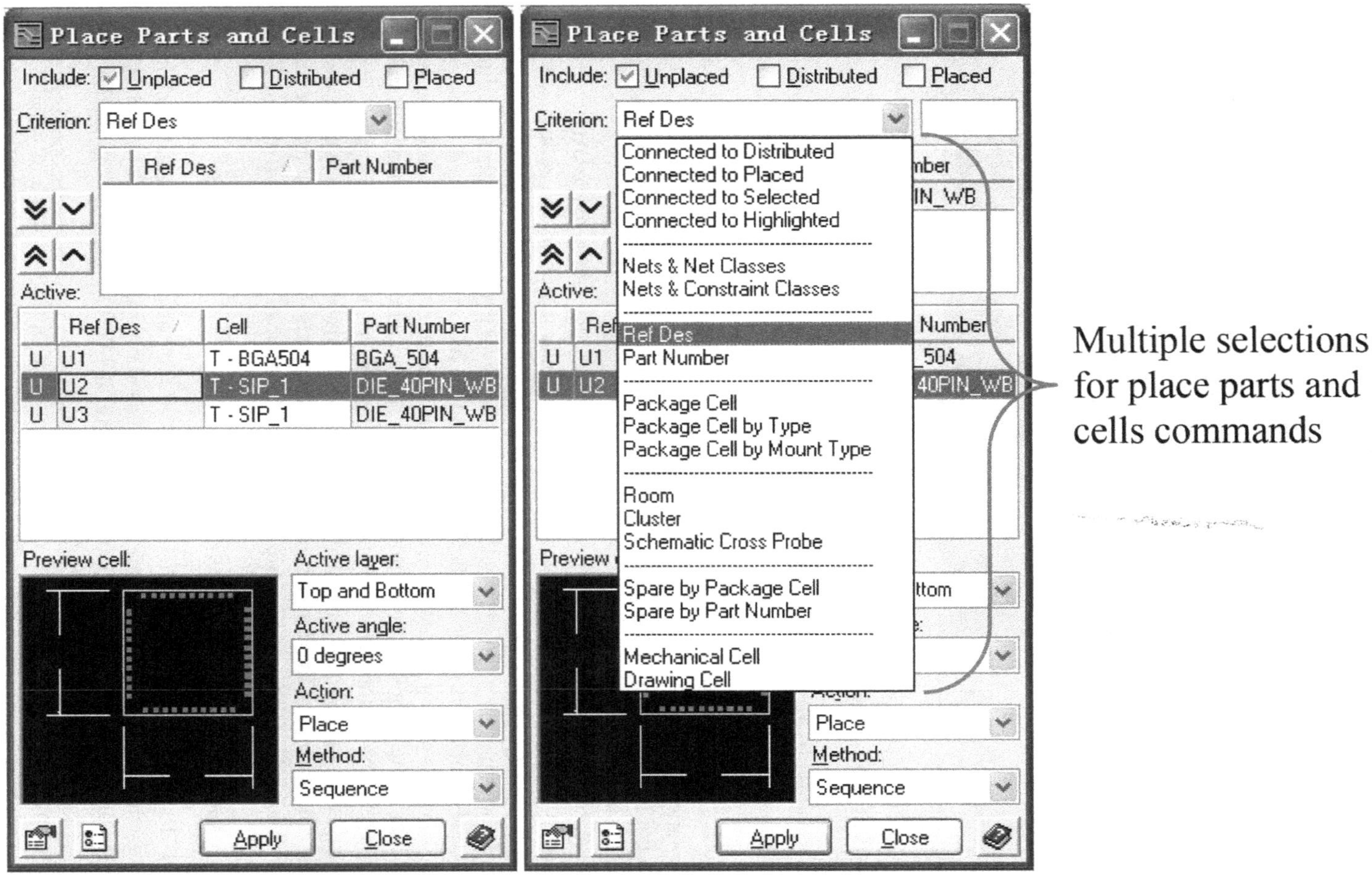

Figure 9.54 Place Parts and Cells window.

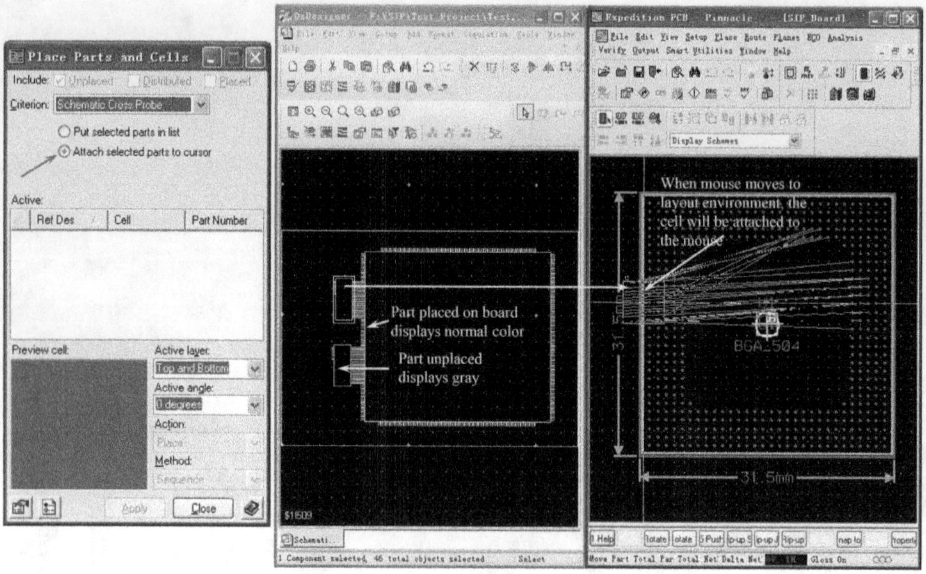

Figure 9.55 Cross probe of schematic and layout.

In the Place Parts and Cells window, there are a variety of options for placement; we have adopted the most commonly used, Ref Des, as the criterion. Designers can select criterion according to the complexity of layout and personal preference.

For example, for more complex SiP design, we often need to guide layout with the schematic, so we can choose Schematic Cross Probe as a criterion, check the Attach Selected Parts to Cursor option, and then minimize the Place Parts and Cells window (do not close this window).

Select one component in the schematic; when the mouse moves to the layout window, the cell of the selected component will automatically attach to the mouse, and the designer only needs to place the cell in the desired position.

In the schematic, we can see whether a component was placed in layout or not through real-time component color display. If a component has not been placed in layout, its color shows as gray; if placed in layout, its color is the normal color (usually the default light blue). The color changes immediately the component is placed in layout, indicating Expedition's strong real-time performance, as shown in Figure 9.55.

For this example SiP design, since there are not so many components, we use Ref Des as the criterion, place the BGA device on the bottom of the substrate, set its coordinates to (0, 0), and place the two bare chips on the top of the substrate, as shown in Figure 9.56.

9.4.2 Automatic Optimization of Net Connections

In SiP design, after device placement is completed, we need net optimization according to the actual net connections; optimization is based on the principles of shortest connection and minimal cross, swapping the general BGA I/O pins, but keeping the position for predefined power supply or GND pins.

Net optimization includes automatic and manual optimization, and usually both are needed in practical design. For the specific net optimization method, see the

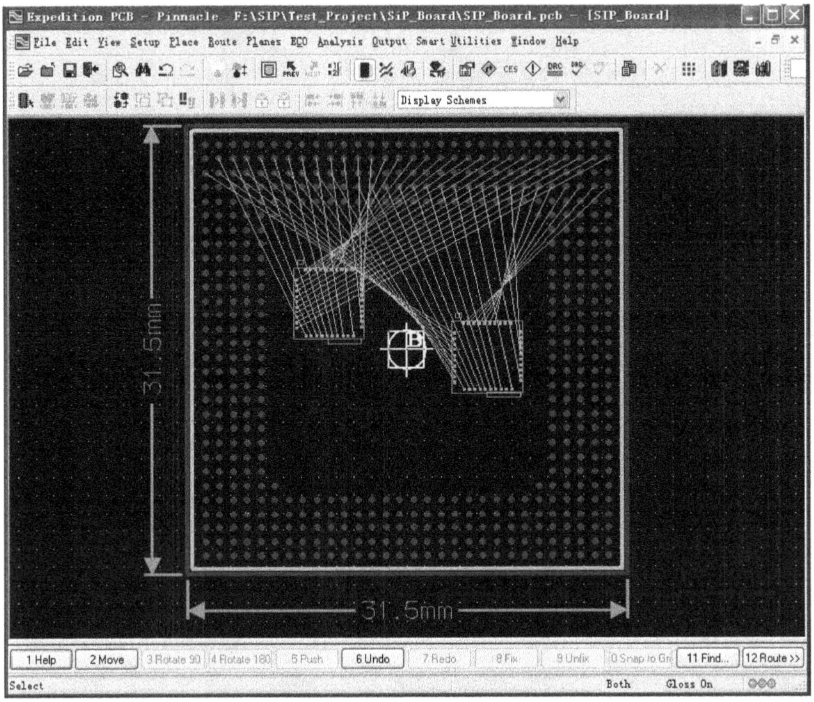

Figure 9.56 Components and net connections after placement.

description in Chapter 13; here we just show the result of net optimization; see Figure 9.57.

If you are not satisfied with the results of automatic optimization, you can adopt manual net optimization, swapping the pins of the BGA package manually, so as to achieve optimal net connections.

9.5 eDxD View

The eDxD View function allows the designer to easily see the schematic design in the layout design tool, and does not require any additional schematic license support.

With this function, the layout designer can check the schematic design or use the schematic for interactive check; it is not necessary to invoke the schematic tool.

To view eDxD View, first we check the Create eDxD view during Forward Annotation option in the Additional Options window during the forward annotation process, as shown in Figure 9.58.

Similarly, if we want to see eExp in the schematic tool, we need to check the Create eExp View during Back Annotation option, and then run back annotation. After successful back annotation, we can view eExp in the schematic, as shown in Chapter 7, Figure 7.14.

After choosing Create eDxD View during Forward Annotation, and running forward annotation successfully, in the Expedition menu, select Window→add eDxD View; we can then open a schematic window in layout design, as shown in Figure 9.59; this is very convenient for interactive check.

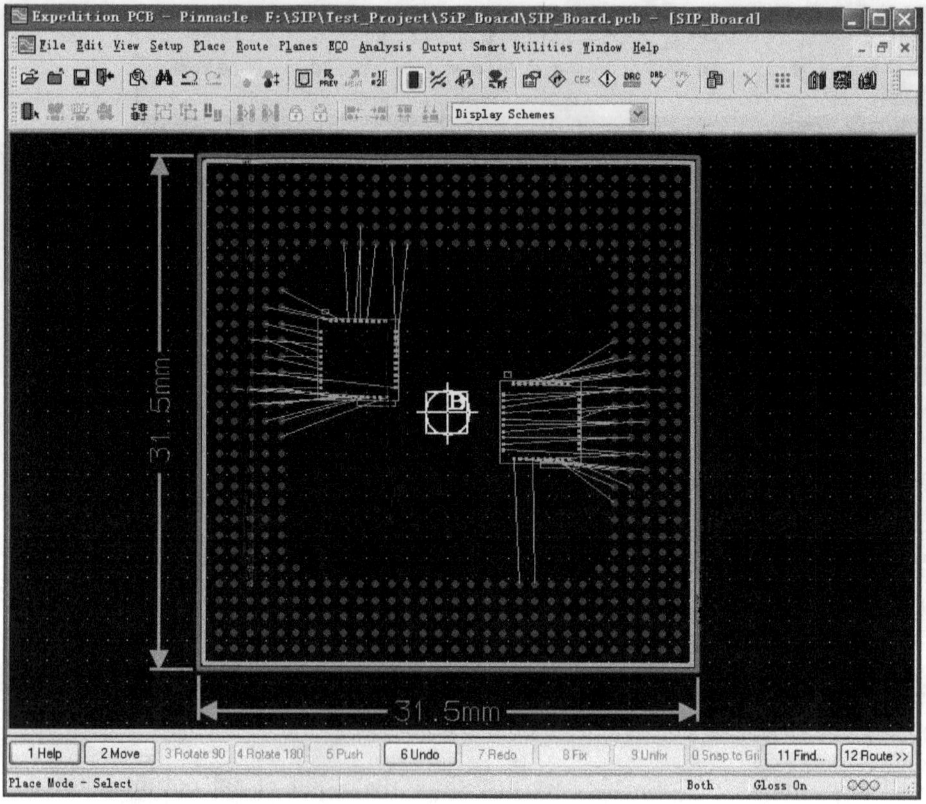

Figure 9.57 Net connections after automatic optimization.

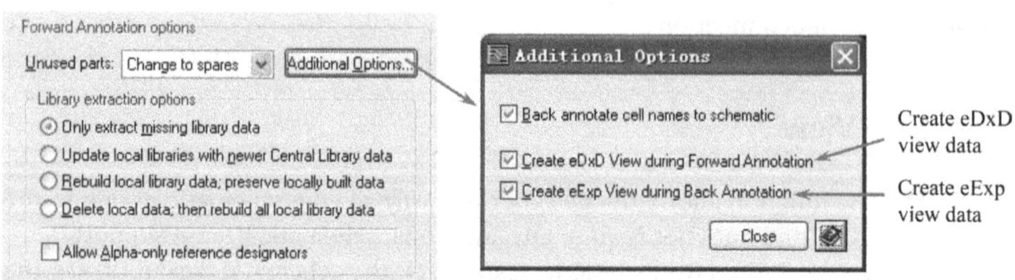

Figure 9.58 Create eDxD and eExp data during forward annotation.

9.6 Input Chinese Characters in Layout

9.6.1 Manually Input Chinese Characters

In Chapter 7 we described how to input Chinese characters in DxDesigner; now we discuss how to input Chinese characters in Expedition.

In draw mode, click the Add Text button A, in the pop-up Properties window String bar directly input the required words.

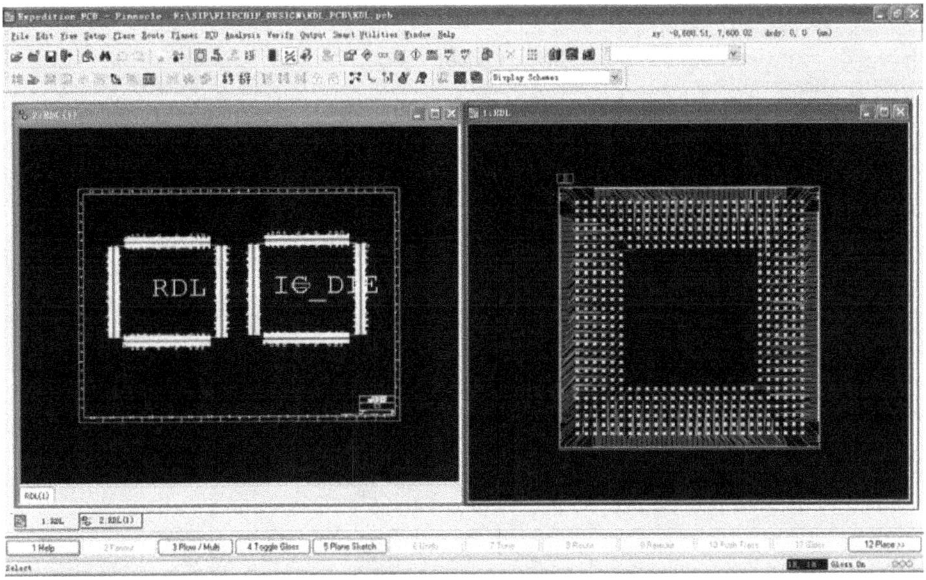

Figure 9.59 View schematic in layout environment.

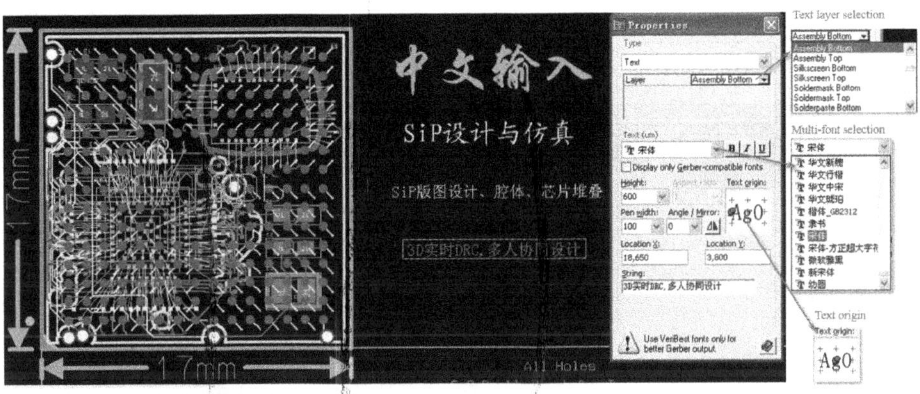

Figure 9.60 Chinese characters input in layout.

These words can be put in any layer of the design, and support a variety of Windows fonts. The designer can define word height, pen width, rotation angle/mirror, text origin, etc., as shown in Figure 9.60.

If it is necessary to edit the text, first use the mouse to click the text, then change it in the Properties window String bar.

9.6.2 Import Chinese Words from DXF File

In addition to manually inputting Chinese characters, Expedition also supports importation of characters from a DXF file. On the menu select File→Import→DXF, in the pop-up DXF Import window choose the file you need to import, and set up as shown in Figure 9.61. Then click the OK button; after import a window will pop up stating that

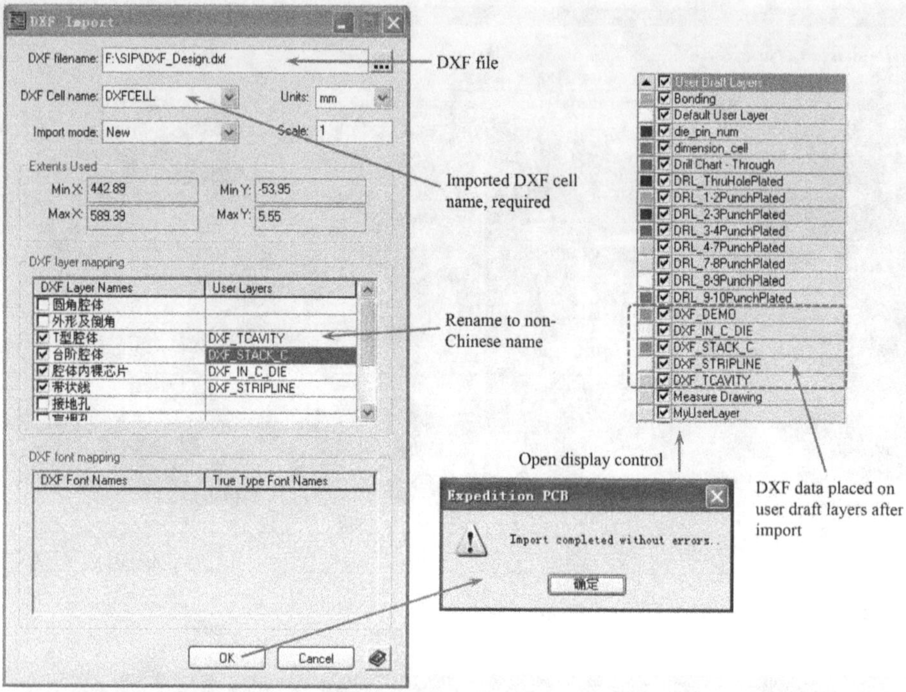

Figure 9.61 DXF file import.

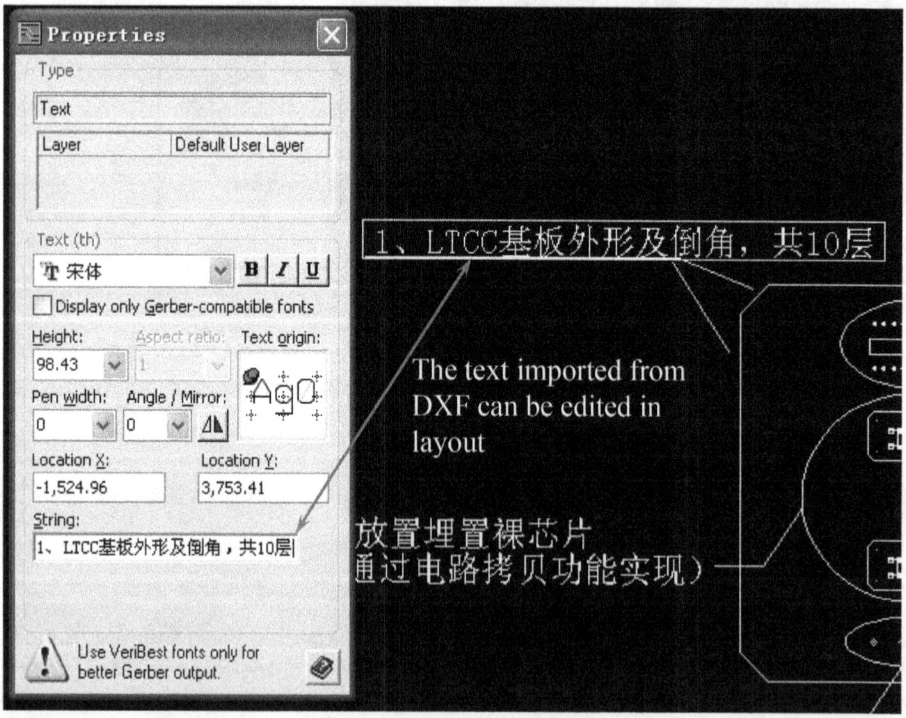

Figure 9.62 Chinese characters imported from DXF can be edited in layout.

import has occurred without errors, then in Display Control open the corresponding layer in User Draft Layers group, and you can see the imported DXF data.

The imported DXF file contains graphics and text; the characters of the text remain in an editable state. Designers can change the type of imported graphics, such as converting them to board outline or cavity, etc.; thus, the structure designed in mechanical software can be inherited by layout, without the necessity to draw it again in layout.

The designer can edit the imported text in layout. Click the text characters; we can see the text content automatically show on the String bar in the Properties window. The designer just needs to edit in the String bar, as shown in Figure 9.62.

10

Constraint Rules Management

10.1 CES – Constraint Editor System

CES(Constraint Editor System)is a unified environment for design rule setting and management; it can be invoked through the schematic tool DxDesigner or the layout tool Expedition, and it maintains data synchronization between schematic and layout.

In design forward annotation, new schematic constraint rules are automatically updated to layout; in design back annotation, new layout constraint rules are updated to schematic, realizing two-way constraint transmission. Figure 10.1 shows the flow of constraint rules transmission.

The designer can start CES from the schematic or layout environment; Figure 10.2 shows the CES tool icon in the schematic and layout environments; the designer can click on the icon to go into CES easily; he or she can also use the menu Tools→Constraint Editor System in the schematic to start CES, or in the layout menu can select Setup→Constraints to start CES.

The CES interface on start up is as shown in Figure 10.3.

CES is based on the spreadsheet work environment; the basic operation method is almost the same as that of Excel, and for the designer it can be very convenient to define and edit all kinds of rules in the CES environment.

First, let's learn some concepts of CES.

1) Constraint: a kind of rule, requirement or attribute. In SiP design, it is usually assigned to the relevant nets or devices, controlling the layout and routing to meet the design goals.
2) Scheme: a collection of design rules, can be applied to the entire design or some specific areas (e.g., rule area).
3) Net Classes: a group of nets with the same rules. Usually they have the same physical constraints, such as trace width, clearance, via specified, differential clearance, route in a specific layer, etc.
4) Constraint Classes: a group of nets with the same constraint rules. Usually they have the same electrical constraints, such as maximum/minimum delay, length matching, custom topology, etc.

SiP System-in-Package Design and Simulation: Mentor EE Flow Advanced Design Guide,
First Edition. Suny Li (Li Yang).

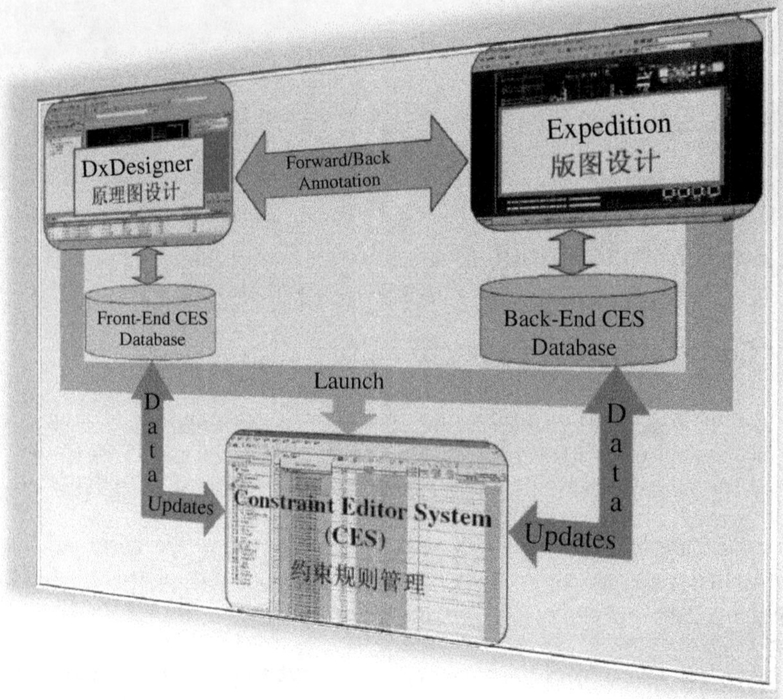

Figure 10.1 The flow of constraint rules transmission.

Figure 10.2 The designer can start CES from schematic or layout.

5) Clearances: used to set the physical space rules in the process of layout design, for example, trace to trace, trace to via, trace to pad, trace to plane, pad to pad, pad to via, pad to plane, etc. clearance rules.
6) Parts: used to set properties of the component parts, such as part type, process, model, power consumption, temperature limit, etc.
7) Noise Rules: noise relevant constraints; can effectively control the crosstalk and noise in layout design, such as crosstalk parallel routing rules.
8) Constraint Template: in CES, multiple constraint templates can be set, in order to quickly reuse the same constraint rules for the same kinds of nets. For example, for nets with the same topological structure as some previously defined in the design, we can first create a constraint template with this topology, and then quickly apply the constraint template to associated nets.

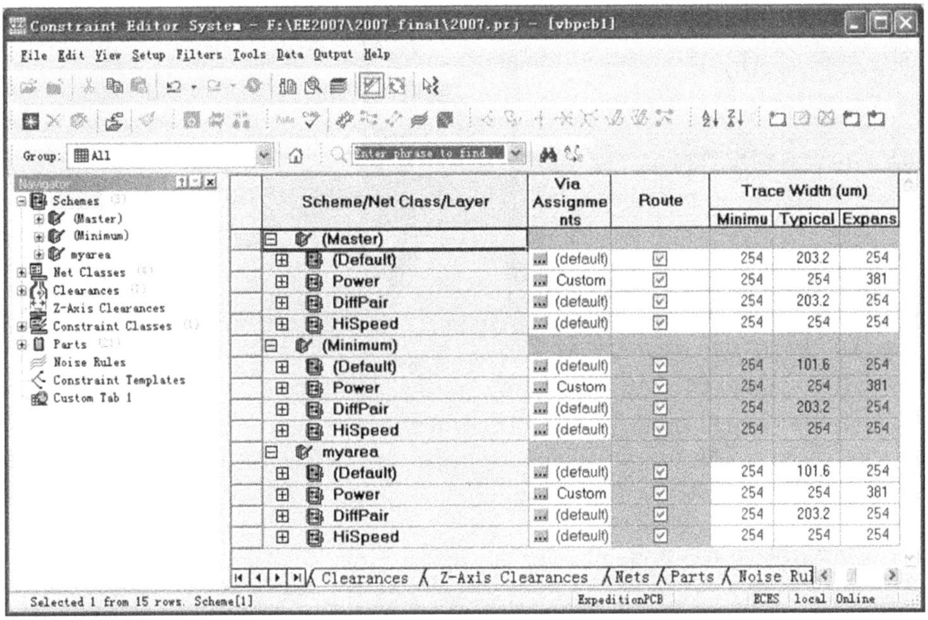

Figure 10.3 CES interface.

10.2 Scheme

A scheme is a collection of physical design rules which can be applied to the whole design or to specific areas; the physical design rules mainly include net classes and clearances.

There are two default schemes in CES, Master and Minimum, and users can also define custom schemes.

1) Master scheme: physical rule settings applied to the whole design.
2) Minimum scheme: not an editable scheme, only used to display the minimum trace width or clearance in the design to allow the designer to easily check the physical rules for minimum measurements.
3) Custom schemes: in some specific area of design, such as BGA routing, design rules will be different from the master scheme rules; this can be realized by user-defined schemes. These schemes automatically inherit the net class definition, layer definition, etc. from the master scheme, but the net class and clearance rules can be different from the master scheme. According to the design requirements, the designer can create multiple custom schemes.

10.2.1 Create Scheme

Right-click on Schemes, select New Scheme in the menu, and then name it according to the functional characteristics; a new custom scheme is created. While creating the scheme, the number following "Schemes" increases by one; the number of schemes in the design is now four, as shown in Figure 10.4.

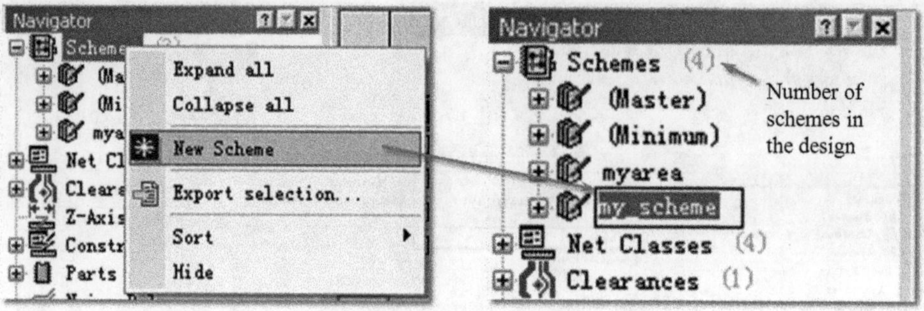

Figure 10.4 Create a new custom scheme.

The custom scheme automatically inherits the master scheme rules definition, and the designer can modify the rules definition in the custom scheme according to demand, as shown in Figure 10.5.

10.2.2 Use Scheme in Layout

After modifying the rules in the custom scheme, close the CES window, and the rules in CES will automatically update to Expedition. In the Expedition design environment, draw a polygon and define the type as Rule Area, then select the custom scheme named "myarea" and assign it to this rule area.

Inside this rule area, the rules defined in the myarea scheme will be valid, and in other areas, the main scheme rules are valid.

The rules in the custom scheme are commonly defined by the layout designer, because the rules take effect only when drawing the rule area in layout and specifying the custom scheme, as shown in Figure 10.6.

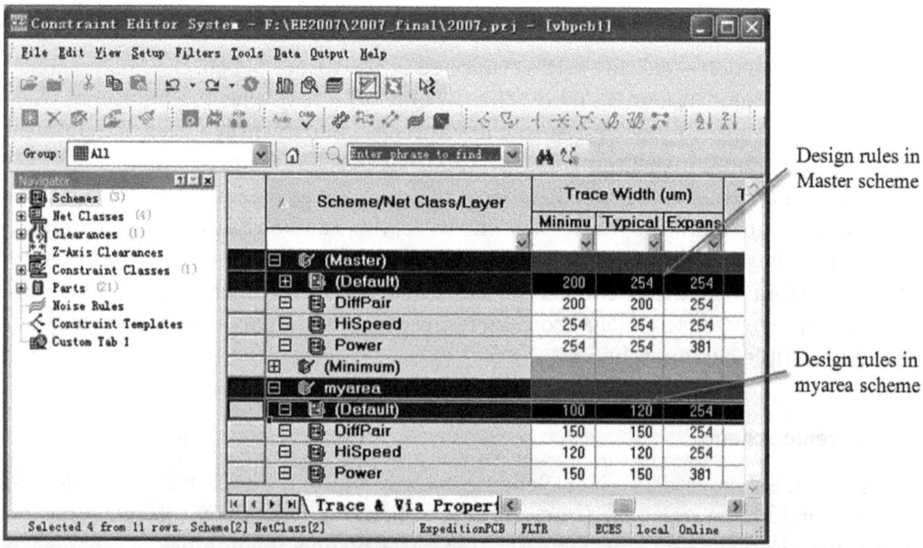

Figure 10.5 Modify the rules in the custom scheme.

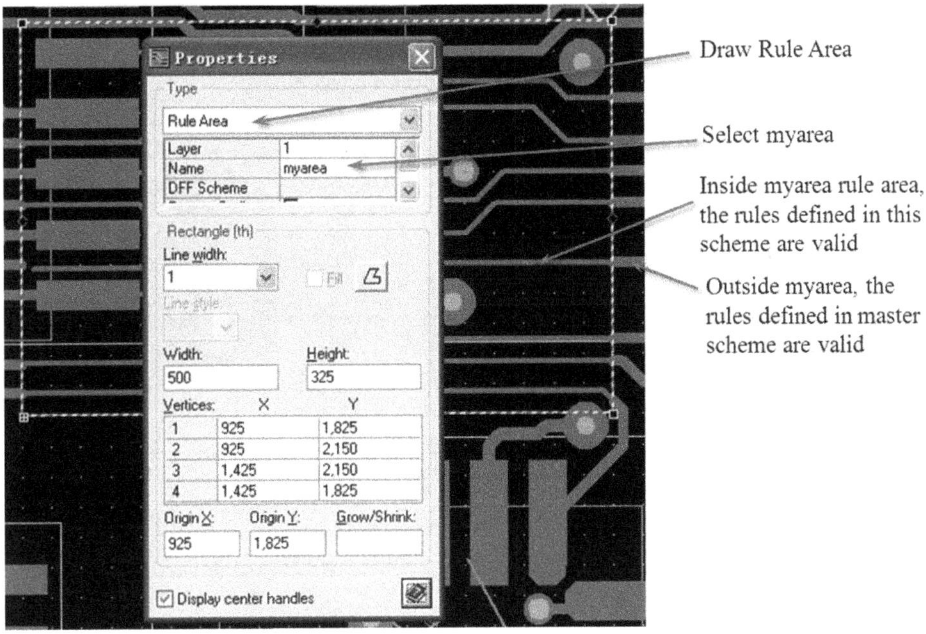

Figure 10.6 Apply myarea scheme in Rule Area.

It is important to note that we should define the rule area first, and then in routing the rule area can play a role. If we define the rule area after routing, we need to edit or move traces after drawing and defining the rule area, and the rule area can then take effect.

10.3 Define Layer Stackup and Parameters

Click the toolbar Stackup Editor icon in CES and enter the layer Stackup Editor; whether we do simulation in HyperLynx, or calculate trace impedance in CES, we should set up the correct layer stackup and physical parameters of the substrate in advance.

In Stackup Editor, the designer can modify the layer name to fit his or her preference, and can also set up the thickness of signal and dielectric layers, as shown in Figure 10.7. If the designer has defined layer stackup parameters in Expedition, the parameters can be automatically inherited by CES. Similarly, if the layer stackup parameters were edited in CES, they can also be updated to layout design.

Stackup Editor contains five tabs: Basic, Dielectric, Metal, Z0 Planning and Customer View, focusing on different setting options, similar to the stackup editor in HyperLynx. If we set the substrate-related physical parameters in CES, they can be passed to the HyperLynx simulation tool automatically, so it is not necessary to set them up repeatedly in HyperLynx.

After setup is finished, click the toolbar Word button ; the set up information can be copied to a Word format picture, and then pasted into a Word document. This is very convenient in the standardization of production document preparation, as shown in Figure 10.8.

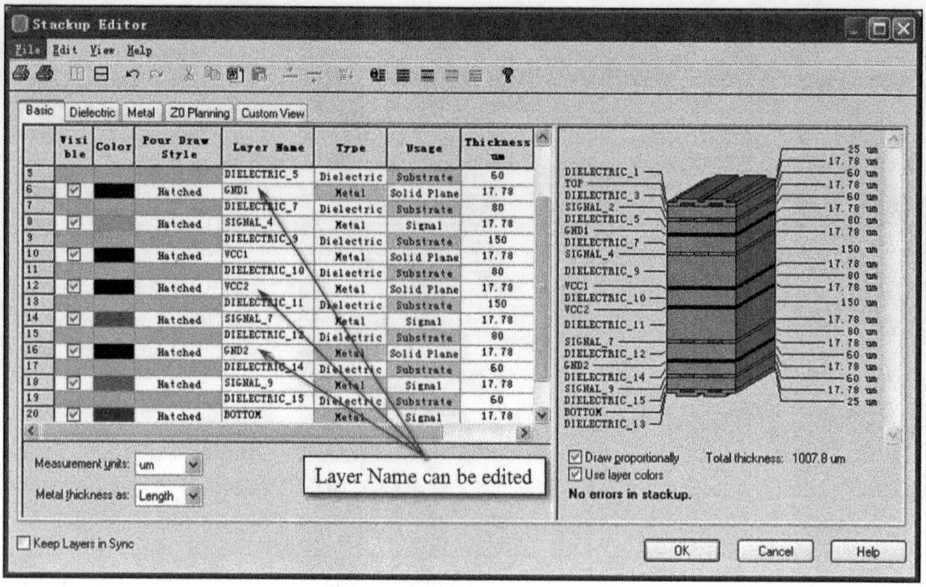

Figure 10.7 Stackup Editor in CES.

10.4 Net Class

10.4.1 Create Net Class and Assign Nets to Net Class

In the top-left-corner list of CES Navigator, all the existing net class names are shown. Right-click on an existing net class name and select Delete to delete it, or press the Delete key on keyboard directly. It is important to note that the designer can't delete the default net class.

Right-click on Net Classes in the pop-up menu, select New Net Class, and then name it according to the functional characteristics; thus a new net class is created. At the same time, we can see that the number to the right of Net Classes increases, to (5) in the figure below, indicating that there is a total of five net classes in the design, as shown in Figure 10.9.

After net class definition, we need to assign nets to the net class. The nets assigned to the net class will follow the rules defined in the net class. Right-click on Net Class Name, select Assign Nets, choose the nets we want to assign in the pop-up window, and click the single arrow > to add the selected nets to the target net class. Clicking the double arrow >> will add all the nets to the target net class, as shown in Figure 10.10.

10.4.2 Define Net Class Rules

After creating a net class and assigning nets to it, we need to define net class rules; the defined rules will apply to all the nets in this net class.

As shown in Figure 10.11, first, in the Trace Width column, set Typical, Minimum and Expansion Trace Width to 150μm, 100μm and 500μm respectively. In the Typical Impedance column, CES will automatically calculate the impedance on the basis of typical width. As we can see, even for the same width, the impedance in different layers is

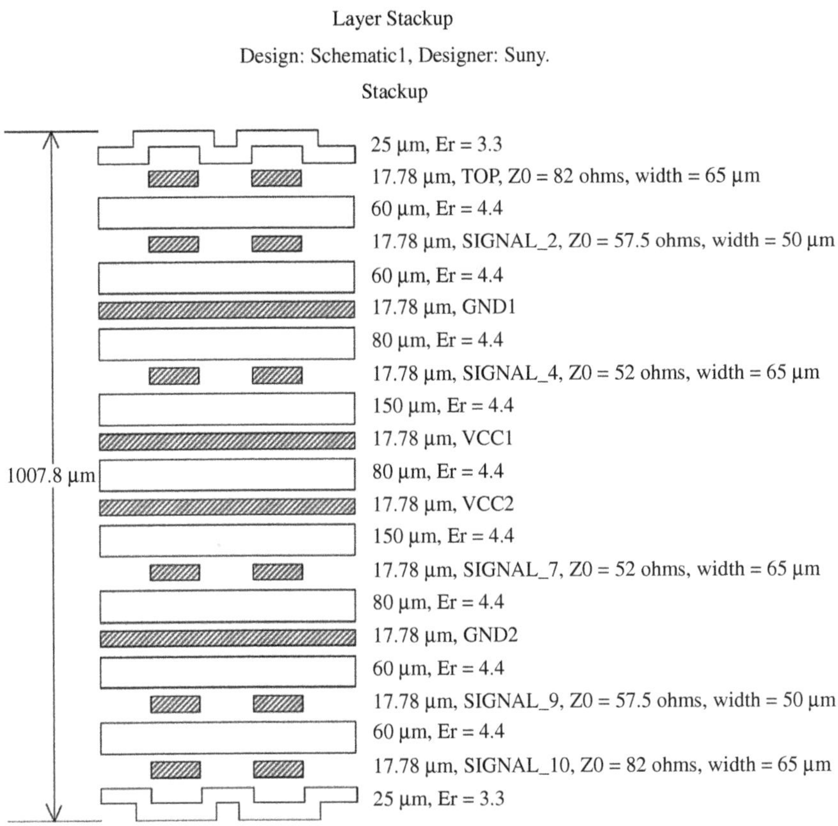

Figure 10.8 The layer stackup information copied to Word.

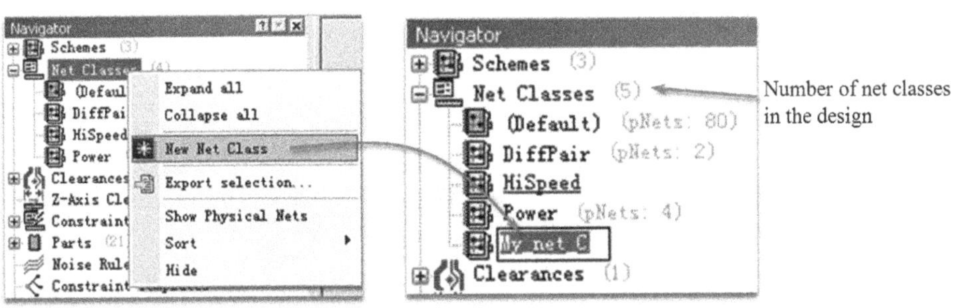

Figure 10.9 Create new net class.

different, so if we want to keep impedance consistent between different layers, different trace widths should be set in different layers.

In Figure 10.11, the Type column shows the layer type, currently signal layer and plane layer; layer type cannot be modified here, it needs to be modified in the Stackup Editor.

The Via Assignments column is used for net class via definition; click the button before "(default)", select Net Class Via in the pop-up window, click the drop-down list, and choose the required via, as shown in Figure 10.12.

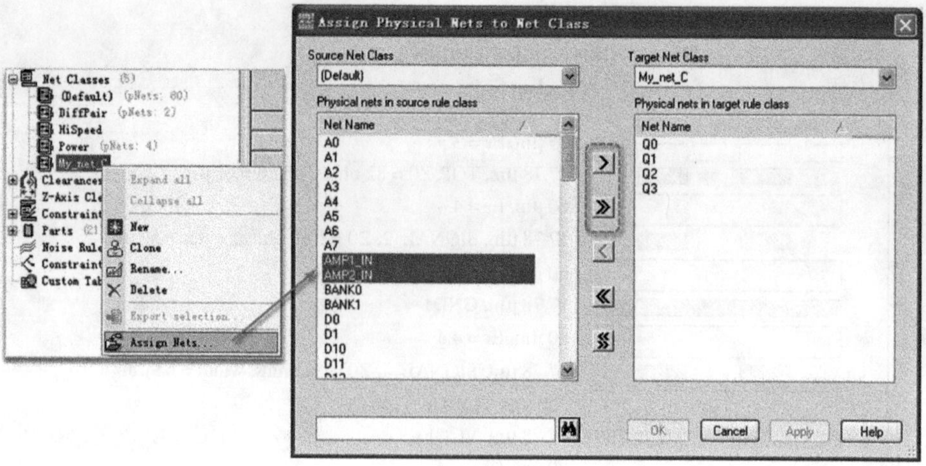

Figure 10.10 Assign nets to net class.

Scheme/Net Class/Layer	Type	Via Assignments	Route	Trace Width (um)			Typical Impedance (Ohm)	Diff Pair Spacing (um)
				Minimum	Typical	Expansion		
(Master)								
Suny_New		(default)	☑	100	150	500		70
TOP	Signal		☑	100	150	500	60.869..	70
SIGNAL_2	Signal		☑	100	150	500	35.619..	70
GND1	Plane		☐	100	150	500	58.535..	70
SIGNAL_4	Signal		☑	100	150	500	35.06..	70
VCC1	Plane		☐	100	150	500	37.273..	70
VCC2	Plane		☐	100	150	500	37.273..	70
SIGNAL_7	Signal		☑	100	150	500	35.06..	70
GND2	Plane		☐	100	150	500	58.535..	70
SIGNAL_9	Signal		☑	100	150	500	35.619..	70
BOTTOM	Signal		☑	100	150	500	60.869..	70

Figure 10.11 Define the trace width rules for net class Suny_New.

The default via is defined in the Setup Parameters dialog Via Definition tab of the Expedition environment; see Chapter 9.

In Figure 10.11, the check item in the Route column indicates whether to allow routing in this layer or not; currently, we do not route in the plane layer, so we do not usually check the plane layer in the Route column. The DiffPair Spacing column is used to define the spacing of two diffpair traces.

10.5 Clearance Rules

10.5.1 Clearance Rule Creation and Setup

In CES Navigator, select Clearances below Net Classes; the same method as was used for net classes is used to define clearance rules. Right-click on Clearances, in the pop-up menu select New Clearance Rule, as shown in Figure 10.13.

Clearance rules are very comprehensive, including trace to trace, trace to pad, trace to via, etc., and pad to pad, pad to via, pad to plane, etc.; the rules can be set separately for each layer, as shown in Figure 10.14.

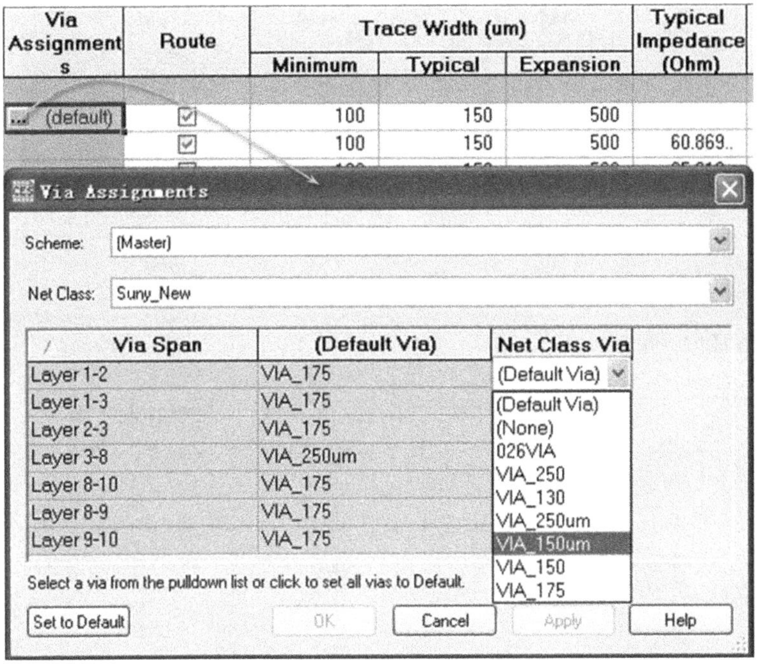

Figure 10.12 Define the net class via.

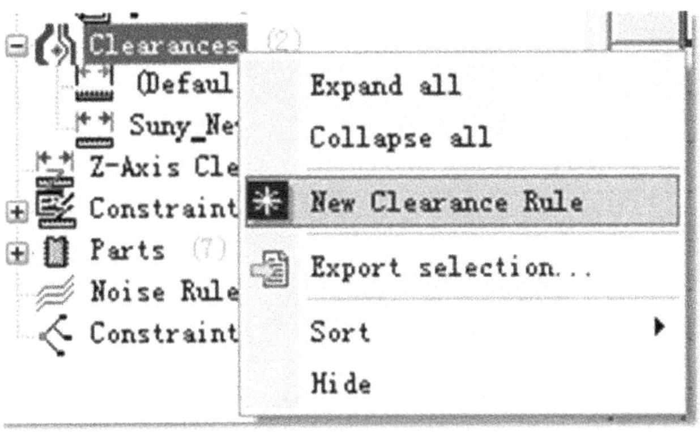

Figure 10.13 Create a new clearance rule.

Set the clearance rules; as there are too many columns, this often needs to be adjusted. In CES each column width can be adjusted by the designer; select and drag the edge of each column to adjust the width; drag left to make the width of the column zero, meaning that this column is automatically hidden; designers can use this method to hide columns not often used, as shown in Figure 10.15. The same method can be used to adjust the height of a row and to hide a row in CES.

If the designer needs to view the hidden column, use View→Reset Column Widths, and for resetting row height we can use the same method, as shown in Figure 10.16.

Scheme/Clearance Rule/Layer	Index	Type	Trace To (um)					Pad To (um)			Via To (um)			Plane To (um)
			Trace (	Pad (	Via (u	Plane	SMD P	Pad (um)	Via (um)	Plane	Via (um)	Plane (u	SMD P	Plane (
(Master)														
(Default Rule)			50	50	50	50	50	100	100	50	50	50	50	50
TOP	1	Signal	50	50	50	50	50	75	100	50	50	50	50	50
SIGNAL_2	2	Signal	50	50	50	50	50	50	100	50	50	50	50	50
GND1	3	Plane	50	50	50	100	50	100	100	100	50	100	50	200
SIGNAL_4	4	Signal	50	50	50	100	50	100	100	100	100	100	50	200
VCC1	5	Plane	50	50	50	100	50	100	100	100	100	100	50	200
VCC2	6	Plane	50	50	50	100	50	100	100	100	100	100	50	200
SIGNAL_7	7	Signal	50	50	50	100	50	100	100	100	100	100	50	200
GND2	8	Plane	50	50	50	100	50	100	100	100	50	100	50	200
SIGNAL_9	9	Signal	50	50	50	50	50	50	50	50	50	50	50	50
SIGNAL_10	10	Signal	50	50	50	50	50	50	50	50	50	50	50	50
(Minimum)														

Figure 10.14 Clearance rules setup.

Scheme/Clearance Rule/Layer	Index	Type	Trace To (um)	Pad To (um)	Via To (um)	Plane To (um)	Embedded Resistor To	EP Mask To (um)
			Trace (um)	Pad (um)	Via (um)	Plane (um)	Trace (um)	Trace (um)
(Master)								
Suny_New			50	100	50	50	50	50
TOP	1	Signal	50	75	50	50	50	50
SIGNAL_2	2	Signal	50	50	50	50	50	50
GND1	3	Plane	50	100	50	200	50	50
SIGNAL_4	4	Signal	50	100	100	200	50	50
VCC1	5	Plane	50	100	100	200	50	50
VCC2	6	Plane	50	100	100	200	50	50
SIGNAL_7	7	Signal	50	100	100	200	50	50
GND2	8	Plane	50	100	50	200	50	50
SIGNAL_9	9	Signal	50	50	50	50	50	50
SIGNAL_10	10	Signal	50	50	50	50	50	50

Figure 10.15 Set up clearance rules and adjusting column width.

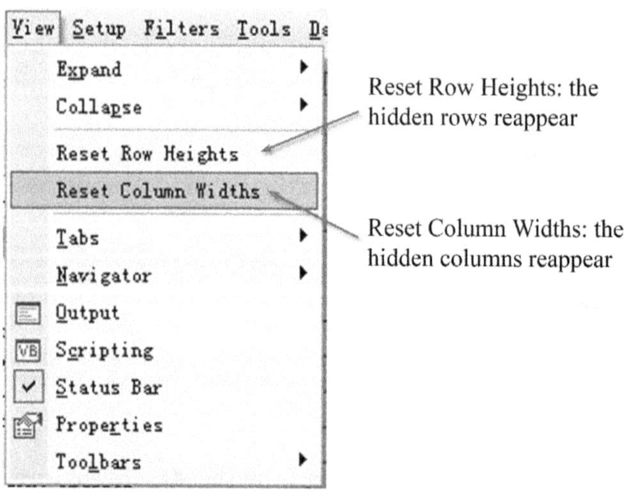

Figure 10.16 Reset column widths and row heights.

10.5.2 General Clearance Rules

In the CES Navigator list, under Clearance rules selected, the designer can click the General Clearance toolbar icon to open the General Clearance Rules definition window. In other cases, the icon is gray and not available.

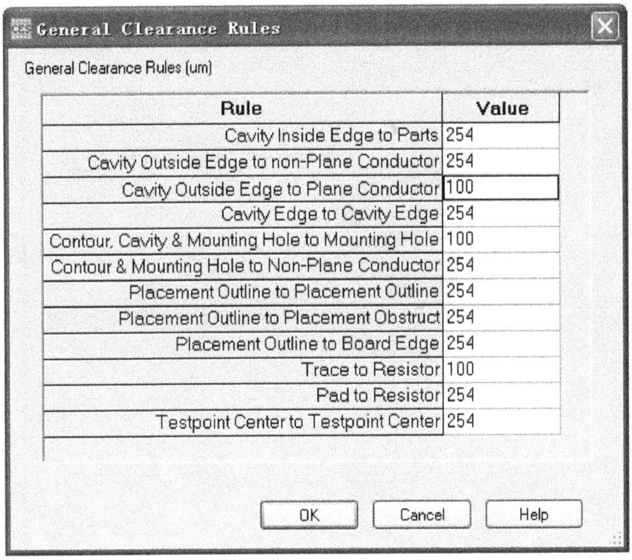

Figure 10.17 General clearance rules definition.

General clearance rules define some common spacing rules, including cavity to parts, cavity to plane conductor, cavity to cavity, placement outline to placement outline, test point to test point, placement outline to board edge, etc., as shown in Figure 10.17.

10.5.3 Net Class to Net Class Clearance Rules

Under Clearance rules selected, click the toolbar icon to open the Class to Class Clearance rules window; these rules are definitions of the clearances between various net classes.

By default, all nets follow the default clearance rule, which is applicable to the internal nets of the same net class and to different net classes.

In many cases, because of the special needs of the design, the designer defines some specific clearance rules, and applies different clearance rules between different net classes.

When nets belong to the same net class, it is necessary to follow one clearance rule; when nets belong to different net classes, it is necessary to follow other clearance rules. This will require specific definitions of net class to net class clearance rules.

In the Class to Class Clearances window, as shown in Figure 10.18, nets belonging to NetClass1 and nets belonging to Default follow the default rules; nets belonging to NetClass1 follow Clearance1 rules.

Nets belonging to NetClass2 follow Clearance2 rules, and nets belonging to NetClass1 and nets belonging to NetClass2 follow the Clearance2 rules.

Defining the class to class clearance rules is very convenient in CES; the designer just needs to click the drop-down list in each unit of the table, and choose the needed clearance rule in the list.

Define a non-default clearance rule for the net class as follows:

1) Create a new clearance rule, such as Clearance1.

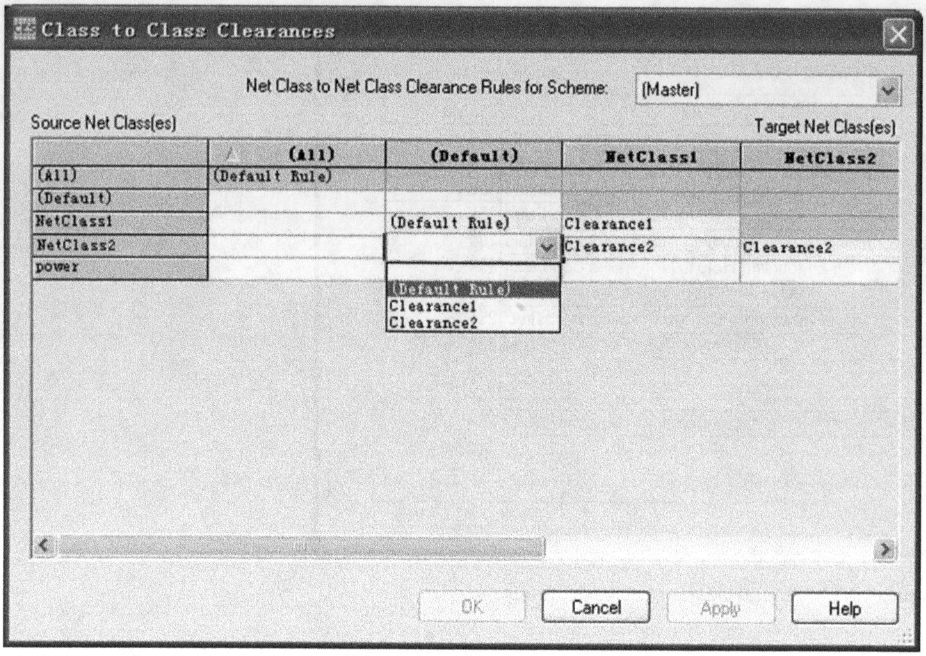

Figure 10.18 Net class to net class clearance rules definition.

2) Define rules in Clearance1.
3) Open the Class to Class Clearances window; first choose a scheme, such as the master scheme.
4) Select the relevant net class in Source Net Classes and Target Net Classes; choose Clearance1 in the clearance rules drop-down list.
5) The rules of Clearance1 are applied between the source net class and target net class; here, the source net class and target net class can be the same.

10.6 Constraint Class

A constraint class is used to define the electrical constraints of nets, including length, delay, topology, crosstalk, overshoot, ringing, etc.

10.6.1 Create Constraint Class and Assign Net to Constraint Class

Creating a new constraint class uses the same method as for net class; right-click on Constraint Classes, select New Constraint Class, and input the class name. Then, use the right-click menu, select Assign Nets, and assign related nets to the constraint class, as shown in Figure 10.19.

10.6.2 Constraint Classes Classification

Electrical constraints are usually all shown in CES. For the convenience of designers viewing and setting up these constraints in CES, constraints are divided into subclasses or groups.

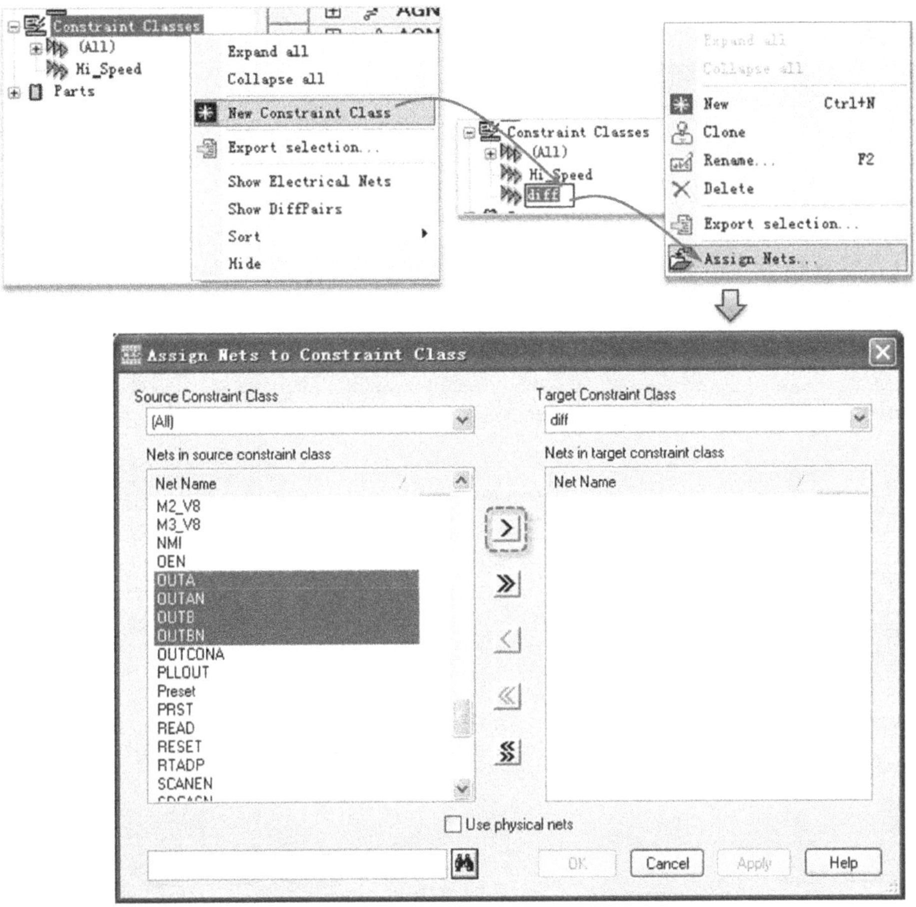

Figure 10.19 Create new constraint class and assign nets to the constraint class.

These groups are: Delays and Lengths, Differential Pair Property, I/O standard, Net Properties, Overshoot/Ringback, Simulated Delays, etc., as shown in Figure 10.20.

1) Delays and Lengths group

   This group is mainly used for definition of net delays and lengths, including minimum and maximum length, length match and tolerance. At the same time, Formulas can be used to set complex length relationships between different nets, as shown in Figure 10.21.

   In Delays and Lengths, we can also import the actual net length from layout design, comparing it with the length set in CES. The detailed description is given in section 10.7.

2) Differential Pair Properties group

   This group is mainly used for differential pair properties settings, including differential pair tolerance, convergence tolerance, distance to convergence, differential pair spacing and differential impedance, as shown in Figure 10.22.

3) I/O group

   This group is mainly used for the network I/O standard setting; designers can select the I/O standard from the drop-down list, as shown in Figure 10.23.

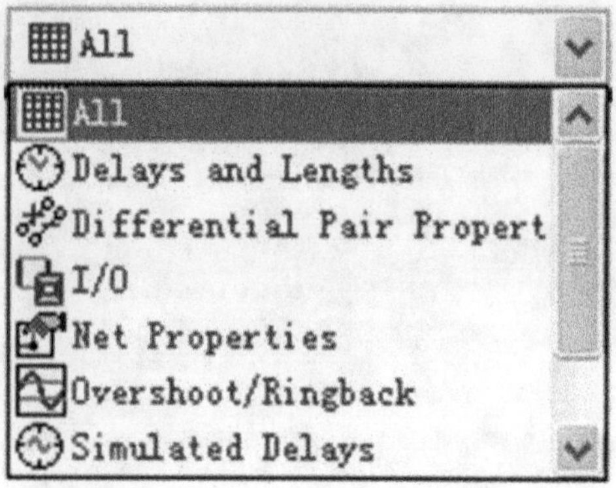

Figure 10.20 Constraint classes classification.

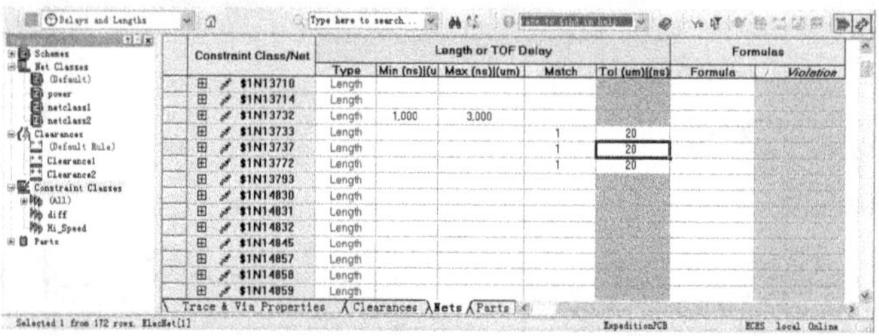

Figure 10.21 Delays and Lengths.

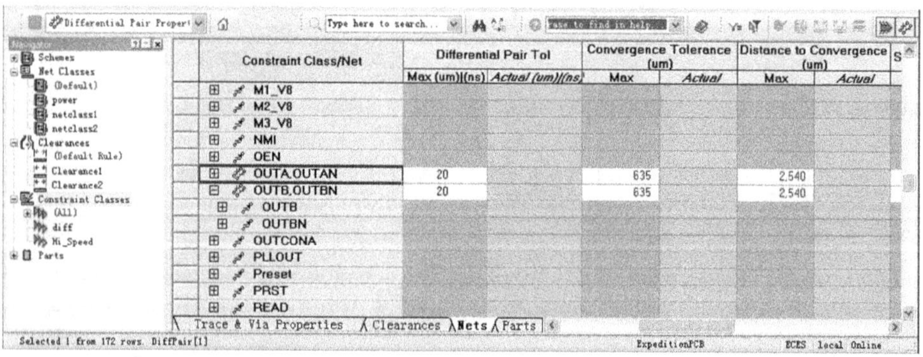

Figure 10.22 Differential Pair Properties.

4) Net Properties group
This group is mainly used for net property setting, as shown in Figure 10.24.
The net properties include net topology (comprising MST, chained, TShape, HTree, star, custom, etc.); analog, which should be checked for analog nets as well as for those nets not automatically recognized as electrical nets by CES; stub length; maximum allowed numbers of vias (#Vias), etc.

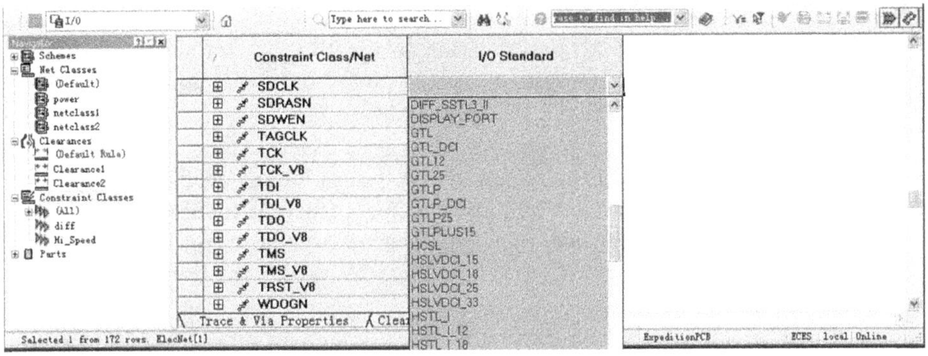

Figure 10.23 I/O standard.

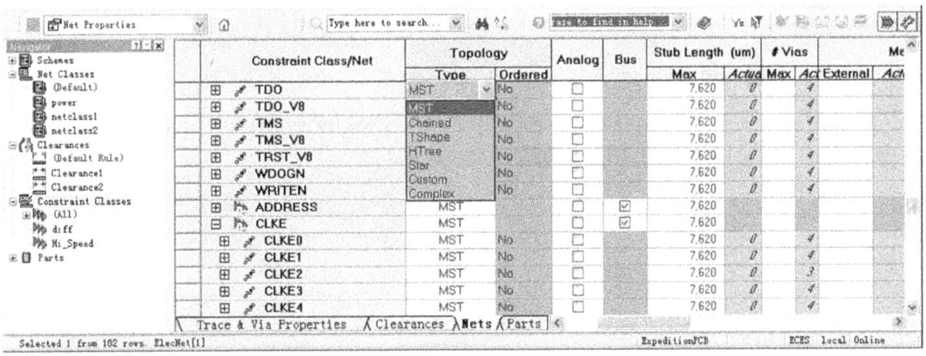

Figure 10.24 Net properties.

5) Overshoot/Ringback group
   This group is mainly used for overshoot and ringback option setting, including static low-level maximum allowable overshoot, static high-level maximum allowable overshoot, dynamic low-level maximum allowable overshoot, dynamic high-level maximum allowable overshoot, high-level, low-level ringback boundary and nonmonotone, etc., as shown in Figure 10.25.

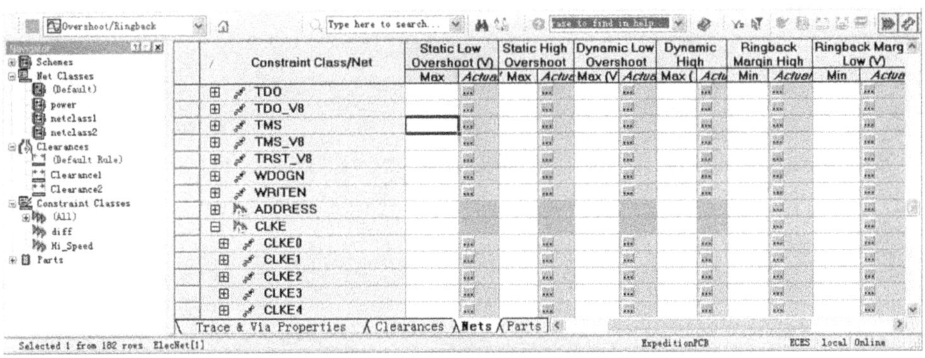

Figure 10.25 Overshoot/Ringback.

6) Simulated Delays group
   This group is mainly related to the simulation delay setting, including the maximum/minimum delay time, the maximum allowable range, and the delay-matching relationship with other nets, as shown in Figure 10.26.
7) Template group
   This group is mainly used for the selection and application of constraint templates.
   First, define rules for one net of a group of nets (such as topology, delay, overshoot, etc.) After setup is finished, right-click the net to create a constraint template, then apply the template to the same type of nets, to quickly improve the efficiency and accuracy of settings.
   At the same time, in the HyperLynx SI simulation environment, generate a constraint template based on signal integrity simulation results, and then in CES, use File→Import→Constraint Template to import the template and apply to corresponding nets; thus, the simulation results will be directly applied to the design rules, using the simulation as a design guide, as shown in Figure 10.27.

10.6.3 Edit Constraint Groups

With Edit Constraint Groups we can adjust and edit the existing constraints groups, and create a new group to manage the constraints.

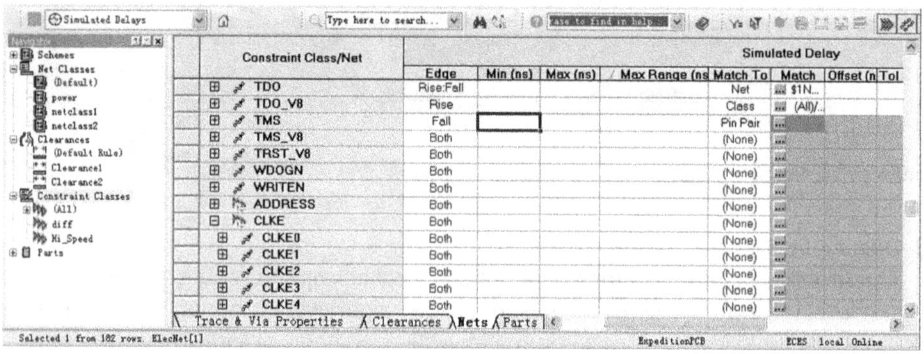

Figure 10.26 Simulated delays.

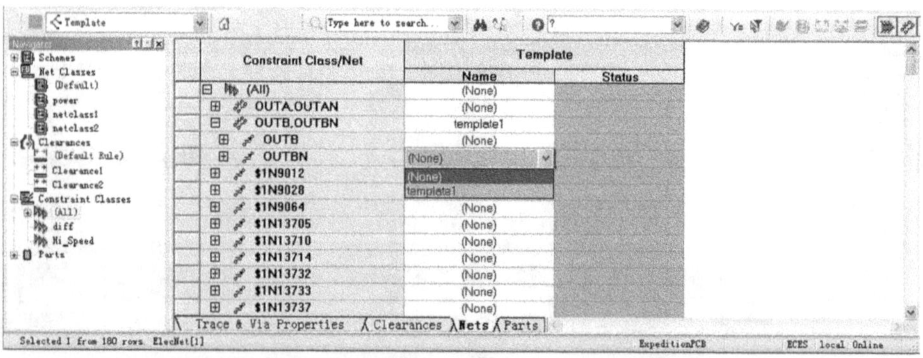

Figure 10.27 Template group.

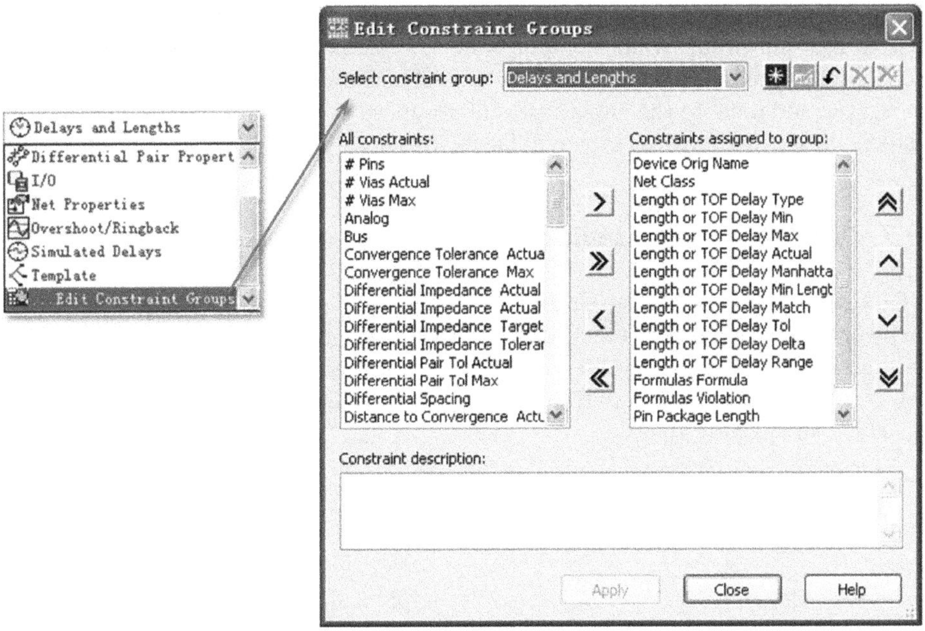

Figure 10.28 Start Edit Constraint Groups.

Start Edit Constraint Groups; the interface is shown in Figure 10.28. In this window, the left-side lists are all constraints, the right-side lists are constraints assigned to corresponding constraint groups.

In addition to editing and adjusting the current constraints group, we can also create new constraint groups, and assign corresponding constraints, so as to realize personalized setting and management of constraints.

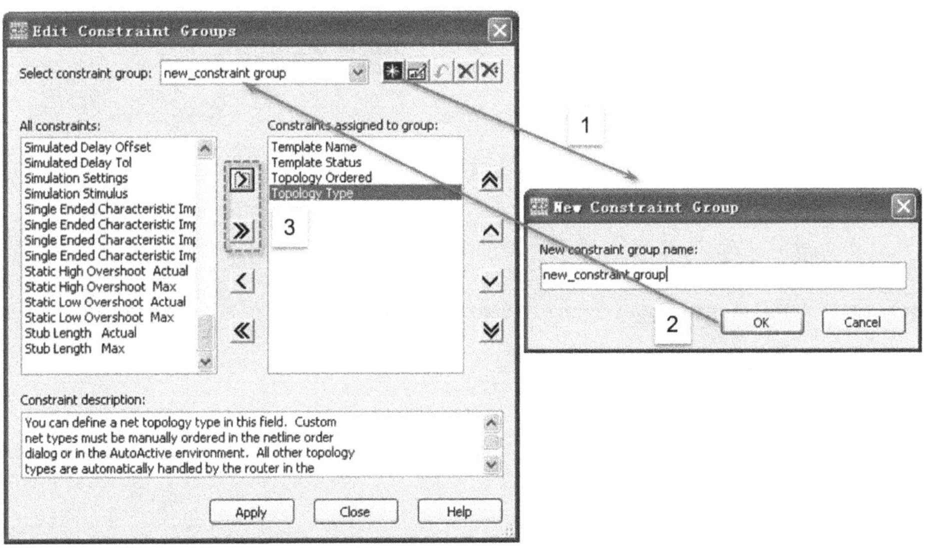

Figure 10.29 Create new constraint group and assign constraints to it.

In the Edit Constraint Groups window, first, click the New button and on the pop-up New Constraint Group window input New Constraint Group name and click OK. Second, select the relevant constraints in the All constraints lists and use the arrow to the right to add them to the new constraint group, so as to realize the management of custom constraint groups, as shown in Figure 10.29.

10.7 Update CES Data with Layout

In CES, select Data→Actuals→Update All; the actual routing length in layout design can be updated to CES, and compared with the length setting in CES; the length within the constraint shows gray. A value exceeding the constraint shows red, indicating there is an error, while a value close to the constraint shows yellow, indicating there is a warning, as shown in Figure 10.30.

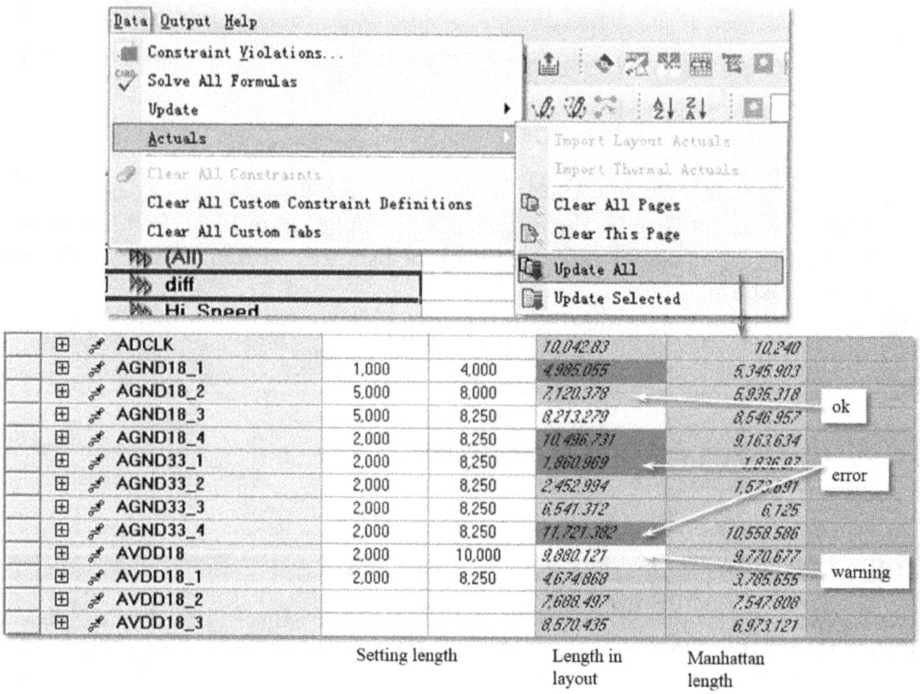

Figure 10.30 Actual routing length compared with the length set in CES.

11
Wire Bond Design

11.1 Wire Bond Overview

Wire bond is sometimes referred to as bond wire. Wire bonding describes the whole process of bonding, bond wire refers to the physical object. In practical terms, there is not usually a strict distinction.

At present, wire bonding is still a dominant method of package and System-in-Package assembly. The process of wire bonding is relatively simple, and the advantage of the process is that it can support a greater number of chips. But it also has its limitations, such as parasitic inductance, slow process, the need to run the process according to a sequence, etc.

Wire bonding plays a key role in the interconnection of package or SiP; using wire bonding chip pins can be electrically connected with bonding pads and traces in substrate. The rationality and accuracy of wire bond design is also crucial for SiP product yield and reliability. Figure 11.1 shows a design screenshot in Expedition (left side); the chip is placed in a multistep cavity, and linked to the substrate by multilevel bonding; the right side is a real photo of wire bond.

Because bond wire itself is a 3D (three-dimensional) design element, the bond wire model definition and rules management also need to be 3D.

Expedition has the most advanced bond wire model, and can meet the most complex wire bond design and manufacture requirements; it supports continuous line, arc curve and spline curve forms of wire bond bend, and complex multilayer wire bond design, as shown in Figure 11.2.

If we want to design wire bond in the Expedition environment, the license needed is Pinnacle, Ascent LX or Ascent, as well as Advanced Technology Pro (AdvPkg), as shown in Figure 11.3.

With the AdvPkg license, Expedition has the following design features:

1) Chip on Board (COB): design bare chip and bonding wire on PCB board.
2) Multi-Chip Module (MCM) design.
3) IC package design.
4) System-in-Package (SiP).
5) Advanced board cavity.
6) Embedded discrete components in inner layers.

SiP System-in-Package Design and Simulation: Mentor EE Flow Advanced Design Guide,
First Edition. Suny Li (Li Yang).

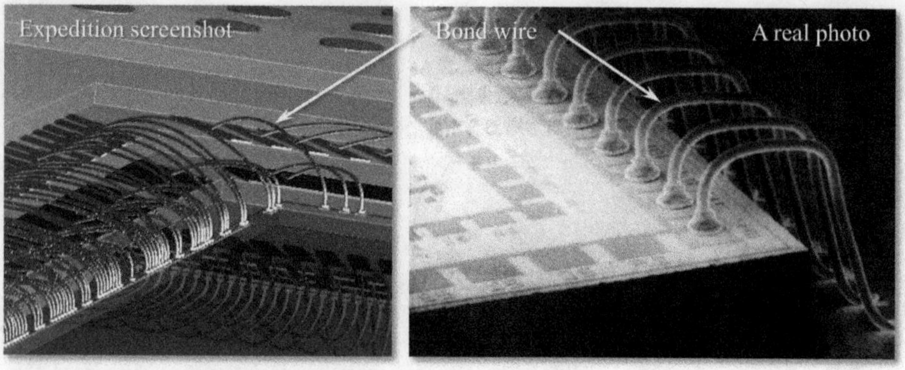

Figure 11.1 Connect chip and substrate through wire bond.

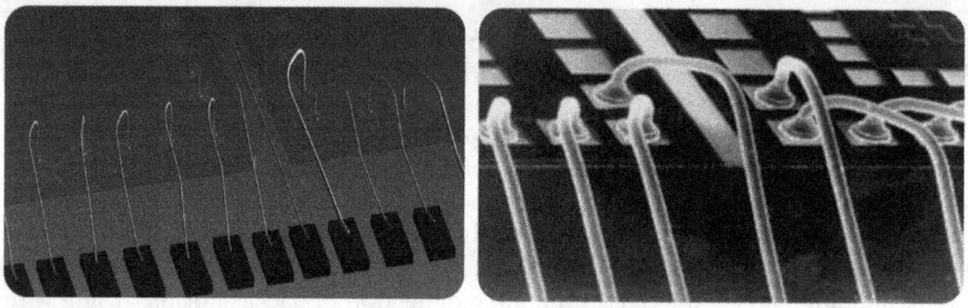

Figure 11.2 Expedition supports the most complex wire bonding design. *(See color plate section for the color representation of this figure.)*

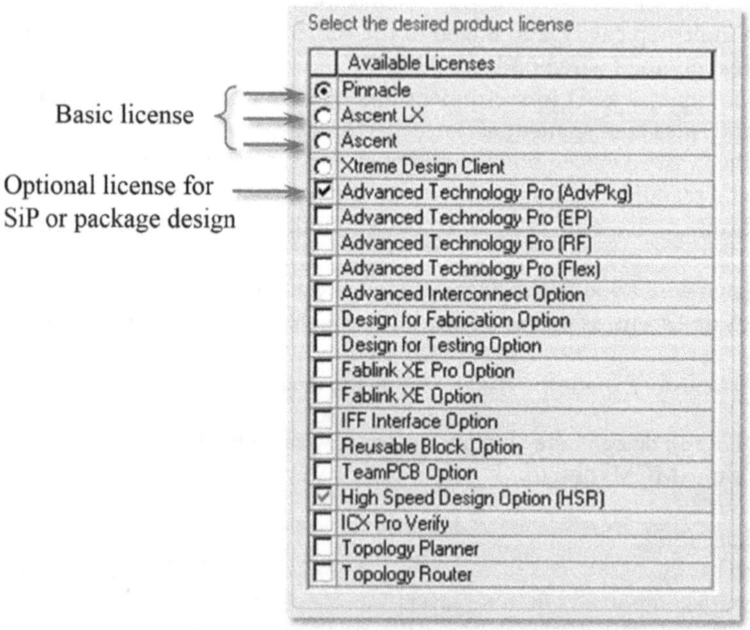

Figure 11.3 Wire bond design requires AdvPkg license support.

7) 3D chip stacking.
8) Bond wire and power ring auto-generation.

11.2 Bond Wire Model

Bond wire can be used to connect chips and substrate electrically. Bond wire-related elements include die pin, bond pad, bare die, etc.; these elements and the bond wire model itself affect the actual bond wire curve shape, either in design or in actual production processing. Figure 11.4 shows bond wire definition and related elements.

In the wire bonding process, there are usually two types of bonding end point, the ball end point and the wedge end point. Both ball and wedge can be used as the wire bond start point or terminal point; the most common bonding method is to use the ball as the start point and wedge as the terminal point, as shown in Figure 11.5.

11.2.1 Bond Wire Model Creation

Expedition supports accurate bond wire model creation, so as to meet the requirements of complex wire bond design and manufacture. Expedition supports three kinds of wire bond bend – straight line, arc and spline curve, and controls bond wire model accuracy using a variety of parameters.

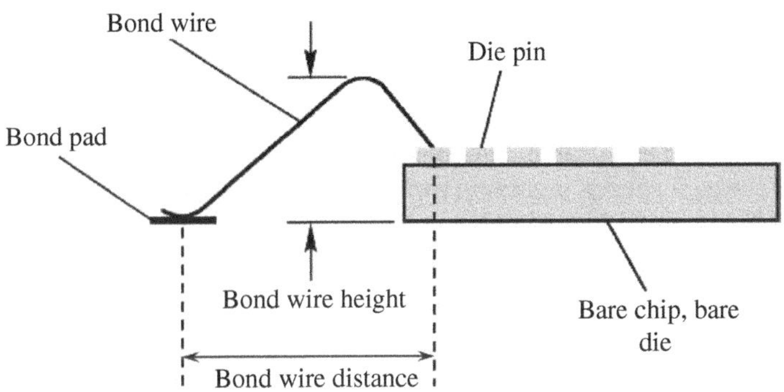

Figure 11.4 Bond wire definition and related elements.

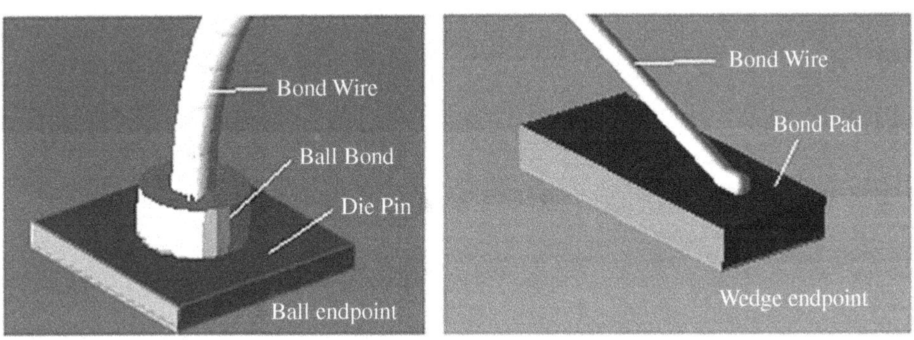

Figure 11.5 Bond wire ball end point and wedge end point.

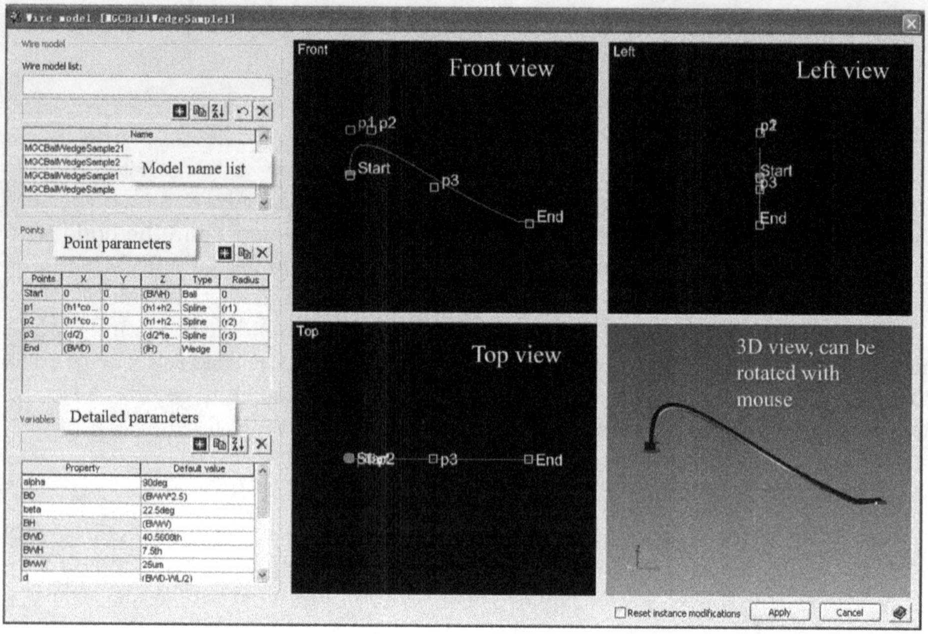

Figure 11.6 Bond wire model editor.

Start the Expedition environment and open an SiP design, then from the menu select Setup→Wire Model Editor to start the bond wire model editor.

In the wire model window, the model name and parameters are listed on the left; on the right are previews of the wire model, including three views (front view, left view and top view) and a 3D view, as shown in Figure 11.6.

1) The wire model list: this is mainly used to create, copy, sort and delete bond wire models. The blank at the top of the wire model list is used for filtering model names, for example, input MGC and only models starting with MGC will be listed, as shown in Figure 11.7.

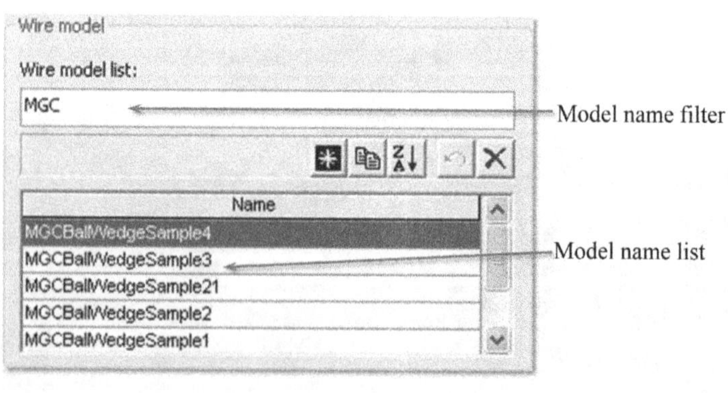

Figure 11.7 Wire Model name list and filter.

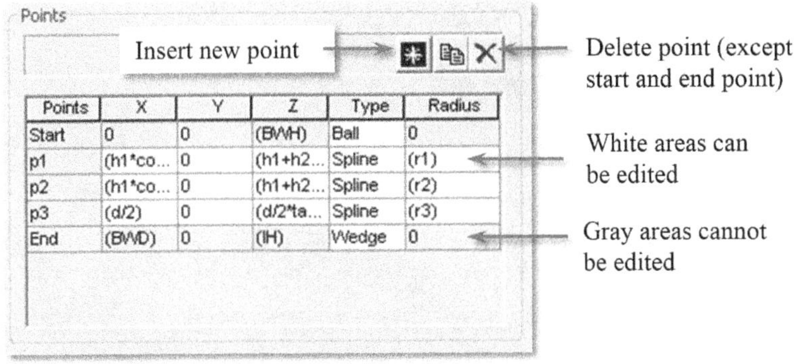

Figure 11.8 Bond wire control points.

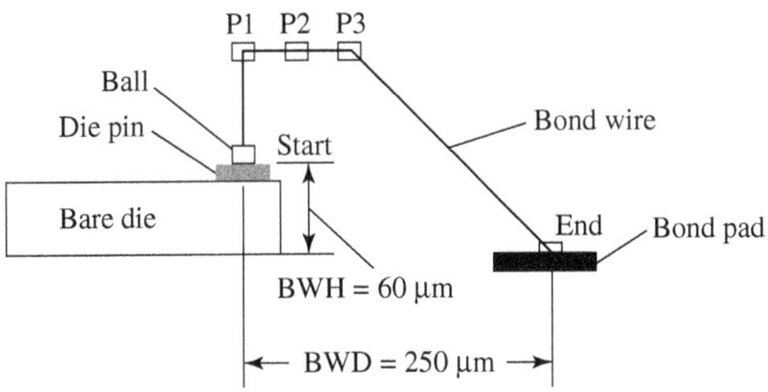

Figure 11.9 JEDEC standard five-point bond wire model.

2) Points: these are mainly used to set the control point parameters of the bond wire model, such as the X, Y, Z coordinates of each point, and the type of each point. The parameters in the white color area can be modified by the designer; the gray color area is the reserved area and cannot be changed.

   The designer can insert a new control point, and can also remove any other points except the start and end point, as shown in Figure 11.8.

   The JEDEC standard bond wire model is controlled by five points, respectively Start, p1, p2, p3 and End, as shown in Figure 11.9.
3) Start point (X, Y) coordinates (0, 0): this means that the deviation value relative to the center of the die pin Z coordinate value is Bond Wire Height (BWH).
4) BWH: this defines the height of the bond wire start point, the value of bare die height plus die pin height, BWH = (part height±die pin delta).
5) There are two types of start point, ball and wedge, as shown in Figure 11.10.
6) The main parameters of ball include ball diameter and ball height; the main parameter of wedge is wedge length.
7) The setting of intermediate points p1, p2, p3 are as shown in Figure 11.11.
   i) The X coordinate stands for bond wire span distance, the X coordinate of p1=(h1*cos (alpha)/sin (alpha)), the designer defines the value of h1 based on point p1's position upward relative to the start point; alpha represents the inclination angle of the start point.

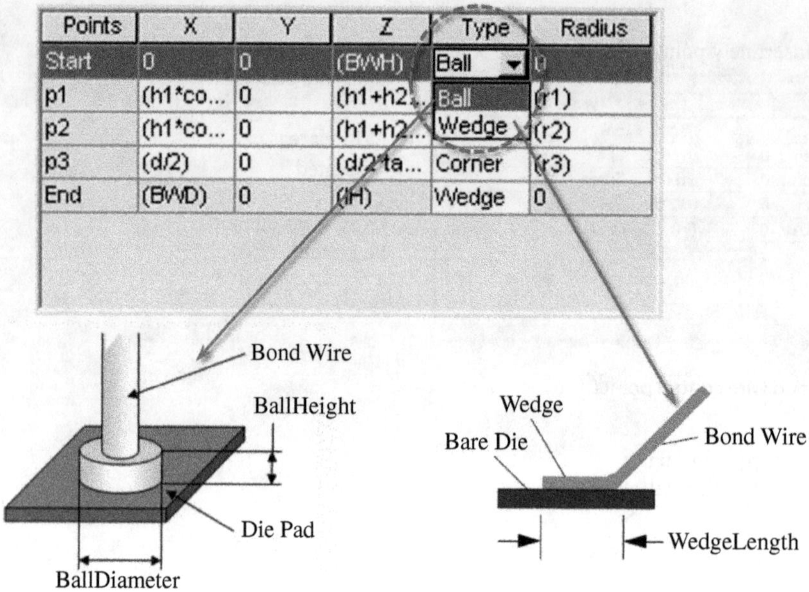

Figure 11.10 Two types of start point and the relevant parameters.

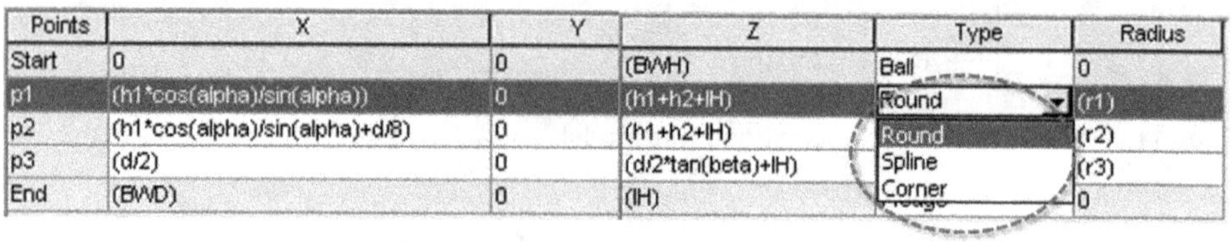

Figure 11.11 Three bend types for intermediate points.

ii) Y stands for Y direction deviation of p1, Y=0; this shows the lateral deviation of p1 is 0.
iii) Z indicates the height of p1, the Z coordinate of p1=h1+h2+IH; h2 stands for bare die height, IH stands for the die pin height relative to bare die.
iv) The operating point, if selected, will become green. At this time, the designer can drag the operating point, and its corresponding coordinate values will change accordingly.

There are three bend types for each intermediate point: round, spline and corner, as shown in Figure 11.11.

Figure 11.12 shows the bond wire shape (the front view and 3D view), when the bend types of the three points p1, p2, p3 are set to corner. It can be seen that this type of model is relatively rough, and cannot simulate real bond wire shape; there will be a large tolerance, and this should be avoided in real design.

Figure 11.13 shows the bond wire shape when p1, p2, p3 bend types are set to round. It can be seen that this type of model is an improvement on the former; the curve is

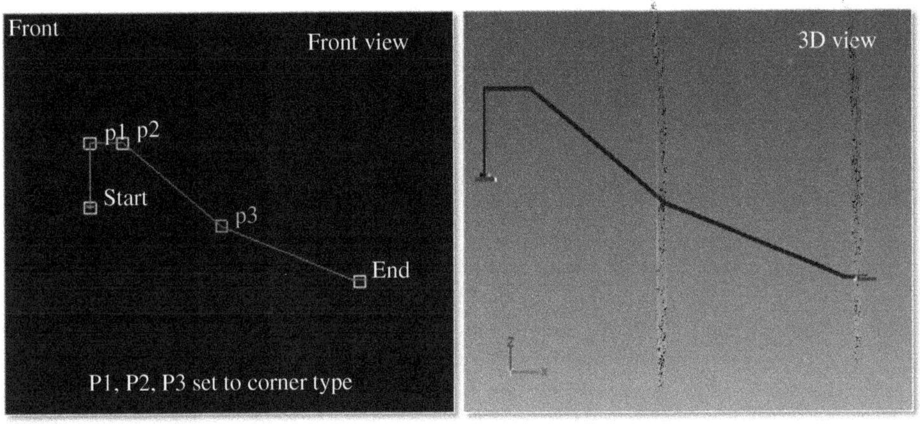

Figure 11.12 The bond wire shape when p1, p2, p3 bend types are set to corner. *(See color plate section for the color representation of this figure.)*

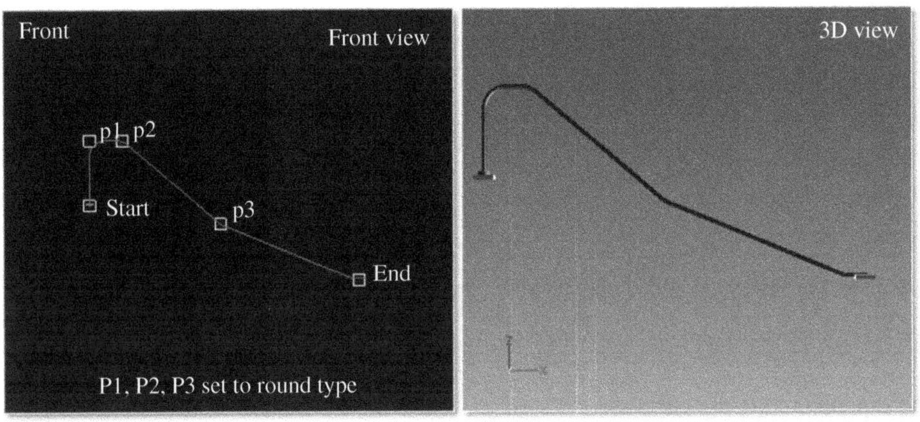

Figure 11.13 The bond wire shape when p1, p2, p3 bend types are set to round.

smoother, but some parts are still not continuous. In addition, the round corner radius cannot adapt to change, so cannot simulate real bond wire shape, but the tolerance is not large, and this can be used in real design.

Figure 11.14 shows bond wire shape with p1, p2, p3 bend types set to spline. We can see that this type of model is better than the previous two and there are many improvements: the corner is smooth and the whole curve is continuous; this kind of model can simulate real bond wire shape, and it is recommended that designers use this in real design.

Using accurate bond wire models, we can effectively consider DFM (Design for Manufacture), thus achieving perfect unity between design and manufacture.

11.2.2 Bond Wire Properties

In the bond wire model window, variables include property names and default values, as shown in Figure 11.15; these are explained below.

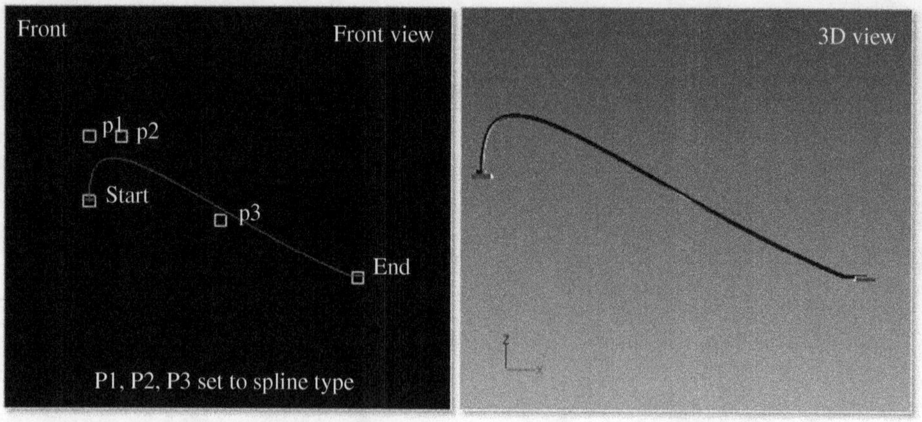

Figure 11.14 The bond wire shape when p1, p2, p3 bend types are set to spline.

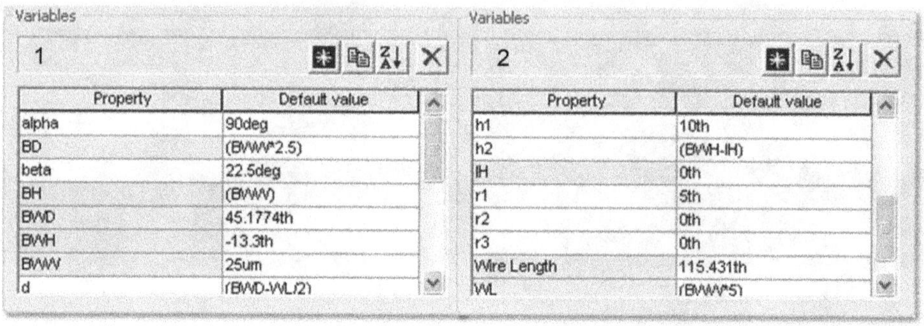

Figure 11.15 Variable properties and default values.

1) BD (Ball Diameter), defines the diameter of the initial ball, the default value is 2.5 times the bond wire diameter.
2) BH (Ball Height), defines the height of the initial ball.
3) BWD (Bond Wire Distance), defines the bond wire span distance, the value change of the X coordinate from the starting point to the end point; this value will change according to the bond pad position in design.
4) BWH (Bond Wire Height), defines the height of the start point; this height is usually related to the design itself, such as chip thickness and die pin height in design.
5) BWW (Bond Wire Width), defines the bond wire diameter; the default value is 25μm.
6) WL (wedge length), defines the length of the wedge connection.
7) Wire length, defines the 3D bond wire length that is the actual length of the bond wire, calculated comprehensively by the software based on other parameters.
8) IH (Inherit Height), the height information inherited from design or library.
9) Alpha, the tilt angle of the start point.
10) Beta, the tilt angle of the end point.
11) d, the span of bond wire minus half of the WL.

12) h1, the height of p1 relative to the start point.
13) h2, the height of the start point.
14) r1, r2, r3, the arc radius of p1, p2, p3 when corner type is selected for round.

Note that bond wire models are defined with properties; some properties will change with changes in the design, so when defining the bond wire model, there is no need to detail the value of all properties. The designer only needs to define a curve type: corner, round or spline; then, according to the requirements of the design (such as multilayer bonding needed to create different bond wire model heights and lengths), to drag the p1, p2, p3 operating points through the 3D view check to get the corresponding bond wire form.

11.3 Wire Bond Toolbar

In Expedition, from the menu select View→Toolbars→Wirebond; the wire bond toolbar is displayed.

We can also use, from the menu, Route→Wirebonds, and start the wire bond submenu to do various bond wire editing operations, as shown in Figure 11.16.

11.3.1 Add Bond Wire Manually

The button is used to add bond wire manually. To carry out this operation, click this button, and then click the die pin needing bonding; we can then manually add bond wire.

Expedition software can automatically show the length and angle of bond wire. Note, here the length refers to the bond wire span, the actual length of the bond wire needs to be viewed in the 3D parameters.

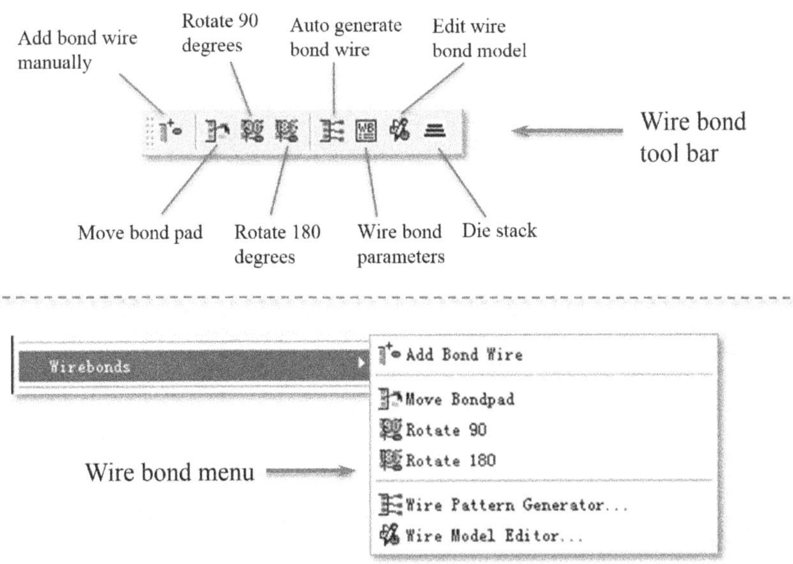

Figure 11.16 Wire bond toolbar and wire bond menu.

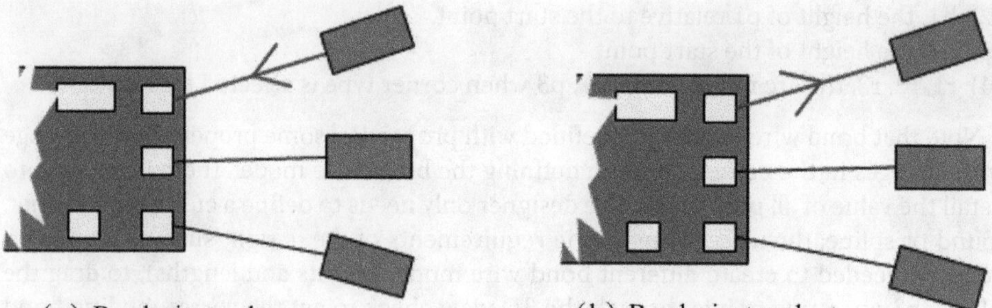

(a) Bond wire too long, move bond pad towards die

(b) Bond wire too short, move bond pad away from die

Figure 11.17 Move mouse according to the arrow direction.

In the process of adding bond wire, if the bonding line shows a red arrow, this suggests that the bond wire length is not within the scope of the setting. An inward arrow indicates the current length is too long and should move in the direction of the chip; an outward arrow indicates the current length is too short and should move away from the chip. In both cases, the designer only needs to move the mouse according to the direction of the arrow hint, as shown in Figure 11.17.

11.3.2 Move and Rotate Bond Pad

1) Move bond pad

   Click the button and then select the bond pad to be moved.

   In the process of moving the bond pad, the designer can view the bond wire span and angle information in real-time. When the span is beyond the setting length set by the rules, the length will automatically show red; when the angle is beyond the setting angle set by the rules, the angle will automatically show red; when both are beyond the range, both are red, notifying the designer that the scope is beyond the rule settings, as shown in Figure 11.18.

2) Rotate bond pad 90°

   Select the bond pad for rotation, and then click button , as shown in Figure 11.19.

3) Rotate bond pad 180°

   Select the bond pad for rotation, and then click on button .

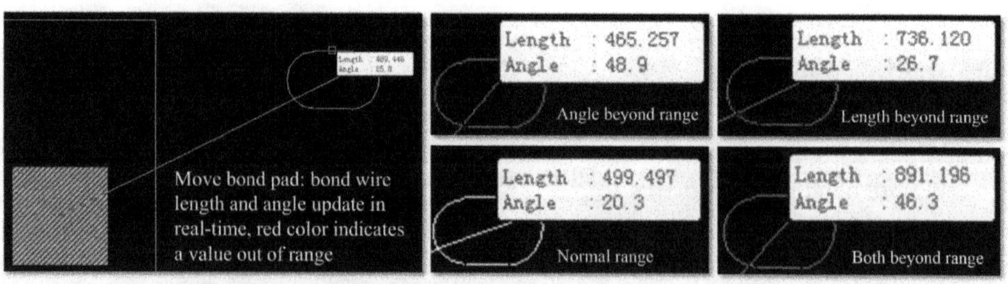

Figure 11.18 Move bond pad: bond wire length and angle update in real-time. *(See color plate section for the color representation of this figure.)*

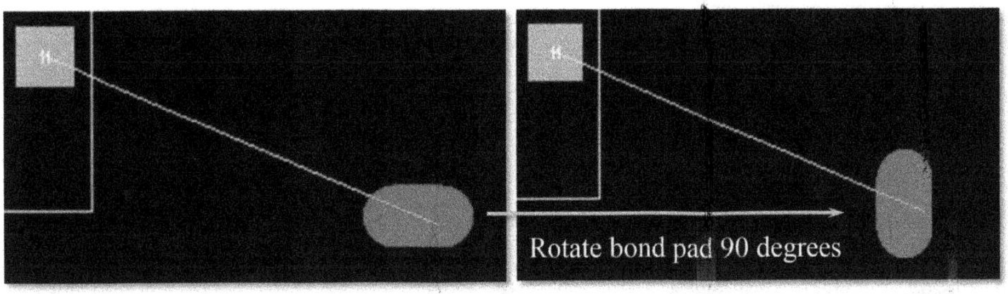

Figure 11.19 Rotate bond pad 90°.

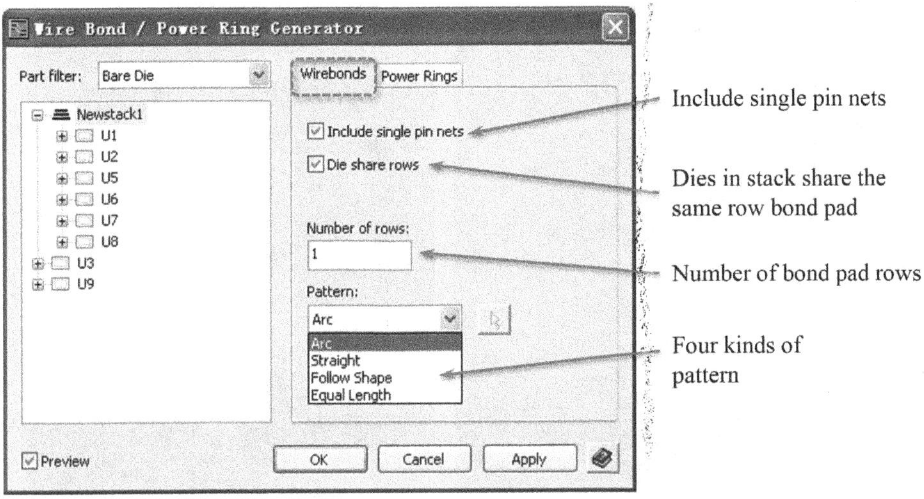

Figure 11.20 Wire bond generator.

11.3.3 Wire Bond and Power Ring Generator

Click the bond wire and power ring generator button.

1) Bond wire tab, automatically generate bond wire
   The generator is similar to the autorouter; it can bond all pins or some pins of a chip, and can also bond die stack automatically, as shown in Figure 11.20.
   The settings of the wire bond generator include the bond pad row number and the arrangement of the pattern. Designers can input the required row number in the Number of rows bar; the pattern options include: arc, straight, follow shape and equal length. Figure 11.21 shows the effect of four different patterns.
2) Power ring tab, automatically generate power ring
   The power ring settings interface includes the type of power ring, clearance from die, ring width, arc height, connected net and layer to add to. There are four types of power ring for selection: Arc – Conductive Shape, Arc – Plane Shape, Rectangle – Conductive Shape, Rectangle – Plane Shape, as shown in Figure 11.22.
   Figure 11.23 shows the different shapes of power ring.

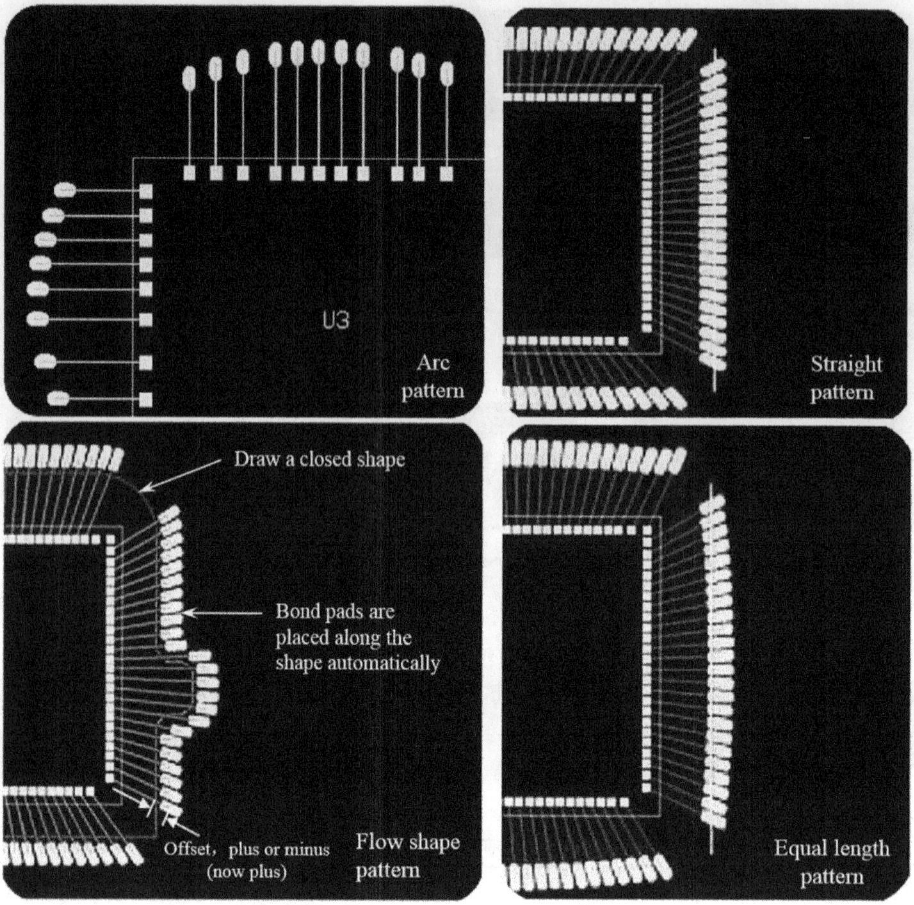

Figure 11.21 Four types of pad pattern.

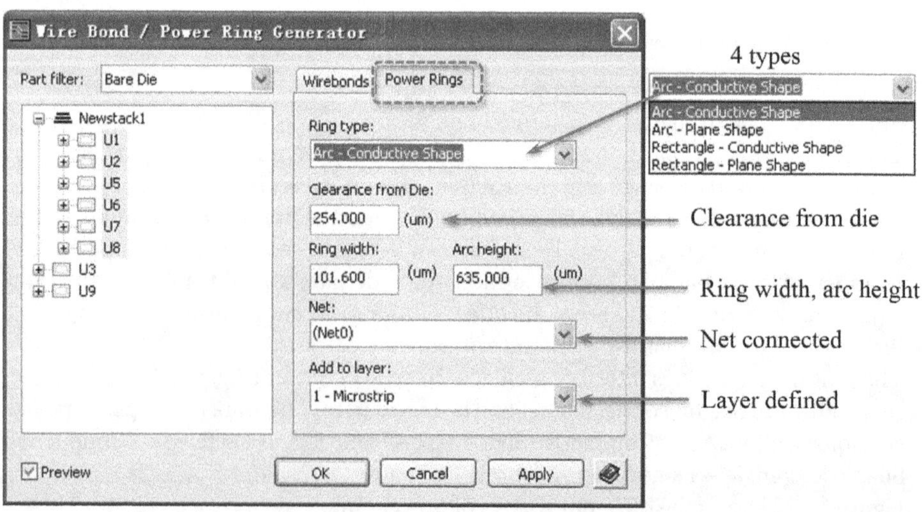

Figure 11.22 Power ring generator.

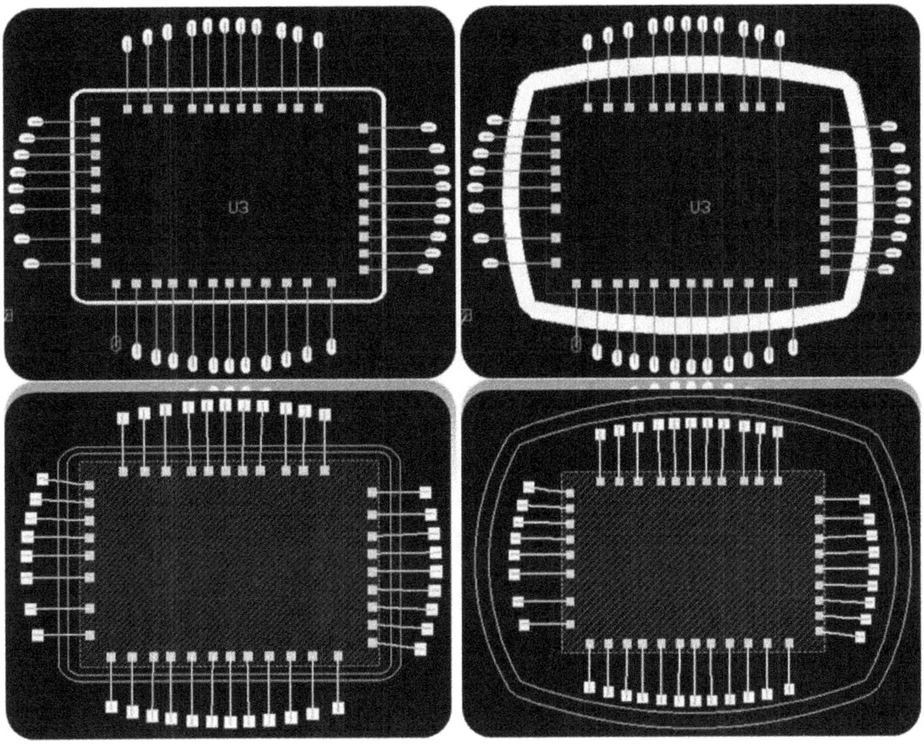

Figure 11.23 Power rings in different shapes.

11.3.4 Wire Bond Rule Setting

The wire bond parameter- and rule-setting button is .

The settings are divided into four levels: design, groups, part and pin.

The design settings are applied to the whole design, group settings are applied to multiple components, part settings are applied to a single chip, and pin settings are applied to single or multiple die pins.

The wire bond parameter and rule settings contain 15 items in total, as shown in the settings matrix in Figure 11.24. The settings vary for different objects, for example, the bondside and row settings are only available for the pin. In this book, we focus on the part and pin settings.

11.3.4.1 Settings for Part

Start the Bond Wire Parameters & Rules window; in the Part Filter bar select bare die; in the part list select a component, such as U3, as shown in Figure 11.25.

Next, we explain the setting items one by one.

1) Wiremodel, in the drop-down list, select the wire model created in wire model editor, as shown in Figure 11.26.
2) Bond padstack, click the drop-down list, choose the bond pad created in Padstack Editor, as shown in Figure 11.27.
3) Align with wire, set the direction of the bond pad consistent with the bond wire, as shown in Figure 11.28.

Rules	Design	Group	Part	Pin
Wiremodel	X	X	X	X
Bond Padstack	X	X	X	X
Align With Wire	X	X	X	X
Wire Offset	X	X	X	X
Die Pin Delta	X	X	X	
Angle	X	X	X	
Bond site margin	X	X	X	

X indicates can be setup, blank indicates no setup

Rules	Design	Group	Part	Pin
Wire To Wire	X	X	X	
Wire To Die Edge	X	X	X	
Wire To Part	X	X	X	
Wire To Metal	X	X	X	
Wire To Cavity	X	X	X	
Min/Max Length	X	X	X	
Bondside				X
Row				X

Figure 11.24 Wire bond parameter- and rule-setting matrix.

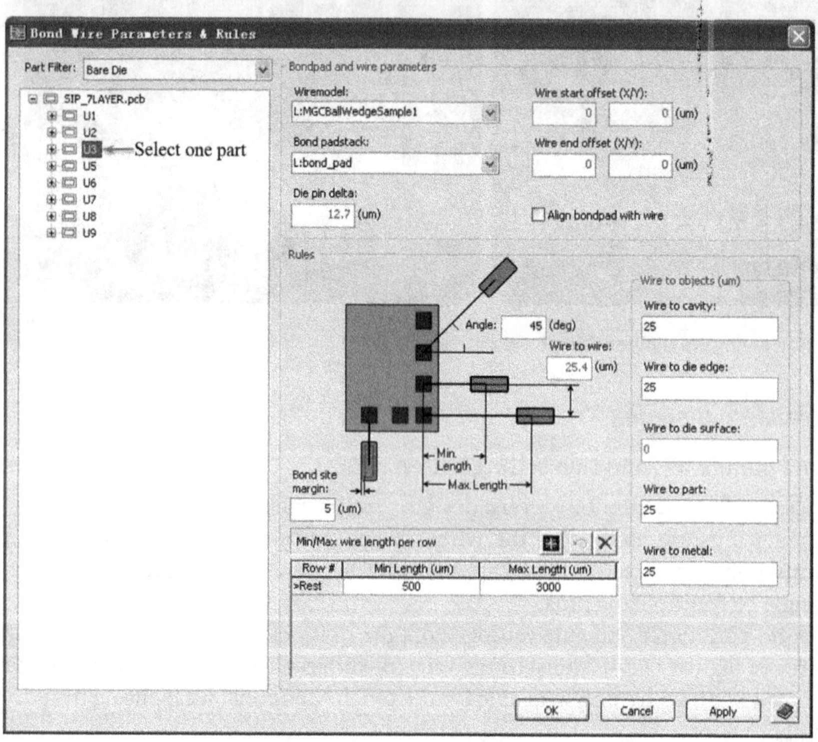

Figure 11.25 Bond Wire Parameters & Rules for part.

4) Wire offset, wire start offset is the definition of the bond wire start point center offset, it includes X and Y coordinates; wire end offset is the definition of the bond wire end point center offset, it also includes X and Y coordinates. Keep the default value 0 in general.
5) Die pin delta, defines the die pin relative height to the chip surface; a positive value indicates that the die pin is higher than the chip surface, a negative value indicates that the die pin is lower than the chip surface.

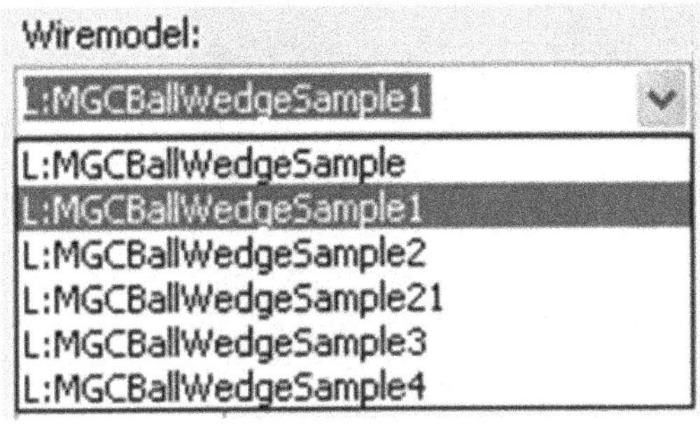

Figure 11.26 Wire model drop-down list.

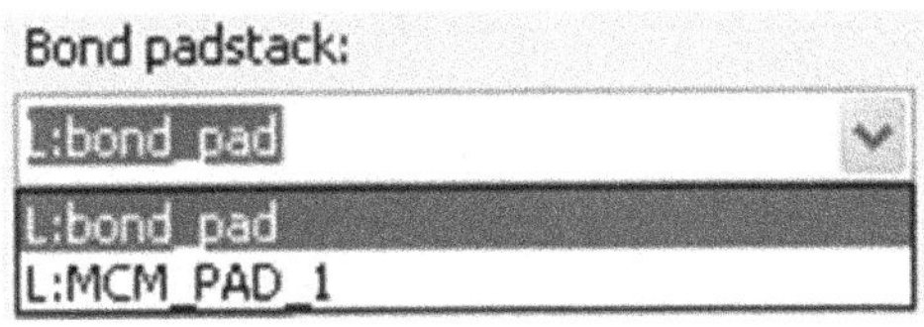

Figure 11.27 Bond padstack drop-down list.

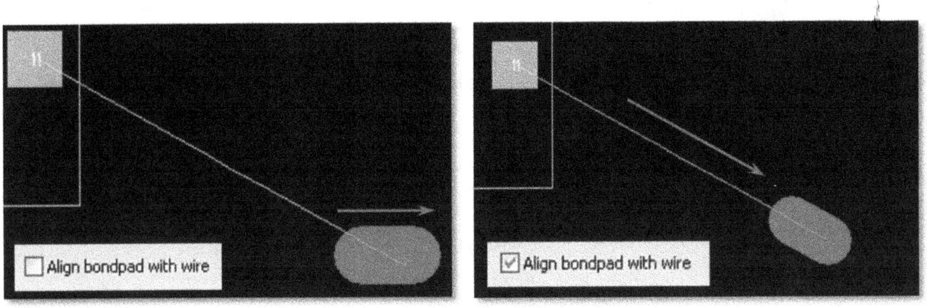

Figure 11.28 Align bond pad with wire before and after contrast

6) Angle, defines the maximum allowable deviation angle; if set to 0 (deg) or 360 (deg), the angle will be unlimited.
7) Bond site margin, defines the distance of bond wire to the edge of the bond pad; the maximum value cannot exceed the width of the bond pad, as shown in Figure 11.29. This setting should be used in cooperation with the bond wire offset.
8) Wire to wire, defines the minimum allowable distance between bond wire. Expedition software can make real-time DRC checks in 3D space according to the value of the setting. As shown in Figure 11.30, in the 2D space we cannot check whether the safety distance from wire to wire is met (wires have crossed each other), but in the 3D space we can see that there is a safe distance between the two bond wires.
9) Wire to die edge, defines the minimum distance between the bond wire and the die edge, as shown in Figure 11.31.
10) Wire to die surface, defines the minimum distance from the bond wire to other chip surfaces in the same chip stack.

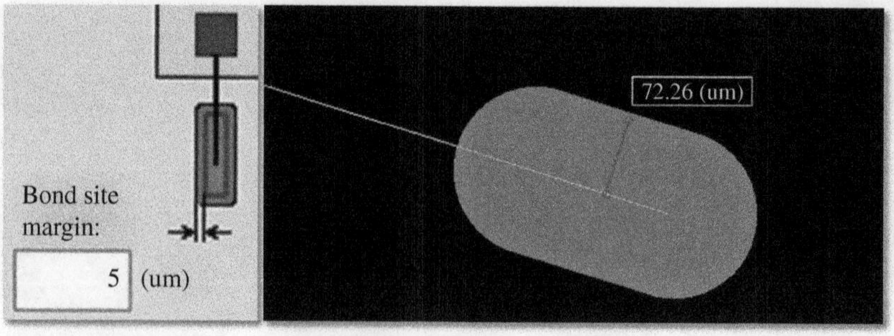

Figure 11.29 Bond site margin definition.

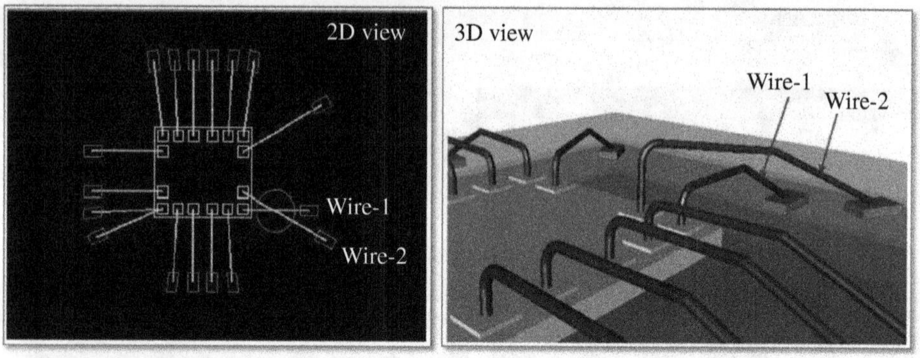

Figure 11.30 3D space DRC check.

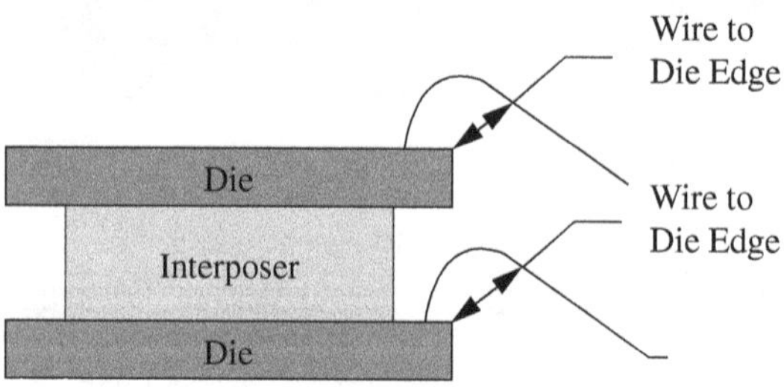

Figure 11.31 Wire to die edge definition.

11) Wire to part, defines the minimum distance between the bond wire and other components, as shown in Figure 11.32.
12) Wire to metal, defines the minimum distance between the bond wire and the metal (trace, plane or pad), as shown in Figure 11.33.
13) Wire to cavity, defines the minimum distance between the bond wire and the cavity, as shown in Figure 11.34.
14) Min/max wire length per row, defines the bond wire minimum and maximum span per row in the bond wire generator. Figure 11.35 shows bond wire length parameters for three rows; ">Rest" means bond wires not defined after the row numbers

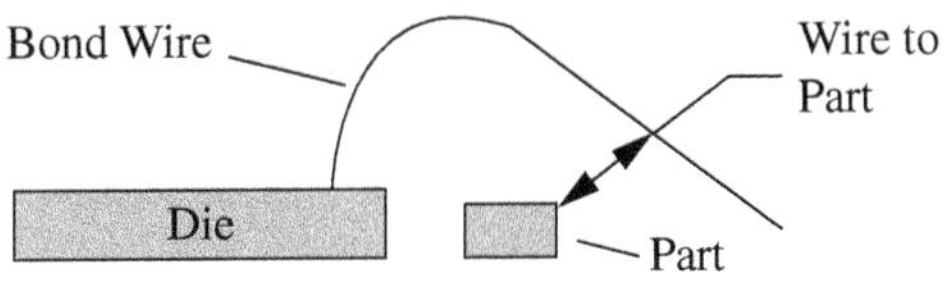

Figure 11.32 Wire to part definition.

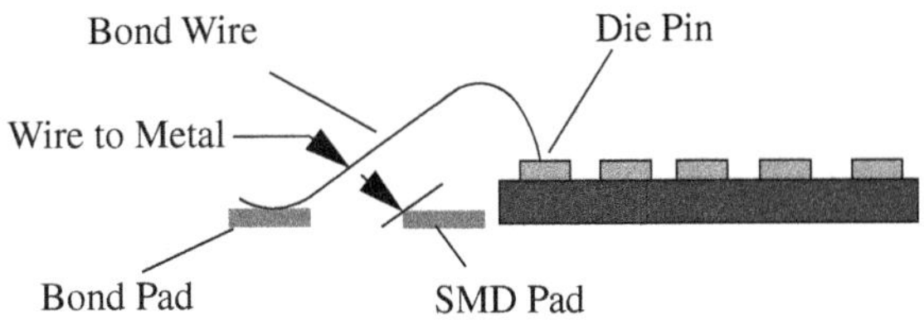

Figure 11.33 Wire to metal definition.

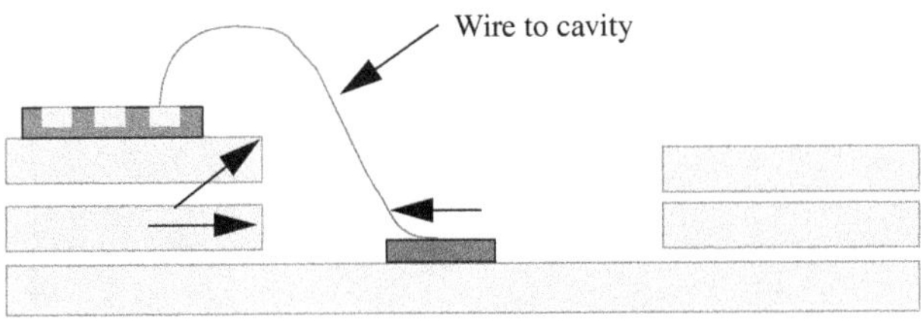

Figure 11.34 Wire to cavity definition.

Min/Max wire length per row

Row #	Min Length (um)	Max Length (um)
1	500	1000
2	1200	1800
3	2000	3000
>Rest	3200	3600

Figure 11.35 Min/max wire length setup.

follow the rules. If the bond wire is bonded by hand, it will be allowed in the length range set in all the rows.

11.3.4.2 Settings for Die Pin

In the die pin set, we can select one of the die pins or, by holding down the < Ctrl > key, select multiple die pins, as shown in Figure 11.36.

1) Bondside, set the direction of the wire bond; if we select Automatic, the outlet direction of the wire bond is chosen by the system automatically; usually the software will select the direction based on the die edge nearest to the die pad as the default. If we

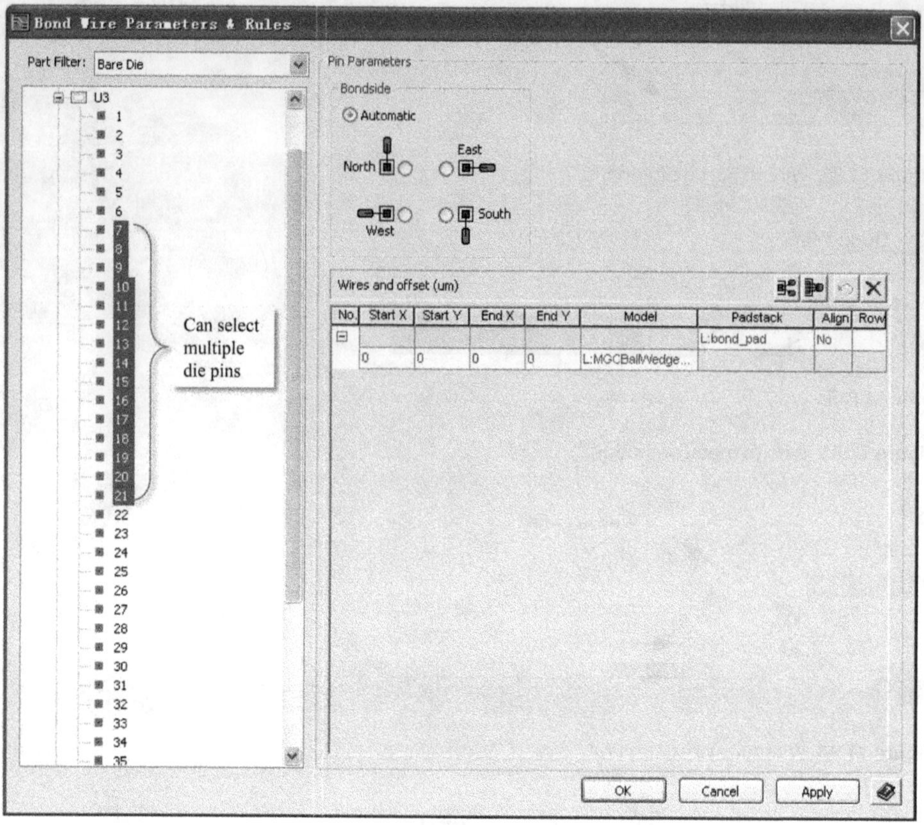

Figure 11.36 The rule settings for die pin.

choose any of north/south/east/west, the software will select the direction as shown in Figure 11.37. Note that this definition is a reference to the chip orientation defined in the library, in a specific design; if the chip is rotated, the orientation will be rotated accordingly.

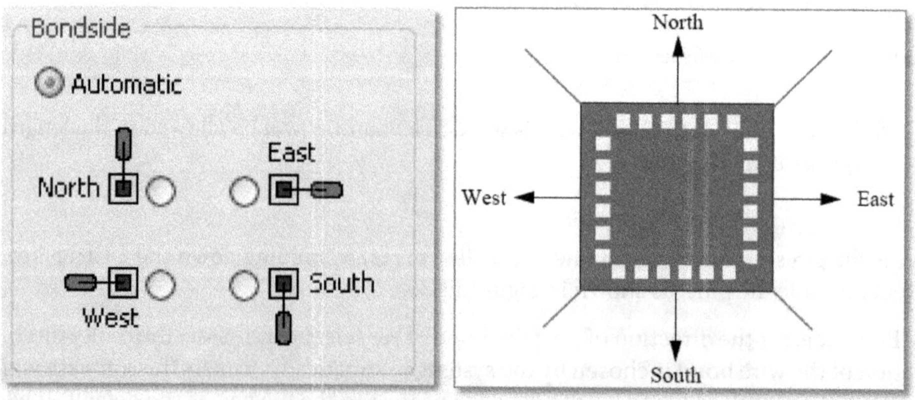

Figure 11.37 Bond wire outlet direction setting.

2) Wires and offset, mainly used to set multibonding wires in one bond pad or to fan out multiple bond pads from one die pin.
3) Start X, Start Y, End X, End Y, these set the X axis and Y axis relative offset value to the center of the start points or the center of the end points.
4) Model, select the bond wire model created in wire model editor.
5) Padstack, select the bond pad created in Padstack Editor.
6) Align, define the direction of the bond pad consistent with the bond wire.
7) Row, set which row the bond pad belongs to; for single row bonding, you don't need to consider this.

11.3.4.3 Add Multiple Bond Wires between Die Pin and Bond Pad

Click the Add wire button to add more bond wires to one die pin; the specific settings are shown in Figure 11.38.

After the completion of the settings, click the die pin to add bond wire; the system will automatically generate three bond wires from a die pin at the same time in accordance with the above settings; the start and end points of each bond wire have corresponding offsets; the designer only needs to specify the location of the bond pad, and then place the bond pad on the substrate. The resulting effect is shown in Figure 11.39.

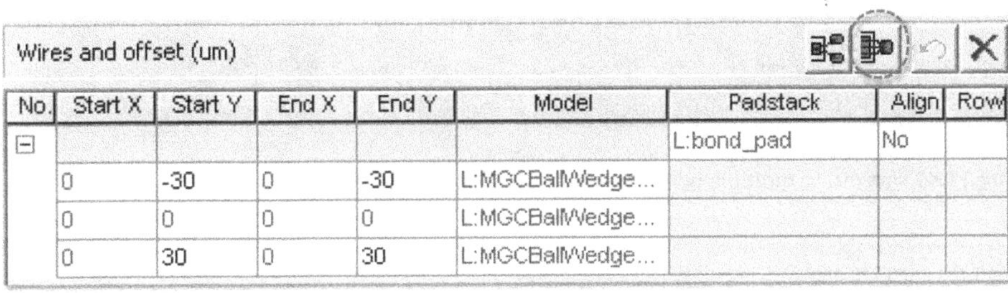

Wires and offset (um)

No.	Start X	Start Y	End X	End Y	Model	Padstack	Align	Row
⊟						L:bond_pad	No	
	0	-30	0	-30	L:MGCBallWedge...			
	0	0	0	0	L:MGCBallWedge...			
	0	30	0	30	L:MGCBallWedge...			

Figure 11.38 Settings for adding three bond wires to one die pin.

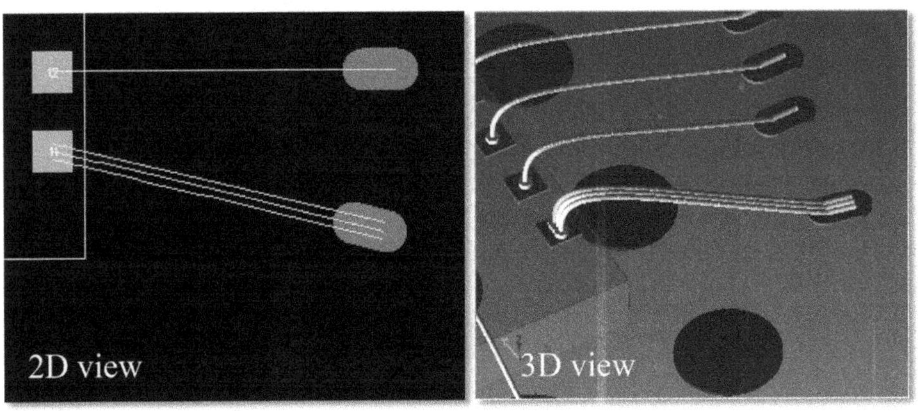

Figure 11.39 Add multiple bond wires between die pin and bond pad.

Wires and offset (um)

No.	Start X	Start Y	End X	End Y	Model	Padstack	Align	Row
⊟						L:bond_pad	Yes	
	0	0	0	0	L:MGCBallWedge...			
⊟						L:bond_pad	Yes	
	0	25	0	0	L:MGCBallWedge...			
⊟						L:bond_pad	Yes	
	0	-25	0	0	L:MGCBallWedge...			

Figure 11.40 The setting for a single die pin fan out to three bond pads.

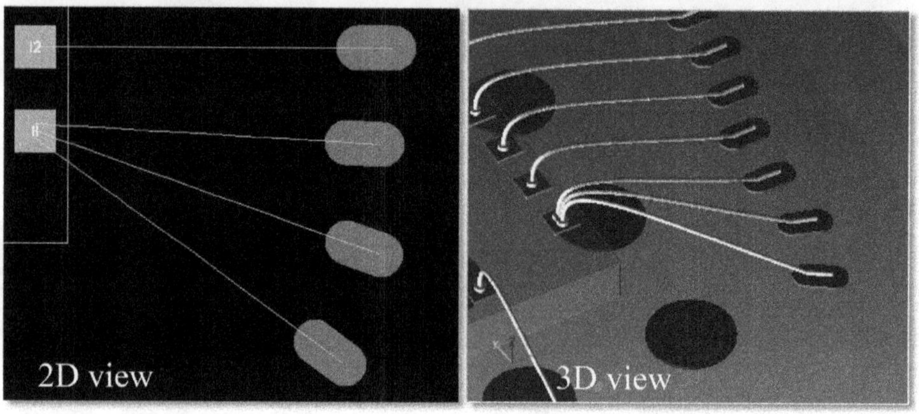

Figure 11.41 Fan out to multiple bond pads from single die pin.

11.3.4.4 Fan Out to Multiple Bond Pads from Single Die Pin

Click the Add Bond Pad button; fan out the bond wire to multiple bond pads from a single die pin; the specific settings are as shown in Figure 11.40.

After completion of the settings, click the die pin to add bond wire; the system will, in turn, fan out three bond wires with their bond pads; the designer only needs to specify the location of each bond pad and place on the substrate; the resulting effect is shown in Figure 11.41.

11.3.4.5 Bond Multiple Die Pins to One Bond Pad

In wire bond design, there is a case where the designer wants to connect two or more die pins from the same net to one bond pad. We first need to set each die pin end point offset value, as shown in Figure 11.42. For example, for the two bond wires, the End point Y coordinates are set to 30 and −30 respectively, otherwise the two bond wires will overlap.

After set up, add the first bond wire according to the normal method. The process for the second bond wire is as follows: first, hold down the < Ctrl > key, click the respective die pin and bond pad, and then execute the Add Bond Wire command, as shown in Figure 11.43.

Die Pin 1

No.	Start X	Start Y	End X	End Y	Model	Padstack	Align	Row
⊟						L:bond_pad	No	
	0	0	0	30	L:MGCBallWedge...			

Die Pin 2

No.	Start X	Start Y	End X	End Y	Model	Padstack	Align	Row
⊟						L:bond_pad	No	
	0	0	0	-30	L:MGCBallWedge...			

Figure 11.42 Set the end point offset value of each bond wire.

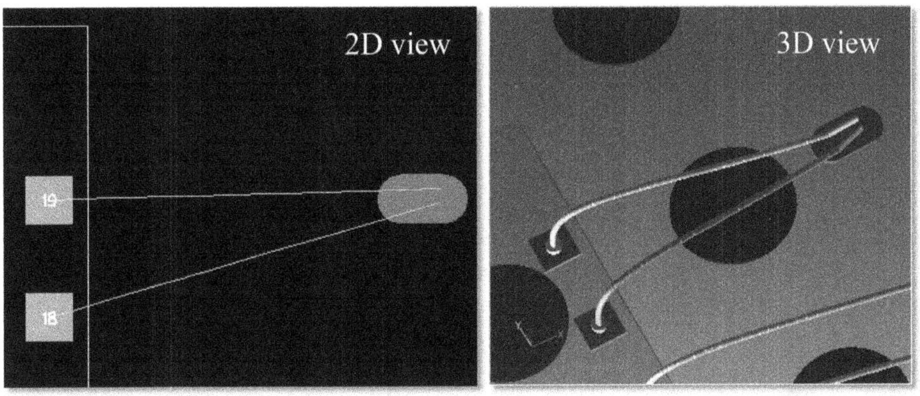

Figure 11.43 Two die pins bonded to the same bond pad.

No.	Start X	Start Y	End X	End Y	Model	Padstack	Align	Row
⊟						DieToDie	No	
	-0.0762	0	0	0	L:MGCBallWedge...			
⊟						L:MCM_PAD_1	No	
	0.0762	0	0	0	L:MGCBallWedge...			

Figure 11.44 Set up die-to-die bonding.

11.3.4.6 Die-to-Die Bonding

In SiP design, in addition to bonding from die pin to bond pad, another kind of situation is die-to-die bonding, from one die pin directly to another die pin. Figure 11.44 shows a die pin fan out with two bond wires, one for bonding to a die pin, and the other to a bond pad.

After set up, bond the first wire according to the normal method; the second bond wire will automatically pop up, and the designer only needs to click on another die pin, as shown in Figure 11.45.

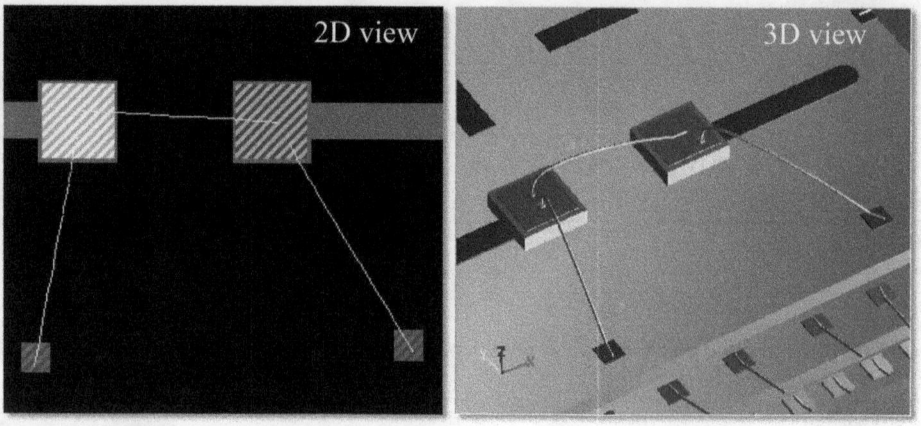

Figure 11.45 Die-to-die bond with ball to wedge.

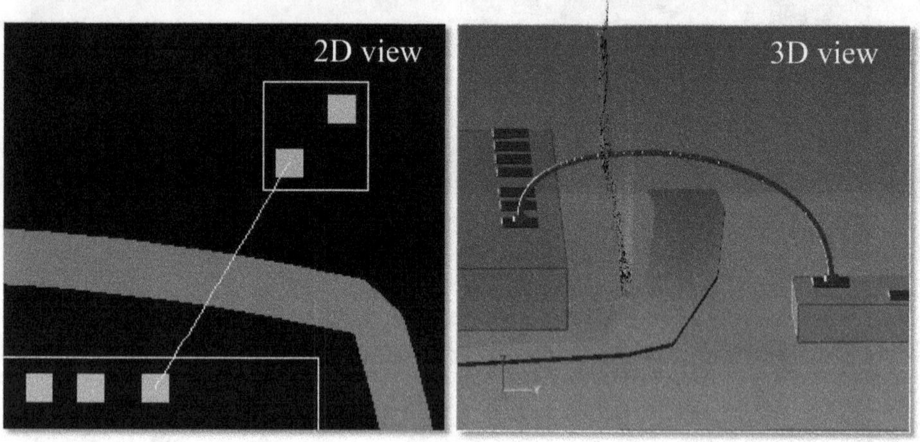

Figure 11.46 Die-to-die bond with ball to ball.

For die-to-die bond connection, selection of ball end or wedge end is mainly based on the bond wire model. For example, in Figure 11.45 the bond wire model is defined with ball to wedge; in Figure 11.46 the bond wire model is defined with ball to ball; in Figure 11.47 the bond wire model is defined with ball to wedge and wedge to wedge.

11.3.5 Wire Model Editor

The wire model editor is a real-time editing and adjustment tool for bond wire models. The designer should note that the functions of the wire model editor started from the Wirebond toolbar are different from the functions of the wire model editor started from Setup→Wire Model Editor, but the same as the functions started from Route→Wirebonds→Wire Model Editor, as shown in Figure 11.48.

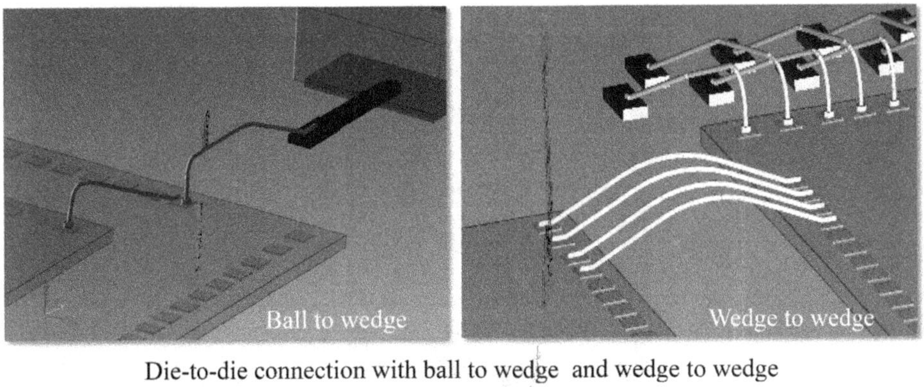

Figure 11.47 Die-to-die bond with ball to wedge and wedge to wedge.

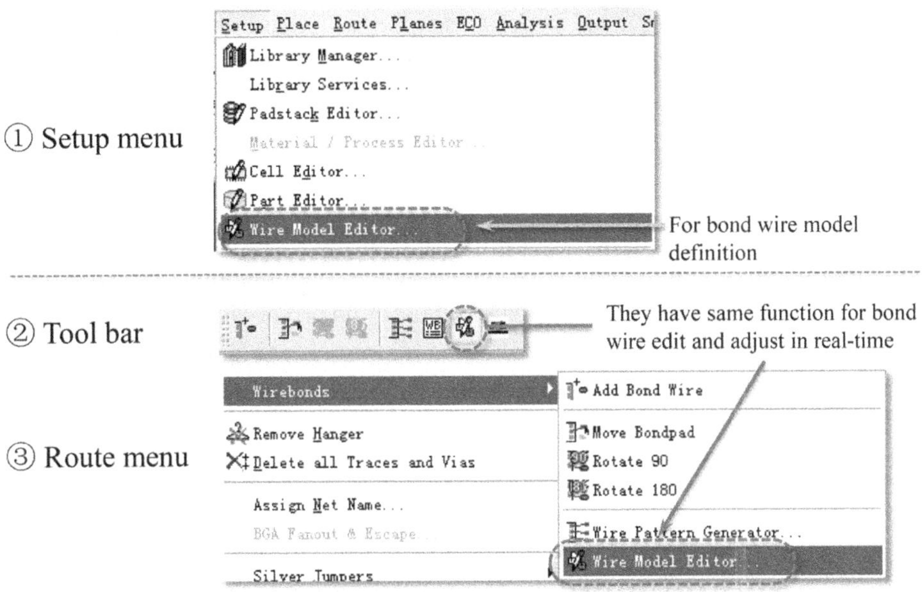

Figure 11.48 The functions of the three wire model editors.

The main difference is that the wire model editor in the Setup menu is used to create and define properties for bond wire model, while the wire model editor started from the Wirebond toolbar and the Route menu is used to inspect, edit and adjust existing bond wire design in real-time.

Before starting the wire model editor, we first need to select one bond pad or die pin connected to the bond wire; we can then see the 3D bond wire and surrounding graphics in the 3D window, otherwise the window display is blank.

The interface of the wire model editor is the same for Setup→Wire Model Editor; the main difference is that the 3D window display also contains 3D graphics of other

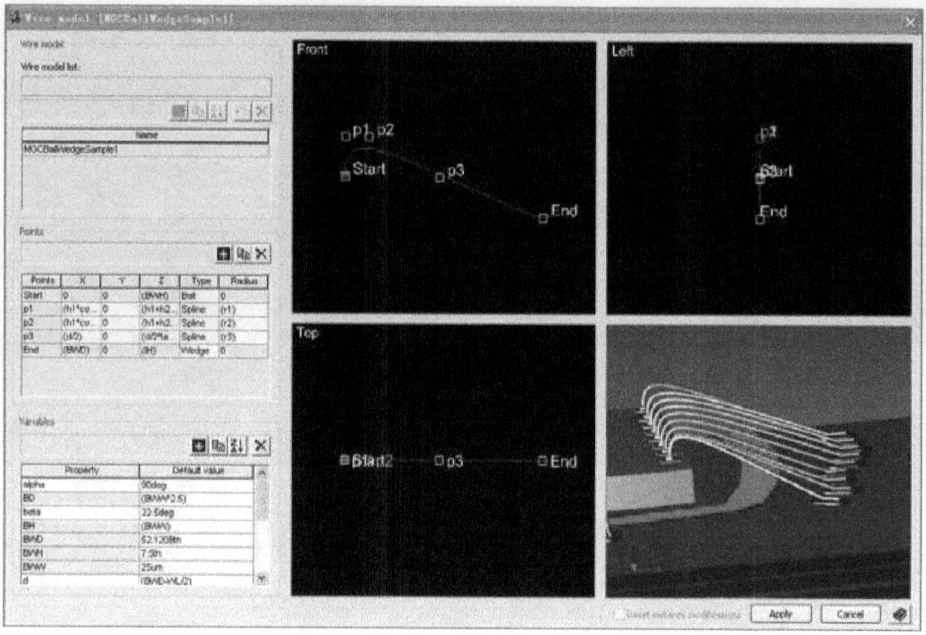

Figure 11.49 The 3D window display of the wire model editor shows bond wire and other elements.

elements near the selected bond wire; in the model list section only the wire model assigned to the selected bond wire is listed, as shown in Figure 11.49.

Select a bond wire (in practice we often select the bond pad or die pin connected with the bond wire), then click the tool button  ; in the Wire model editor window, we can operate by dragging points p1, p2 and p3 to adjust the bond wire shape in real-time, and see the effect in the 3D window.

As shown in Figure 11.50, the Z coordinates of p1, p2 and p3, which affect the bond wire longitudinal curve and height, can be adjusted. In the process of adjustment, we can view the resulting 3D figure in real-time and then click the Apply button to directly use the adjustment in design. This kind of longitudinal adjustment applies to multiple row bonding, as well as bond wire across components, etc. In actual design, we also need longitudinal adjustment to avoid conflict.

As shown in Figure 11.51, the Y coordinates of p1, p2 and p3, which affect the bond wire lateral curve and deviation, can be adjusted. In the process of adjustment, we can view the 3D result in real-time, and then click the Apply button to directly use the adjustment in design. This adjustment is applicable for bond wire interlace, and bond wire around components, etc. In actual design, we also need lateral adjustment to avoid conflict.

By adjusting the (X, Y, Z) coordinates of the bond wire control points p1, p2 and p3, we can design any complex bond wire curve; this kind of complex curve is often used in actual design. Using accurate simulation via the wire model editor, we can

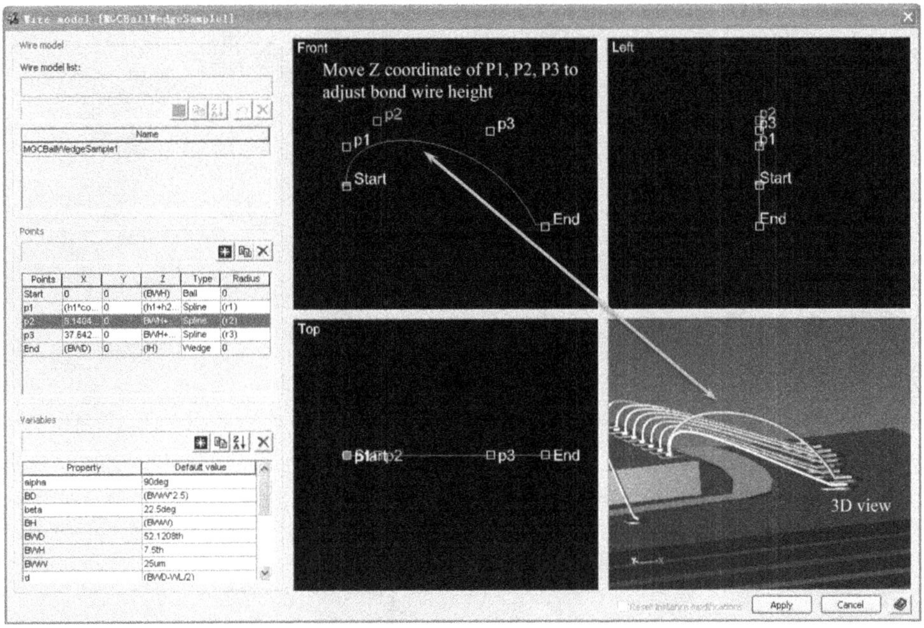

Figure 11.50 Bond wire curve height adjustment (Z coordinate).

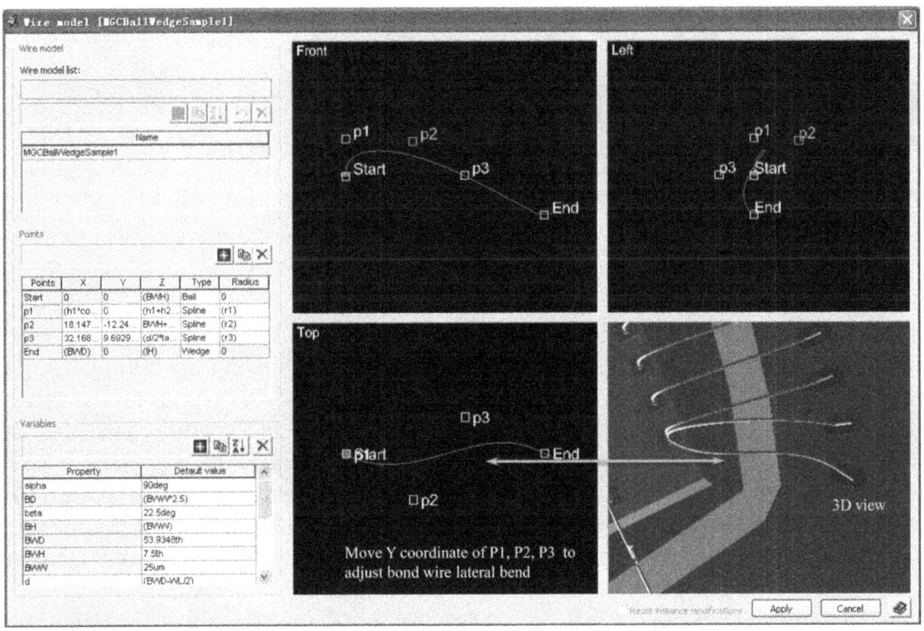

Figure 11.51 Bond wire lateral (Y coordinate) adjustment.

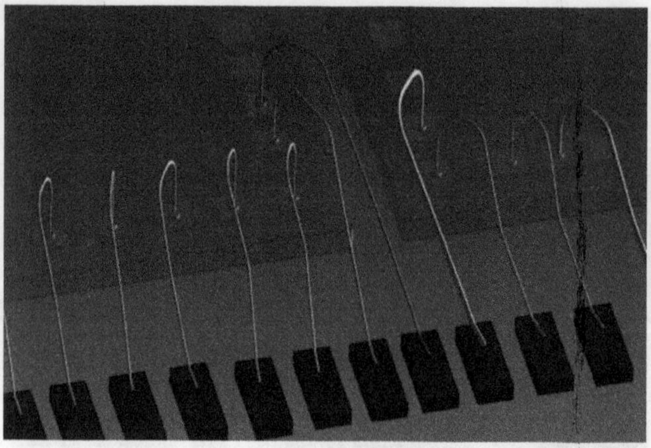

Figure 11.52 Bond wire with complex curve in Expedition design.

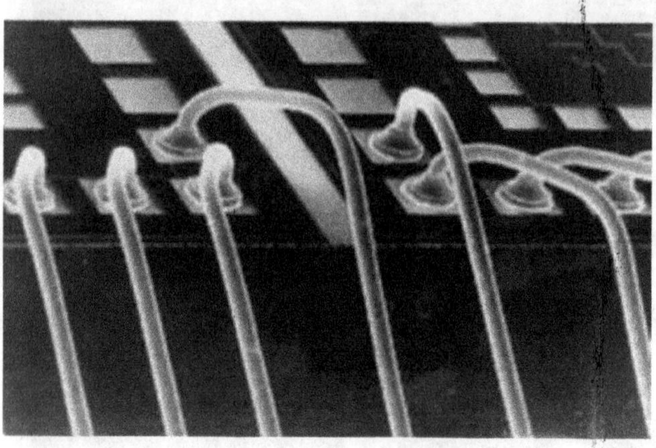

Figure 11.53 Real photo of complex curve bond wire.

effectively predict whether there will be problems in practical production, thus improving the manufacturability of the wire bond, and achieving the aim of Design For Manufacture (DFM).

Figure 11.52 shows a complex bond wire curve designed in Expedition; because of chip placement and die pin positioning, design needs wire bond with complex curves.

In actual design, we will encounter this kind of complex curve bond wire. For example, the real photo in Figure 11.53 shows the complex curve bond wire for the specific layout and die pin distribution. Comparing Figures 11.52 and 11.53, we can see that designing complex curve bond wire in Expedition is effective and of value.

Double-click the left mouse button in the real-time 3D window; the 3D window will automatically jump out and center in the screen, and the figure can be viewed more clearly. The 3D window is always updated in real-time, that is, any adjustment in 2D design will be updated to the 3D window in real-time, as shown in Figure 11.54.

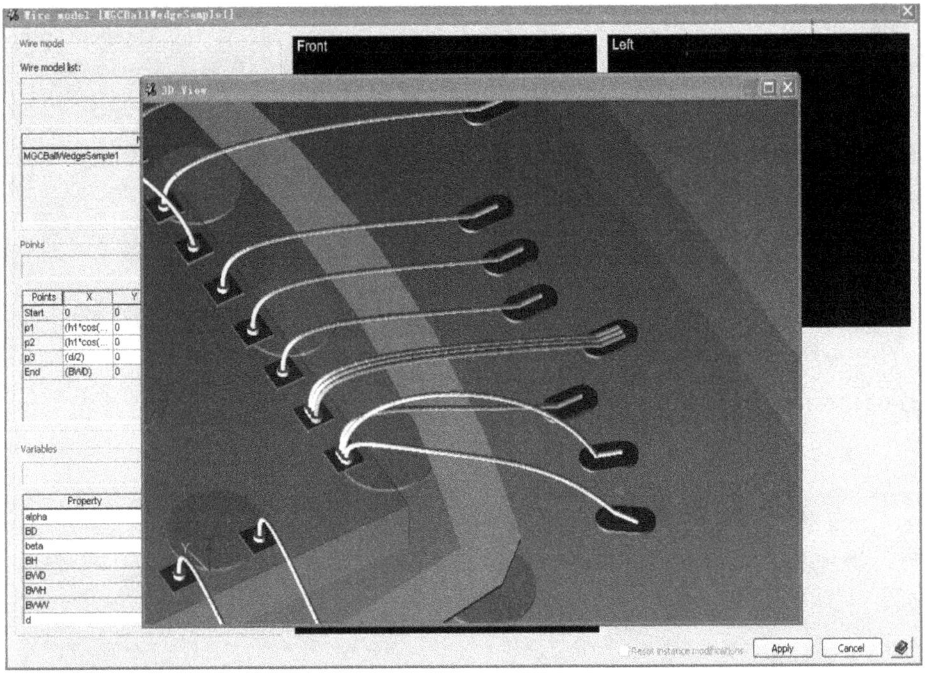

Figure 11.54 Expand 3D window with double-click of left mouse button. *(See color plate section for the color representation of this figure.)*

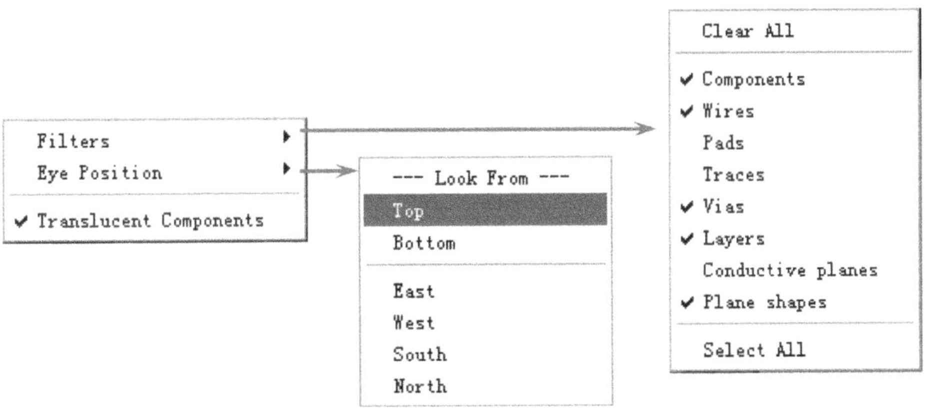

Figure 11.55 Right-click menu of 3D window.

In the 3D window, we can hold the left mouse button to rotate the design. If we click the right mouse button, the right-click menu will appear, as shown in Figure 11.55.

With this menu, we can carry out the following operations:

1) Using Filters we can open or close elements for selective viewing, as shown in Figure 11.56.

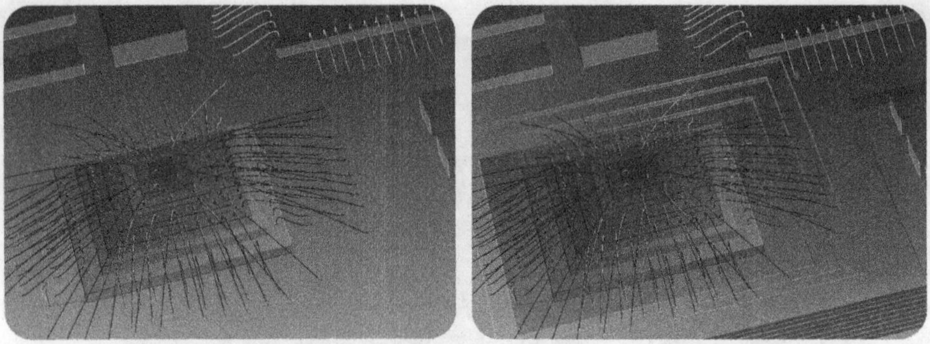

(a) Display die and bond wire (b) Display die, bond wire and substrate

Figure 11.56 Selective view.

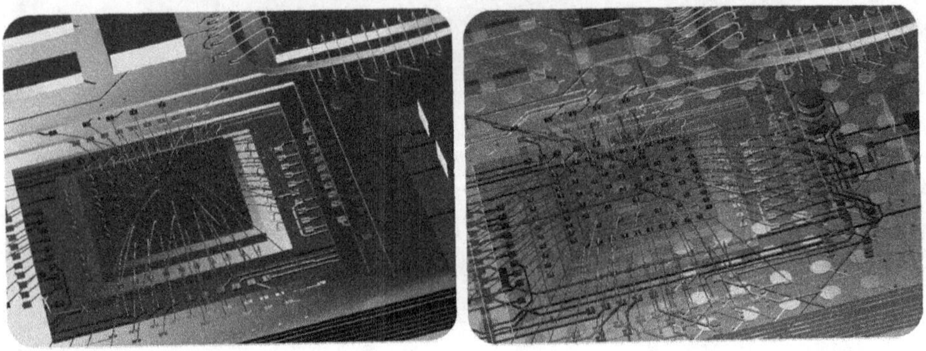

(a) Opaque mode display (b) Translucent mode display

Figure 11.57 Opaque and translucent display modes comparison.

2) Using the Eye Position we can view design from six different viewpoints, the options being top, bottom, east, west, south and north.
3) Using the Translucent Components menu, we can toggle between opaque and translucent, convenient for viewing surface or internal elements, as shown in Figure 11.57.

12

Cavity and Chip Stack Design

12.1 Cavity

Cavity, one of the most common and basic technics in ceramic package substrate design, has become more and more important in package or SiP deign. Currently, with technic improvements, some projects also use cavity in plastic substrate; for example, Loongson processor plastic substrate uses cavity structure.

Cavity is a 3D structure; to simulate cavity more realistically, what is needed is package or SiP design software supporting professional 3D substrate design.

12.1.1 Cavity Definition

A cavity is a blind hole in substrate that does not traverse all layers (in Expedition, we call a hole traversing all layers a contour). Cavity is created by removing substrate material to get an open void area in the internal electrical layers. Cavities can start from one of the outer layers or be embedded within the design stackup. Cavities can be single-step or multistep; the multistep cavity is "cavity in cavity". Figure 12.1 shows an open multistep cavity.

From Figure 12.1 we can see that Cavity 2 is in Cavity 1, and the die is placed in Cavity 2. Some bond wires connect die pins to the surface layer, some bond wires connect die pins to the step layer between Cavity 2 and Cavity 1, other bond wires connect die pins to the bottom of Cavity 2. All bond wires connect to different layer traces with their respective bond pads.

Below are listed several typical cavity structures.

Figure 12.2 displays an eight-layer stackup, including the electrical layer, substrate and prepreg.

Figure 12.3 displays an eight-layer stackup with 1–3 layer cavity profile; the start layer of the cavity is layer1, the end layer is layer 3, which is the bottom of the cavity, exposed to the outside; components can be installed on this.

Figure 12.4 displays an eight-layer stackup with a multistep cavity profile: 1–3 layer cavity, 3–5 layer cavity and 5–7 layer cavity respectively; this kind of multistep cavity is popular in ceramic package and SiP applications. In this case, we can put the bond pad in the step of the cavity.

Figure 12.5 displays an eight-layer substrate with embedded cavity; with such cavity we can embed the components into the substrate.

SiP System-in-Package Design and Simulation: Mentor EE Flow Advanced Design Guide,
First Edition. Suny Li (Li Yang).

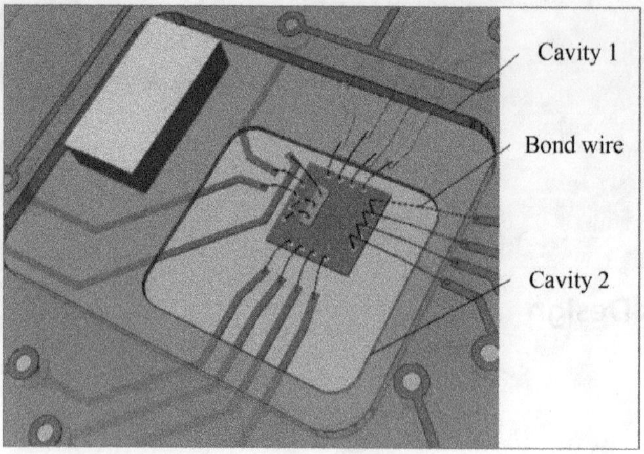

Figure 12.1 Open multistep cavity.

Figure 12.2 The eight-layer stackup.

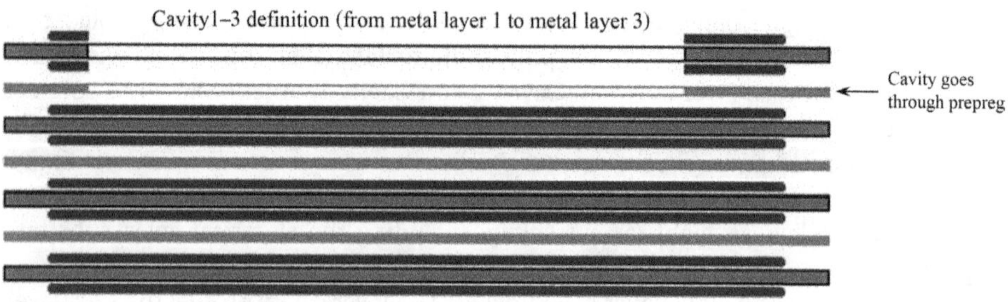

Figure 12.3 The eight-layer stackup with 1–3 layer cavity profile.

12.1.2 Cavity Creation

On the Expedition tool bar, click the button and toggle to draw mode. Then click the Property button, in the Type list select Cavity, then set Start Layer=1, Allow Metal=No, End Layer=2, Allow Metal=Yes, as shown in Figure 12.6.

Then, draw the cavity graphics, and in the General tab of Display Control, check Cavity display below Board Items; we can see the 1–2 layer cavity in Figure 12.7.

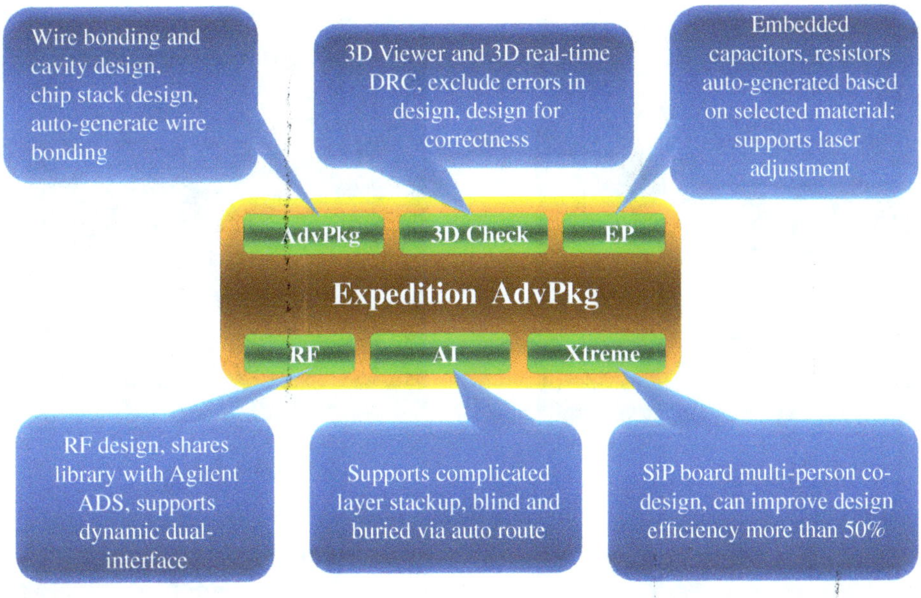

Figure 1.10 The six core design functions of Mentor SiP layout.

SiP System-in-Package Design and Simulation: Mentor EE Flow Advanced Design Guide,
First Edition. Suny Li (Li Yang).

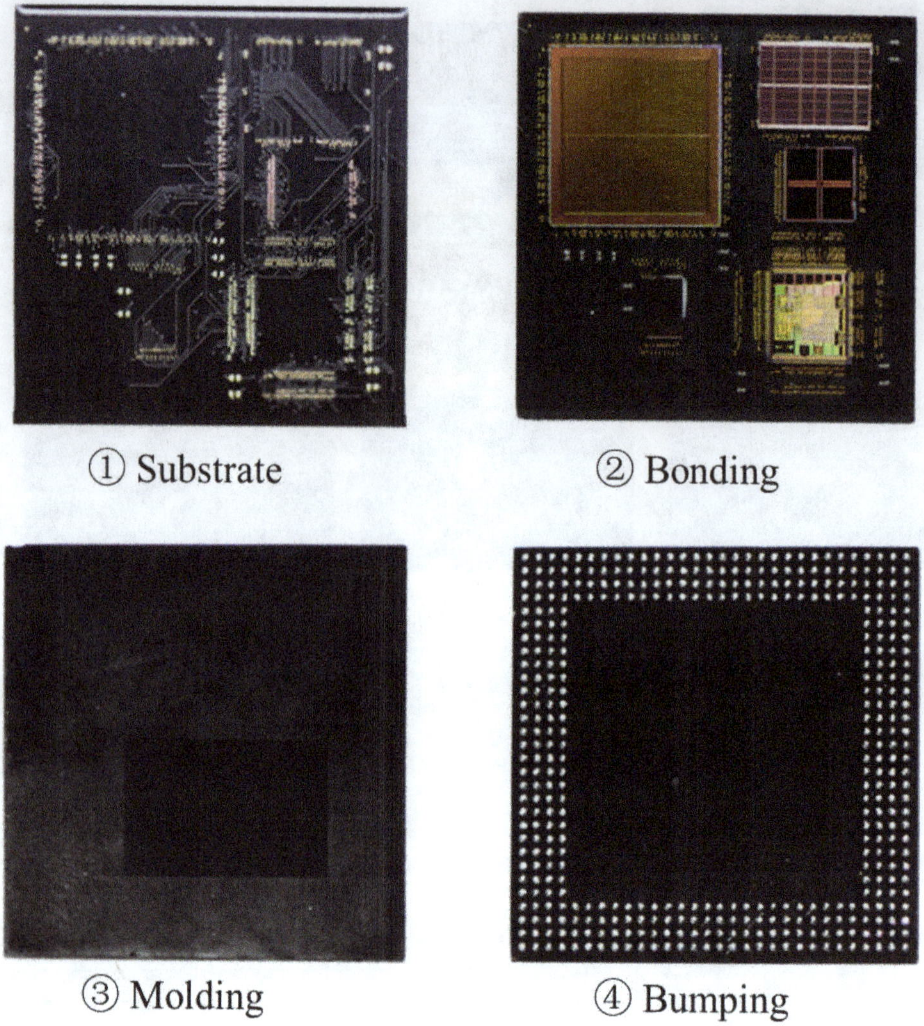

Figure 1.20 One of the finished customer projects using the Mentor SiP platform.

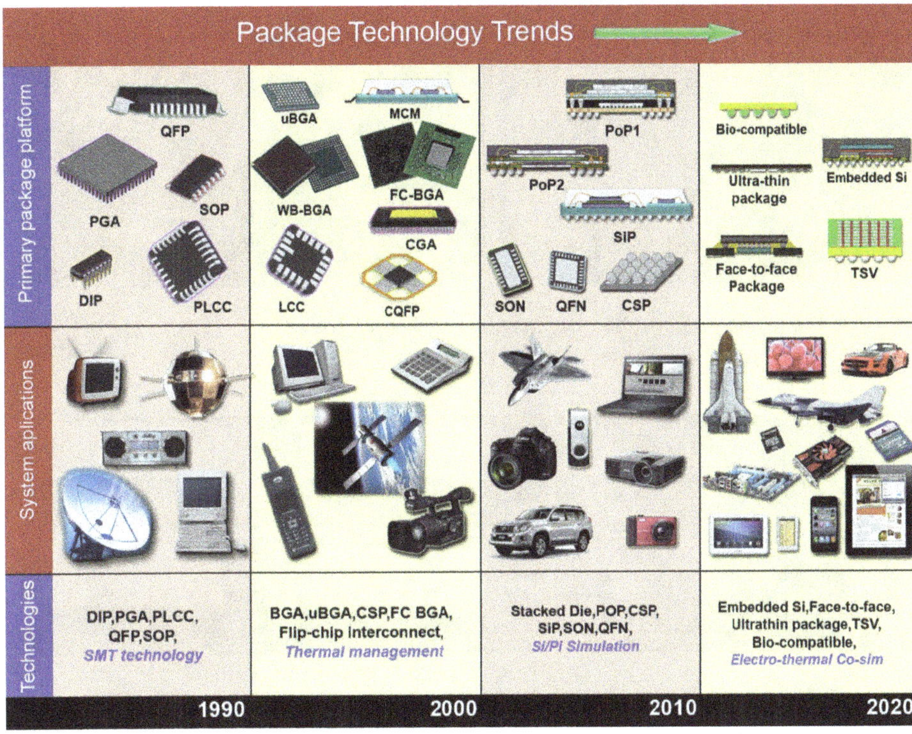

Figure 2.4 Development trends in electronic packaging technology.

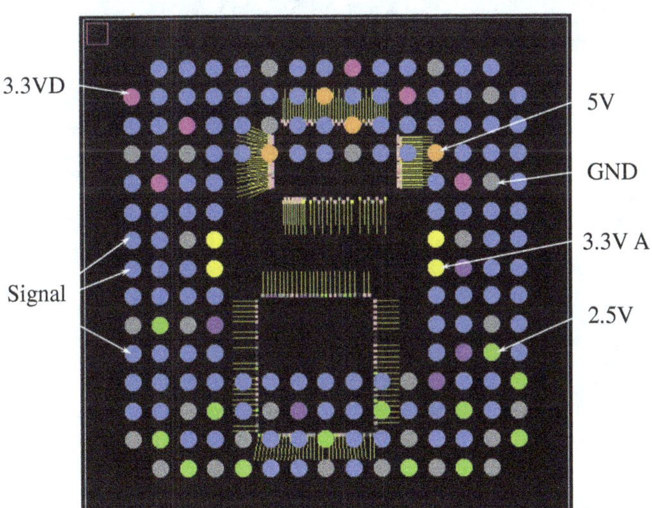

Figure 5.4 Typical distribution of power, GND and signal pins.

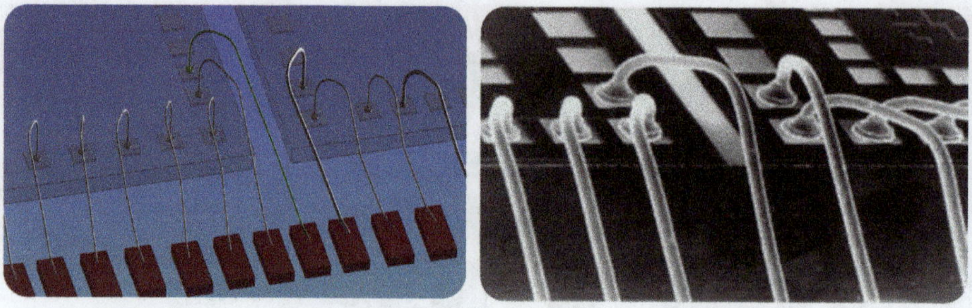

Figure 11.2 Expedition supports the most complex wire bonding design.

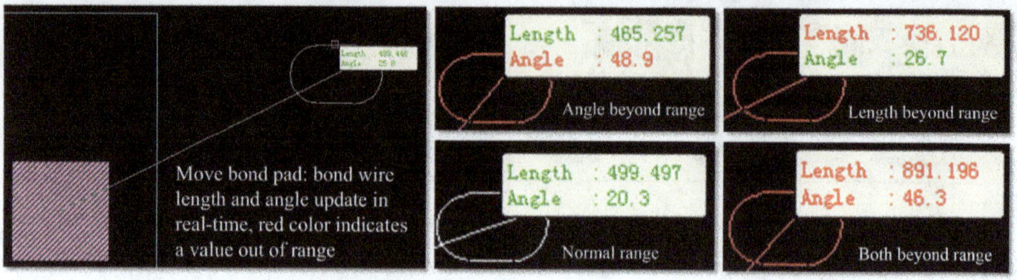

Figure 11.18 Move bond pad: bond wire length and angle update in real-time.

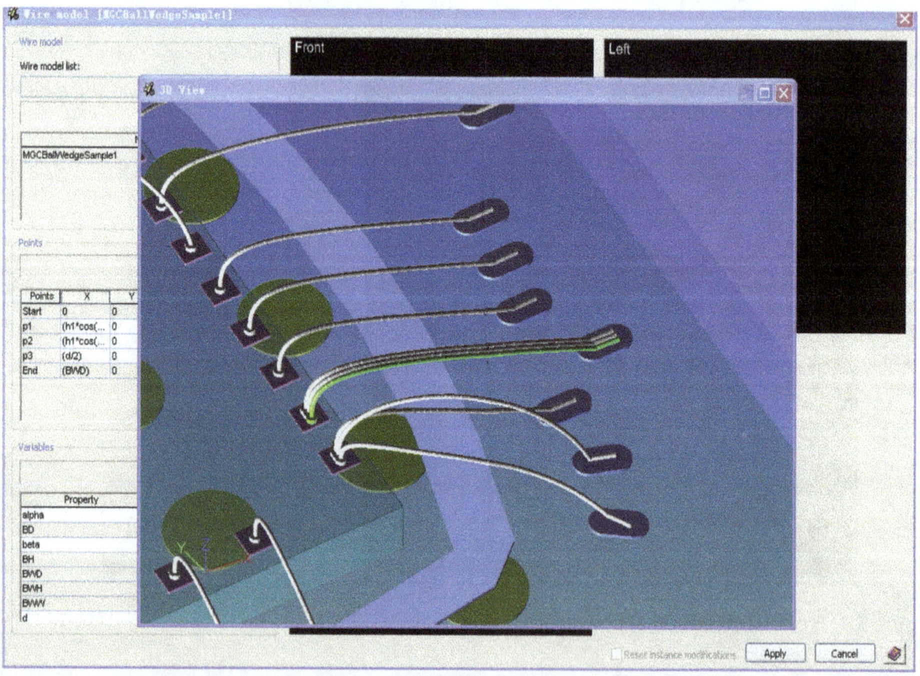

Figure 11.54 Expand 3D window with double-click of left mouse button.

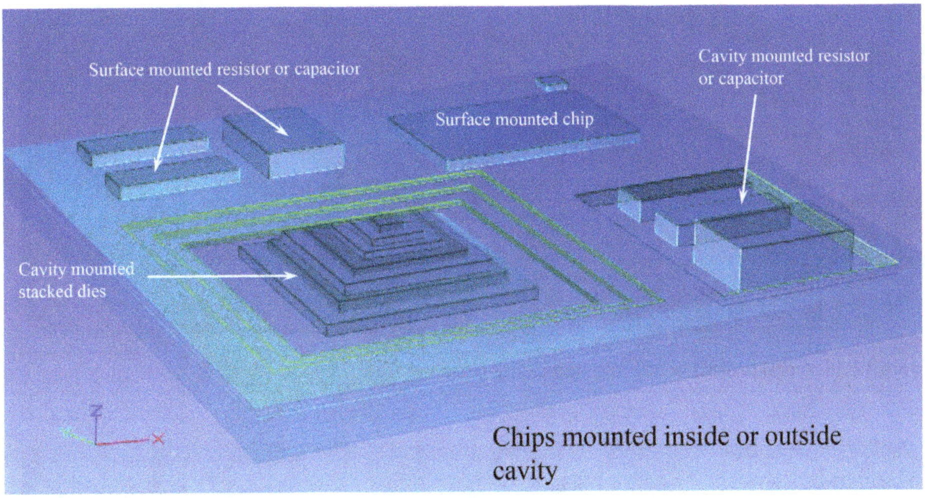

Figure 12.12 Place components into cavity.

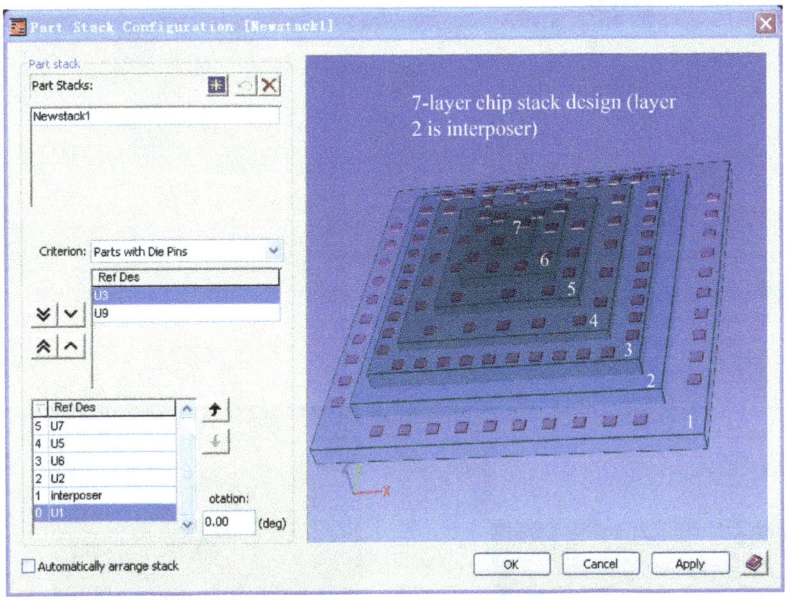

Figure 12.25 Chip stack configuration window.

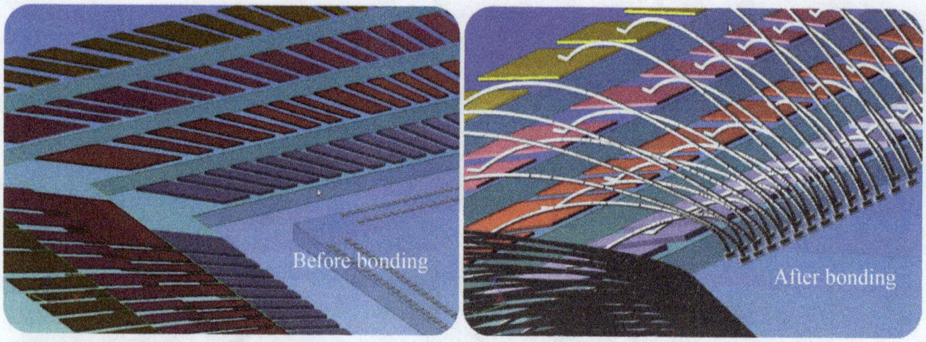

Figure 12.35 A real project with multistep cavity bonding.

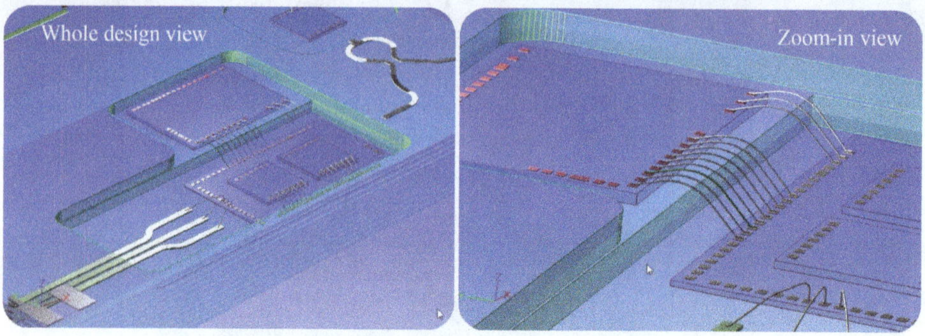

Figure 12.36 A real project with mixed design technologies.

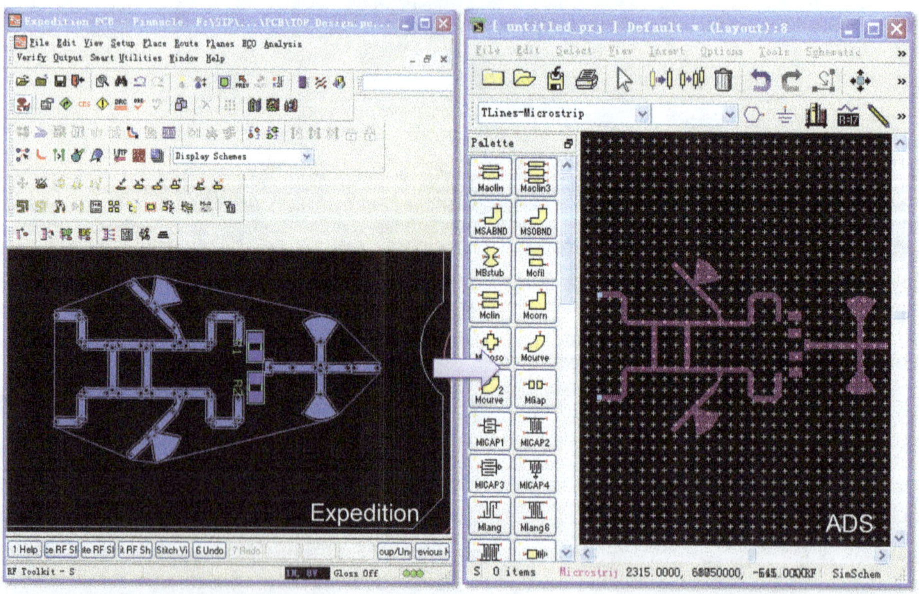

Figure 16.54 Transmit RF layout data from Expedition to ADS.

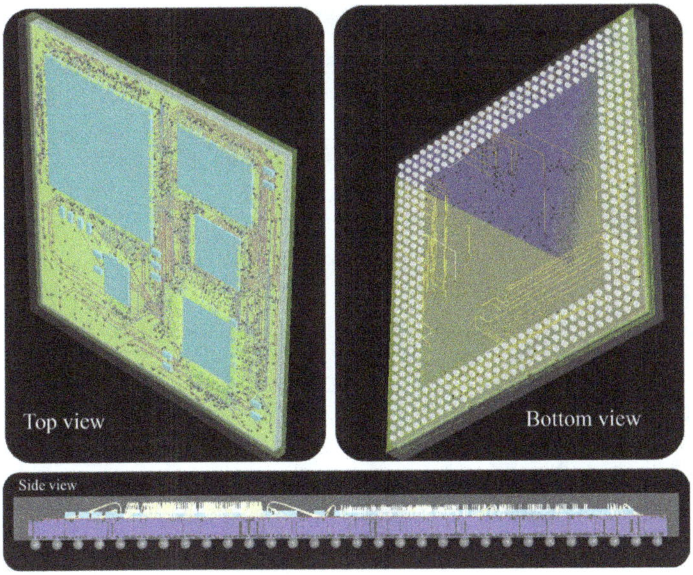

Figure 18.19 View SiP product from three directions in 3D Viewer.

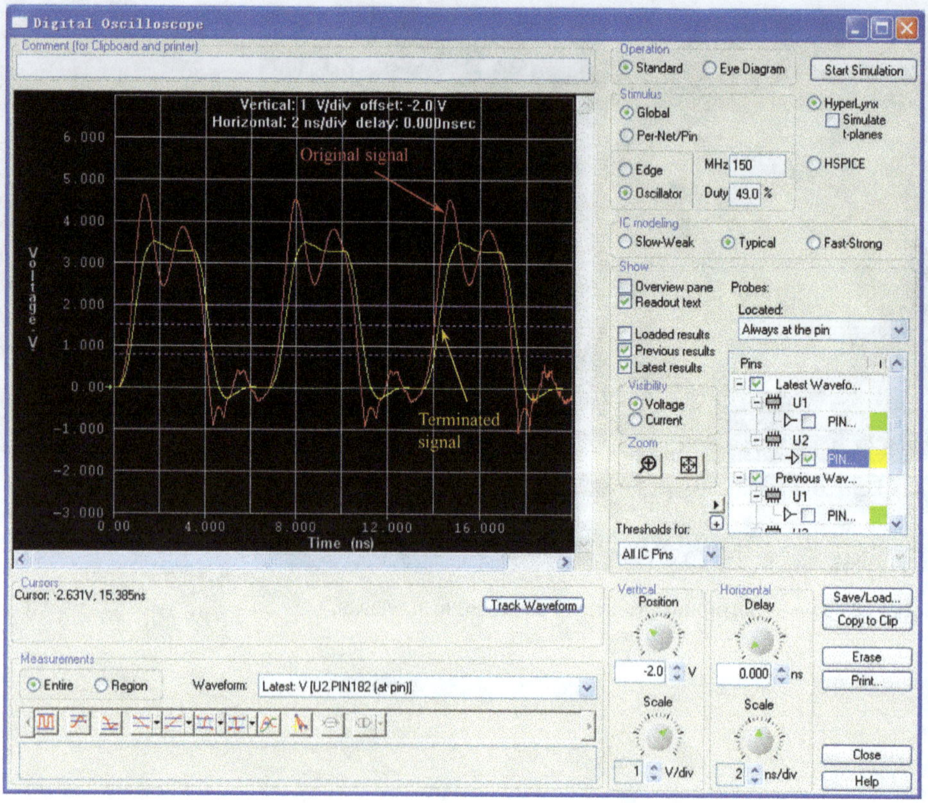

Figure 21.12 Comparison of original signal and terminated signal.

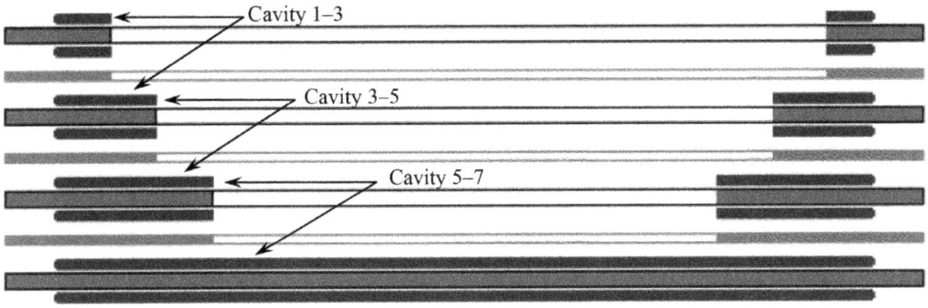

Figure 12.4 The eight-layer stackup with multistep cavity profile.

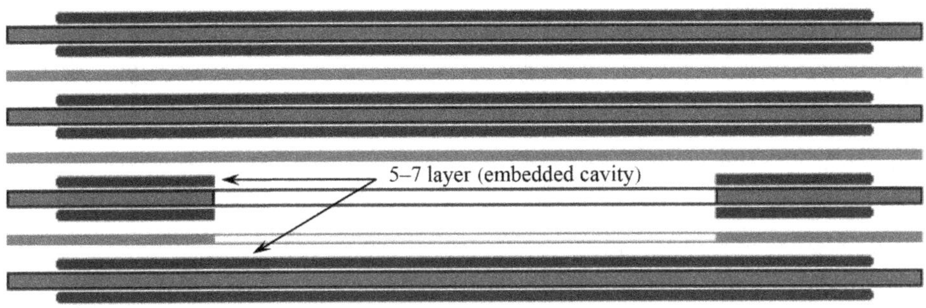

Figure 12.5 The eight-layer stackup with embedded cavity profile.

Then we can draw a multistep cavity; include 1–2 layer cavity, 2–3 layer cavity and 3–4 layer cavity.

First, we draw a new cavity inside the 1–2 layer cavity, edit the property to 2–3 layer, then draw a new cavity inside the 2–3 layer cavity, and edit the property to 3–4 layer. Here, we can also draw a 1–4 layer cavity beside the multistep cavity for comparison. All properties are set up as shown in Figure 12.8.

When cavity drawing is finished, open the 3D view (described in Chapter 18); we can see there are two kinds of cavity, as shown in Figure 12.9.

Cavity structure is popular in ceramic package and SiP applications.

In ceramic package design, complicated chips, especially when multilevel bonding is needed, will be placed inside cavity; large-scale chips such as SOC, DSP and FPGA in ceramic package need multistep cavity support. Currently, in the wire bond process, we put the bond pads in different layer steps of the multistep cavity.

There are two advantages for complicated chip bonding inside multistep cavity: 1) the height of bond wire can be decreased effectively so as to decrease the thickness of package or SiP, 2) the length of the outer ring bond wire in multilevel bonding can be decreased.

12.1.3 Place Component into Cavity

After finishing cavity creation, we can place a component into the cavity.

Figure 12.6 Cavity properties definition.

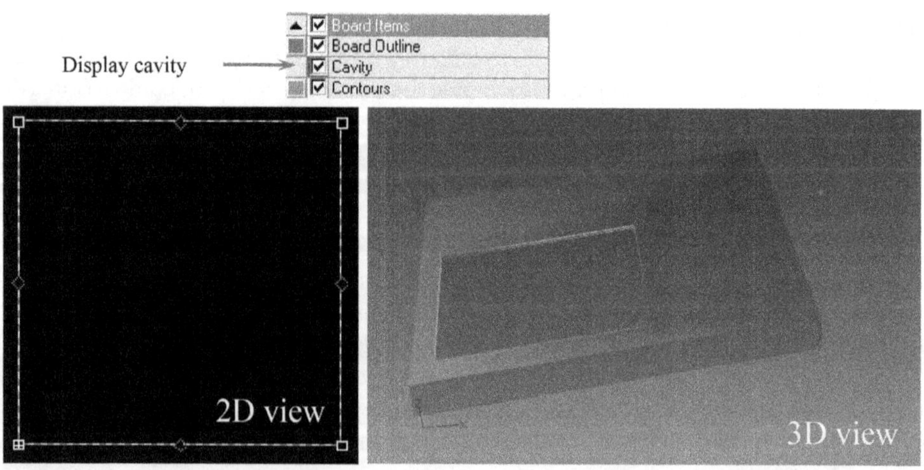

Figure 12.7 1–2 layer cavity structure.

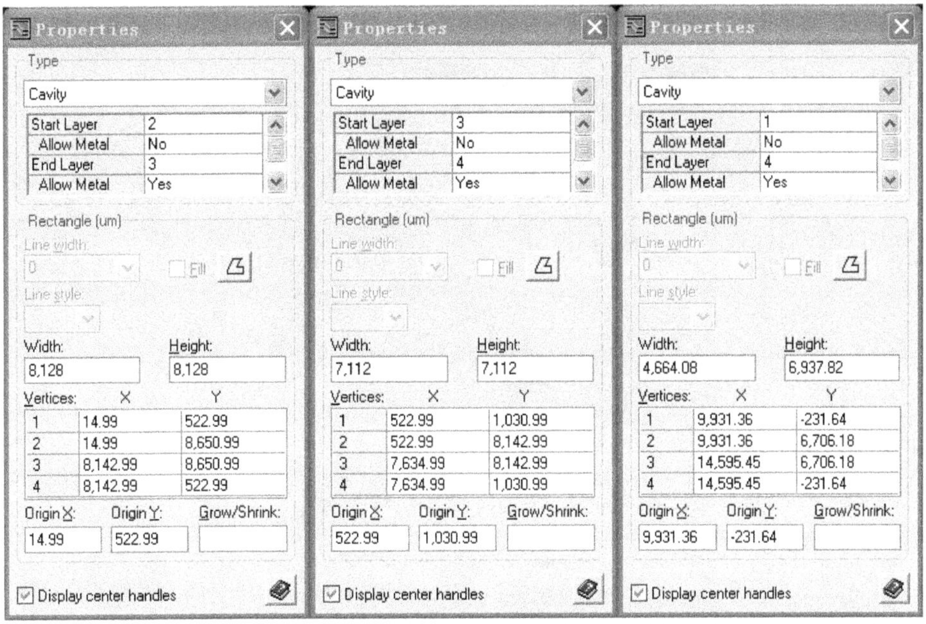

(a) 2–3 layer cavity properties (b) 3–4 layer cavity properties (c) 1–4 layer cavity properties

Figure 12.8 Properties setup for different cavities.

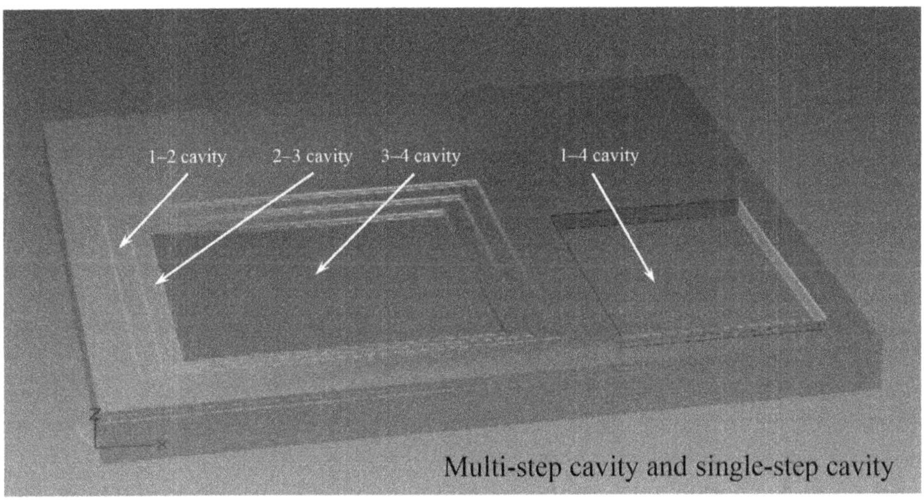

Figure 12.9 Multistep cavity and single-step cavity comparison.

Currently, two steps are needed: 1) move component over the cavity; 2) release the mouse button; if there is no clearance violation, the component will drop down into the cavity automatically, as shown in Figure 12.10.

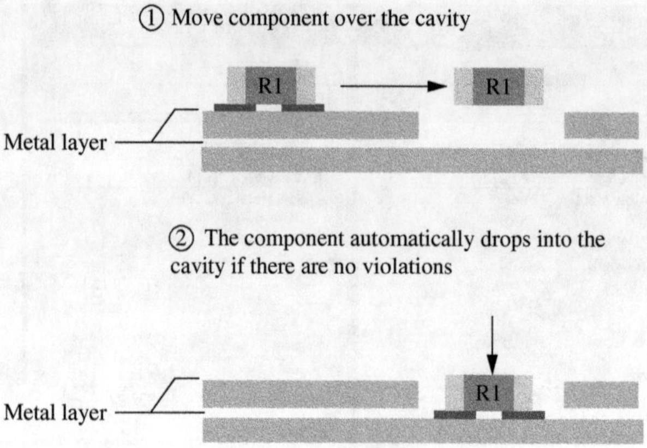

Figure 12.10 Place component into cavity.

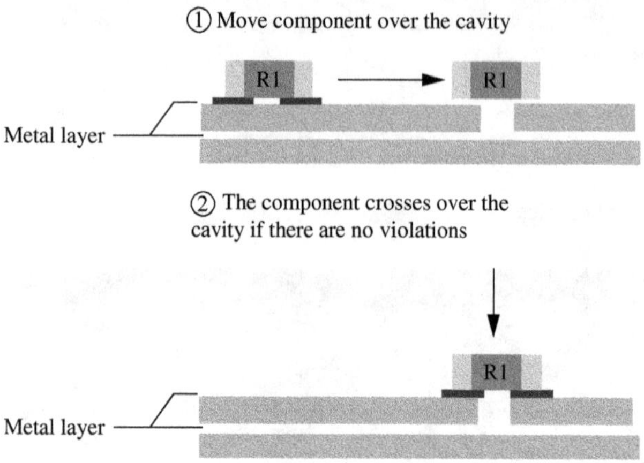

Figure 12.11 Place component crossing over the cavity.

In the Expedition design environment, the most obvious sign of a component drop into cavity is the component pad color change, from the color of the top layer pad to the color of the layer in which the cavity bottom is located.

During the process of placing a component into cavity, if a clearance violation occurs, we cannot place the component into cavity properly; the related design rules definition must be checked.

One exception is when the designer wants to place the component to cross over the cavity; in this situation, the component is bigger than the cavity. When there is no clearance violation, the component will not drop into the cavity; instead it will cross over the cavity, as shown in Figure 12.11.

After the components are placed into cavity, open the 3D viewer; we see a 3D view, as shown in Figure 12.12.

12.1.4 Bonding in Cavity

After chip placement is finished, the next step is bonding.

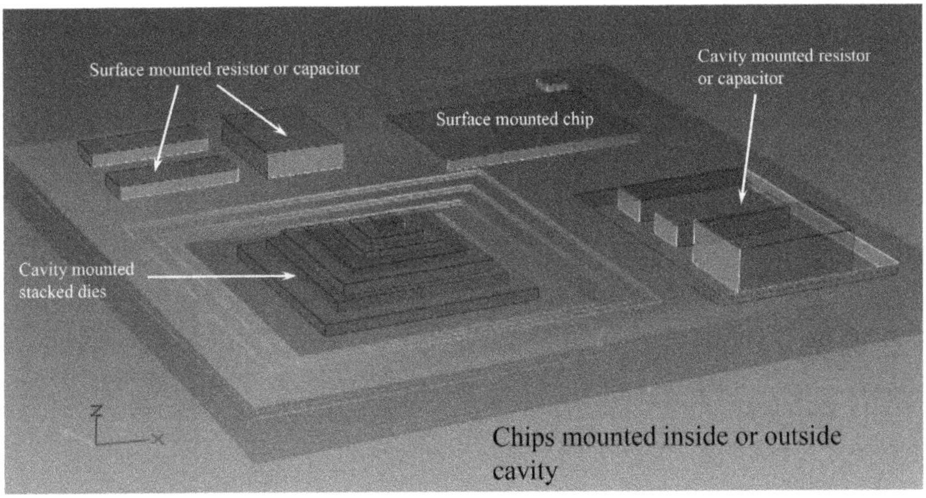

Figure 12.12 Place components into cavity. *(See color plate section for the color representation of this figure.)*

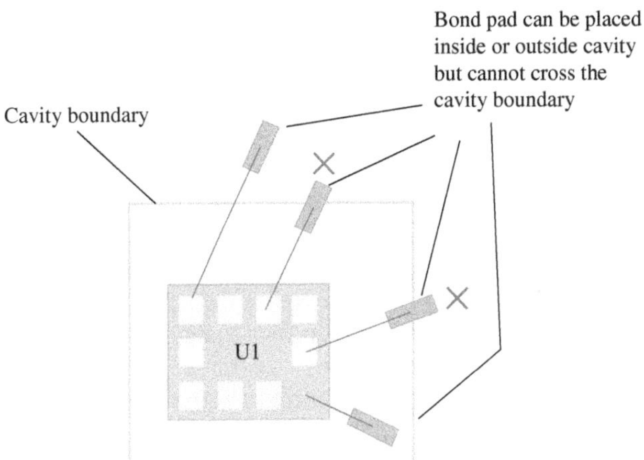

Figure 12.13 Bond pad cannot be placed across the cavity boundary.

First, we need to set design rules for bond wire and cavity, for example, we need to set wire to cavity clearance; please refer to Chapter 11.3.4, Wire bond rules setting. In CES, we can also define related design rules for wire bond and cavity; please refer to Chapter 10.5.2, General clearance rules.

One thing the designer must be careful about is that the bond pad cannot be placed at the edge of the cavity(it is not allowed in the manufacturing process); the bond pad can be placed inside or outside cavity; see Figure 12.13.

For multistep cavity, the bond pad can be placed in the steps of cavity. If there is no violation of the DRC rules, the bond pad can be placed in any step of the multistep cavity.

When the bond pad moves from one step to another, the bond wire profile will adjust accordingly(through the spline curve fitting, consider the height, length and stress effect); we can see this phenomenon in the Expedition environment; see Figure 12.14.

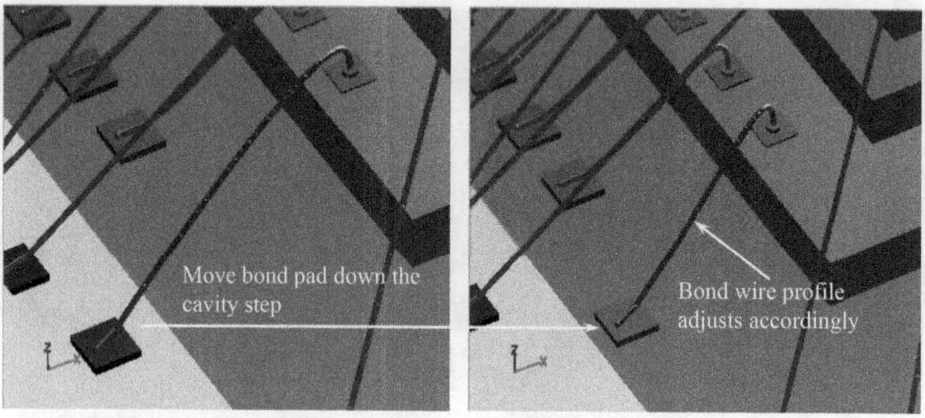

Figure 12.14 Bond wire profile will adjust with bond pad move.

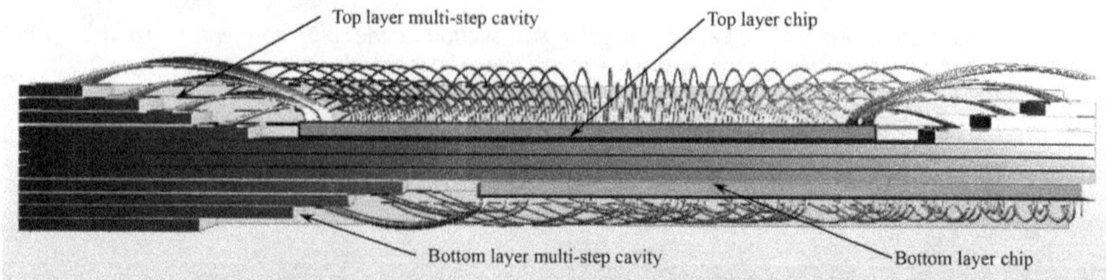

Figure 12.15 Two-sided multistep cavity and chip-bonding profile of a real project.

Expedition supports two-sided multistep cavity; we can create the cavity in the top layer and bottom layer of the substrate, then place the chip into the cavity and bond; Figure 12.15 shows the profile of one real customer project.

From Figure 12.15, we can see that there are multistep cavities in both top and bottom layers; the chip is installed and bonded in the cavity. In the top layer, there are four level bond wires, bonding to the related steps of the multistep cavity; with this method, we can shorten the length of the bond wire effectively, especially for the outer-ring bond wires, so as to decrease the thickness of the package. At the same time, this kind of cavity structure is beneficial for the airproofness of ceramic package or SiP.

12.1.5 Embedded Cavity Design and Embedding Chip into Substrate

With the increase in the complexity of system design and continual enhancement of system functionality, more and more components are integrated within systems to meet various functional demands.

At the same time, because the demand for system miniaturization is becoming increasingly important, the surface of substrate available for mounting chips is becoming smaller and smaller. More and more chips are required to be mounted in a limited surface, which is a contradiction.

To solve this problem, apart from reducing the volume of the component, another solution is to embed components into the substrate to save the limited surface space.

There are two methods for embedding components; one is discrete embedding and the other is planar embedding.

Discrete embedding is where discrete components are placed directly into the substrate. Because the component volume influences the substrate thickness and area, we can only embed small chips or passive devices into the substrate. The advantage of discrete components is that they have high accuracy and support larger resistor, capacitor and inductor values.

Planar embedding is using printing technics or press-in dielectric thin film in substrate for creating embedded resistors or capacitors.

Planar embedding has almost no influence on the thickness of the substrate, but it also has limitations. For example, because of limitations of the dielectric constant and the area of the capacitor material, the planar capacitor value is generally smaller than the discrete capacitor value.

For planar resistors, the power will be restricted because of the limitation of embedded resistor volume.

At the same time, the degree of accuracy of planar components is not as good as their discrete counterparts; sometimes we need to use lasers to adjust and improve the precision accuracy of planar components.

Expedition supports both embedded methods; here we discuss only discrete embedding technics; see Chapter 15 in relation to planar embedding technics and embedded resistor and capacitor design.

Next, we learn how to embed discrete chips into the substrate in Expedition step by step.

In the example below, we plan to embed one chip into layer 4–6 of the substrate.

First, we need to be clear which layer is ready for embedding the component; we also need to consider the height of the embedded component and the thickness of each layer of the substrate so as to calculate the internal space for the embedded component.

Second, we draw an open-type cavity in the substrate (see Figure 12.16), and we draw a 1–6 layer cavity; the size of the cavity is based on the size of the chip; we also consider

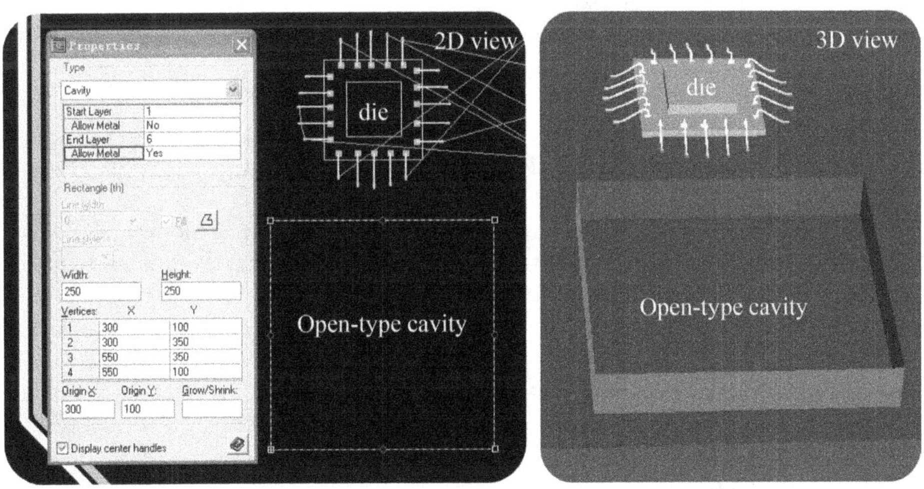

Figure 12.16 Draw an open-type cavity in the substrate.

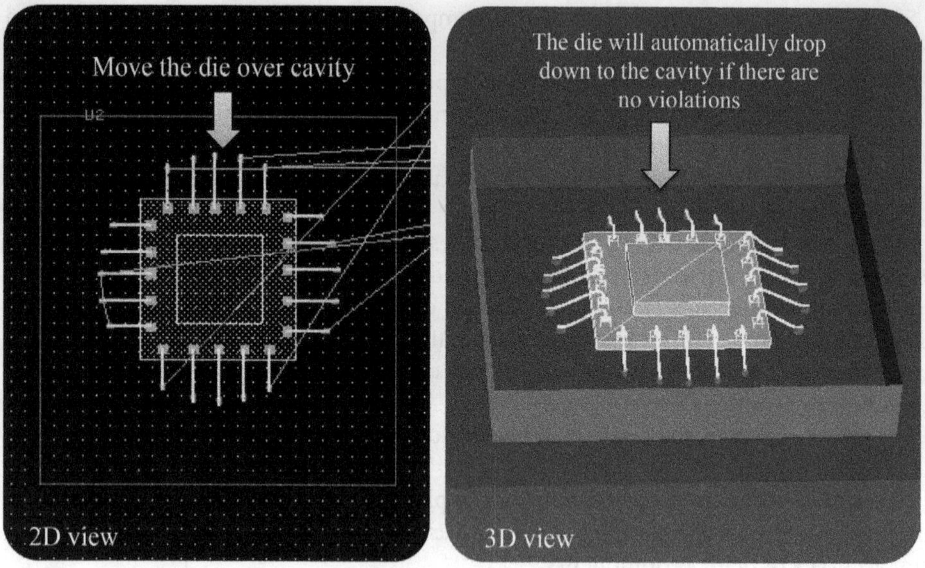

Figure 12.17 Place the chip into the open cavity.

the bond wire, the bond pad size and the clearance of cavity to metal. Thus, after finishing the cavity, the chip will drop into the cavity automatically when we move it over the cavity.

For a more real effect, select from the menu Window→3D Viewer to open the 3D viewer window. On the right side of Figure 12.16 we can see the sideward 3D view of the cavity and chip.

Third, toggle to place mode, move the chip over the cavity; if there are no violations of the DRC rules, the chip will drop down into the cavity automatically; see Figure 12.17.

In the Expedition environment, by default the finished bond wire and bond pad will move together with the chip and keep the relative position. If the designer wants to move the chip only and to preserve the bond pad location, he or she needs to select the chip, then click the right mouse button and choose "Move chip only" in the pop-up menu.

The chip will move and the bond pad will keep its location; the bond wire will adjust automatically for the correct connection, but of course, the prerequisite is that there is no DRC violation.

Fourth, after the chip drops down into the cavity, toggle to draw mode again, select the cavity and open the cavity properties window, change the start layer of the cavity to 4, giving an embedded 4–6 layer cavity, as shown in Figure 12.18.

Fifth, open the 3D Viewer again; we can see that the cavity has closed, it crosses layers 4–6, the chip is located in the inner layer of the substrate, the bottom of the chip is located in layer 6, the bond pad is also located in layer 6, the trace connect to bond pad will also start from layer 6. See Figure 12.19. In Figure 12.19(a) display is set to translucent, in (b)display is set to opaque.

Sixth, after the chip is embedded in the substrate, we can route around the chip. Apart from layers 4–6, which were occupied by the cavity, we can route in layer 1, layer 2 and

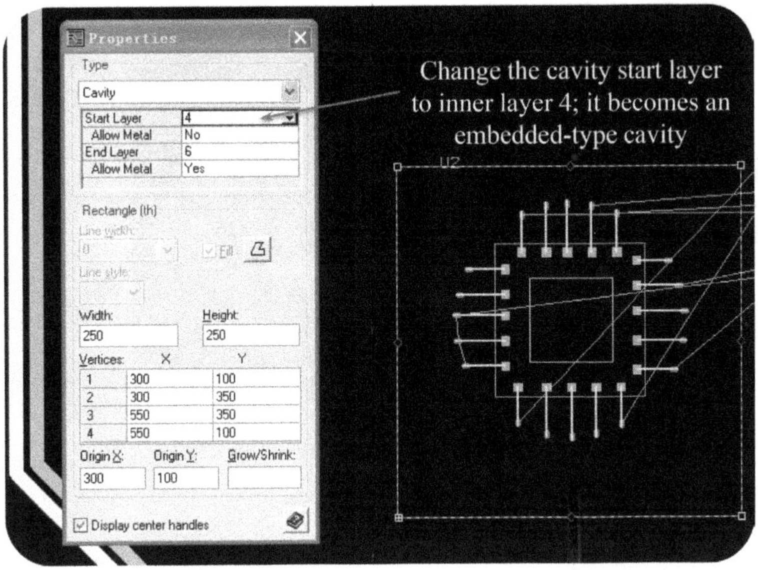

Figure 12.18 Change the start layer of the cavity to layer 4.

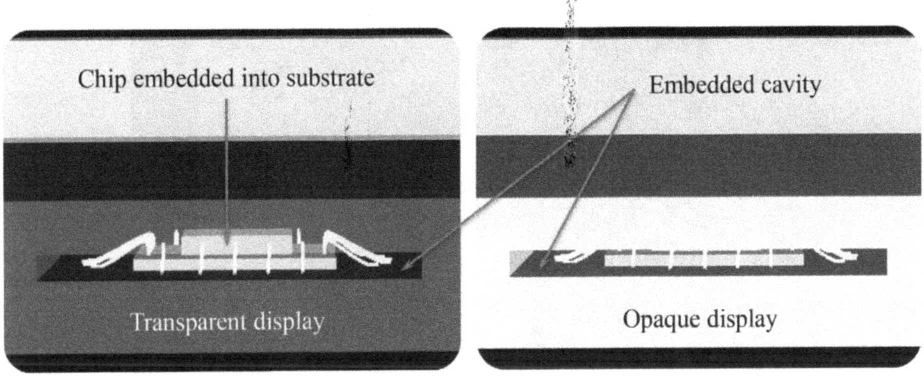

Figure 12.19 Chip was embedded in the inner layer of the substrate.

layer 3 over the chip. In layer 6, we can route from the bond pad, in other layers we can route as for ordinary substrate. Figure 12.20 shows the position of the chip, cavity and trace in each layer of the substrate.

Now we are finished embedding the discrete chip into the substrate in the Expedition environment.

12.2 Chip Stack

12.2.1 The Concept of Chip Stack

In SiP design, for the sake of saving space in reducing the substrate area, we often use chip stack (sometimes called die stack or part stack). Chip stack is where many chips are put in one stack. Figure 12.21 shows a real photograph of a chip stack.

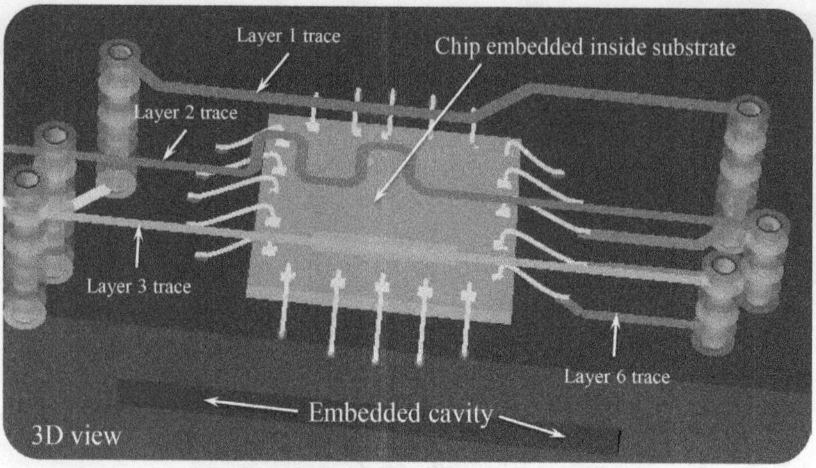

Figure 12.20 The position of the embedded chip and trace in each layer.

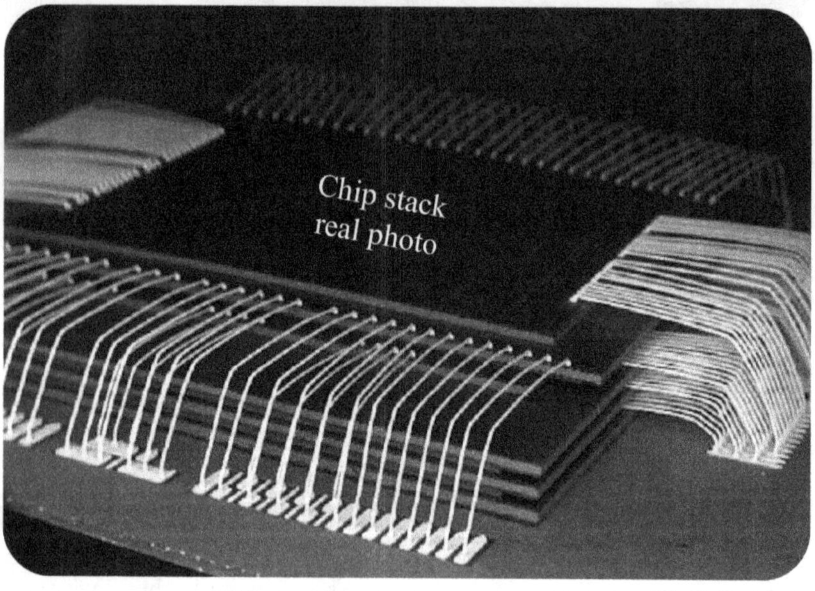

Figure 12.21 Real photo of chip stack.

In chip stack, we can insert an interposer for electric isolation or raise the upper-level chip to avoid conflict with the lower-level bond wire.

Figure 12.22 displays a 3D view of chips, interposer and bond wires in a chip stack.

There are two usual kinds of chip stack: chip stack pyramid and chip stack cantilever; the interposer for these two kinds of chip stack is different; for chip stack pyramid, only a thin interposer or sometimes no interposer is needed; for chip stack cantilever, a relatively thick interposer is needed to raise the upper-level chip so as to avoid conflict with the lower-level bond wire. See Figure 12.23.

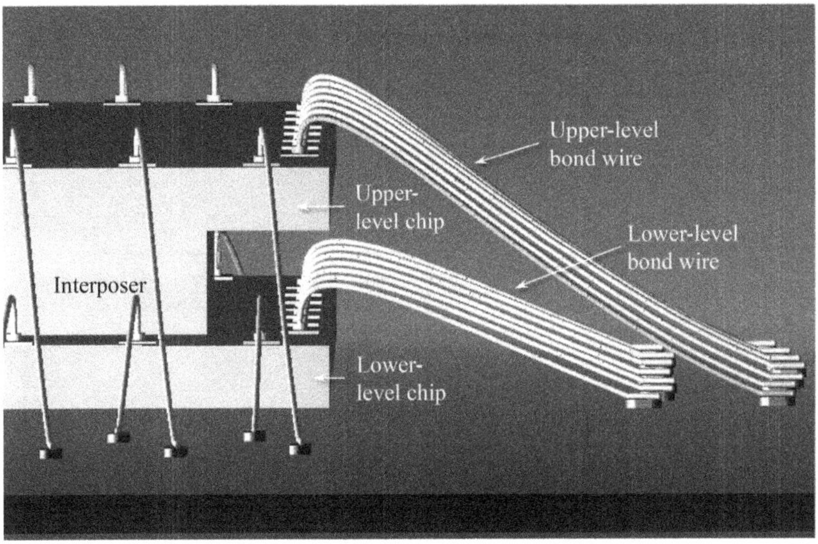

Figure 12.22 Chips, interposer and bond wires in chip stack (Expedition 3D Viewer).

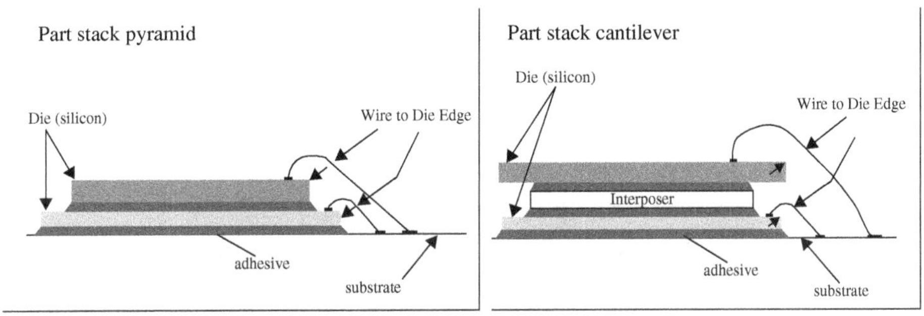

Figure 12.23 Chip stack pyramid and chip stack cantilever.

12.2.2 Chip Stack Creation

In chip stack, an interposer is sometimes needed, so we should generate the interposer before chip stack creation. Select menu: Setup→Cell Editor→Mechanical to create the interposer. The important parameters for the interposer are height and size; the height of the interposer is set in cell properties; see Figure 12.24.

The size of the interposer must be drawn in cell editor; click the Editor Graphics button , enter the cell editor workspace, draw a placement outline of the interposer according to the dimension requirements, then save and exit cell editor.

After interposer creation, click the Part Stack Configuration button in the wire bond tool bar, as shown in Figure 12.25.

First, create a new chip stack, such as Newstack2, in the Criterion bar select Parts with Die Pins; we can see all the bare dies in the list below; select one of them, click the single downward arrow , this die is placed in the chip stack; we can also use the double downward arrow to add all the dies in the list to the chip stack.

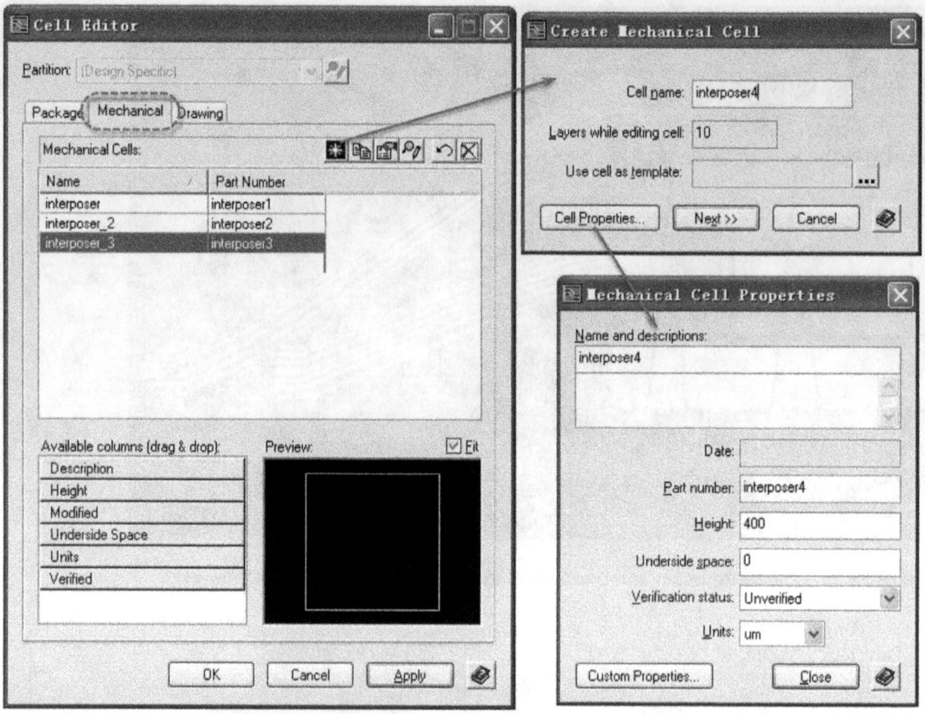

Figure 12.24 Interposer creation windows.

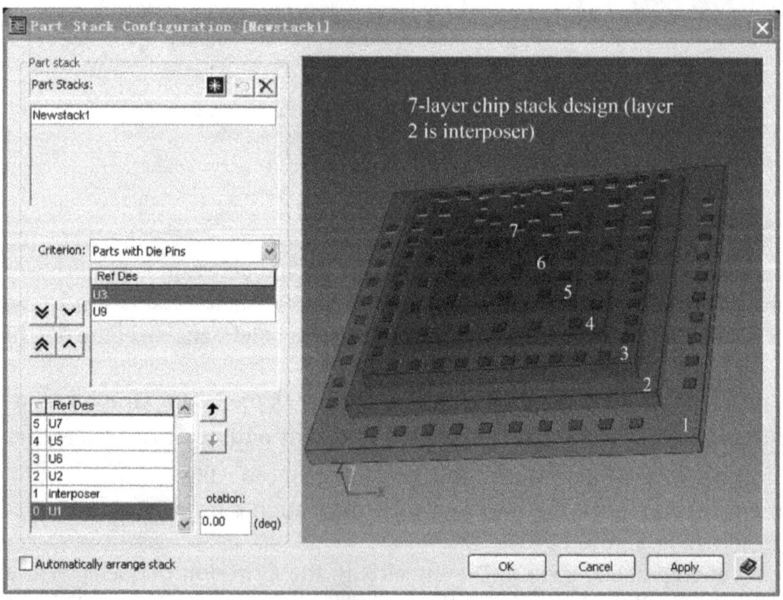

Figure 12.25 Chip stack configuration window. *(See color plate section for the color representation of this figure.)*

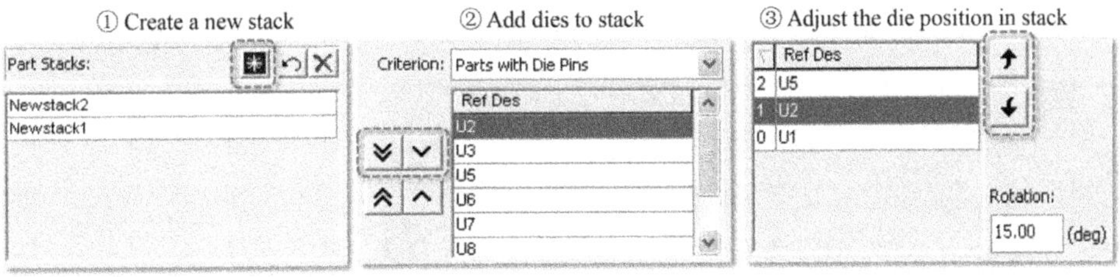

Figure 12.26 Chip stack operations.

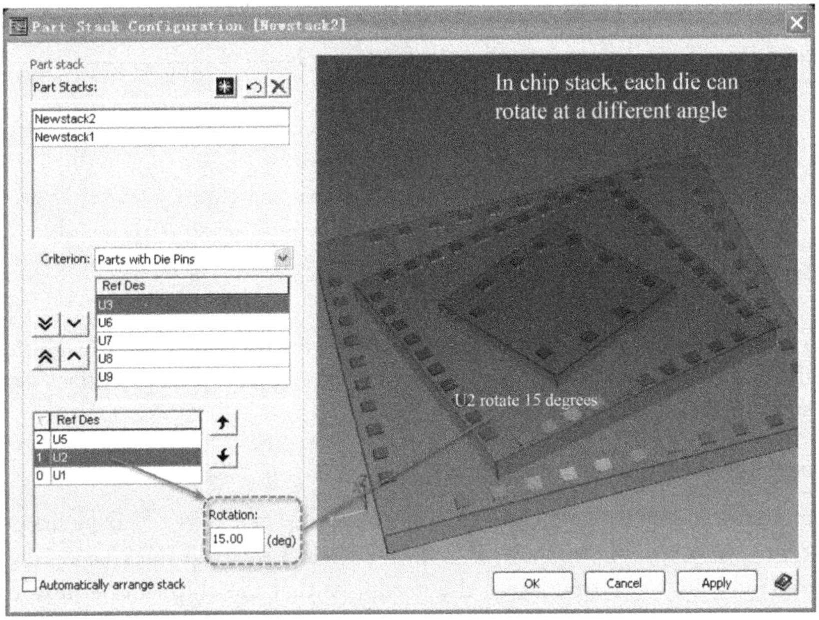

Figure 12.27 Chips in stack with different rotation angles.

In the right-most bar, 0, 1, 2 represent the layers of chip in stack, typically 0 means the lowest layer, 1 indicates the layer above layer 0, etc. The chip position can be adjusted with the up arrow or down arrow, as shown in Figure 12.26.

Rotation indicates the rotation angle of the selected chip in the stack; in the same chip stack, different components can have different rotation angles, as shown in Figure 12.27.

12.2.3 Stack Chips Side by Side

In the design of chip stack, two or more smaller chips can be placed side by side on top of a larger chip, that is, multiple chips are stacked on the same plane; this is known as stacking chips side by side, as shown in Figure 12.28.

To stack chips side by side, simply use the upward arrow or downward arrow ; this adjusts chips stacked in the same layer. In Figure 12.29, U5 and U7 are both on layer 2, naturally they are side by side.

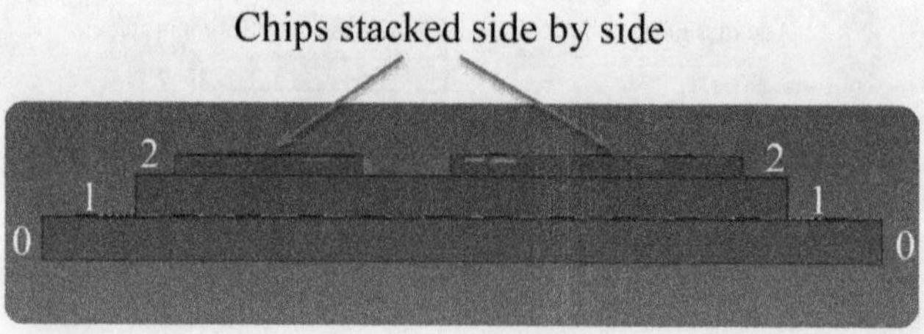

Figure 12.28 Stacking chips side by side.

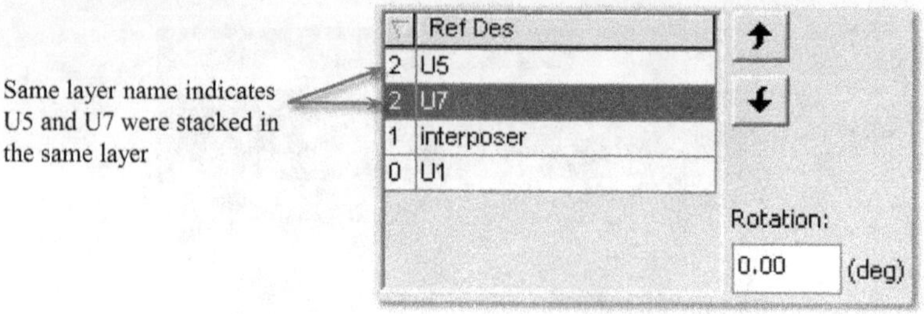

Figure 12.29 Set chips stacked side by side in the same layer.

When two or more chips are stacked in the same layer, the bond wire start height is the same; if the design tool cannot accurately describe this situation, this will affect the production and processing of SiP.

12.2.4 Adjust Relative Position of Chips in Stack

If we need to adjust the relative position of chips in a chip stack, in the Expedition environment, first switch to place mode, and then select the chip stack, using the ungroup command button to dissolve the group, so that the chip can be moved separately, as shown in Figure 12.30. Otherwise, the chip stack will be moved as a whole group.

After the chip position adjustment is complete, in place mode reselect all chips in the chip stack, then use the group command button again to make group them. After chip stack setup is complete, as shown in Figure 12.31, the stack can be placed or moved like a normal component.

12.2.5 Chip Stack Bonding

Chip stack bonding is almost the same as bonding for a single chip; we need to consider bond wire height and span. In the Wire Bond/Power Ring Generator window, select the chip stack, such as Newstack1, and set the appropriate bonding rules; as shown in Figure 12.32, autobonding can be carried out. Because the 3D chip stack is more complex than a single chip, so chip stack autobonding currently takes longer than single chip bonding.

If we are not satisfied with the results of automatic bonding, or in cases where the 3D space DRC relations are too complex and the software cannot bond automatically, the designer must carry out manual bonding.

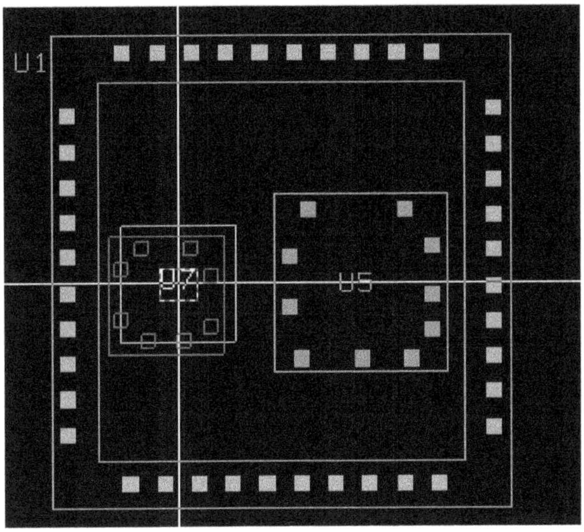

Figure 12.30 Adjust relative position of chips in stack.

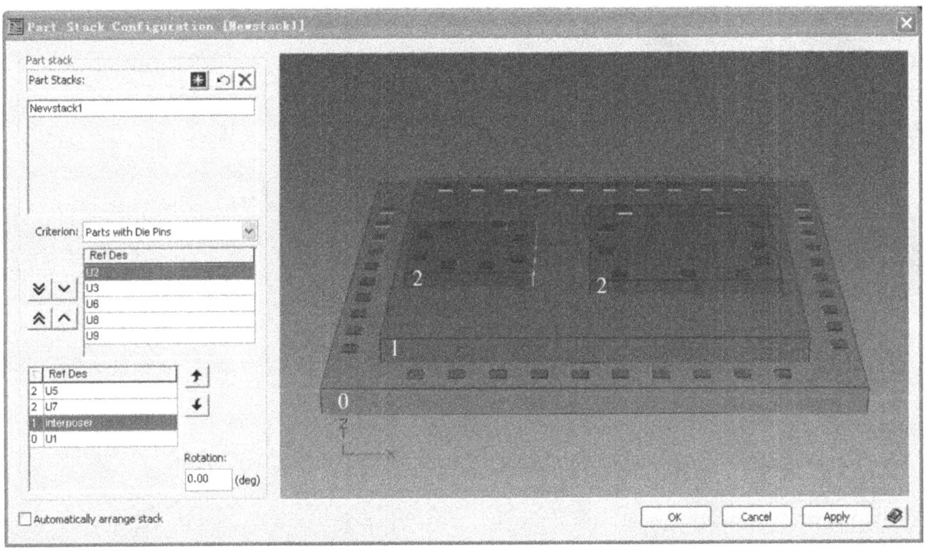

Figure 12.31 Chip stack setup complete.

Bonding by hand is flexible for the designer; it can be based on chip-specific factors such as the location or height of the chip and allows adjustment of the height, span and profile of bond wire to better meet the needs of production and processing. The effect of manual chip stack bonding is shown in Figure 12.33.

Chip stacking has effectively reduced the area of the SiP substrate, but increased the height of package or SiP, and sometimes the top-level chip bond wire is too long.

In order to effectively reduce the thickness of SiP or package and reduce the upper-level chip bond wire length, it is usual to place chip stacks into a cavity. In most cases,

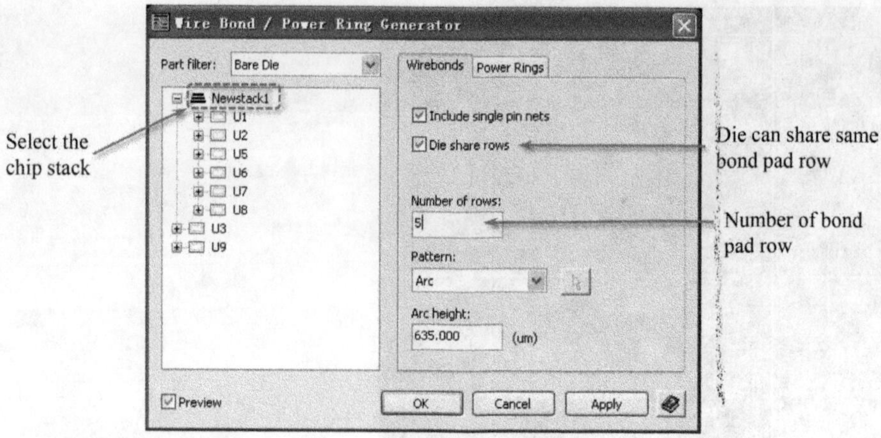

Figure 12.32 Chip stack automatic bonding setup.

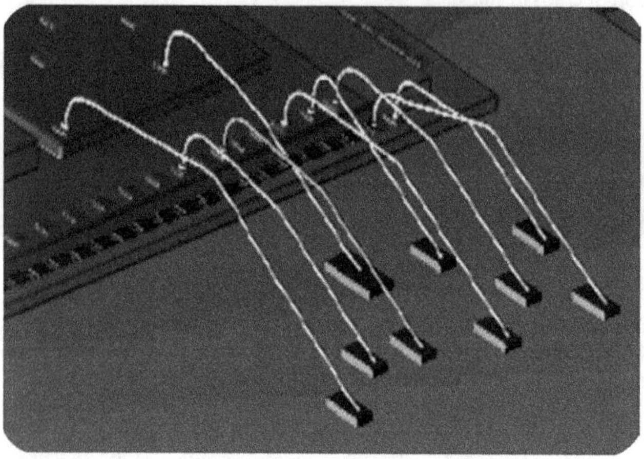

Figure 12.33 The effect of manual bonding of chip stack.

multistep cavity is used to place chip stacks, with the bond pad placed in a different step of the cavity, as shown in Figure 12.34.

Figure 12.35 shows a real customer multistep-cavity design project in Expedition. Designed with four-step cavity, the chip was placed in the cavity; the chip had two rows of die pins; the outside die pins were bonded to steps 3 and 4 of the cavity, the inside die pins were bonded to steps 1 and 2 of the cavity, thus effectively reducing the height and span of bond wire.

Figure 12.36 shows a customer example of mixed-technology design in Expedition, containing multistep cavity, chip stack (side by side), die-to-die connection (containing two kinds of connection: ball to wedge and wedge to wedge), and RF circuit (see Chapter 16 RF Circuit Design). This type of design is common in LTCC microwave projects. Mixed-technology design is well supported in the Expedition environment.

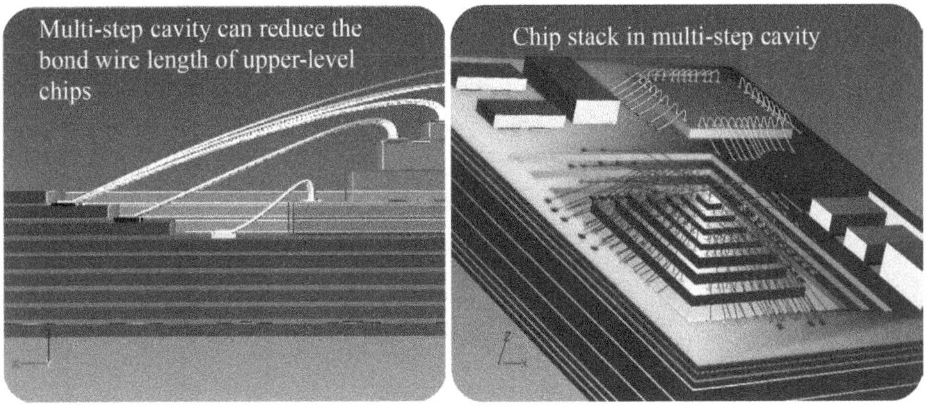

Figure 12.34 Chip stack is placed in the cavity for bonding.

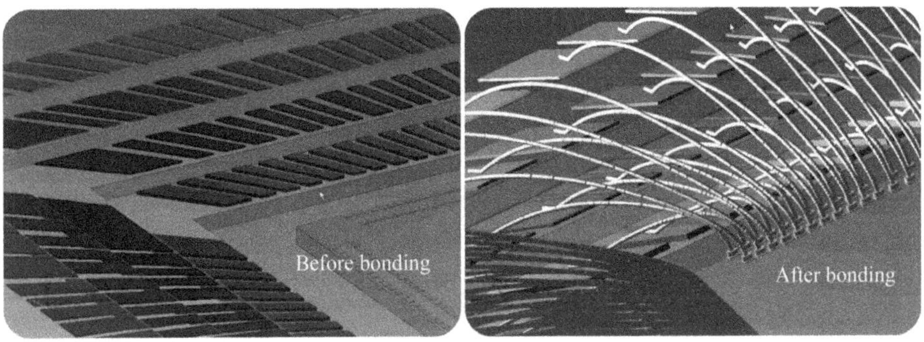

Figure 12.35 A real project with multistep cavity bonding. *(See color plate section for the color representation of this figure.)*

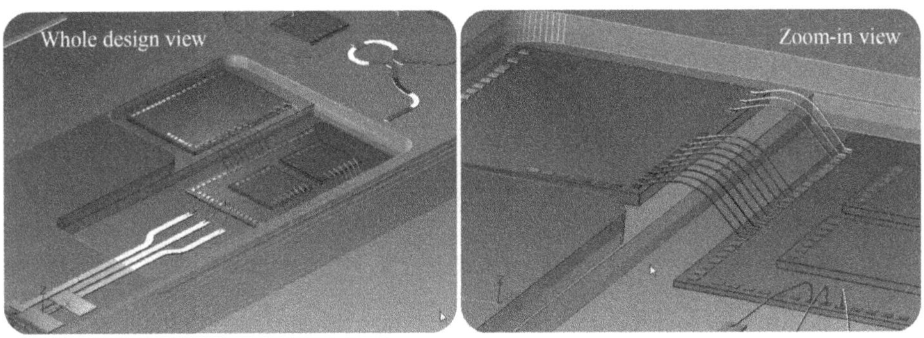

Figure 12.36 A real project with mixed design technologies. *(See color plate section for the color representation of this figure.)*

13

Flip Chip and RDL Design

13.1 The Concept and Characteristics of Flip Chip

Flip chip, abbreviated FC and also called invert welding, is a new type of micro-assembly technology. In recent years, flip chip has become a frequently used packaging format in the field of high-end devices, high-density package and SiP.

Today, flip chip is becoming a widely used encapsulation technology. As the requirements of such technology increase, a series of difficult new challenges are presented to design and production, and designers need to face these challenges and solve the related problems, providing reliable support for design, assembly, testing and the whole process involved in this complex technology.

In traditional packaging technology, the active chip face is upward, with the back stuck to the substrate, and wire bonding or tab technology is then used for electrical interconnection. With flip chip the active side faces down towards the substrate, and a solder bump array is used to realize the chip and substrate interconnection, with the chip or die directly connecting to the package or SiP substrate.

On the one hand, flip chip greatly shortens the length of the signal interconnection, reduces the delay, and effectively improves the performance, which is important for high-speed design. On the other hand, because of the use of planar array connection, flip chip interconnection can provide higher I/O density. At the same time, flip chip occupies a very small area, almost consistent with the chip size. Compared with wire bond and tab, flip chip can achieve the smallest, thinnest encapsulation.

Flip chip has become mainstream for high-I/O chip packaging technology for high-performance CPU, GPU, FPGA and Chipset. Because flip chip I/O terminals are distributed throughout the chip surface, there are advantages for packing density and signal speed. Flip chip can be used with SMT-like processing, which is the direction in which SiP and high-density package are developing.

Flip chip has distinct advantages: 1) flip chip pin length is short, with minimal parasitic parameters; 2) flip chip pins are connected by array, which greatly improves performance and reduces the packaging area; 3) flip chip can support more pins to meet the growing demand for high I/O.

Figure 13.1 shows the flip chip diagram. We can see from the diagram that flip chip connects die to package substrate using bumps, with bottom under fill glue used for reinforcement. Compared with bond wire connection, flip chip has no dangling wires,

SiP System-in-Package Design and Simulation: Mentor EE Flow Advanced Design Guide,
First Edition. Suny Li (Li Yang).

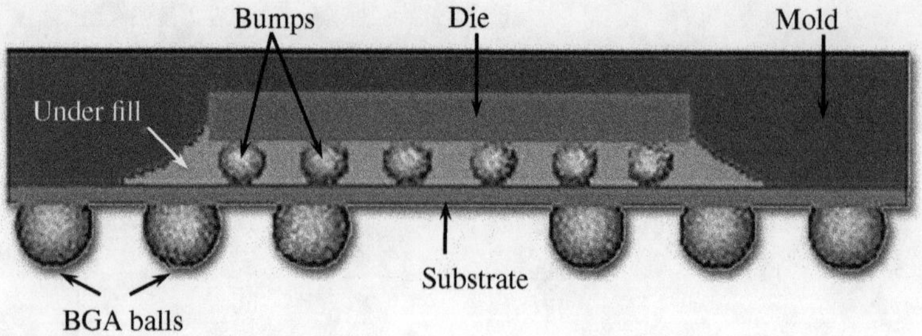

Figure 13.1 Flip chip diagram.

the connection is shorter, the parasitic effect is small, and it can support higher-frequency signals.

Of course, flip chip also has its limitations: 1) if flip chip needs to generate bumps on wafer, the process is relatively complex; 2) if the chip is not specially designed for flip chip, it is necessary to design and process an RDL layer; 3) flip chip is more easily affected by temperature change, and more consideration must be given to the chip and SiP substrate CTE match (Coefficient of Thermal Expansion), and there are also higher requirements on thermal analysis.

To sum up, the advantages of flip chip are increased opportunities, decreased package volume, and higher speed. Its shortcomings are its significant challenges, the need for more complex processing, and the increased demands for material CTE matching and thermal analysis. If all these problems are solved, the application of flip chip will greatly increase.

13.2 The RDL Concept

RDL, an abbreviation for Redistribution Layer, involves making a layer on the active chip side, for chip pin redistribution.

With RDL, chip pins can be rearranged to any reasonable position on the chip. Using RDL technology, the die pads located in the chip periphery to support traditional wire bond technology can be reassigned to the "redistribution pins" of the chip anywhere on the surface; these pins are hereinafter referred to as RDL pins, as shown in Figure 13.2.

Through such redistribution, the location of the pin changes, and the designer can be more flexible about chip placement in SiP design.

Usually, RDL uses single-layer aluminum, single-layer copper or multilayer copper. For most RDL design, we can choose single layer, as shown in Figure 13.3.

Another advantage of RDL is the cost control in relation to the IC; the chip lifecycle can be extended through RDL technology. Instead of expensive IC chip re-design, RDL usually minimizes the cost of IC overhead.

13.3 RDL Design

From the perspective of SiP or package design, if the IC chip is developed specifically for the flip chip, we don't need to design and process an RDL layer. In this case, the SiP

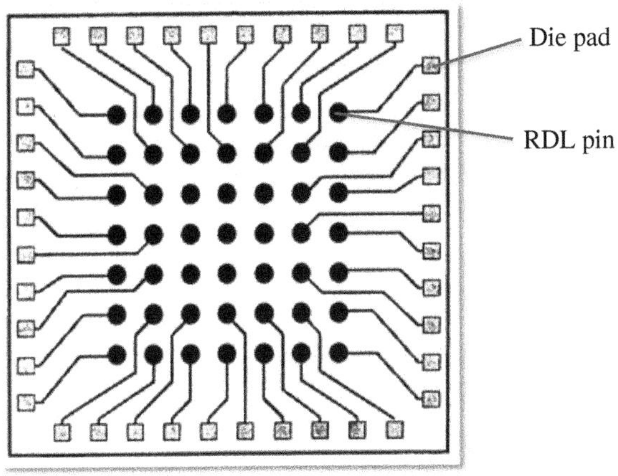

Figure 13.2 RDL diagram (chip-top view).

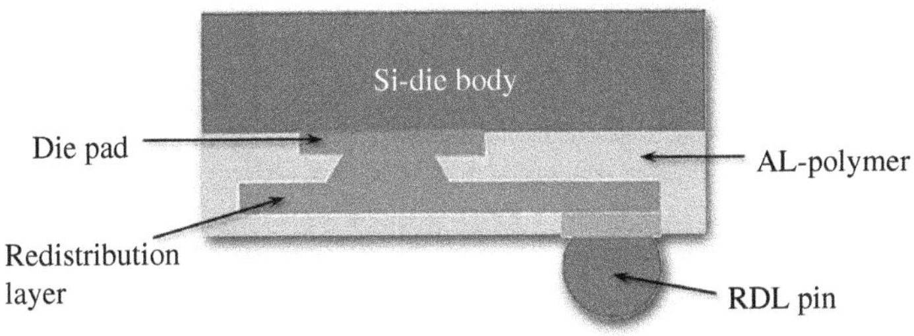

Figure 13.3 RDL diagram (side view).

designer only needs to design the flip chip substrate. If the IC chip is not specifically developed for flip chip technics, but only supports the wire bonding process, the SiP or package designer needs to design and process an RDL layer.

Using a practical design, we introduce the design method for RDL and flip chip in the Expedition environment.

13.3.1 Create Bare Die and RDL Library

First, in the Expedition environment, a bare die cell, RDL cell and BGA package cell must all be created.

Looking at the package structure, the bare die cell and RDL cell are placed on the top of the substrate surface, the BGA cell is placed on the bottom side of the substrate, and the signal path flows from die pad→RDL layer trace→RDL pin to package substrate trace→BGA pad, as shown in Figure 13.4.

The 3D view of the bare die cell, RDL cell and BGA cell locations on the package substrate are shown in Figure 13.5.

After the cells have been created and the positions determined, we need to define the RDL padstack shape and size; it is based on the number of die pads, the bare die chip area, the process ability of flip chip, etc.

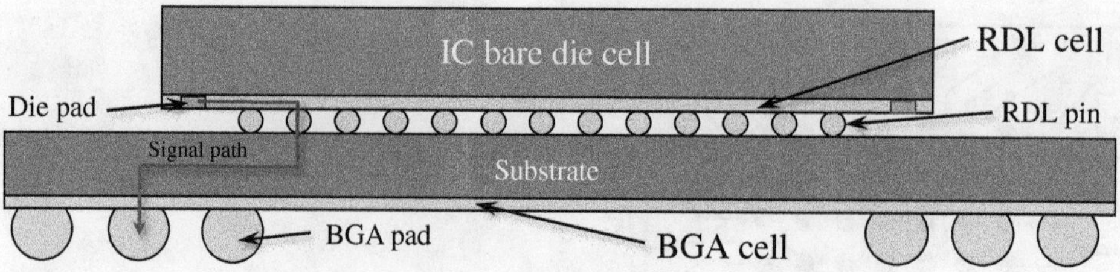

Figure 13.4 Bare die cell, RDL cell and BGA cell positions on substrate (side view).

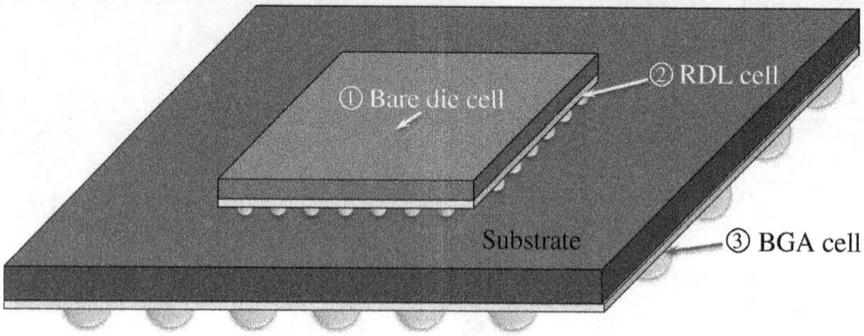

Figure 13.5 Bare die cell, RDL cell and BGA cell locations on substrate (3D view).

In this case, we select a 150 μm square pad named DIE150um_FC as the RDL pad, and for pin type select pin-SMD.

See Figure 13.6 for RDL padstack creation and parameter setup.

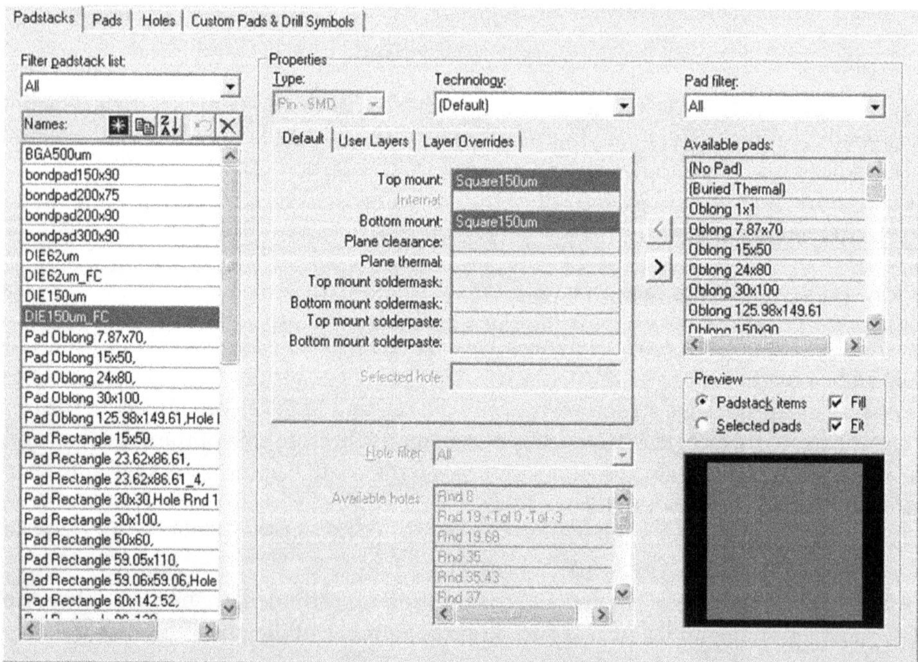

Figure 13.6 RDL padstack creation and parameters setup window.

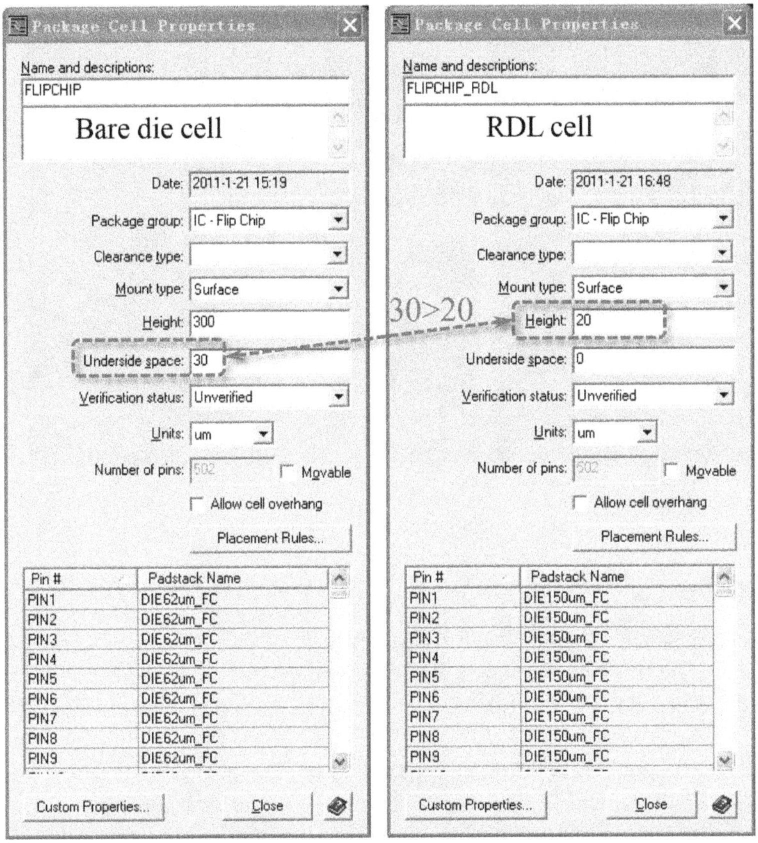

Figure 13.7 Bare die cell and RDL cell properties setup.

For the IC die itself, a pin-SMD-type padstack must also be created. In this example, we create DIE62um_FC for the IC die padstack.

After the RDL padstack and die padstack have been created, the next step is to create the bare die cell and RDL cell, as shown in Figure 13.7. Here, note that the bare die underside space value should be bigger than the RDL cell height value. so that in component placement, the bare die cell and RDL cell can be placed in the same position, so as to avoid DRC conflict.

After completion of properties setup, the die cell can be created according to the data provided by the IC die maker, using the method previously described, and using Die Wizard to import the data provided by the die manufacturer. See the detailed description in Chapter 6. The RDL cell-creation method is almost the same as for BGA, and again reference can be made to Chapter 6. In this example, the bare die cell and RDL cell created are shown in Figure 13.8.

After the cells have been created we need to create a schematic symbol; if the bare die cell and RDL cell have exactly the same pin number, we can use the same symbol; for the symbol creation method please also refer to Chapter 6.

After symbol creation, we need to create an IC_DIE part and RDL part to map the corresponding symbol and cell, and these parts can then be used in RDL design.

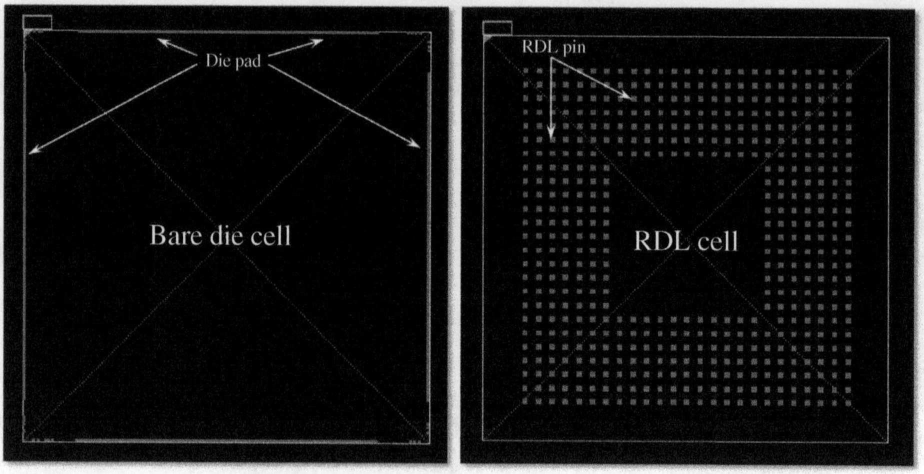

Figure 13.8 The bare die cell and RDL cell.

13.3.2 RDL Schematic Design

RDL schematic design is very simple; we only need to place the RDL part and IC_DIE part in the schematic. In the process of placing them, please make sure to check Add Nets and Add Net Names; in this way, the same net will be automatically connected.

Because in this case the pin numbers of RDL part and IC_Die part are the same, we use the same symbol for them, and the pin names of these two parts are identical.

In schematic design, we only need to select the Add Nets and Add Net Names options when placing the RDL part and IC_DIE part in the schematic; the net name generated by the pin name is completely consistent, and the net with same name will connect to them automatically, without the need for any manual connection.

The RDL schematic created is shown in Figure 13.9.

After completion of RDL schematic design, review the design and run packager; after packaging, we can enter the RDL layout design environment.

13.3.3 RDL Layout Design

In the layout design environment, the first thing we should do is set the RDL layer stackup. Because the RDL design only needs one layer in this case, the layer stackup can be set as shown in Figure 13.10.

The thickness and the parameters of each layer can be adjusted according to the actual conditions; the setting here does not have an impact on production and processing, but will affect the transmission line impedance calculation, and will also affect the result of subsequent software simulation.

Then, carry on with device layout. The bare die cell and RDL cell are both placed on the top layer in the cell-creation process; we have set the bare die cell underside space value (30 μm) greater than the RDL cell height value (20 μm), so we can place them in the same position, and even if the two cells overlap, there will be no conflict, as shown in Figure 13.11.

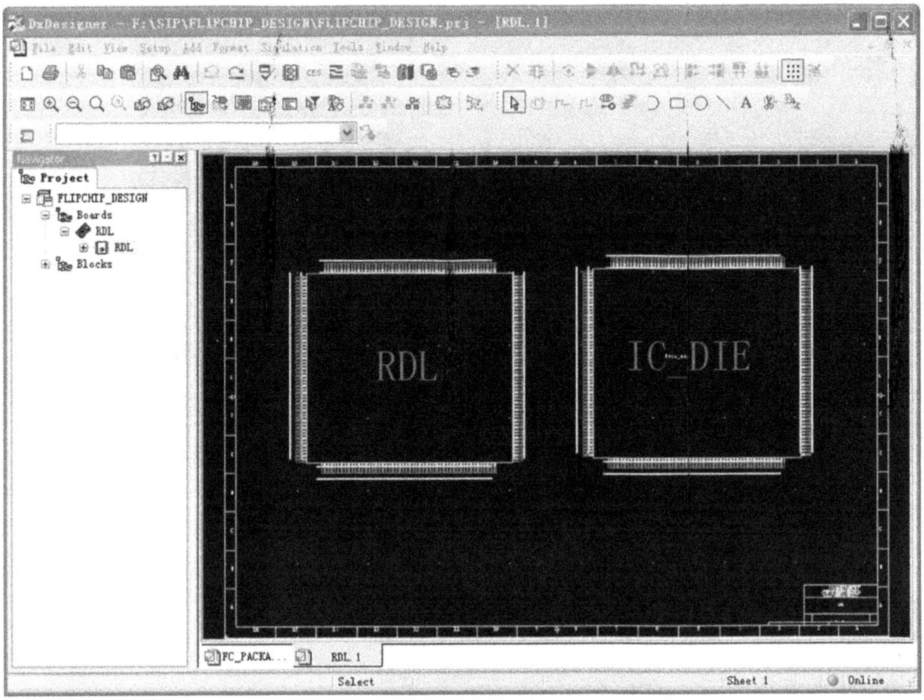

Figure 13.9 RDL schematic design.

General | Via Definitions | Via Clearances | Layer Stackup | Buried Resistors & Rise Time

Layer Type	Thickness (um)	Resistivity (OhmMeters)	Dielectric Constant	Description
Dielectric	76.2		3	polyimide
Signal	17.78	1.72E-008	1	AL_Redistribution
Dielectric	76.2		3	polyimide

Figure 13.10 RDL layer stackup setting.

After cell placement, open the net line, because in schematic design we did not consider the position of the pins in layout, and so the net connections were crossed, as shown in Figure 13.12; it is now necessary to optimize the net connections.

For net optimization in Expedition, first, select from the menu Setup→Part Editor→Pin Mapping, and set all the pins of RDL part swappable.

The specific method is as follows: first, select the pins to be swapped; we can select multiple adjacent pins by holding down the < Shift > key, or hold down the < Ctrl > key to select non-adjacent pins. Then click the swap button ; swappable pins in the same group have the same shape shown in the left column, as shown in Figure 13.13.

In this case, all the pins of the RDL cell are set to the same swappable group. Start Place→Automatic→Swap by Part Number; the Automatic Swap by Part Number interface appears as shown in Figure 13.14. Select the part for which optimization is needed, and use the arrow to the right to send the part to the Included bar on the right-hand

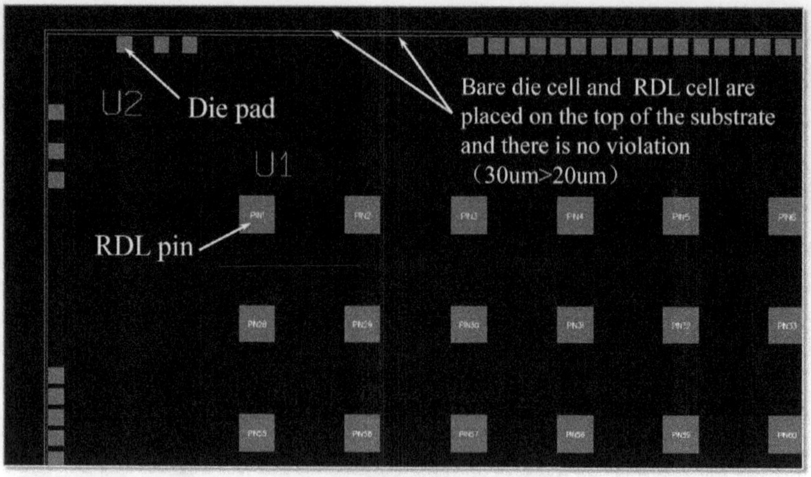

Figure 13.11 Bare die cell and RDL cell placement.

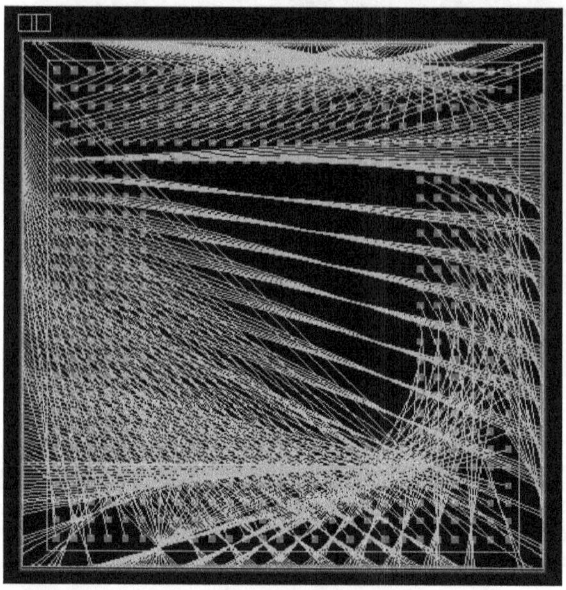

Figure 13.12 Net connections before auto-optimization.

side; in Swap items, select Pins, and then click the Apply button; the software starts automatic pin optimization.

Depending on the size and pin numbers of the chip, the time taken for automatic optimization will be different. Usually automatic optimization will be completed within minutes.

After completion of net optimization, we can view the percentage of route space savings in the Savings bar and Total Savings bar, as shown in Figure 13.14.

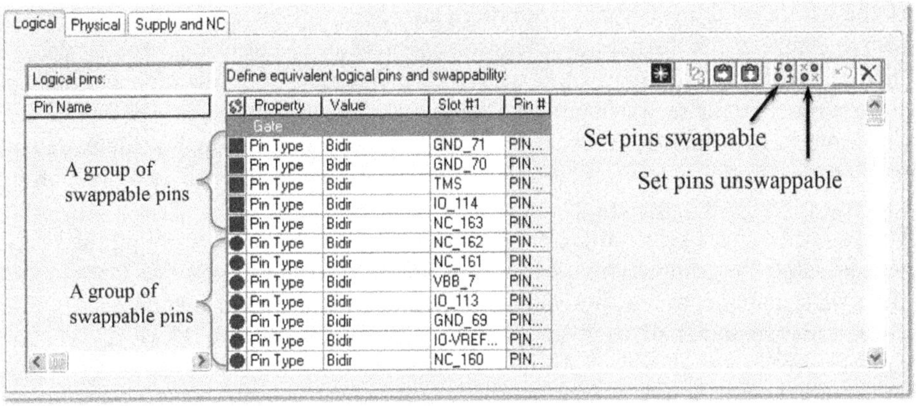

Figure 13.13 Set part pins swappable.

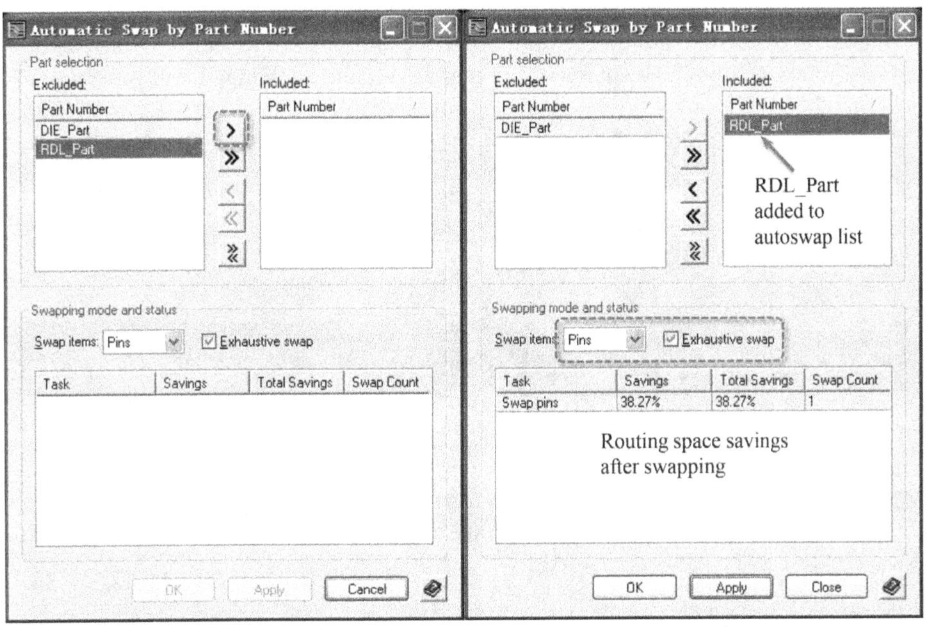

Figure 13.14 Automatic net optimization interface.

Normally, after automatic optimization, the net connections become smoother, but it is difficult to achieve perfection, and the designer may need to do manual optimization according to the actual design situation.

In route mode, click the Swap Pins button to manually swap the pins, and handle according to the prompt information located in the lower-left corner of the status bar.

First, the status bar tip shows `Select first pin to Swap.` ; select the pin of the RDL cell it is necessary to swap; at this time, all swappable pins will be highlighted, and the prompt information in the status bar changes to `Select second pin to Swap.` ; select a suitable pin; the prompt information changes to `Confirm Swap.` ; this time we only need to click the left mouse button at any position in layout, and pin swapping is complete. As shown in Figure 13.15, the net lines change from crossing to parallel.

Figure 13.16 shows a comparison of the results of automatic optimization and manual optimization; it can be seen that after the completion of automatic optimization, the net connections have greatly improved compared to the initial situation, but there are still some nets crossing. Using both automatic optimization and manual optimization by pin swap, optimal net connection can be achieved.

The essence of net optimization, either automatic or manual, is swapping the swappable pins; after the completion of net optimization, the result must be back annotated to the RDL schematic, to ensure synchronization of schematic and layout.

There are two kinds of back annotation, as shown in Figure 13.17. 1) Via the Setup→Project Integration Back Annotation button. If the project needs back annotation, this button's color will turn to orange, and the words behind the button will read "Back Annotation Required"; click this button for back annotation. 2) From the menu select ECO→Back Annotate. After the completion of back annotation, a prompt window will confirm the success of the back annotation.

Comparing the schematic before and after back annotation, we can see that the pin-mapping relationships have changed, pin names remain unchanged and pin numbers

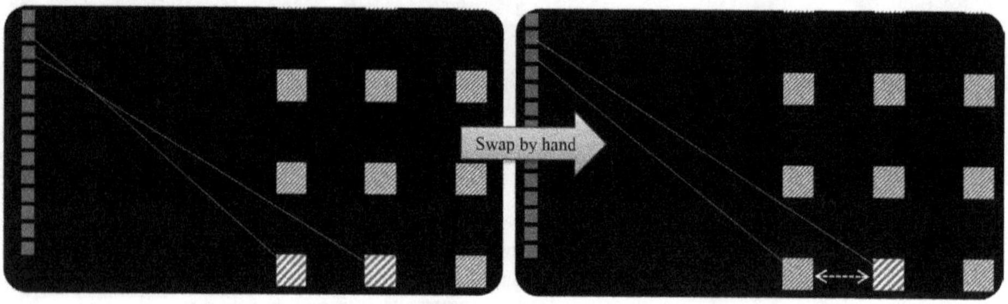

Figure 13.15 Manually swap pins of RDL cell to optimize the net connection.

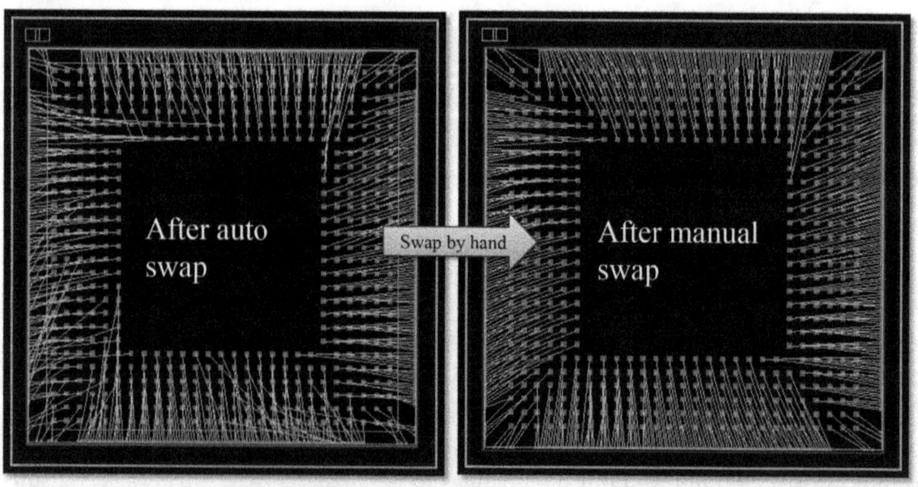

Figure 13.16 Automatically optimized net connection and its manual counterpart.

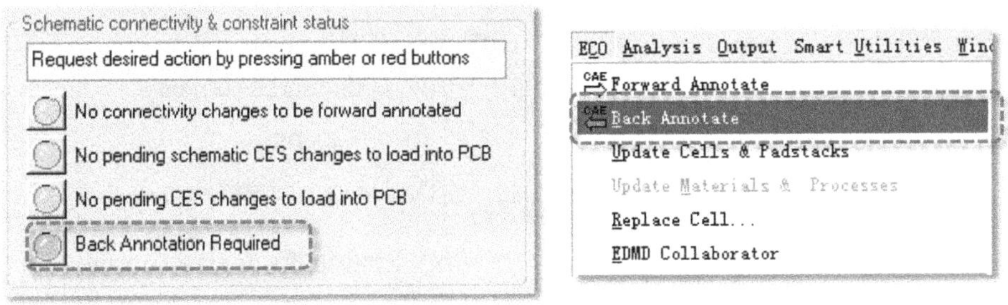

Figure 13.17 Two methods for back annotation.

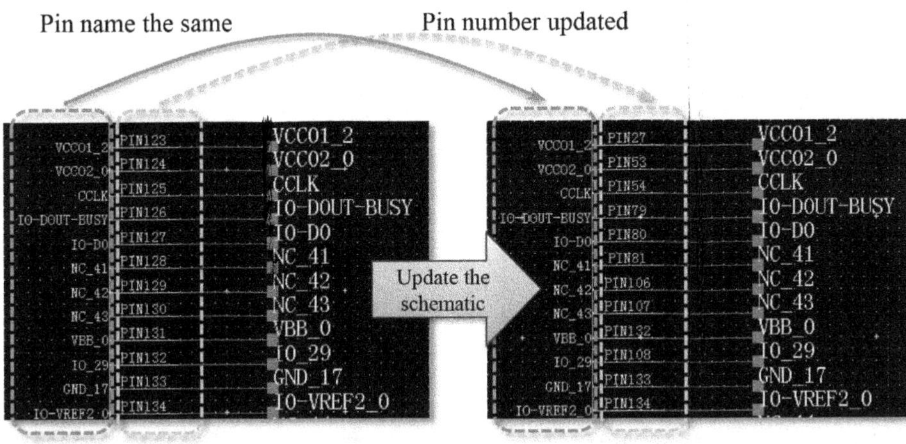

Figure 13.18 The schematic updated after back annotation.

have been updated, so as to achieve a reasonable net allocation in layout. This rationality is unpredictable in the schematic design phase, so it is only achievable through physical connection optimization in layout, followed by back annotation to the schematic. The comparison of the pin-mapping relationship before and after back annotation is shown in Figure 13.18.

After the completion of net optimization, enter CES for route rule setting, mainly RDL layer line width and clearance, etc. The method is referred to in Chapter 10's description of constraint rule management. After design rule setting, we can start routing automatically or manually; usually the effect of manual routing is a better and more reasonable result.

Either in manual or automatic routing, the 45° route function can be enabled or disabled.

In Setup→Editor Control→Plow→General Options, toggle the Allow 45 degree corners option to enable or disable the 45° route function, as shown in Figure 13.19.

After setup, route automatically or manually; the results of the two different route functions are shown in Figure 13.20.

After routing, DRC check is needed. For a detailed description of DRC check please see Chapter 19, Design Review. After DRC check, we can output the file for RDL

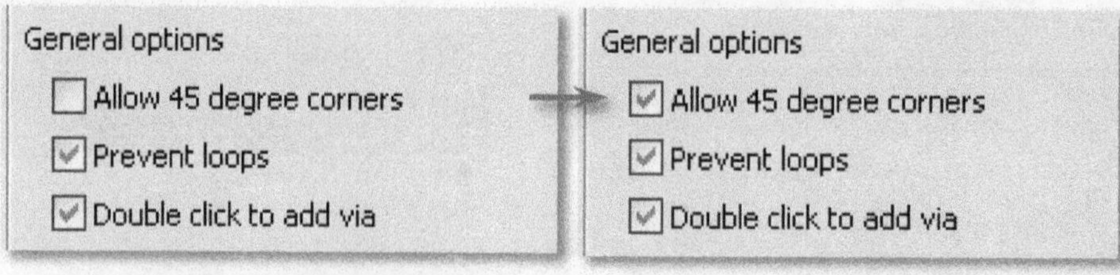

(a) Disable 45-degree routing (b) Enable 45-degree routing

Figure 13.19 Disable or enable the 45° route function.

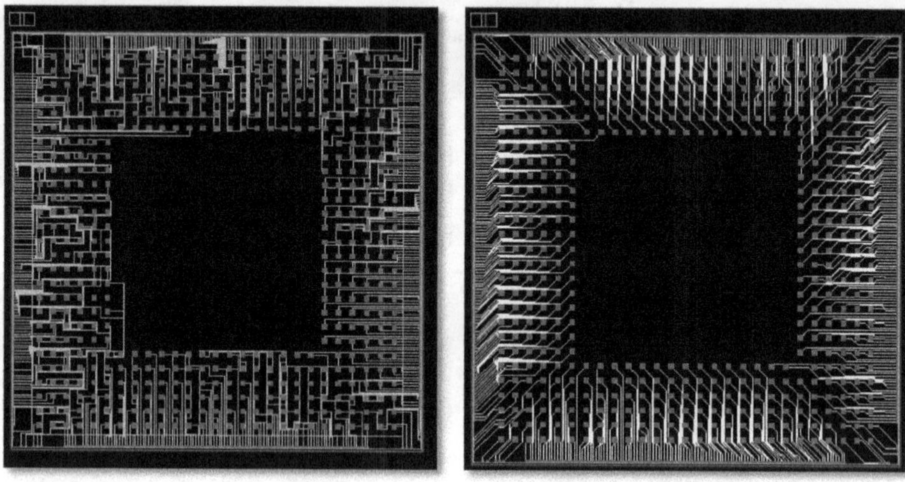

(a) Routing result with 45-degree routing disabled (b) Routing result with 45-degree routing enabled

Figure 13.20 The results of the two types of routing.

production processing. The production data supported by Expedition includes GDSII, DXF, Gerber, ODB++, Neutral File, etc.

13.4 Flip Chip Design

After the completion of RDL design, we can start flip chip design.

In RDL design, bare die pins have been connected to RDL cells; in flip chip design, we need to connect RDL cell pins to BGA cell pins.

It is important to note that if the IC chip manufacturers have defined the IC pins for flip chip array, we don't need to design and process an RDL layer. If this is the case, Chapter 13.3, RDL design, can be omitted, and flip chip design can start directly from this section.

In this book, we discuss the complete design process, from IC die→RDL→FlipChip.

13.4.1 Flip Chip Schematic Design

In the same project as was used for RDL design, select from the menu File→New board, and then right-click to rename the board to FC_PACKAGE; also rename the schematic of this board to FC_PACKAGE. Then, directly copy the RDL symbol along with its net connection from the RDL schematic to the FC_PACKAGE schematic; to make it more intuitive, the symbol can be marked "FC".

It is important to note that here we copy the RDL symbol to the FC_PACKAGE schematic, rather than calling the RDL part and symbol directly from the central library, mainly because the RDL symbol must inherit the pin-mapping relationship that has been previously optimized in RDL schematic design.

Then, call the corresponding BGA Package Part & Symbol from the central library and place in the schematic; here it is also necessary to check the Add Net and Add Net Name items. If the net name is the same as the net name of the FC symbol, the nets will connect automatically. If there are new nets on the BGA package symbol, such as redefined power and ground pins, etc., we need to manually add nets and net names.

As shown in Figure 13.21, in the FLIPCHIP_DESIGN project there are two designs: RDL and FC_PACKAGE. The corresponding schematics and pages are shown below the design. If we have any spare design in this project, it can be placed in block, and when needed we can copy its contents to the schematic, or, using the right-click command Create Board, transform the block to a new board.

Please note, as for RDL design, the net connections of FC_PACKAGE design must also be optimized in Expedition layout and back annotated to the schematic.

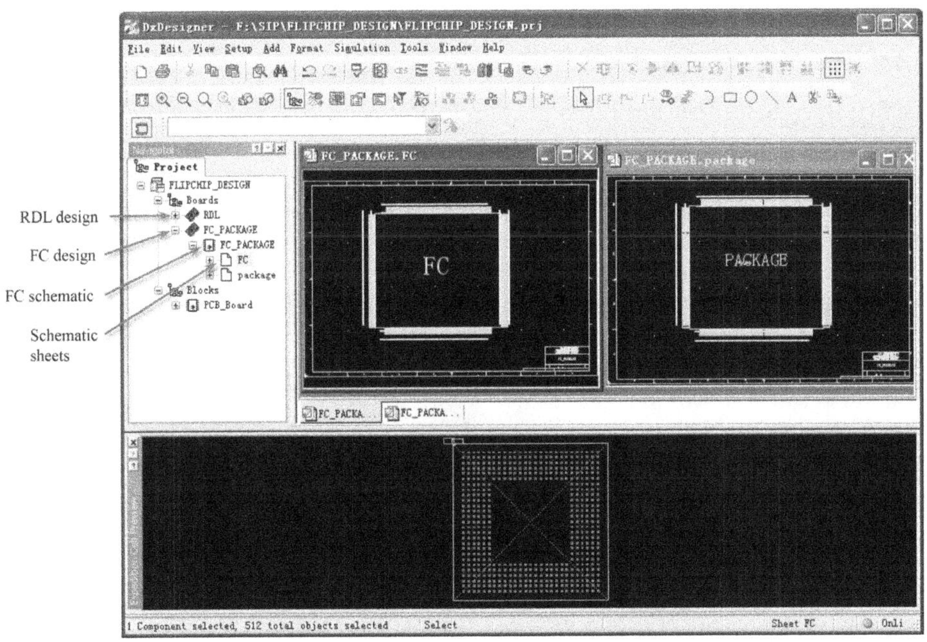

Figure 13.21 FC_PACKAGE schematic design.

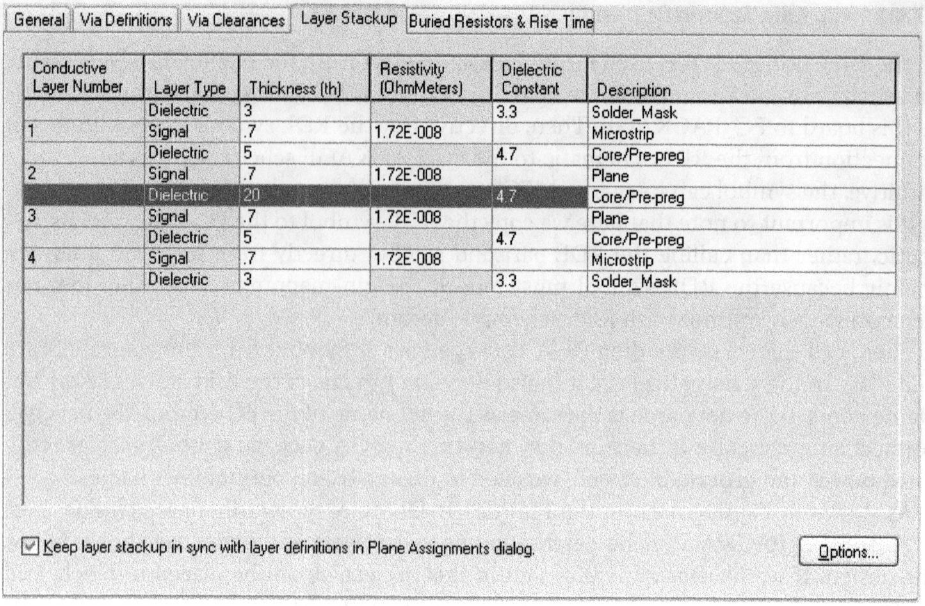

Conductive Layer Number	Layer Type	Thickness (th)	Resistivity (OhmMeters)	Dielectric Constant	Description
	Dielectric	3		3.3	Solder_Mask
1	Signal	.7	1.72E-008		Microstrip
	Dielectric	5		4.7	Core/Pre-preg
2	Signal	.7	1.72E-008		Plane
	Dielectric	20		4.7	Core/Pre-preg
3	Signal	.7	1.72E-008		Plane
	Dielectric	5		4.7	Core/Pre-preg
4	Signal	.7	1.72E-008		Microstrip
	Dielectric	3		3.3	Solder_Mask

Figure 13.22 FC_PACKAGE substrate layer stackup setting.

After the completion of schematic design and review, package the schematic, then go into the Expedition layout environment.

13.4.2 Flip Chip Layout Design

The first thing to be done after entering Expedition is the layer stackup and via settings; set BGA substrate to four layers, with layer stackup 1+2+1; the layer stackup settings are shown in Figure 13.22.

FC_PACKAGE via settings are shown in Figure 13.23, with the middle layer for the laminate process and the surface layer for the buildup process.

After layer stackup and via setup, adjust the board size to the BGA cell size, and change the board origin to the center of the substrate, then place the parts. Because there are only two parts, place the FC at the top of the substrate, and the BGA at the bottom side, then set the origin of both cells to (0, 0). After the completion of placement, the result is shown in Figure 13.24.

After placement, click the button CES in the Expedition toolbar, enter the CES constraint editing system, and set up the related design rules. These include physical rules such as line width and clearance and electrical rules such as delay and differential, as shown in Figure 13.25. For the detailed setting method, please see Chapter 10 on constraint rules management.

After design rule setup, we can start routing.

Before routing, we need to carry out net optimization; the method is the same as for RDL layout design, described above.

First, select from the menu Setup→Part Editor→Pin Mapping and set all the BGA pins to swappable. At this time the flip chip pins need to be set to unswappable.

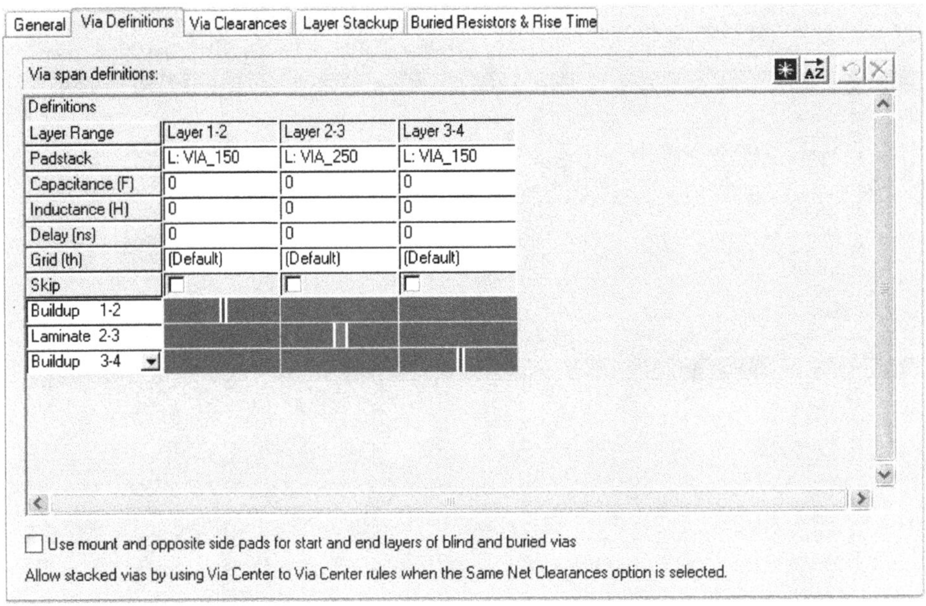

Figure 13.23 FC_PACKAGE via settings.

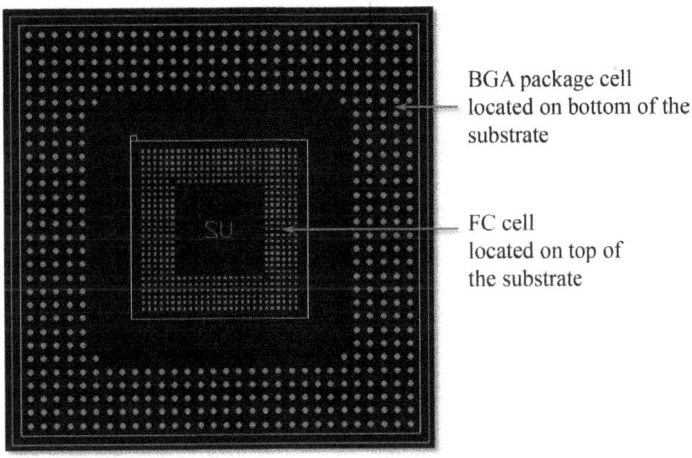

Figure 13.24 FC_PACKAGE placement.

Then, select Place→Automatic→Swap by Part Number, move the BGA part to the right side of the window for optimization, set Swap items to pins, and then click the Apply button; this starts automatic pin optimization.

Depending on the number of pins and the computer performance, the time required for net auto-optimization will differ. After the completion of auto-optimization, the net connections will be smoother, but as it is difficult to achieve perfection, the result may need manual optimization according to the actual design situation.

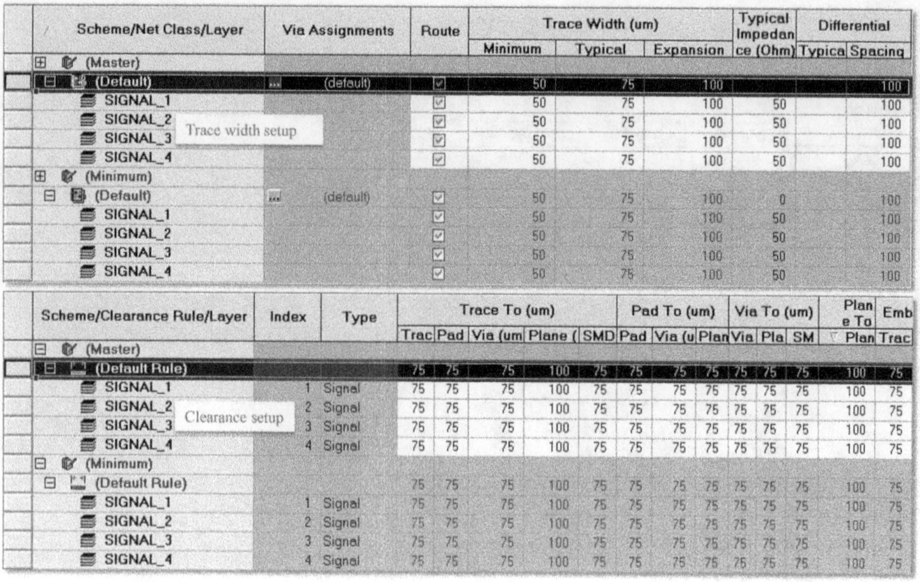

Scheme/Net Class/Layer	Via Assignments	Route	Trace Width (um)			Typical Impedance (Ohm)	Differential	
			Minimum	Typical	Expansion		Typica	Spacing
(Master)								
(Default)	(default)	✓	50	75	100			100
SIGNAL_1		✓	50	75	100	50		100
SIGNAL_2		✓	50	75	100	50		100
SIGNAL_3		✓	50	75	100	50		100
SIGNAL_4		✓	50	75	100	50		100
(Minimum)								
(Default)	(default)	✓	50	75	100	0		100
SIGNAL_1		✓	50	75	100	50		100
SIGNAL_2		✓	50	75	100	50		100
SIGNAL_3		✓	50	75	100	50		100
SIGNAL_4		✓	50	75	100	50		100

Scheme/Clearance Rule/Layer	Index	Type	Trace To (um)					Pad To (um)			Via To (um)			Plane To	Emb
			Trac	Pad	Via (um	Plane (	SMD	Pad	Via (u	Plan	Via	Pla	SM	Plan	Trac
(Master)															
(Default Rule)			75	75	75	100	75	75	75	75	75	75	75	100	75
SIGNAL_1	1	Signal	75	75	75	100	75	75	75	75	75	75	75	100	75
SIGNAL_2	2	Signal	75	75	75	100	75	75	75	75	75	75	75	100	75
SIGNAL_3	3	Signal	75	75	75	100	75	75	75	75	75	75	75	100	75
SIGNAL_4	4	Signal	75	75	75	100	75	75	75	75	75	75	75	100	75
(Minimum)															
(Default Rule)			75	75	75	100	75	75	75	75	75	75	75	100	75
SIGNAL_1	1	Signal	75	75	75	100	75	75	75	75	75	75	75	100	75
SIGNAL_2	2	Signal	75	75	75	100	75	75	75	75	75	75	75	100	75
SIGNAL_3	3	Signal	75	75	75	100	75	75	75	75	75	75	75	100	75
SIGNAL_4	4	Signal	75	75	75	100	75	75	75	75	75	75	75	100	75

Figure 13.25 Design rules setup in CES.

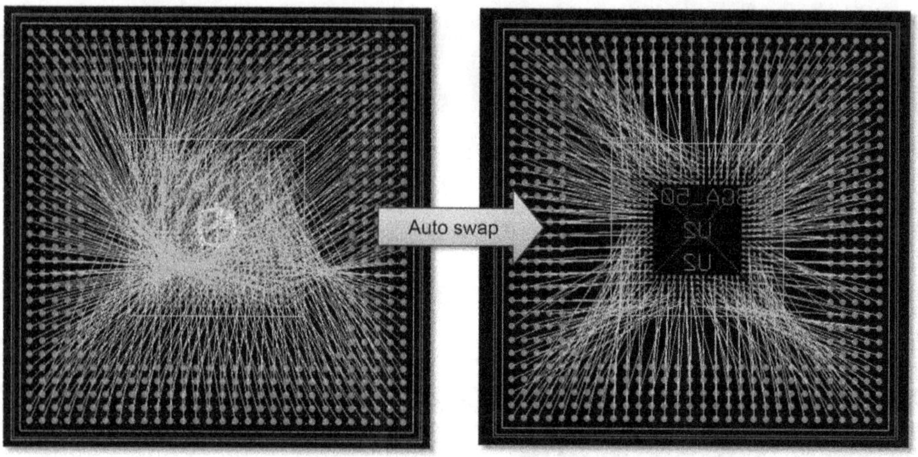

Figure 13.26 Comparison of before and after auto-optimization.

Figure 13.26 shows a comparison of before and after auto-optimization. It can be seen that most of the crossovers have been optimized, but there is still room for improvement.

If the auto-optimization result is not satisfactory, using manual optimization the designer can swap BGA pins according to local net connections and manual routing strategies. For the detailed method for swapping pins, please refer to the RDL design section above.

To begin routing after completion of net optimization, we can choose automatic route or manual route; usually the manual route result will be better and more reasonable. In both manual and automatic routing, we can enable or disable the 45° route function; the

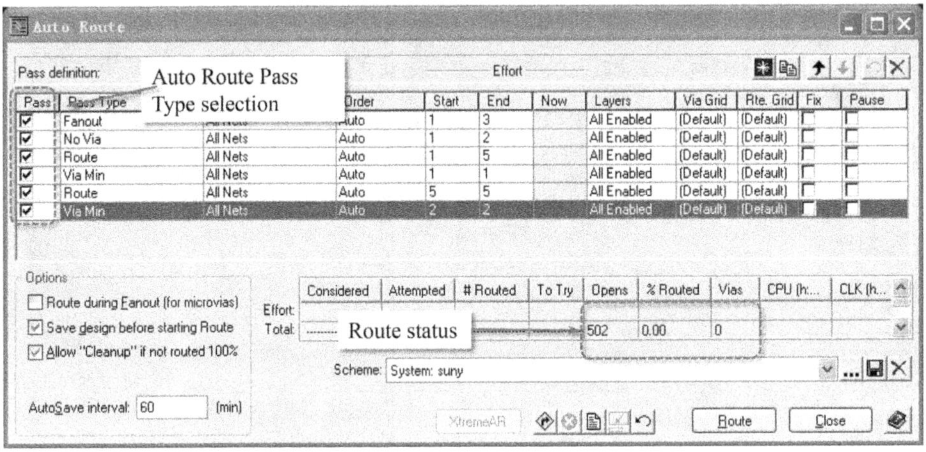

Figure 13.27 Start Auto Route.

method is the same as for RDL design. In this case, we choose automatic route with 45° route allowed.

In route mode, click the Automatic Route button ; the Auto Route window will pop up; select the corresponding pass options, click the Route button and start the auto route.

According to the complexity of the design, the time required for auto route will differ. In the auto route process, information on open net numbers, percentage routed and use of vias will be updated in real-time, as shown in Figure 13.27.

If each pass type finishes in auto route, the check item in front of Pass Type will be removed automatically. After the completion of auto route, the percentage of route, routed nets and via numbers are no longer up to date; if routing is not 100% complete, the net connections must be further optimized.

After further optimization, re-route the design; if 100% completeness is still not reached, manual route must be carried out, until the completion rate reaches 100%.

Figure 13.28 shows auto route completion status.

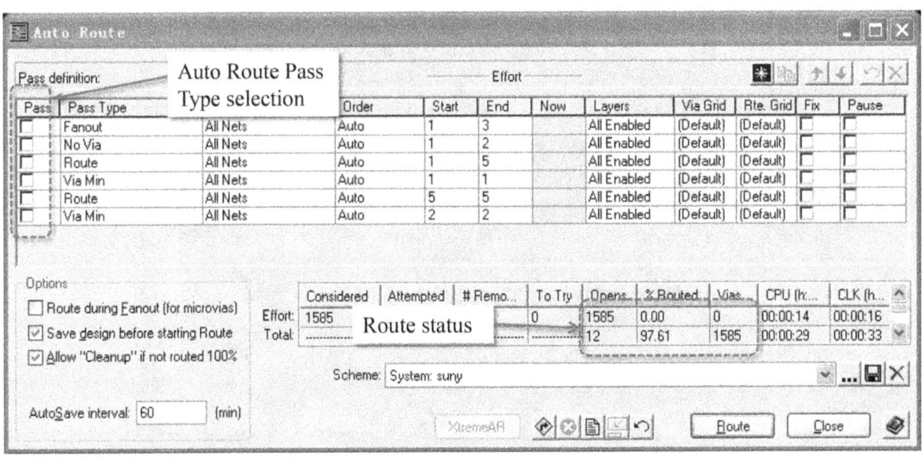

Figure 13.28 Auto route completion status.

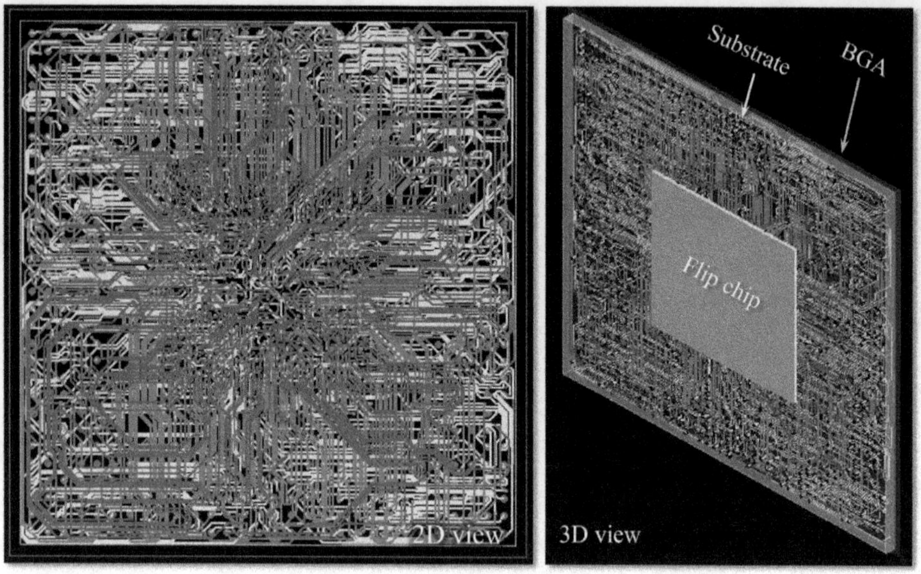

Figure 13.29 2D view and 3D view after route completion.

In this design, the auto route-complete percentage is 97.61%, but without manual optimization the net connection relationship is not optimal.

Usually, after manual optimization, the route-complete percentage will reach 100%. Figure 13.29 shows the board after the completion of auto route, for both 2D view and 3D view (view in 3D Viewer; for specific usage of 3D Viewer please see Chapter 18).

Figure 13.30 shows the traces in the flip chip area 3D view; we can see the elements include the FC pad, surface trace, inner trace, 1–2 via, 2–3 via, etc. From the 3D view we

Figure 13.30 Flip chip area trace and via in 3D view.

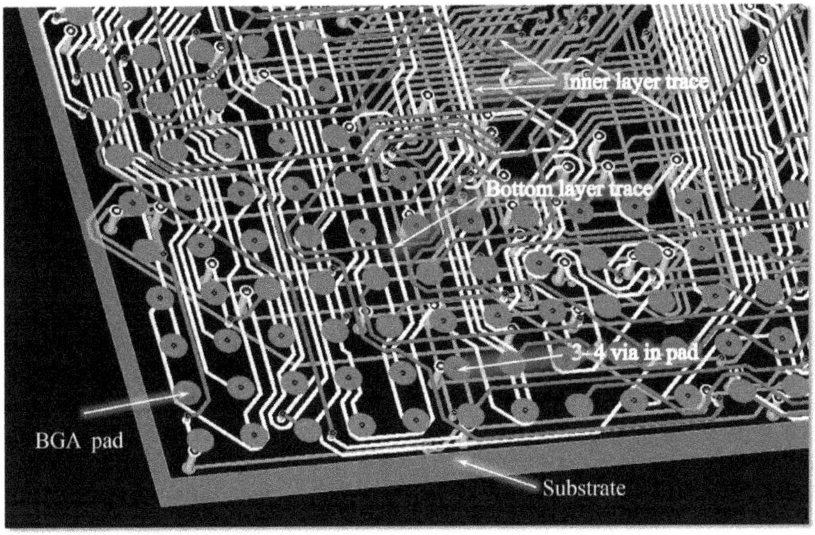

Figure 13.31 Trace and via 3D view in BGA area.

can intuitively see how the traces go to the other layers through a via, and at the same time we can visually check the via and layer stackup settings.

Figure 13.31 shows the 3D view in the BGA area; the figure shows the bottom view of the board, and we can see the BGA pad, bottom-layer trace, inner-layer trace, and via in BGA pad.

Now, flip chip design is almost complete. Subsequent DRC check is needed, as well as the output of the production file. Due to space limitations, only the flip chip design flow is introduced in this chapter; plane layer copper generation, DRC check and production output files are described in subsequent chapters.

Because the BGA package will be applied to PCB design, or placed on a PCB testing board, in the same project we create a new PCB_Board and copy the BGA package along with the nets to the PCB_Board schematic, then add the related testing circuit to finish the PCB board test design, as shown in Figure 13.32.

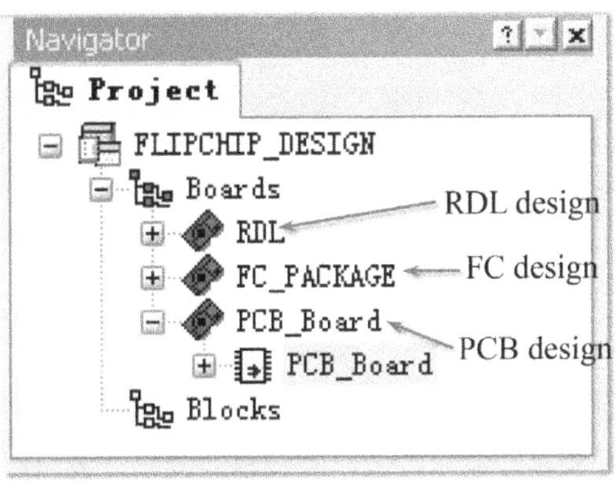

Figure 13.32 Add testing PCB_board to the same project.

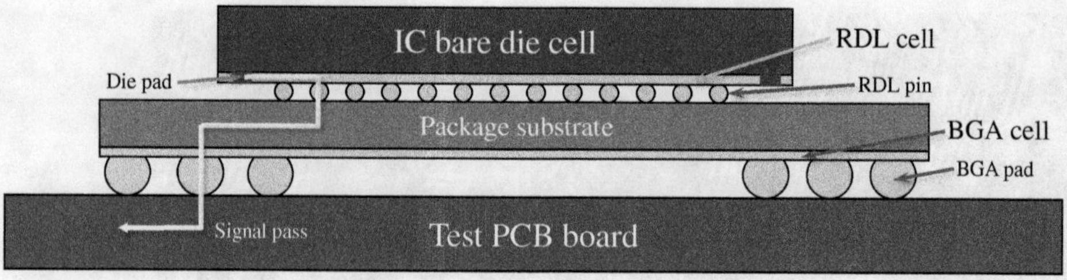

Figure 13.33 Diagram for IC bare die→RDL→BGA package→PCB board.

Finally, let's see the system application structure of the flip chip design described in this chapter.

The signal path flow goes from the die pad→RDL layer trace→RDL pin→Package substrate trace→BGA pad→PCB board, as shown in Figure 13.33.

14

Route and Plane

14.1 Routing

14.1.1 Introduction to Routing

In Expedition, routing is usually divided into three kinds: manual, semi-auto and auto; manual routing is also known as interactive routing.

Manual routing refers to the functions: route, plow, place via and connect to corresponding pad by hand.

Semi-auto route includes some of the functions of auto route; semi-auto route operations include: fanout, tune, route, re-route, auto-finish, gloss, etc.

Auto route refers to using the auto route dialog to set up auto route schemes; the software then automatically executes the schemes to complete routing. Auto route can be used for the whole design or part of the design.

In SiP layout, the best solution is to use manual route together with auto route and semi-auto route. For example, complete the key traces via manual route, with the aid of semi-auto route to speed up the manual route process, then use the auto-route function to finish the rest of the traces.

In Expedition, click the Route mode button and switch to route mode; the Route toolbar will automatically pop up. The route toolbar can be located below the menu, or the designer can drag it to the leftmost or rightmost area of the design window, or suspend it in any position on the design window. Changing the route toolbar display mode can be done by dragging it using the mouse, as shown in Figure 14.1 for the two route toolbar display modes.

14.1.2 Manual Route

In the route toolbar, click the Plow or Multi-plow button to activate manual route. When a single pad is selected the system will execute single plow, when multiple pads are selected the system will execute multi-plow. Plow or multi-plow can start from bond pads, component pins, existing traces or vias at any location, as shown in Figure 14.2 and Figure 14.3.

In multi-plow, we can adjust the trace spacing using the < F6 > and < F7 > keys or the soft key [6 Converge In][7 Converge Out] at the bottom of the design window.

SiP System-in-Package Design and Simulation: Mentor EE Flow Advanced Design Guide,
First Edition. Suny Li (Li Yang).

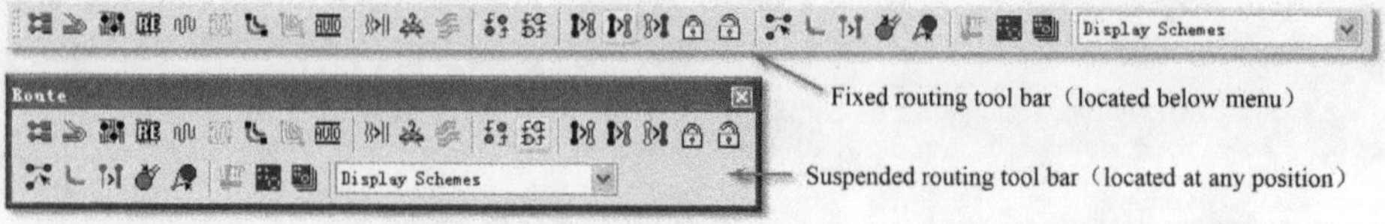

Figure 14.1 The two route toolbar display modes.

Figure 14.2 Single plow.

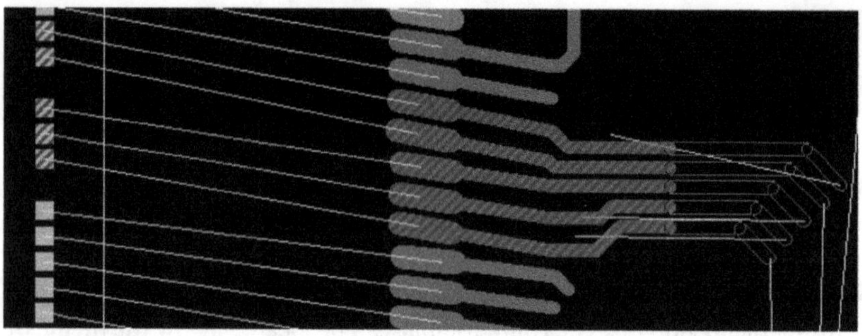

Figure 14.3 Multi-plow.

In addition to using the tool button in the toolbar to start the Plow command, we can also choose the menu commands Route→Interactive→Plow, choose the soft key [3 Plow / Multi] at the bottom of the design window, or press the < F3 > key. With stroke 852 [+], we can also invoke plow or multi-plow.

14.1.3 Plow Modes

When manual plow is used, there are four plow modes to choose from: forced plow, route plow, angle plow and dynamic plow; each mode has its own characteristics and the designer can switch from one to the other.

After invoking the Plow command, we can use the < F3 > key or the soft key [3 Toggle Plow] to toggle plow mode. When toggling, which plow mode is activated depends on which mode is selected in the Editor Control dialog, as shown in Figure 14.4.

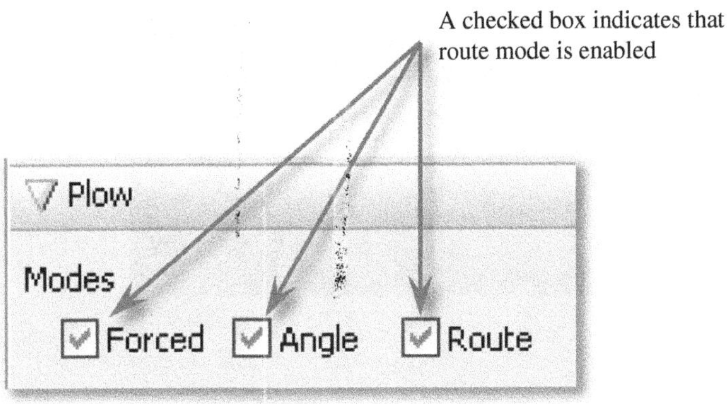

Figure 14.4 Plow mode selection in Editor Control.

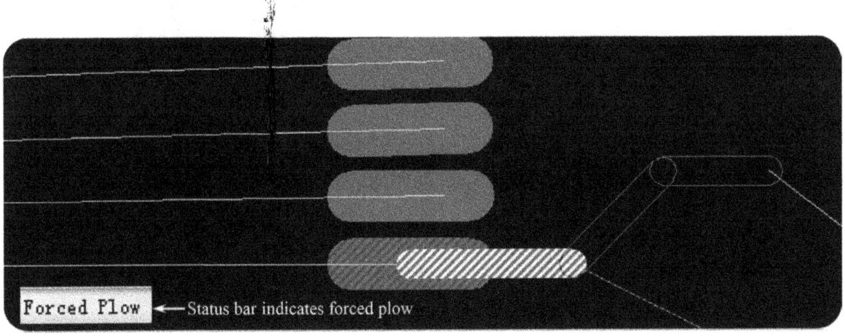

Figure 14.5 Forced plow.

1) Forced plow
   Forced plow is the default plow mode, and the one in which push ability is the strongest, but via cannot be added automatically in this mode; in forced plow mode a dotted white circle suggests to the designer where via might be added; the designer can then add via manually by clicking the mouse button, with the expected trace in the form of a hockey stick. Designers can see the current plow mode in the status bar display at the lower-left corner, as shown in Figure 14.5.
2) Route plow
   Route plow is a more flexible plow mode, it uses a semi-automatic routing algorithm; in the routing process, the software can automatically insert via and switch layers.
   Route plow is suitable for long pad-to-pad routing. The expected trace is displayed in the form of a straight line but the finished trace can automatically be optimized into the form of a hockey stick; the actual routing effect is almost the same as for forced plow, as shown in Figure 14.6.
3) Angle plow
   Angle plow allows any angle routing; the routed trace, which is a semi-fixed trace, can't be pushed, as shown in Figure 14.7. Semi-fixed trace can be moved by the designer, but can't be pushed by other trace or via.

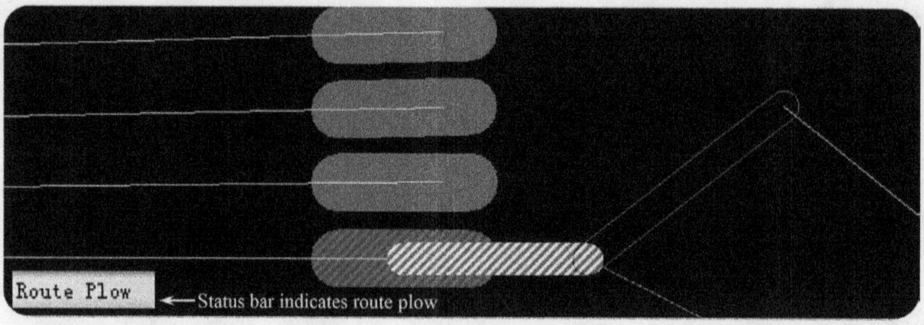

Figure 14.6 Route plow.

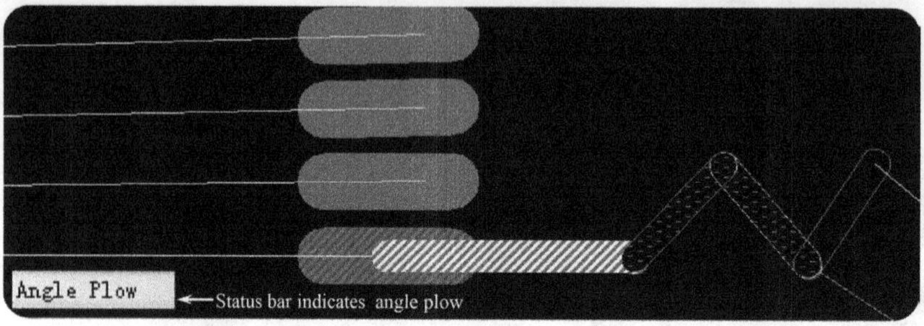

Figure 14.7 Angle plow.

4) Dynamic plow
   Dynamic plow is similar to forced plow, but it has a special startup mode. Instead of using hotkeys to switch, the designer drags the mouse to activate this mode. We can invoke dynamic plow in any one of the above three plow modes; hold the left mouse button and drag the mouse to route; the system will automatically enter dynamic plow mode. The characteristics of dynamic plow are the very strong push ability and that the results of routing and push show in real-time.

14.1.4 Gloss Modes

When routing manually, Expedition will dynamically optimize the route path, remove unnecessary bending, and smooth the trace; this function is called "gloss".

There are three modes of gloss to choose from: gloss on, gloss local and gloss off. Each has its own characteristics and can be toggled on and off using the soft key. In the process of routing, we can toggle gloss mode by pressing the <F4 > key or the soft key [4 Toggle Gloss] at the bottom of the layout window.

1) Gloss on
   Gloss on is the default gloss mode; it can automatically remove the redundant angle and bending in the routing process, and will automatically optimize the pad entry, following the pad entry parameters set in the Editor Control dialog.
2) Gloss local
   Gloss local is a more intelligent gloss mode; in this mode Expedition can better understand the intention of the designer in relation to keeping the bend the designer

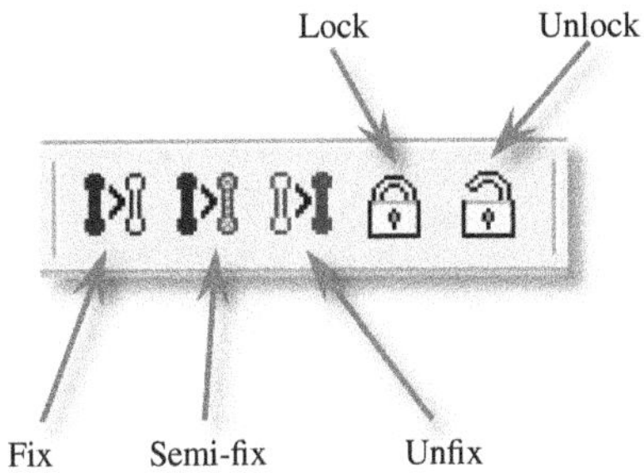

Figure 14.8 Fix and Lock tool bar.

wants. In the route process, if the designer feels that gloss on is too flexible, he or she can switch to gloss local mode.

3) Gloss off
   In gloss off mode, routing is carried out entirely in accordance with the intention of the designer. In this mode the trace is semi-fixed, and cannot be pushed, but can be moved and adjusted by the designer.

14.1.5 Fix and Lock

In the route process, sometimes it is necessary to fix or lock the trace for special protection, to avoid it being pushed or accidentally deleted by the designer in the subsequent route process.

As shown in Figure 14.8, the Fix and Lock toolbar includes fix, semi-fix, unfix, lock and unlock tools.

1) Fixed trace and via cannot be pushed or moved, and cannot be deleted, resulting in effective protection for the trace and via.
2) Semi-fixed trace and via can be moved manually, but cannot be pushed by other trace, and thus incorporates part of the fixed effect. In the route process, pushing by other trace or via doesn't affect semi-fixed trace and via.
3) Fixed or semi-fixed trace can be released with the Unfix command. Semi-fixed trace can be fixed with the Fix command; fixed trace cannot switch to a semi-fixed state. If you need to switch fixed trace to a semi-fixed state, first unfix, then semi-fix the trace.
4) Lock is the highest level of trace protection. Fixed or semi-fixed trace can be locked, and can then only be released by unlock. Locked trace cannot be fixed, semi-fixed or unfixed.
5) Fixed trace and via are displayed with dot filling; semi-fixed trace is displayed with short-line filling, Locked trace and via are displayed as hollow, as shown in Figure 14.9.

14.1.6 Layer Switching

When routing manually, there are several methods to switch route layer.

1) Switch layer with layer pairs
   In the process of manual routing, when the mouse stops at a particular location, a white-dotted circle suggests that a via can be placed here. Double-click the left

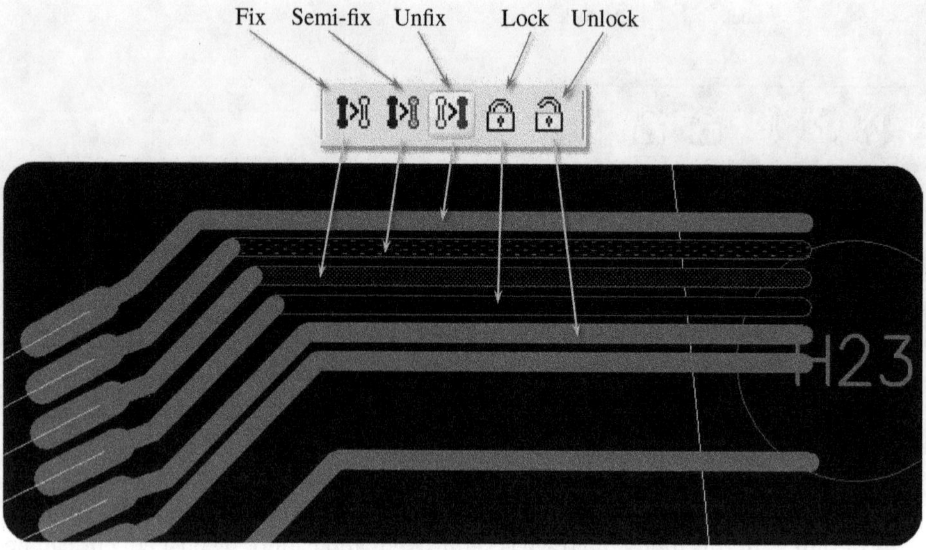

Figure 14.9 Traces in states fix, semi-fix, unfix, lock and unlock displayed together.

mouse button in this position, and the via will be placed. Using the via, the route can go through from the current layer to a corresponding paired layer, as shown in Figure 14.10.

"Corresponding paired layer" refers to one of the two layers set in a layer pair. If there is no paired layer set up, we cannot place the via by double-clicking the mouse button. The current layer and its paired layer are together called a layer pair. In Editor Control, select from the menu Setup→Editor Control→Route→Layer Settings to start the layer pair-setting interface, as shown in Figure 14.11.

After layer pair setting, we can view the layer pair information in the status bar on the lower-right corner of the layout window, for example, 1H, 6V : 1H indicates that the current layer is layer 1 and its default route direction is horizontal; 6V indicates that the paired layer is 6, and its default route direction is vertical.

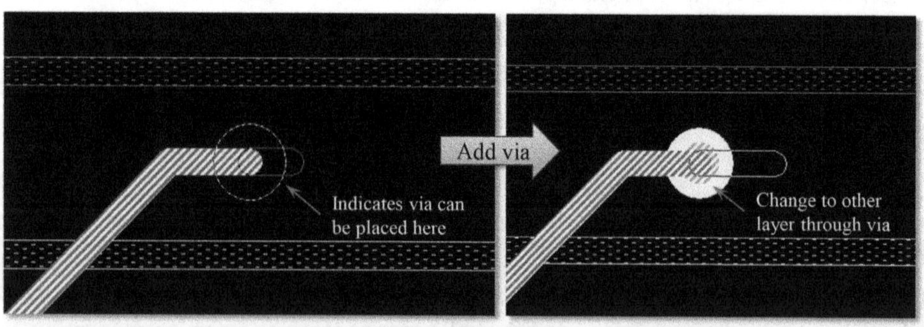

Figure 14.10 Before and after a via is placed.

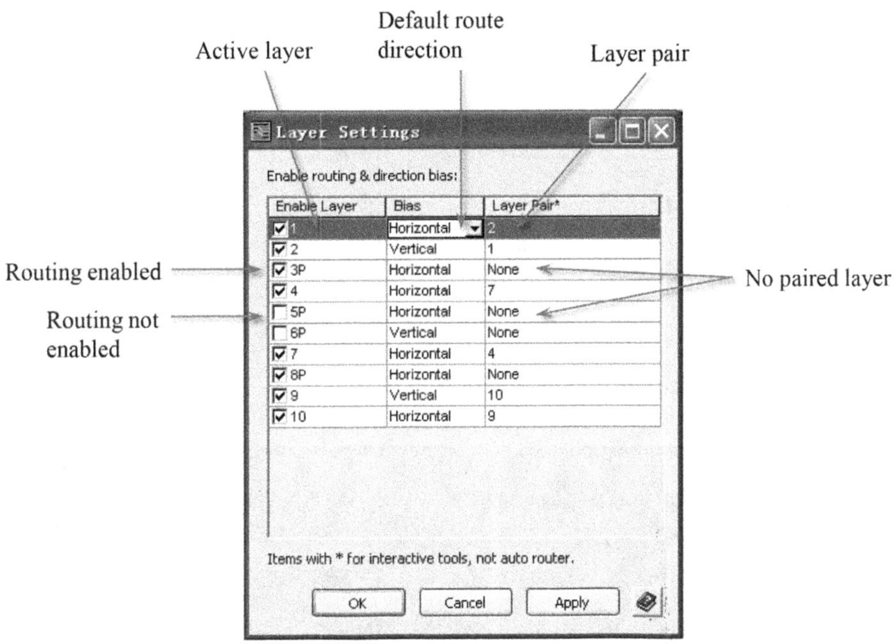

Figure 14.11 Layer pair and related route settings.

2) Switch layer using up and down arrow
During the route process, another layer-switching method is to use the up and down arrows to switch layer. For example, route in the first layer, press the down arrow to switch to the second layer, and then press the down arrow again to switch to the third layer. In the same way, use the up arrow to switch upwards, etc. Please note that layers where routing is not allowed will automatically be avoided in the upward and downward switching.
3) Random switch
If the designer wants to switch to any layer randomly in the route process (excluding layers that do not allow routing), he or she can open the Display Control dialog box, and in the route process use the mouse to click the layer directly; the software will then automatically place the via and switch in this layer and continue to route in this new layer. This switching mode is not affected by the limitations of layer pairs, and has strong via and trace push ability, as shown in Figure 14.12.

14.1.7 Move Trace and Via

To move trace and via, the designer must be in route mode and use the Select command (not the Route command); the lower-left corner status bar displays `Route Mode - Select` or `Select` .

Select the trace or via and hold the left mouse button, then move the mouse; the trace and via can be moved, except fixed or locked trace and via. The peripheral traces or vias, if not fixed, semi-fixed or locked, will be pushed automatically.

We can use the following methods to select trace.

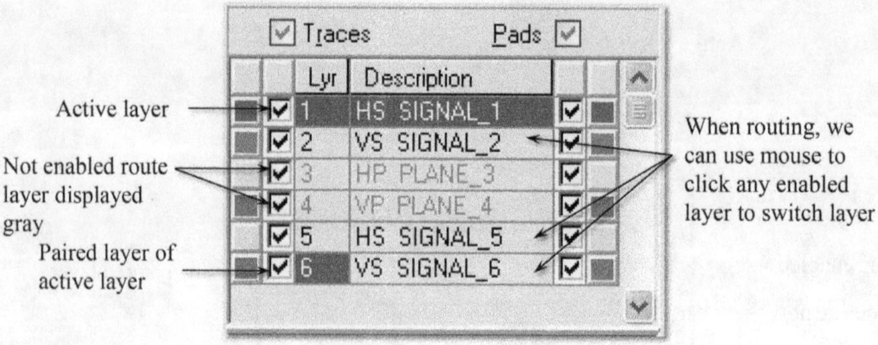

Figure 14.12 Using Display Control for layer switch.

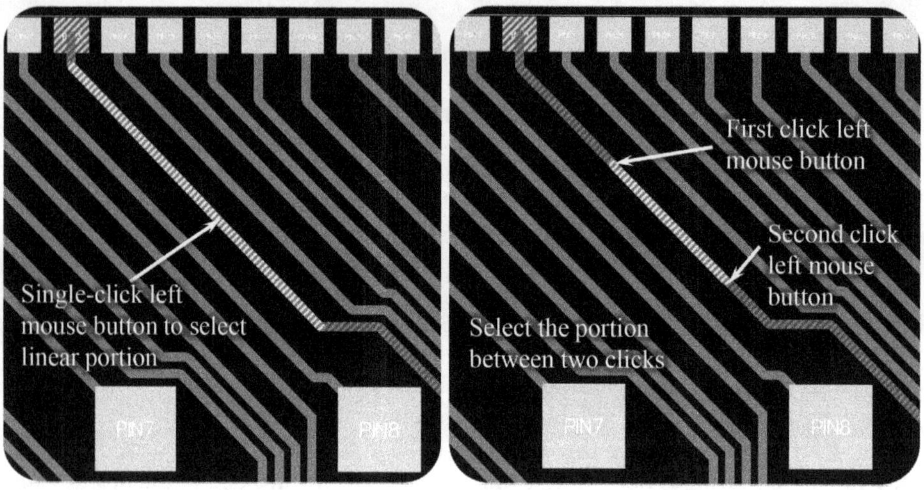

Figure 14.13 Single-click selection and selection clicking start and end points.

Using the left mouse button to single-click the trace, we can select a straight section connected to the click position. If the designer wants to select a particular part of the trace, first, he or she can click the left mouse button at the start point, and then click the end point; the segment between the two points is selected, as shown in Figure 14.13.

After selection, we can drag the trace or via to move it, and only the selected part will be moved while the other parts remain in position.

14.1.8 Circuit Copy

In SiP design, if there are multiple channels or similar circuits in the design, in order to save time and to maintain the consistency of related circuits, usually we adopt the method of circuit copy.

The premise of circuit copy is that there are corresponding circuits in schematic design, that is, in Expedition we cannot produce a circuit out of thin air. Circuit elements that can be replicated include components, traces, vias, planes, even cavities, etc. Select from the menu Edit→Circuit Move and Copy, and the Selection Filter window will automatically pop up. This window is divided into an Object tab and a Layer tab. The object tab is for circuit element selection, and the layer tab for layer selection. Usually we keep the default settings, as shown in Figure 14.14.

Then, use the left mouse button to box select the circuit for copying, and press the < F7 > key or click the soft key [7 Copy Circuit] at the bottom of the design window. The Instantiation Wizard window will pop up automatically. In this window, the software system displays the components and corresponding nets for copying, while the designer can also adjust the components and nets according to the real design situation, as shown in Figure 14.15.

Click the OK button in the Instantiation Wizard window; a virtual box enclosing the replicated circuit will pop up and attach to the cursor; the designer can move the mouse to the required position and then click the left mouse button to place the duplicated circuit at the location, as shown in Figure 14.16.

After the circuit is placed, the Instantiation Wizard window will pop up again; to find components and nets once more, follow the same method as above; the designer can also adjust the mapping relationship according to the specific design situation.

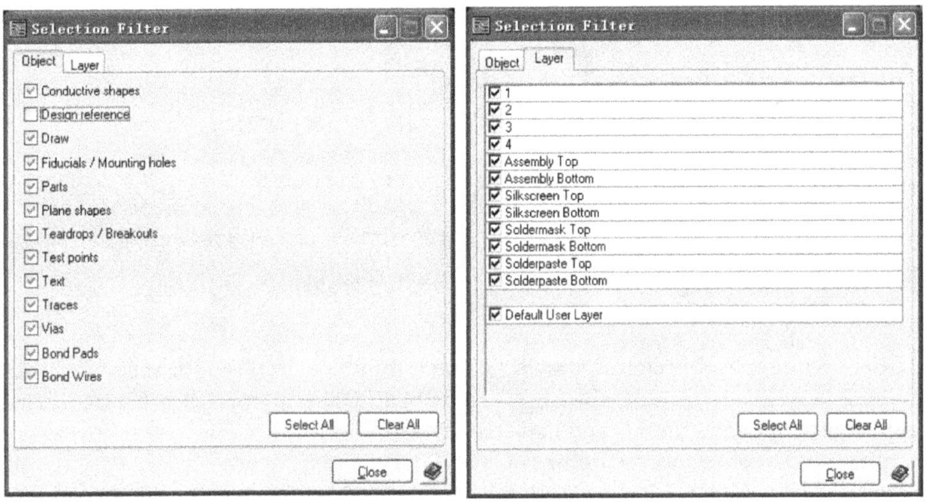

Figure 14.14 Selection Filter window.

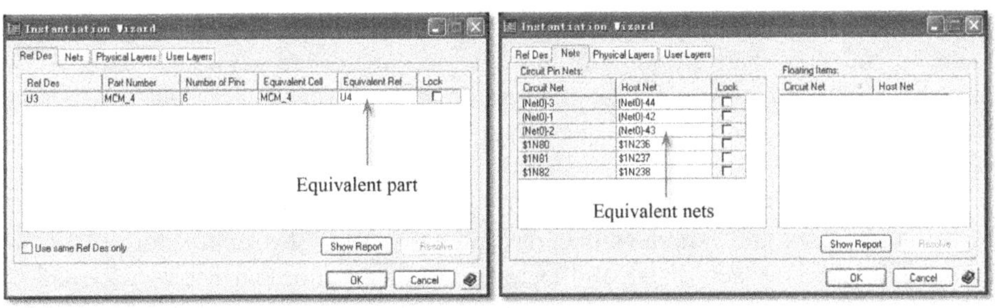

Figure 14.15 Component and net mapping in circuit copy.

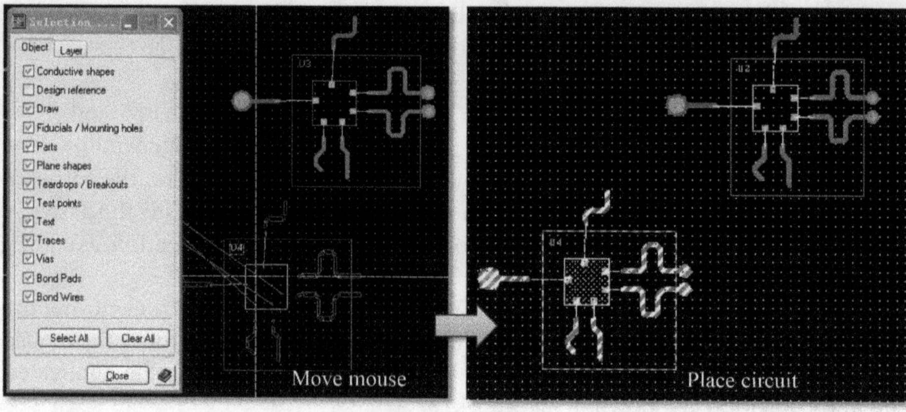

Figure 14.16 Copy circuit and place to specified location.

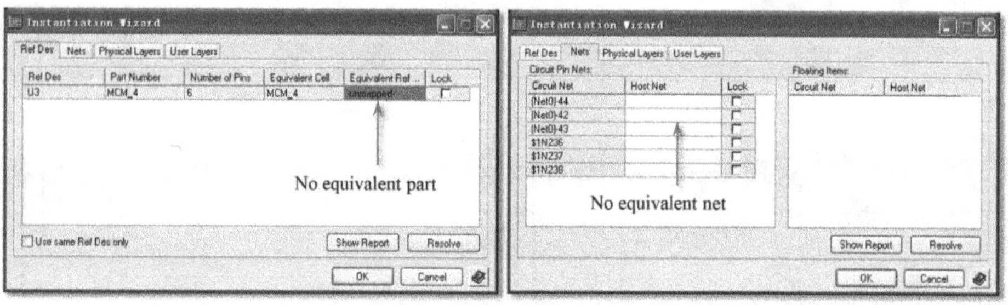

Figure 14.17 There are no equivalent components and nets in the design.

If there is no replicable circuit, that is, no corresponding circuit in the schematic that can be replicated, the window shown in Figure 14.17 will pop up, indicating that there are no equivalent components and nets to copy.

In circuit copy, elements including substrate cavity, bonding wire, trace, via, etc. can be replicated, as shown in Figure 14.18 (2D view and 3D view).

14.1.9 Semi-Automatic Route

Between manual route and auto route, there are some commands or operations, such as fanout, route, re-route, tune, gloss etc., which are called semi-automatic route commands. To use the semi-automatic route operations, first select elements such as nets, pins and vias, and then start semi-automatic routing with the button on the toolbar.

For example, first choose some pads, and then click the Fanout button; if the design rules are met, the system will fan out and add vias, as shown in Figure 14.19.

In manual route, the auto finish and toggle curve commands can also be regarded as semi-auto route operations. When editing trace manually, trace and via shoving is actually semi-auto re-route.

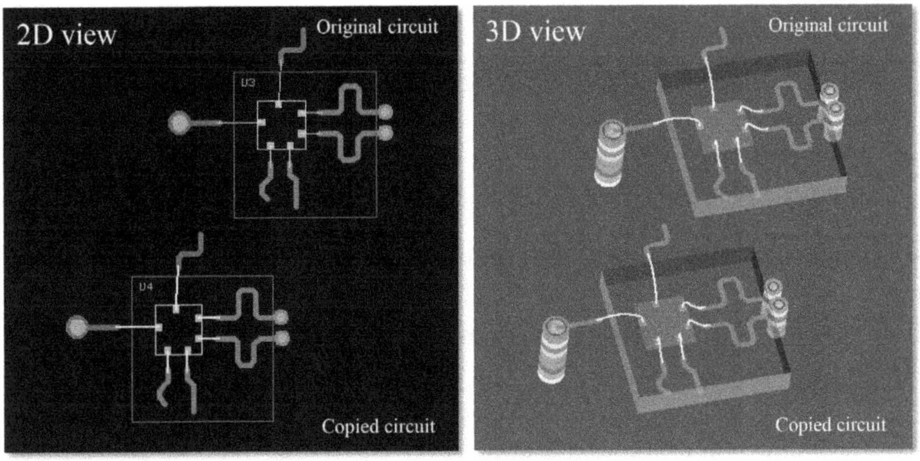

Figure 14.18 Circuit copy realization (including substrate cavity).

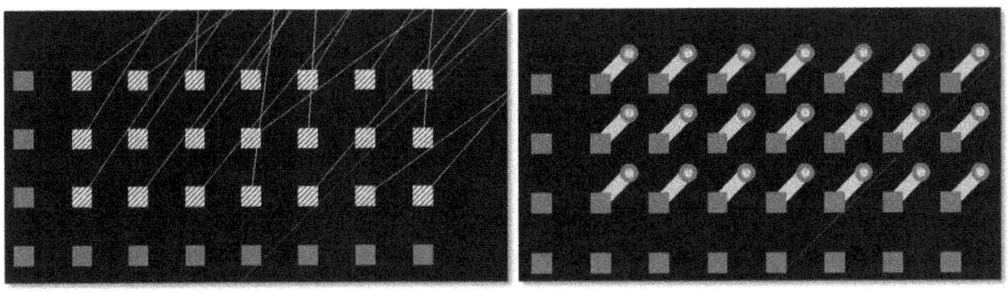

(a) Select the pad needing fan out (b) Execute fan out

Figure 14.19 Select pads then execute fan out.

14.1.10 Auto Route

Expedition can auto route for the whole design, a specific layer, parts of nets and special nets. We can start auto route any time in the design process. Auto route can be used alone and can also be used interactively with manual route.

The auto route algorithm is basically the same as the algorithm in semi-auto route. In auto route, we can use different methods to achieve better effects in less time.

Select Route→AutoRoute or click the tool button in route mode to start the Auto Route dialog box, as shown in Figure 14.20.

Next, we introduce each column of the auto route options, as shown in Figure 14.21, one by one.

1) Pass check box: check to use this pass type.
2) Pass Type: includes fanout, no via, route, via min, tune delay, tune crosstalk, smooth, etc.; using the drop-down list to choose, the designer can create a new pass type or delete an existing pass type.
3) Items to Route: the designer can choose all nets in the design for auto route, or can choose parts of nets; there are a variety of net types for selection. When the designer

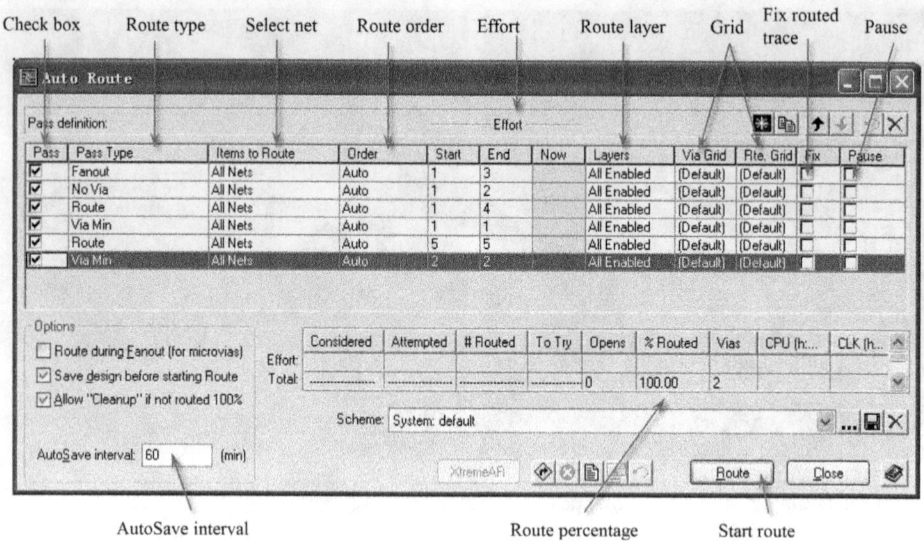

Figure 14.20 Auto route dialog box.

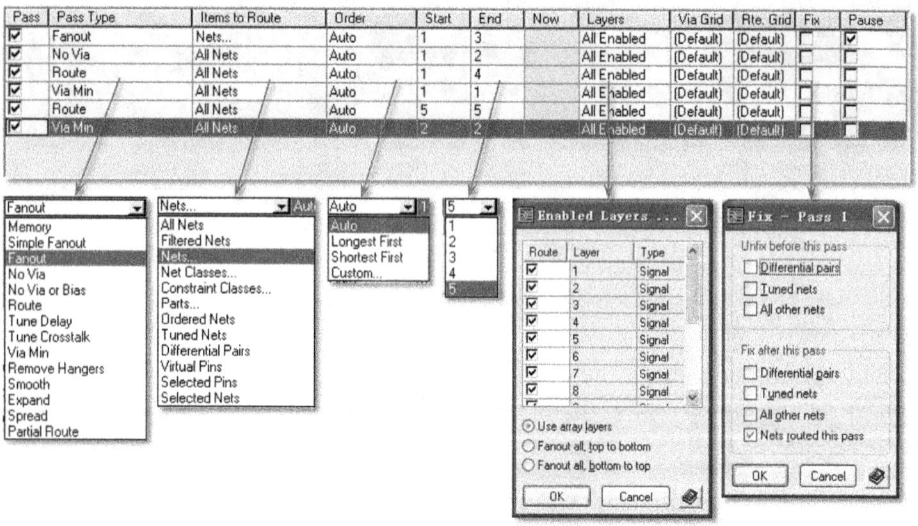

Figure 14.21 The auto route options.

chooses nets, the window shown in Figure 14.22 will pop up; the designer needs to send the nets to the Included bar on the right-hand side for inclusion in auto route. The auto router will only run on the included nets.

4) Order: route sequence, including auto selection, longest, shortest or custom order specified by the designer.
5) Effort: routing algorithm number, a number representing the complexity of the routing algorithm, the bigger the number, the higher complexity of the algorithm; the

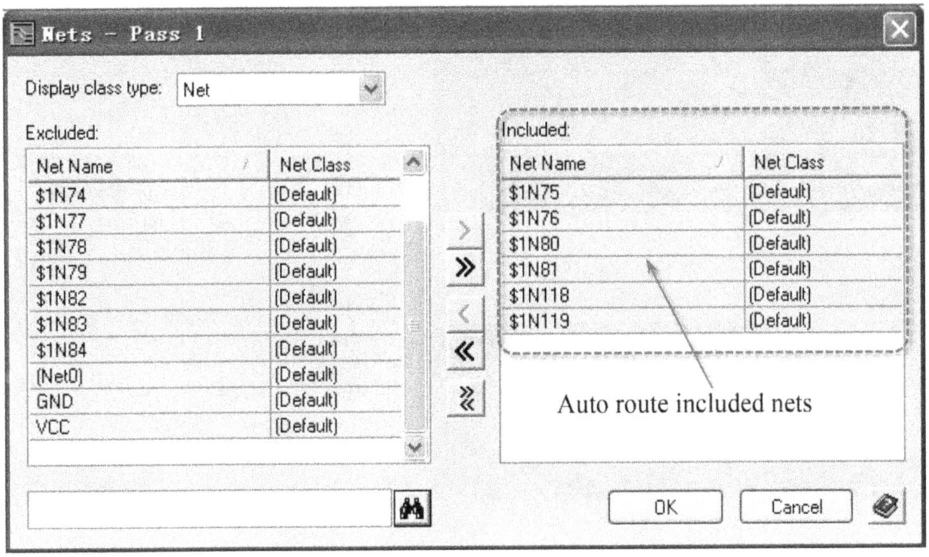

Figure 14.22 Nets included for auto route.

completion rate and route time will also be higher. Start, End and Now respectively stand for route start, route end and the current algorithm complexity.

6) Layers: route layers, can be chosen by the designer in the pop-up window to auto route all layers or only some layers.
7) Via Grid and Rte. Grid: these stand for via and route grid, which by default uses the same grid as for manual route, namely the grid set in Editor Control. None indicates that we are using gridless auto route.
8) Fix: stands for fix trace; it shows whether, after completion of the current routing, the trace should be auto-fixed or not; in addition, the fix option can be used to cancel fixed traces before the start of the current route.
9) Pause: this indicates whether auto route should be paused after completion of the current pass type.

14.1.11 Route Differential Pairs

Differential signaling has been widely used in high-speed circuit design. Differential signaling has strong anti-interference ability, can effectively restrain EMI, and has precision sequence positioning and other technical characteristics to appeal to circuit designers. LVDS (Low Voltage Differential Signaling) has adopted differential signaling technology.

In SiP substrate design, differential signaling is commonly used. We often call a pair of differential signals a differential pair. In the following section, we will describe how to realize differential pair setup and routing in Mentor EE Flow.

First, let us learn how to set up a differential pair in the schematic.

In the schematic design environment DxDesigner, select a net, such as DIFF1_N, and in the Properties interface, set diff pair attribute value to DIFF1_P (choose the DIFF1_P net from the drop-down list). Then, switch to net DIFF_P; we can see the diff pair

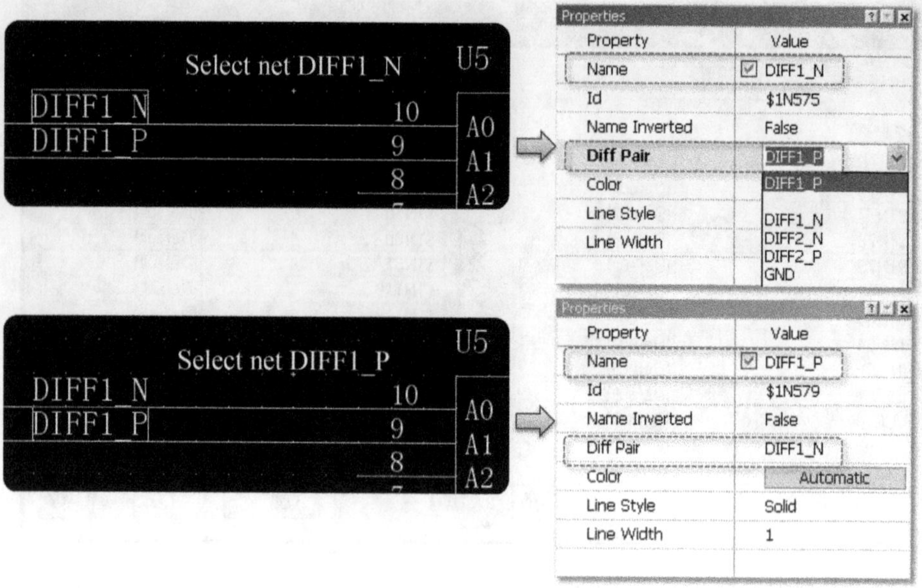

Figure 14.23 Set up diff pairs in the schematic.

attribute of DIFF_P has become DIFF1_N. At this point, diff pair set up is finished in the schematic, as shown in Figure 14.23.

Now, let's learn how to set up diff pairs in CES.

CES supports manual and auto diff pair setup; below, we introduce these two methods.

1) Set up diff pair manually
   In CES, select two nets, and then click the toolbar button "Selected Nets Diff Pair" to create a diff pair with the selected nets. If we want to remove diff pairs, choose the diff pair name, then click the "Remove Diff Pair(s)" button; the system will automatically display a hint window asking whether to remove the differential pair or not; click OK to confirm, and the diff pair will be removed, as shown in Figure 14.24.
2) Set up diff pair automatically
   If there are many diff pairs in the design, we can use the "auto assign diff pairs" method. In the CES toolbar, click the Auto Assign Differential Pairs button and enter wildcards for diff pair names in the pop-up window, such as * _N and * _P, then click button ; the nets meeting the conditions will be placed into the list automatically, as shown in Figure 14.25.

After the Apply button is clicked, the system finishes automatic diff pair setup, as shown in Figure 14.26.

After diff pair setup in CES, the designer can route diff pairs in the Expedition design environment. Choose any net of the diff pair; the system will select the corresponding one automatically; two nets of a diff pair are routed together, with the trace spacing meeting the diff clearance rules set in CES.

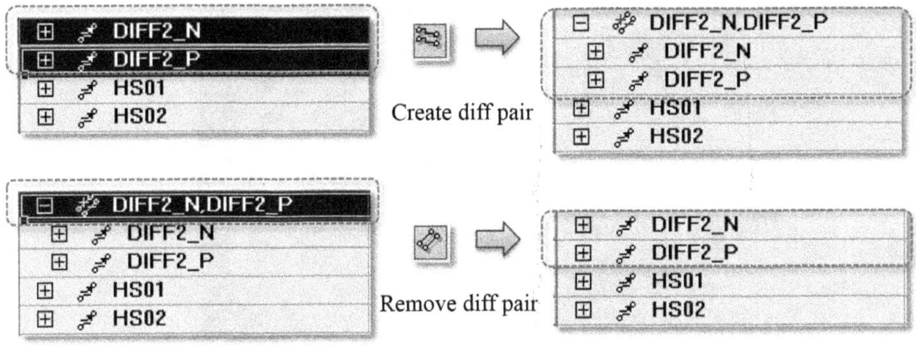

Figure 14.24 Create and remove diff pairs in CES manually.

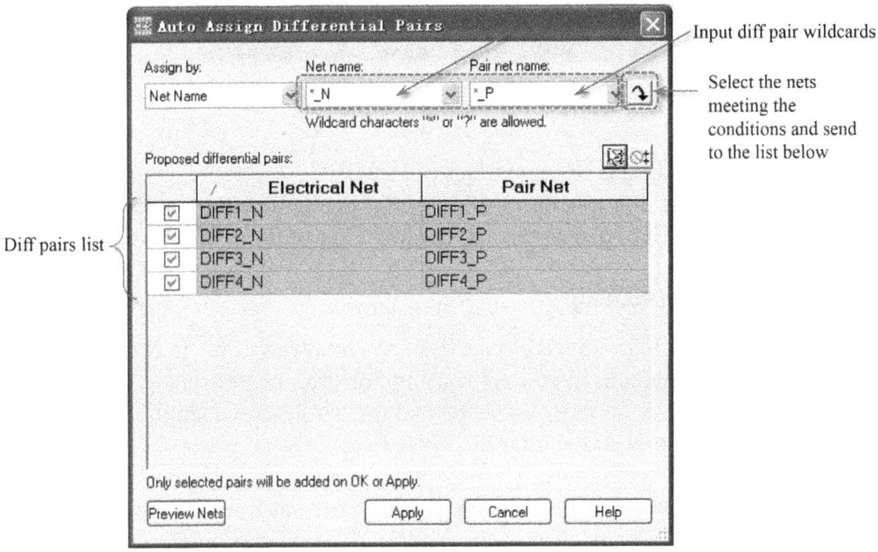

Figure 14.25 Set up diff pairs in CES automatically.

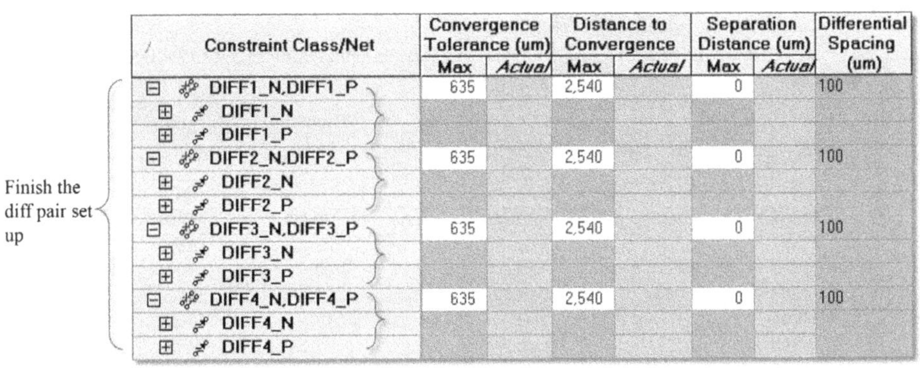

Figure 14.26 Finish automatic diff pair setup.

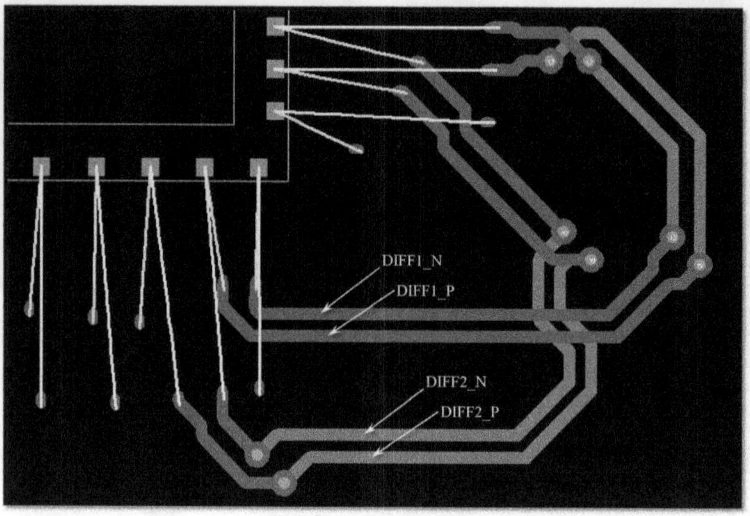

Figure 14.27 Completed diff pair routing.

Completed diff pair routing is shown in Figure 14.27.

This book only contains a simple description of diff pairs; for more detail on diff pair set up in Expedition and CES, please refer to the help files of those applications.

14.1.12 Length Control in Routing

In SiP substrate design, if high-speed circuit design is involved, it is often necessary to carry out high-speed signal length control routing, for example, ensuring that a set of signals are of equal length as far as possible, or that the whole signal length is limited to a certain length, for example, 30000 μm±2%.

The common practice is as follows.

The first step is to set length rules in CES. In this case, set the minimum length of high-speed nets HS01~HS06 to 29400 μm (30000–2%), and the maximum length of HS01~HS06 to 30600 μm (30000+2%), as shown in Figure 14.28. It is important to note

Constraint Class/Net	Net Class	Type	Min (ns)|(um)	Max (ns)|(um)	Match
⊟ (All)	(Default)	Length			
⊞ DIFF1_N,DIFF1_P	(Default)	Length	Minimum length	Maximum length	
⊞ DIFF2_N,DIFF2_P	(Default)	Length			
⊞ DIFF3_N,DIFF3_P	(Default)	Length			
⊞ DIFF4_N,DIFF4_P	(Default)	Length			
⊞ HS01	(Default)	Length	29,400	30,600	
⊞ HS02	(Default)	Length	29,400	30,600	
⊞ HS03 — A set of high speed nets	(Default)	Length	29,400	30,600	
⊞ HS04	(Default)	Length	29,400	30,600	
⊞ HS05	(Default)	Length	29,400	30,600	
⊞ HS06	(Default)	Length	29,400	30,600	

Figure 14.28 Set up length rules in CES.

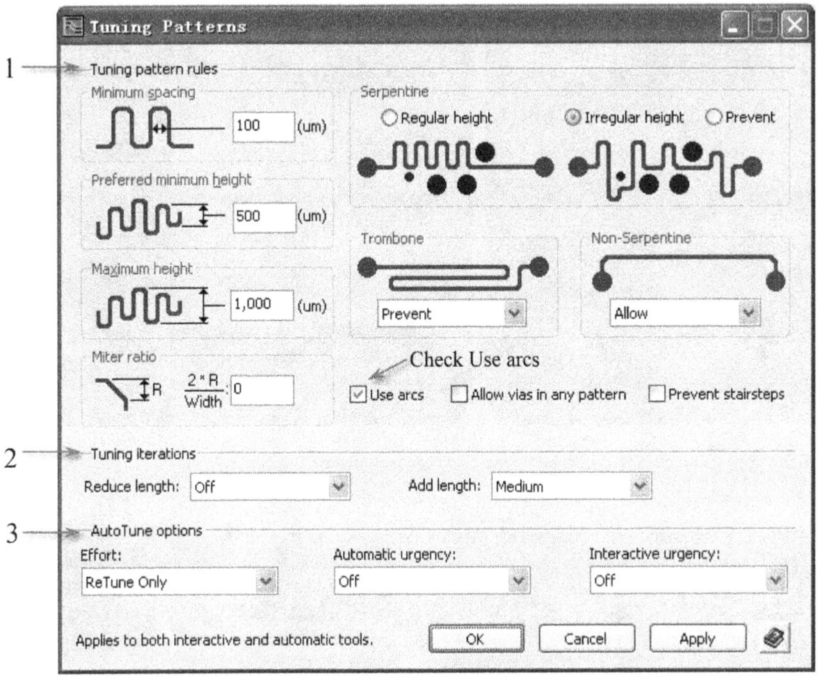

Figure 14.29 Tuning rules setup.

that since all the nets have been limited to the minimum and maximum length range, it is not necessary to set match group attributes for these nets.

If we want to set up a set of high-speed nets of equal length without a length-range limit, we can set a match group for these nets (input the same value in the match column).

After the completion of setup in CES, exit CES; the rules will automatically update to the Expedition environment. Then, set up the high-speed route rules in the Expedition environment.

In Editor Control, select Route→Dialogs→Tuning, and the Tuning Patterns window will pop up for tuning rules setting. These rules are mainly divided into three categories: tuning pattern rules, tuning iterations and autotune options, as shown in Figure 14.29.

Each tuning rule has a different influence on the result of tuning. For example, when the Use arcs option is checked, the tuning automatically uses arc corner; when the Use arcs option is not checked, the tuning uses 45° corner automatically, as shown in Figure 14.30. The left side of the figure shows the Use arcs option and the right side shows the result when this option is not used.

Using different setting combinations, various tuning effects can be obtained. For instance, the tuning effects with different setting combinations for tune spacing, height, miter ratio and arc use are shown in Figure 14.31.

Figure 14.31 shows the results of different spacing and height tuning, height change tuning, arc tuning, diff pair tuning, etc.. Designers can set up tuning patterns according to the design needs.

For detailed interpretation of tuning rules settings, please refer to Table 14.1.

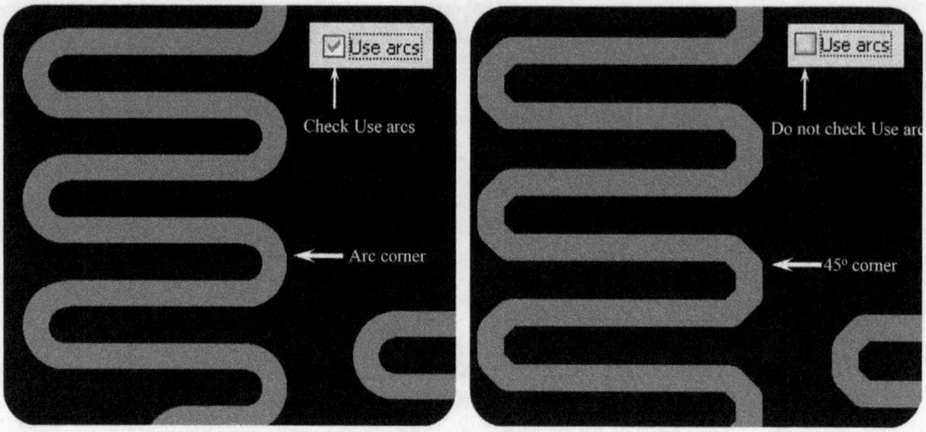

Figure 14.30 Tuning results with the Use arcs option selected and not selected.

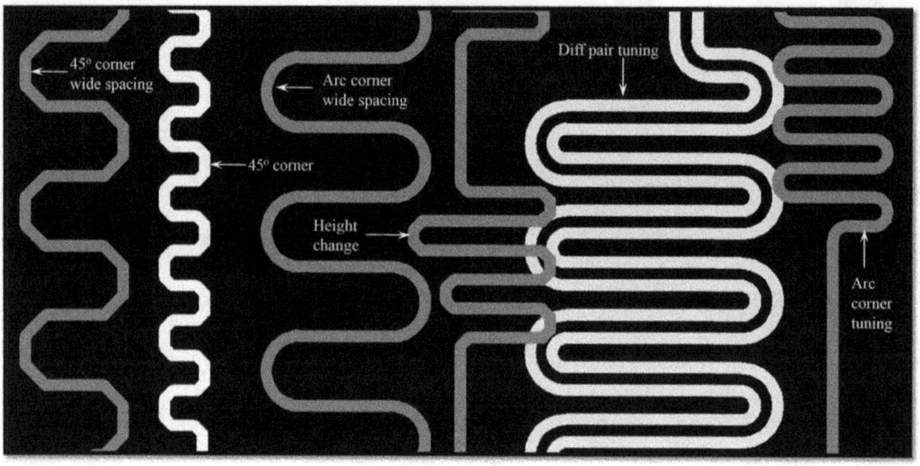

Figure 14.31 Tuning effects with different setting combinations.

Table 14.1 Detailed tuning rules settings.

◆ Tuning pattern rules section – Defines the tuning pattern, spacing and height rules, and the availability of each pattern type.	
Minimum spacing	(Serpentine only) Defines the minimum spacing (edge to edge). Default is 0. You can also enter the value as a factor of the trace width using x. Example: A minimum spacing value of 2x and a trace width of 10 defines a minimum spacing of 20.
Preferred minimum height	(Serpentine only) Defines the minimum height. Default is 0. To compress the pattern and reduce trace segments, enter a high value.
Maximum height	(Serpentine only) Defines the maximum height. Default is 0. To compress the pattern and reduce trace segments, enter a high value. Note: Setting minimum spacing and maximum height to 0 results in tuning using the values entered in CES.

Table 14.1 (Continued)

Miter ratio	(Serpentine only) Defines the miter ratio for the pattern. To have the software calculate the miter length, enter 0. To modify the length of a tuning diagonal, enter a miter ratio (2*R)/Trace width, where R is the miter height. Valid values are 0.5 to 100.0 (with increments of .5). Example: A miter height of 30 and a trace width of 10 results in a ratio of 6 (2*30/10=6).
Serpentine	Defines the serpentine tuning pattern availability and type. Regular height – Selected, allows a uniform height for the serpentine. Uses the spacing and height rules. Irregular height – Selected, allows a free-form height for the serpentine. Uses the spacing and height rules. Prevent – Selected, prevents serpentine patterns.
Trombone	Defines the availability of the trombone tuning pattern. Allow – Selected, allows a trombone pattern. Prevent – Selected, prevents a trombone pattern. Note: This pattern is unavailable for diff pair tuning.
Non-Serpentine	Defines the availability of a non-serpentine tuning pattern. Allow – Allows a non-serpentine pattern. Prevent – Prevents a non-serpentine trace. Prefer – Attempts tuning with a non-serpentine pattern before attempting tuning with serpentine or trombone patterns.
Use arc	(Serpentine only) Selected, replaces mitered corners of the pattern with arcs. Note: If the arc radius is larger than (minimum spacing+trace width)/2, the arc radius reduces to maintain the minimum spacing, which results in a half-round bend.
Allow vias in any pattern	(Non-serpentine only) Selected, allows the router to add vias to non-serpentine tuning patterns. Note: During interactive tuning, the maximum number of vias is determined by the max vias-added value in the plow section of the Editor Control Dialog – Route tab.
Prevent stair steps	(Serpentine only) Selected, prevents stair steps in the pattern.
◆ Tuning iterations section – Defines the options that control reduction and increase of trace lengths (to meet delay constraints) while running a tune delay pass in the autorouter. You can select the Reduce length and Add length options simultaneously.	
Reduce length	Defines the reduce length attribute for tuning. This option operates as a "pre-tune" pass and attempts to resolve delay violations by reducing the length of the longest net to match the shortest net (of a matched set). This generally makes the Add length tuning for the automatic router easier. Off – Prevents any reduction in trace length. Low, Medium and High – Define the amount of time the automatic router should spend reducing trace length.

(Continued)

Table 14.1 (Continued)

Add length	Defines the add length attribute for tuning. This option uses the tuning pattern rules to add trace length to the shortest net of a matched set. Off – Prevents any increase in trace length. Low, Medium and High – Define the amount of time the automatic router should spend resolving delay constraints. Exhaustive – Tunes the trace (to completion) with no time limit.
♦ Auto Tune options section – Defines Auto Tune options, including when, and for what nets, tuning occurs.	
Effort	Defines what nets to tune. This applies for Automatic urgency and Interactive urgency options. ReTune Only – Limits retuning to existing traces and nets that are out of tolerance. This option does not tune untuned nets. Tune & ReTune – Enables tuning of all nets. This option also allows tuning to push and shove traces and vias to achieve optimal tuning.
Automatic urgency	Defines when automatic tuning occurs relative to an auto route pass. Off – Automatic urgency is off. You must use the Auto Route Tune pass to tune nets. + At End of Pass – Tuning occurs at the end of the auto route pass. Auto Route Pass Type, Tune. + At End of Effort – Tuning occurs at the end of effort for Target Area Route and Auto Routing or Auto Route Pass Type, Straight Line Interconnects. + On Netline Routed – Tuning occurs when the automatic router routes a net. This does not require the Tune pass.
Interactive urgency	Defines when interactive tuning occurs. Off – Interactive urgency is off. + On Idle – Tuning occurs when the Interactive command completes. Typically, tuning occurs less frequently than for other options except for Off. + On Netline Routed – Tuning occurs when you route a netline using plow or interactive route. + While Clicking – Tuning occurs when you release the mouse button, or at the end of dragging or digitizing point to point in plow. + On Drag – Tuning occurs while you drag (every mouse move). With this option, tuning occurs most frequently of all options.

After tuning rule setup, we can start the tune process; there are three types: automatic tune, interactive tune and manual tune.

Automatic tune is included in Auto Route Pass Type; first select Route Pass Type, then select nets, such as HS01~HS06, then select Tune Delay Pass Type, as shown in Figure 14.32.

Then click the Route button, and the system will begin to auto route and auto tune. According to the different tuning settings, we can get different results. Figure 14.33 shows the serpentine tuning form and the trombone tuning form.

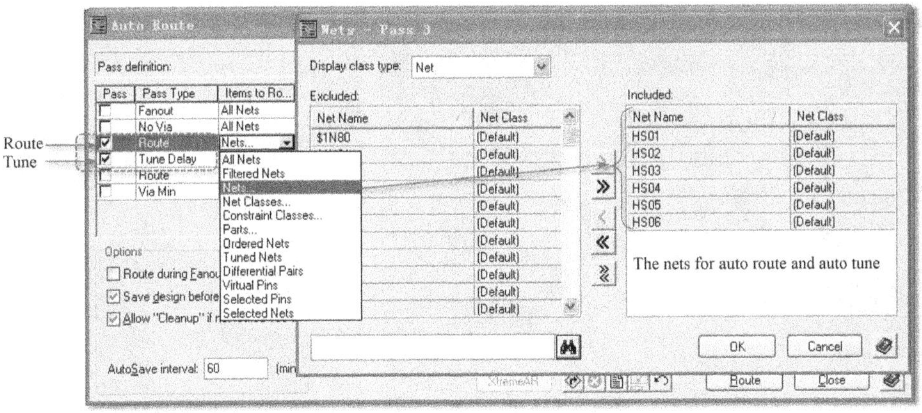

Figure 14.32 Automatic tune setup.

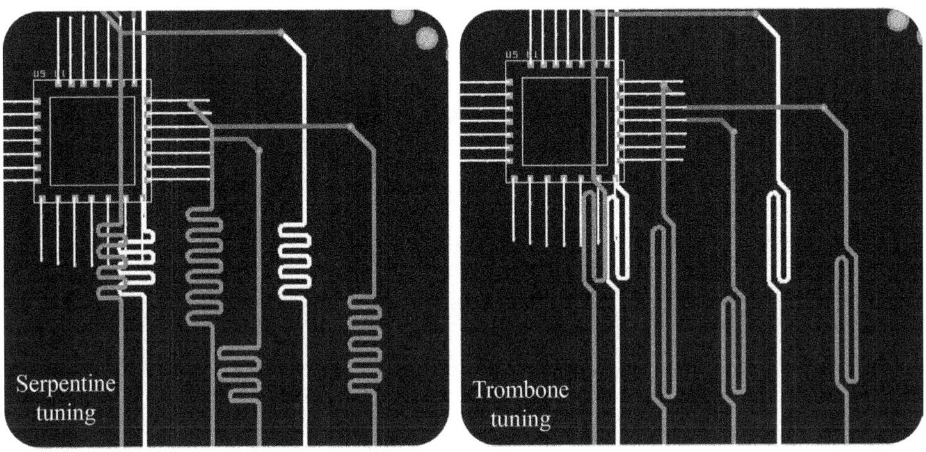

Figure 14.33 Serpentine tuning form and trombone tuning form.

Interactive tune can be carried out when routing is finished; we can select a single net or multiple nets, and then click the tune button to start interactive tune. The tuning effect is almost the same as for automatic tune.

Manual tuning is done by choosing a routed net and then clicking the Manual Tune button . The designer controls the tuning position and length manually.

With the Tuning Meter open, the tuning window will automatically show the length range of the setting in CES and the current tuning length, for the designer's reference. When tuning length is within the setting range, the ruler is green; when tuning length is close to the setting range, the ruler shows yellow for warning; when tuning length is greater than the setting range, the ruler shows red for alert, as shown in Figure 14.34.

After the completion of tuning, in order to compare the actual tuning length and the length set in CES, in CES select from the menu Data→Actuals→Update All; the actual tuning length can be imported to CES, as shown in Figure 14.35.

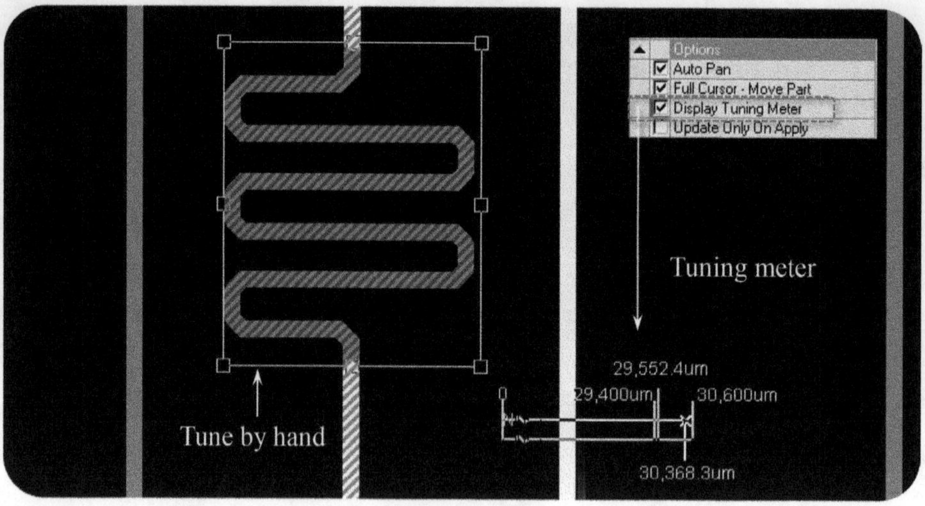

Figure 14.34 Manual tune effect.

Constraint Class/Net	Net Class	Type	Min (ns)|(um)	Max (ns)|(um)	Actual (u	Manhattan (
(All)	(Default)	Length				
DIFF1_N,DIFF1_P	(Default)	Length	length setting			
DIFF2_N,DIFF2_P	(Default)	Length	Minimum length	Maximum length		
DIFF3_N,DIFF3_P	(Default)	Length				
DIFF4_N,DIFF4_P	(Default)	Length				
HS01	(Default)	Length	29,400	30,600	29,400	17,526
HS02	(Default)	Length	29,400	30,600	29,558.423	17,764.923
HS03	(Default)	Length	29,400	30,600	29,571.934	13,725.415
HS04	(Default)	Length	29,400	30,600	29,617.649	17,816.83
HS05	(Default)	Length	29,400	30,600	29,638.093	18,714.21
HS06	(Default)	Length	29,400	30,600	29,592.145	19,971.437

Figure 14.35 Import the actual tuning length.

The Actual column displays the actual trace length of the net; if the actual length is close to the setting range, the bar will display yellow; if the actual length is beyond the scope of the setting, the bar will display red for alert.

In Figure 14.35, Manhattan length refers to the sum of X coordinates difference DX and Y coordinates difference DY. Under normal circumstance, the actual route length can't exceed the Manhattan length, but for tuning nets, the situation may be just the opposite.

14.2 Plane

14.2.1 Plane Definition

Plane, also known as copper pour, is used to connect the same net by copper or other metal filling; it is often used for power net or ground net connection.

When applying plane, we can set a layer for plane usually called plane layer; we can also draw the plane shape in the signal layer. The same net pads and vias will be connected by the plane, the other nets will be isolated automatically.

In Expedition, plane is divided into positive plane and negative plane, as shown in Figure 14.36.

Positive plane, in the generated Gerber file, shows the actual copper area. When applying plane around traces and pads or when a copper pattern is needed, we need to use positive plane.

Negative plane, in the generated Gerber file, is displayed in reverse image, that is, the copper does not display, as negative plane does not support copper pattern.

14.2.2 Plane Setting

In Expedition, plane setting has two aspects. On the right side of the Route toolbar there are two buttons: Plane Classes Parameters and Plane Assignments.

14.2.2.1 Plane Classes Parameters

Click the button to open the Plane Classes and Parameters window; in this window, the designer can create new plane classes and define related parameters for each plane class, as shown in Figure 14.37.

To create a plane class, click the New Plane Class button at the top of the window, and then set parameters for this plane class; we can create multiple plane classes in Expedition for one design and set different parameters for each class.

Parameter setting for plane classes includes three tabs: Thermal Definition, Clearances/Discard/Negative and Hatch Options.

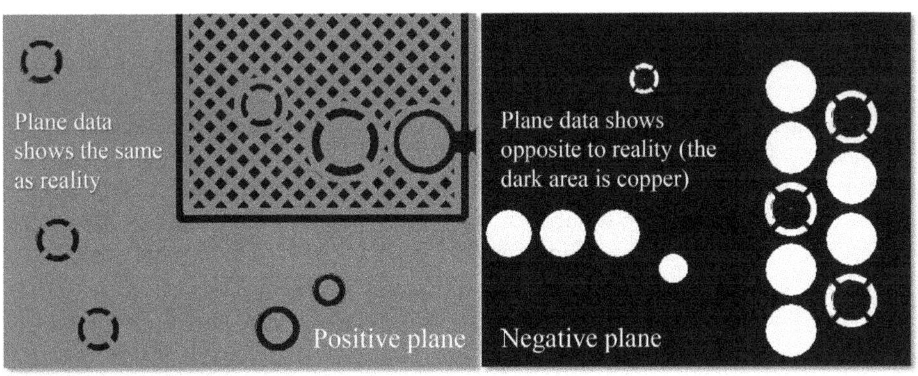

Figure 14.36 Positive plane and negative plane.

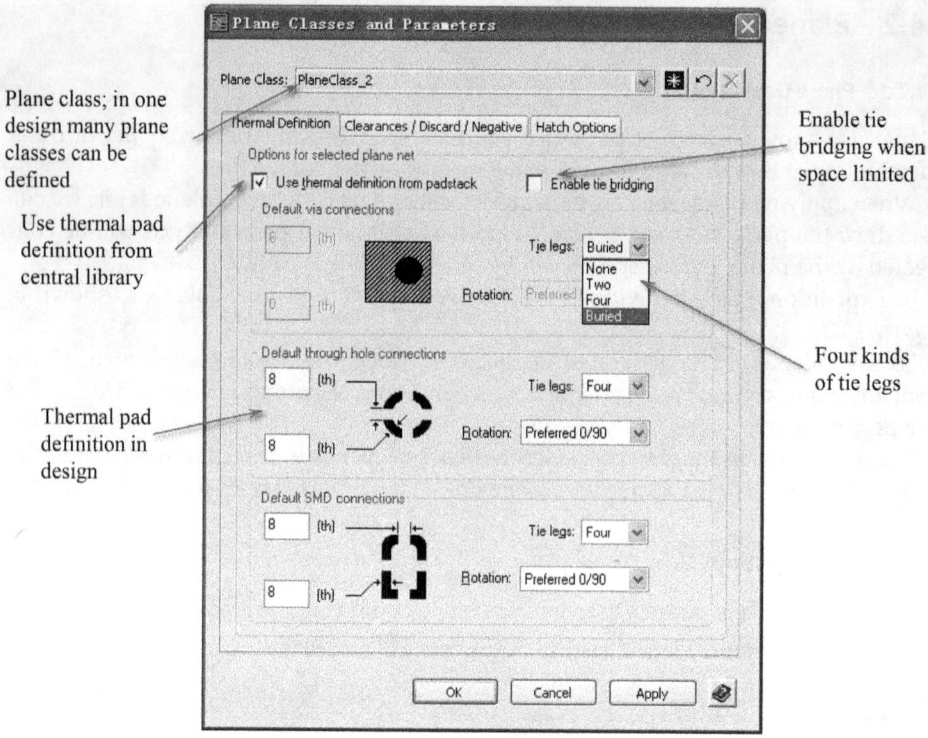

Figure 14.37 Thermal definition in the Plane Classes and Parameters window.

1) Thermal definition tab
   The Thermal Definition tab is mainly used for the definition of thermal connections, including via connections, through-hole pad connections and SMD connections. When "Use thermal definition from padstack" is checked, if the thermal connections were defined in the central library, the central library definition will be used. If the connections were not defined in the central library, the design definition will be used here. If this item is not checked, then this plane class will use the thermal connection defined here.
2) Clearances/Discard/Negative tab
   The Clearances/Discard/Negative tab is mainly used for the plane clearance and plane island settings. Default clearances are set for mounting holes and other objects to plane data; if the values are different from the CES definitions, use the larger values.
   Discard plane area options are used for isolated copper settings; it can help to effectively avoid problems with isolated copper, where it is only connected to one pad or where the copper area is less than a certain restricted area, so as to avoid unexpected EMC problems.
   Negative clearance pads have two modes, Anti-Pad and Donut, and the designer can choose according to the actual design needs. The Negative plane fill distance beyond route border option usually needs a bigger value; its value should be greater than the distance from the route border to the board outline, as shown in Figure 14.38.

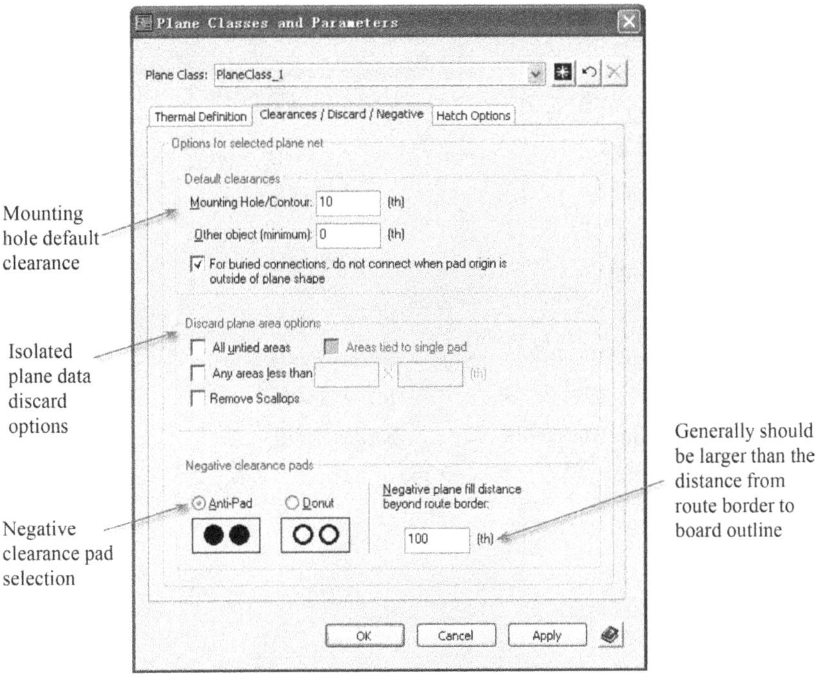

Figure 14.38 Clearances/Discard/Negative tab.

We can see from the right side of Figure 14.39 that if the negative plane fill distance beyond route border is set to 0, in the negative plane the copper will be generated between route border and board outline. The copper will spread to the edge of the board, a situation where the short circuit phenomenon can easily occur, and therefore this should be avoided in design. If a reasonable value is set, such as setting it to 100th, as shown on the left side of Figure 14.39, this will not be the case.

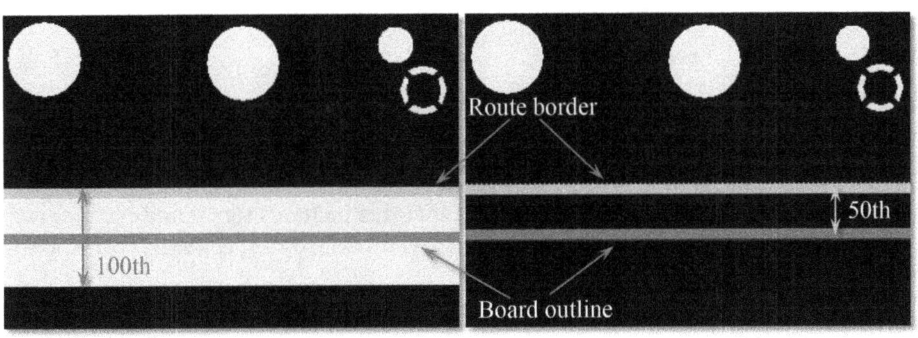

(a) Negative plane fill distance beyond route border set to 100th

(b) Negative plane fill distance beyond route border set to 0th

Figure 14.39 Two different settings for the Negative plane fill distance beyond route border option.

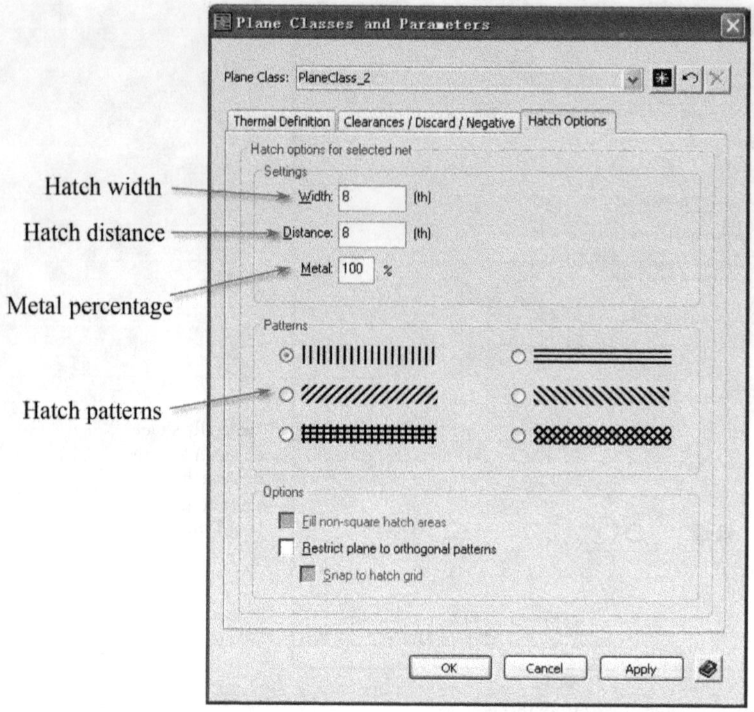

Figure 14.40 Hatch Options tab.

3) Hatch Options tab

   The Hatch Options tab is mainly used to set the copper pattern for the positive plane layer. If the width is set to the same value as the distance, or width > distance, solid copper will be generated; when width < distance, latticed copper will be generated.

   It is important to note that the latticed copper pattern is only applicable to positive plane and can't be generated in the negative plane, as shown in Figure 14.40. After the completion of plane class parameter setting, click OK or Apply, and all settings will be applied to the design.

14.2.2.2 Plane Assignments

The Plane Assignments window is shown in Figure 14.41.

1) Layer\Net is used to list all the layers in the design and the plane nets contained in each layer.
2) Layer Usage column is used to set layer type; this includes signal and plane types, with signal layers for routing. For the plane layer we need to specify the corresponding plane nets, such as GND or VCC.
3) Plane Type is used to set the plane type; it includes positive and negative. Signal layer can only be positive.
4) Plane Class is used to assign plane class; options for plane class are created in the Plane Classes and Parameters dialog window. Figure 14.41 shows Default plane class and PlaneClass_1 to PlaneClass_4.

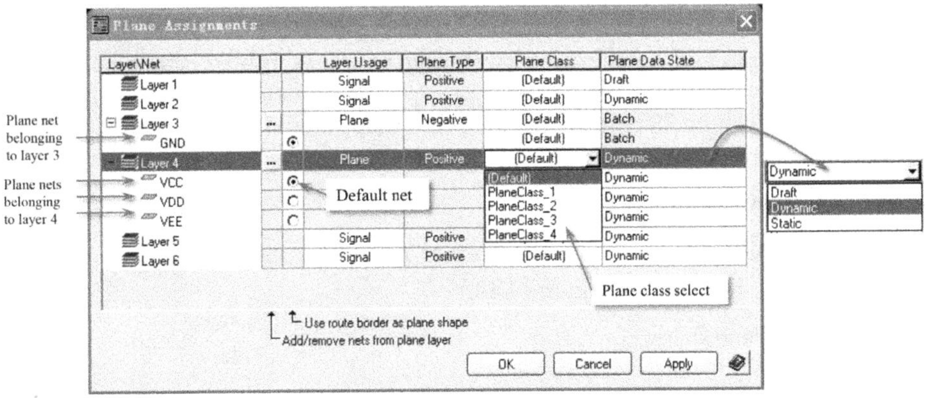

Figure 14.41 Plane Assignments window.

5) sPlane Data State is used to set the plane data state; there are three options: draft, dynamic and static respectively. Dynamic is usually used in design; in dynamic state, plane data can update in real-time, without any special command.

In the Add/remove nets from plane layer column, click the button ...; the nets windows will pop up. The Excluded bar on the left side of the window lists the nets that have not been assigned to the plane layer; the Included bar on the right lists the nets that have been assigned to the plane layer. Use the leftward or rightward arrow in the middle to assign selected nets to the plane layer or to remove selected nets from the plane layer, as shown in Figure 14.42.

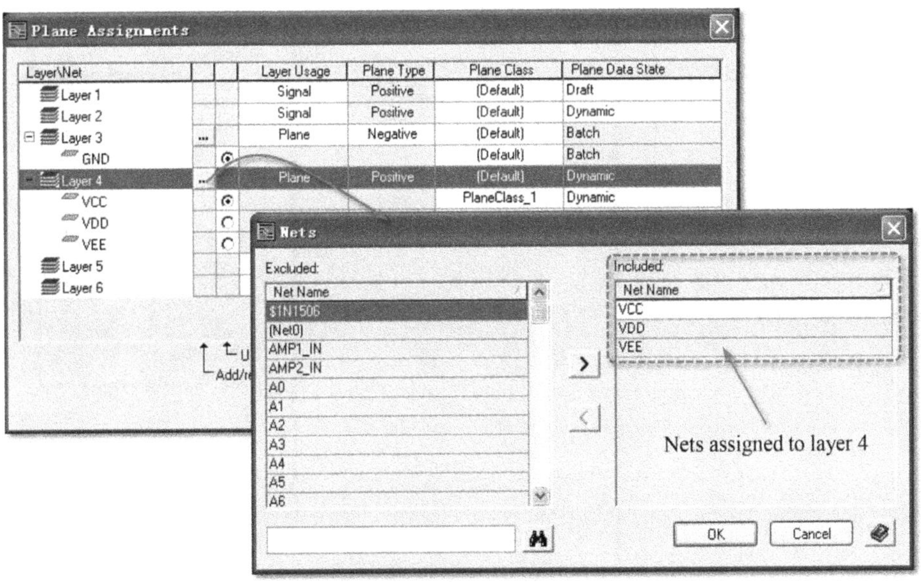

Figure 14.42 Assign nets to plane layer.

In the Use route border as plane shape column, indicates that the corresponding net is the default net in the plane layer, using route border as its plane shape, so it is not necessary to draw the plane shape. In this case, VCC is the default net of Layer4; other nets such as VDD and VEE require that the plane shape boundary be drawn.

14.2.3 Draw Plane Shape

In draw mode , click the properties button ; in the pop-up window select the following options, as shown in Figure 14.43.

1) Type: Select plane shape.
2) Layer: Select 4P, P stands for plane layer.
3) Net: Select VDD; the options for nets are VCC, VDD and VEE, assigned in the Plane Assignment window.
4) Obstruct type: Select none. If trace or via is selected, the system will not allow the designer to route or place via in the plane shape area.
5) Lock option is not selected normally; if selected, the plane shape will be locked.
6) Isolate Plane option is usually not selected; it indicates that the plane shape will automatically merge with nets of the same plane shape when overlapping occurs. If this option is selected, the plane shape will stay isolated from other plane shapes. Isolated spacing refers to plane to plane clearance in CES.

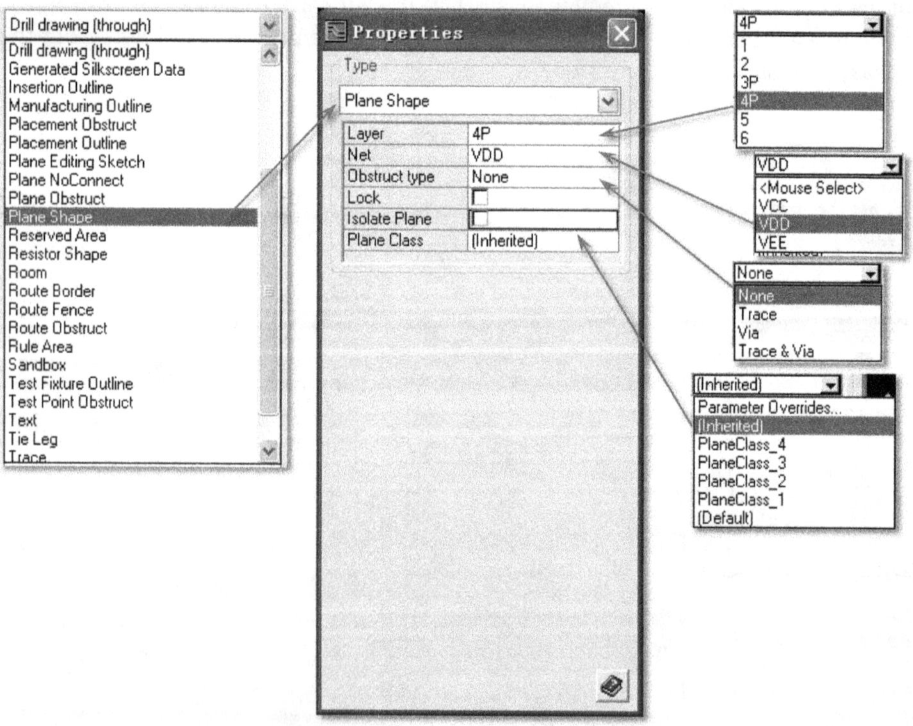

Figure 14.43 Plane shape properties window.

7) Plane Class: the user can select from the drop-down list. "Inherited" indicates that the plane shape inherits the plane class defined in the Plane Assignment window for such net.

After properties setting, click button to draw the plane shape; the selected VDD net will automatically be highlighted at this time. Draw a polygon according to the design needs. Usually, the polygon must include all the VDD net pads and vias, and exclude other plane net pads and vias, as shown in Figure 14.44.

After completing plane shape drawing for VDD, follow the same rules to draw the VEE plane shape. Similarly, the VEE plane shape must include all the VEE net points and exclude other plane net points, as shown in Figure 14.45.

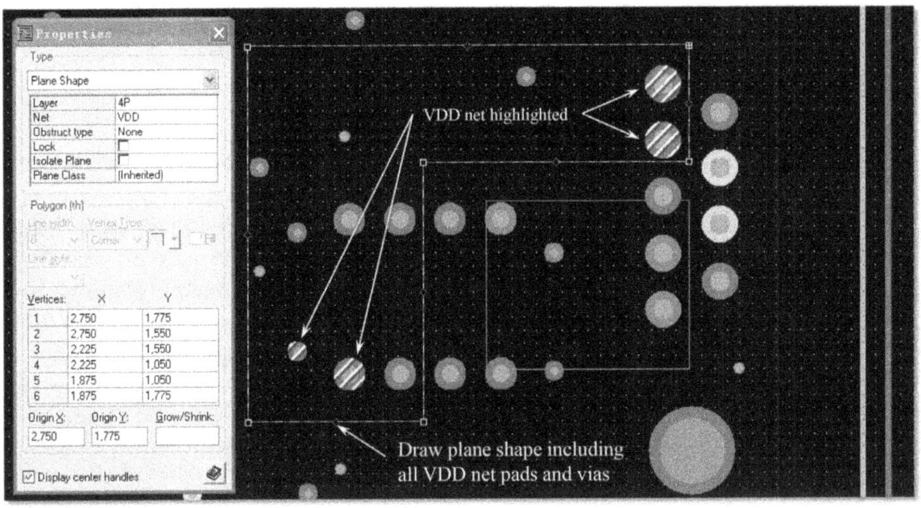

Figure 14.44 Draw VDD plane shape.

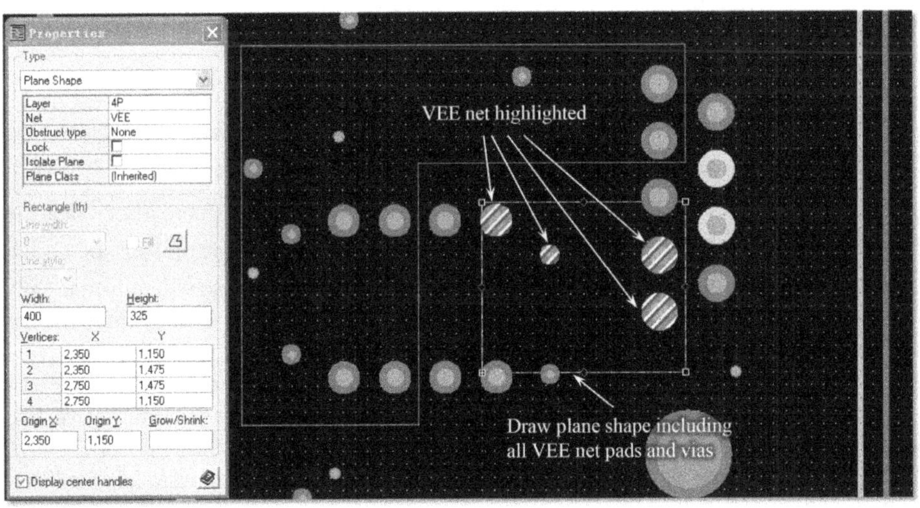

Figure 14.45 Draw VEE plane shape.

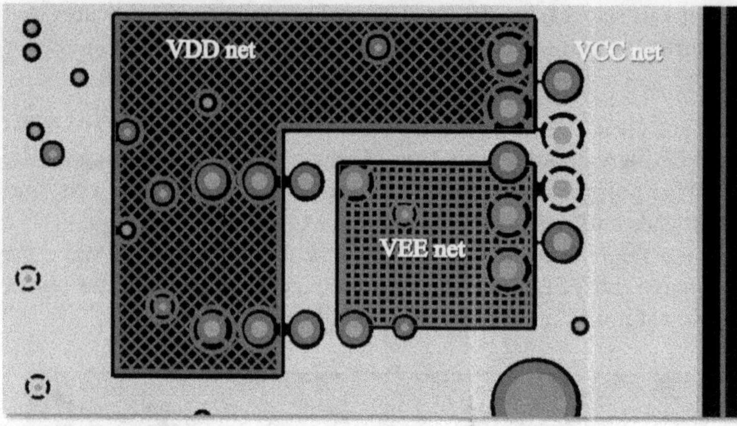

Figure 14.46 Actual plane data.

For the VCC net, because it has been set as a default net in this layer, the route border is its boundary, so it is not necessary to draw its plane shape.

After the completion of plane shape drawing, open plane data in Display Control, and select the Fill/hatch option; we can see the actual effect of the plane data drawing.

Because the plane data state selected is dynamic, in Expedition the plane data is updated in real-time; whether components are moved, or the designer carries out routing or places vias, dynamic plane data will automatically update in real-time.

Figure 14.46 shows actual plane data for the VCC\VDD\VEE nets. For VCC the plane data uses the route border as the boundary, shown with solid hatch; the VDD and VEE plane shapes were drawn by the designer with different hatch patterns based on different plane class definitions.

14.2.4 Modify Plane Shape

After the completion of plane shape drawing, if we want to modify the plane shape, we first need to switch to draw mode. At any time, if the plane shape is modified, the plane data will automatically update.

It is usual to adopt two methods to modify plane shape: one is to drag the vertex of the plane shape; another is to draw graphics, and then use the addition and subtraction buttons [+] [-] to merge or subtract graphics.

The specific method is as follows. First, select the original plane shape, then click the addition button [+] or the subtraction button [-] and select the graphics to be added or subtracted (the graphics need to be drawn in advance). The system will automatically add or subtract the graphics from the plane shape, to give a new plane shape with the same properties as the original one, as shown in Figure 14.47.

In addition, we often need to edit the corner vertex of the plane shape. Expedition supports three types of vertex: corner, round and chamfer. After selecting the vertex, using the drop-down arrow to switch, as shown in Figure 14.48.

For positive plane shape, since the plane status is usually set to dynamic, after editing the plane shape, the plane data will automatically update in real-time. Figure 14.49 shows a comparison of the VDD net before and after plane-shape editing.

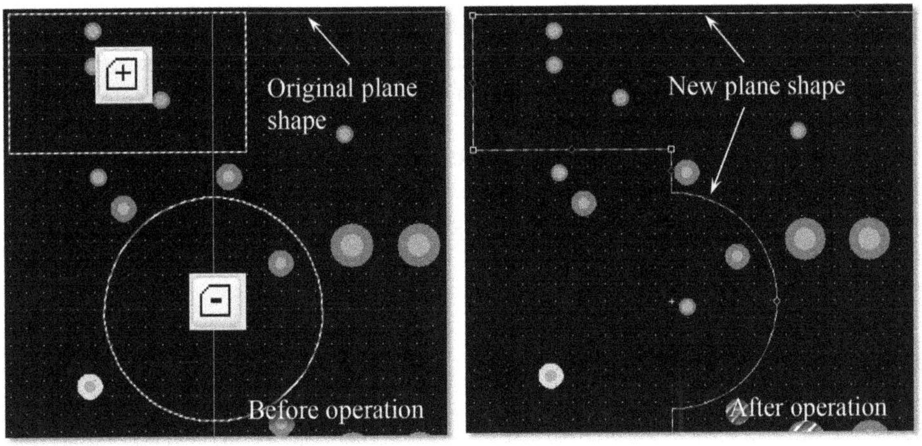

Figure 14.47 Modify plane shape by adding or subtracting graphics.

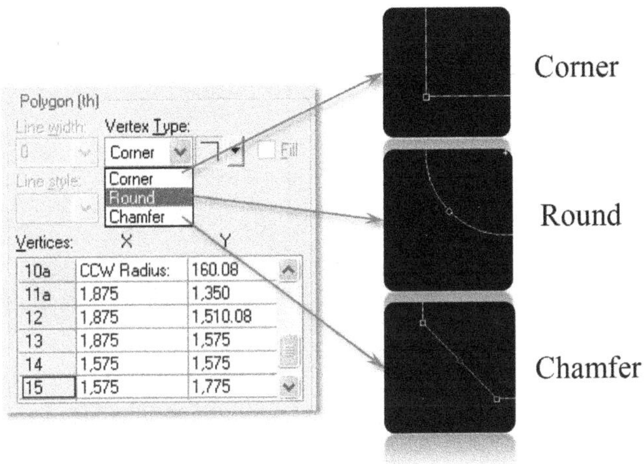

Figure 14.48 Three types of vertex.

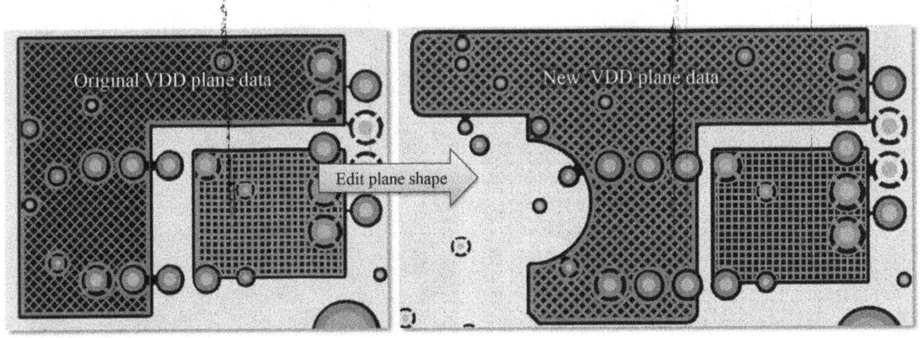

Figure 14.49 Comparison of VDD net before and after plane-shape editing.

14.2.5 Generate Negative Plane Data

For negative plane layers, since the plane data state can only be set to batch, after negative plane setup, we need to run Planes→Generate Negative Planes to generate negative plane data, as shown in Figure 14.50.

After running the Generate negative planes command, if there is no error message, the negative plane data is generated. Figure 14.51 shows a comparison of before and after negative plane data generation; the thermal pad shape indicates that the pad is connected to plane data; the solid pad shape indicates that the pad is isolated from plane data.

14.2.6 Delete Plane Data

To delete positive plane data, we can select the relevant plane shape in draw mode and directly press the < Delete > key; the plane shape and plane data will be deleted together.

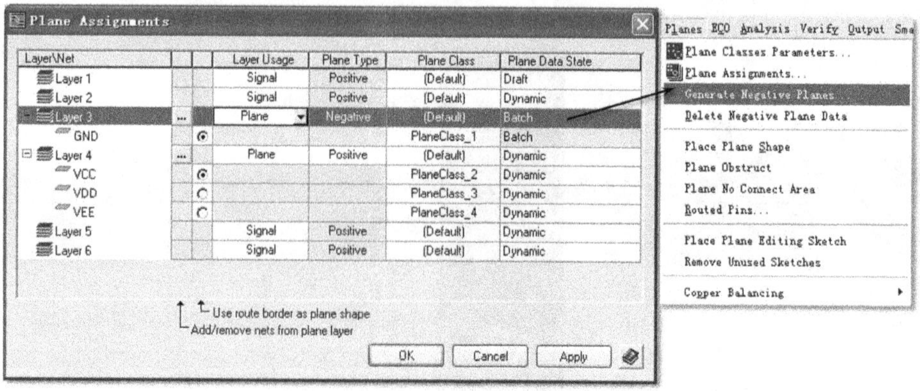

Figure 14.50 Generate negative plane data.

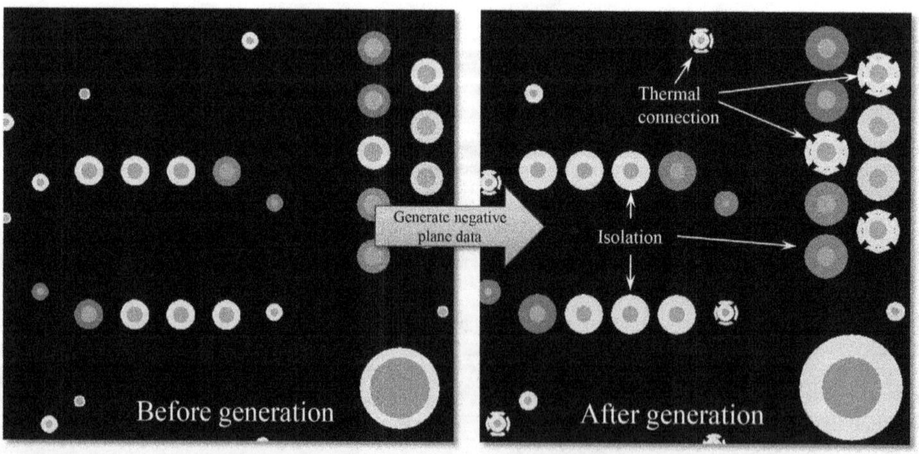

Figure 14.51 Comparison before and after negative plane data generation.

To delete negative plane data, first select from the menu Plane→Delete negative plane data, and then delete the plane shape in draw mode.

14.2.7 Verify Plane Data

After plane data generation, we can use batch DRC to verify it.

Usually, the plane data problems that can arise, and the solutions for those problems, are as follows:

1) Plane data did not cover all relevant pins. Solution: Modify the plane shape to include all the necessary pins, or connect the pins by trace.
2) Plane islands. Solution: Change the Discard plane area options in the Plane Classes and Parameters window; throw away all untied areas.

In addition, because of technics and DFM requirements, we may need to delete plane areas only connected to a single pad or plane areas smaller than a certain limit, or we may need to apply plane data optimization to avoid curve edge that is too complex. These all can be set in the Discard plane area options of the Plane Classes and Parameters window, as shown in Figure 14.52.

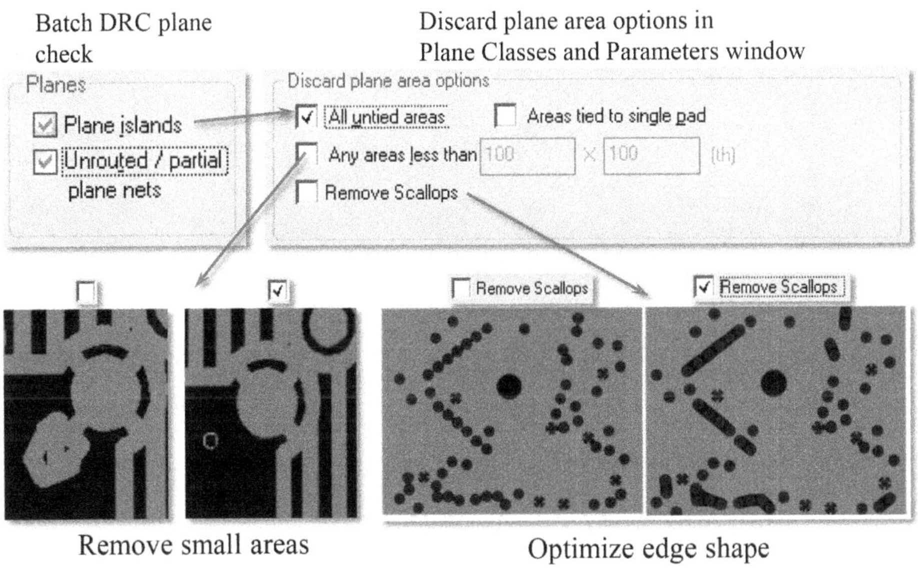

Figure 14.52 Batch DRC check items for plane and discard options in Plane Classes and Parameters.

15

Embedded Passives Design

15.1 The Development of Embedded Technology

With advances in electronic technology, high-speed, high-density design of electronic products increased rapidly. The requirements of electronic product miniaturization, low power consumption, high performance and multiple functions also become more and more imperative, and the increased attention that high-density package and SiP are receiving shows their importance.

With function and performance continually increasing in electronic systems, the layout of components in PCB, packaging and SiP substrate is becoming more difficult. With improvements in IC chip integration and the fast increase in I/O numbers, the number of passive devices will also continue to increase rapidly. Embedding a large number of passive devices in printed circuit board or SiP substrate can shorten the connections between components, improve electrical properties, reduce the package area and reduce welding spots, so as to improve reliability and effectively reduce production cost.

15.1.1 Discrete Embedded Technology

Discrete embedded technology refers to embedding discrete components such as chips, resistors, capacitors and inductors, usually installed on the surface of the substrate, *into* the substrate, using special design and processes, so as to save surface space and shorten interconnections.

In Expedition, discrete embedded technology is implemented using substrate cavity, as shown in Figure 15.1. Components are embedded into a closed cavity inside the substrate; after the completion of substrate production, the internal components are invisible. This technique can be used to embed specific chips into substrate, and to form a base board or PCB with intellectual property rights (IPR) protection.

This chapter focuses on planar embedded technology; for discrete embedded component design please refer to Chapter 12.

15.1.2 Planar Embedded Technology

Planar embedded technology refers to those passive components such as resistors, capacitors and inductors, which, using a combination of design and processing (use of etching or printing methods to generate substitute passive components in the substrate

SiP System-in-Package Design and Simulation: Mentor EE Flow Advanced Design Guide,
First Edition. Suny Li (Li Yang).

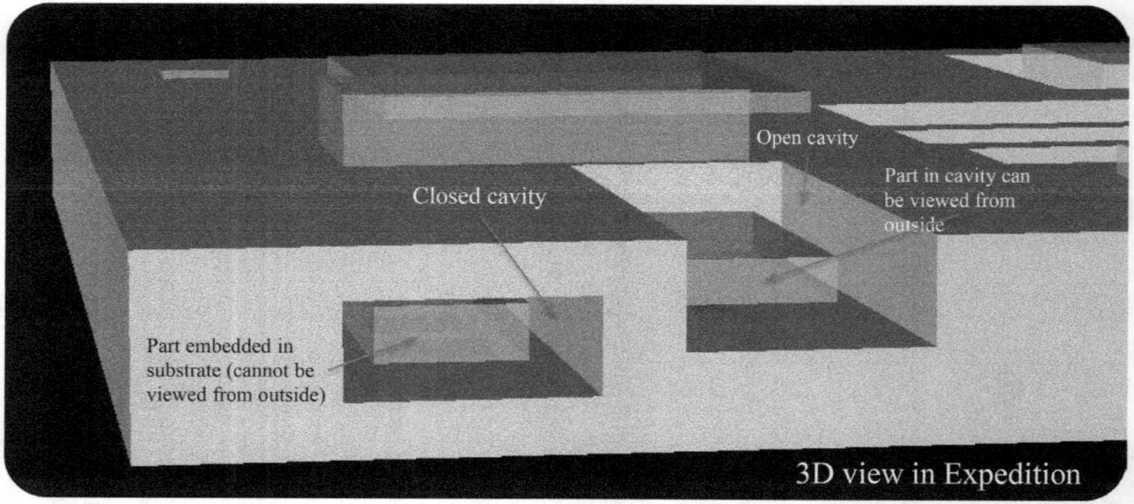

Figure 15.1 Discrete devices embedded with closed cavity in Expedition.

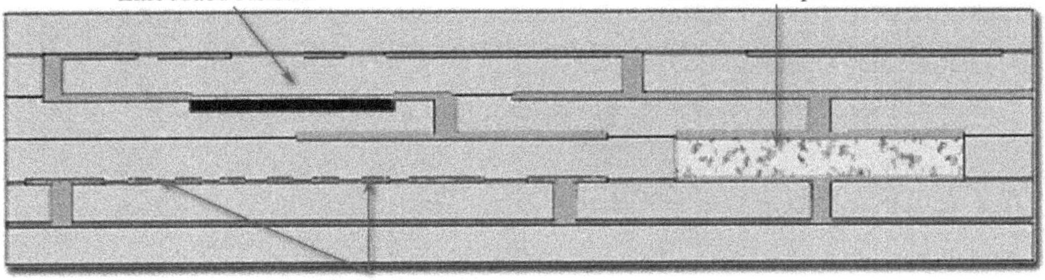

Figure 15.2 Embedded resistors, capacitors and inductors in substrate.

surface or inner layer, and then using laminate or buildup board-production processes) are embedded into the substrate, replacing surface-welded passive components, thus improving active chip layout and the degree of routing freedom.

3) Embedded resistor technology
   Embedded resistor technology usually adopts high-resistivity materials to make various shapes and resistances of planar resistor. Resistance material includes nickel phosphorus alloy and non-metallic materials such as carbon, graphite, diamond powder, etc. It can also be metal powder and nonmetal filler (e.g., silicon powder, glass powder) blended with resin, dispersing agent, etc. to create pulp compounds.
4) Embedded capacitor technology
   Embedded capacitor technology usually adopts the dielectric membrane method, which has two process methods, thick film or thin film. Generally, higher dielectric constant materials are used.
5) Embedded inductor technology
   Embedded inductor technology usually adopts etching of copper foil or copper plating to form screw or bending shapes, or uses via connection to form multilayer spiral structures. Its characteristics depend on the substrate parameters and graphics' shapes. Now this process can support a relatively small inductance value, only about dozens of nH, which is mainly used in high-frequency modules.
   Figure 15.2 shows embedded resistors, capacitors and inductors in substrate.

In Expedition, embedded resistors and capacitors are produced via the automatic synthesis tool, while embedded inductors are generated via the RF design tool. In this chapter we focus on embedded resistor and capacitor design methods.

15.2 Process and Material for Embedded Passives

In Library Manager, select Tools→Material/Process Editor, and open the Material/Process Editor window. The left side of the window is the navigator, which is divided into two categories: processes and materials. "Processes" mainly refers to the process flows for generating embedded resistors and capacitors. "Materials" refers to the many kinds of materials used to generate embedded resistors and capacitors, as shown in Figure 15.3.

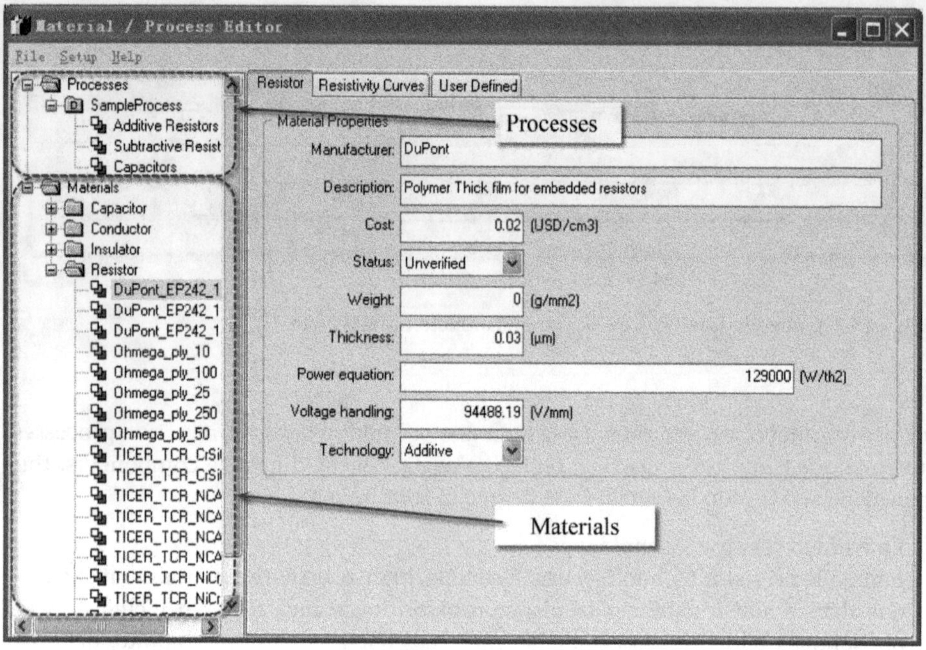

Figure 15.3 Material/ Process Editor window in the central library.

Process management includes the following processes:

1) Additive resistor process, often used for thick-film resistor generation.
2) Subtractive resistor process, often used for thin-film resistor generation.
3) Capacitor process, includes interdigitate capacitor, mezzanine capacitor and printed capacitor.

The designer can also create new processes and set different process parameters to meet the design requirements.

Materials management includes the following:

1) Capacitor materials, used to generate the dielectric for embedded capacitors.
2) Conductor material, used to generate the metal conductors for embedded capacitors.
3) Insulator materials, used to generate the insulating medium, such as FR4 in the substrate, etc.
4) Resistor materials, used to generate embedded resistors.

Under each type of material, the designer can create new material and add material-related parameters, such as custom-made resistor material, so as to edit and control all kinds of materials flexibly.

15.2.1 Embedding Processes

Here we can create multiple processes according to differing production line capacities or manufacturers. The new processes include additive resistor processes, subtractive

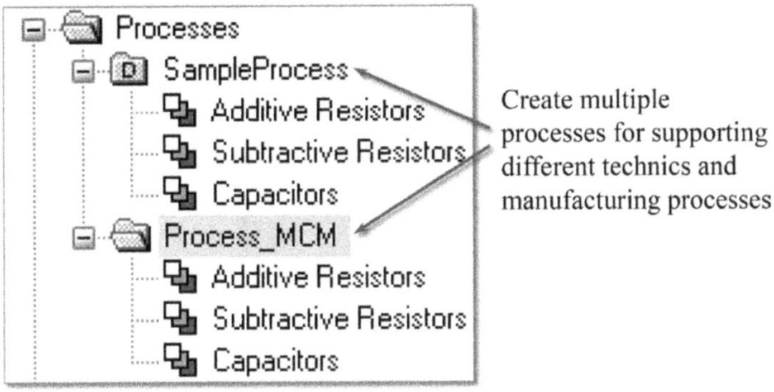

Figure 15.4 Expedition supports multiple processes.

resistor processes and three kinds of capacitor process (interdigitate capacitor, mezzanine capacitor and printed capacitor), as shown in Figure 15.4.

Usually, the rule definition for these processes requires the participation of production departments. Alternatively, before rule setup, the designer should communicate fully with the relevant technology personnel, and set the rules in compliance with process specifications. After setting, the rules can be applied to automatic synthesis tools.

15.2.1.1 Additive Resistors

As shown in Figure 15.5 for the common parameters definition of additive resistor process.

1) Common tab
   a) Material selection area: by default this contains several DuPont materials; the designer can add custom material. It is important to note that the material we can add in should be defined in advance in materials, and must support the same process as the current dialog box. For example, Material_1 is defined in materials as embedded resistor material and as supporting the additive process, thus it can be added in at this time.
   b) Allowed forms: these are the allowed resistor shapes; they include the following four shapes: rectangle, top hat, folded and serpentine, as shown in Figure 15.6.
   c) Overglaze: defines the Delta value of the protection glaze coating for resistor materials.
   d) Process parameters: includes settings for laser-trimming resistors, etc.
      - Minimum trim width, defines the minimum trim width and it can be understood as the width of the laser beam.
      - Process tolerance, defines the tolerance of synthesis resistor and actual resistor. If we define the process tolerance value as 1, the synthesis resistor and actual resistor will be completely equal; if this is defined as 0.8, the synthesis resistor will equal 80% of the actual resistor, and then laser trim (use laser to cut off the extra section) can be used to adjust the resistor value to the actual need.
      - Component tolerance, usually the default component tolerance value is 0%, and is not considered in the process. If the designer defines a non-zero value,

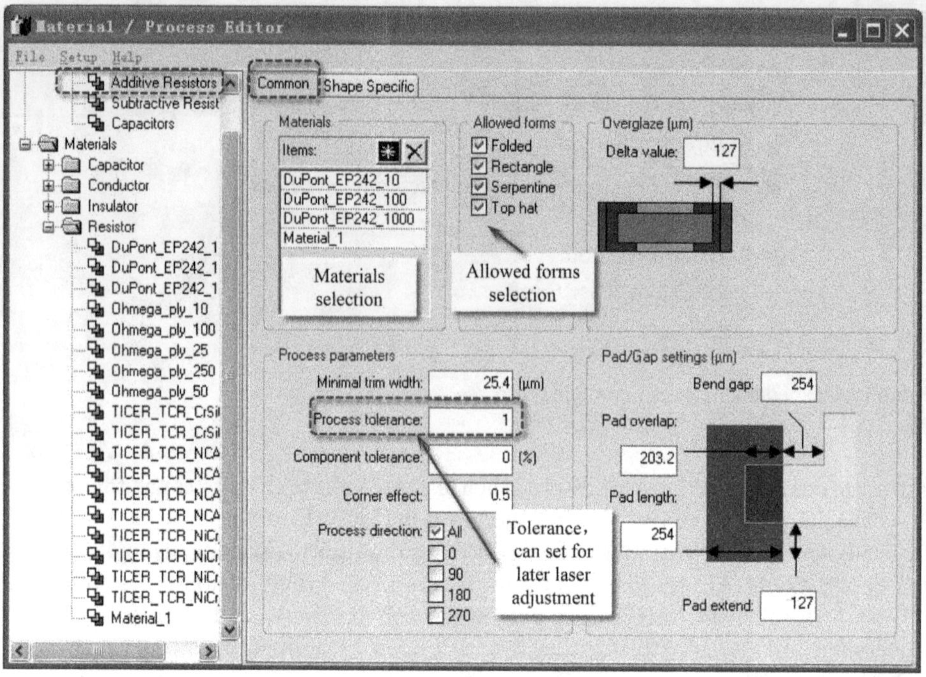

Figure 15.5 Common parameters definition for additive resistors.

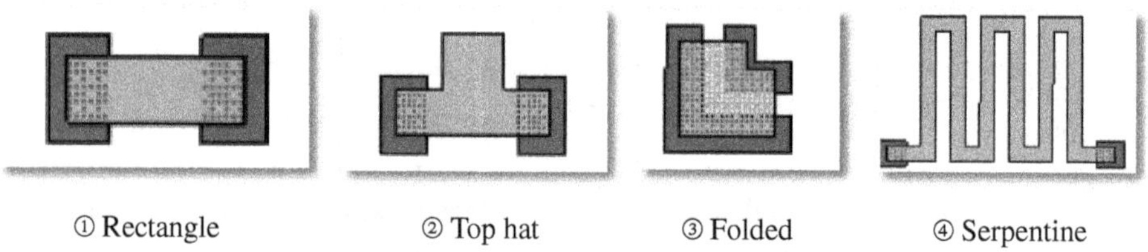

Figure 15.6 Expedition supports four types of resistor shape.

this value and the process tolerance value multiply to define the range of resistor laser trim. We suggest here that the designer keep the default value.

- Corner effect, this mainly defines the resistance-value change caused by a round corner. For the additive resistor process, the default value is 0.5; for the subtractive resistor process, the default value is 1.
- Process direction, defines the angle of the process, includes all, 0°, 90°, 180° and 270°.

e) Pad/Gap settings: mainly define the size of metal pad and the spacing of pad and resistor materials, etc.

- Bend gap, defines the distance between the pad and the bend part of the resistor material; primarily effective for the top hat and serpentine shapes.
- Pad overlap, defines the overlap size of resistor materials and pad.
- Pad length, defines the pad's longitudinal length.

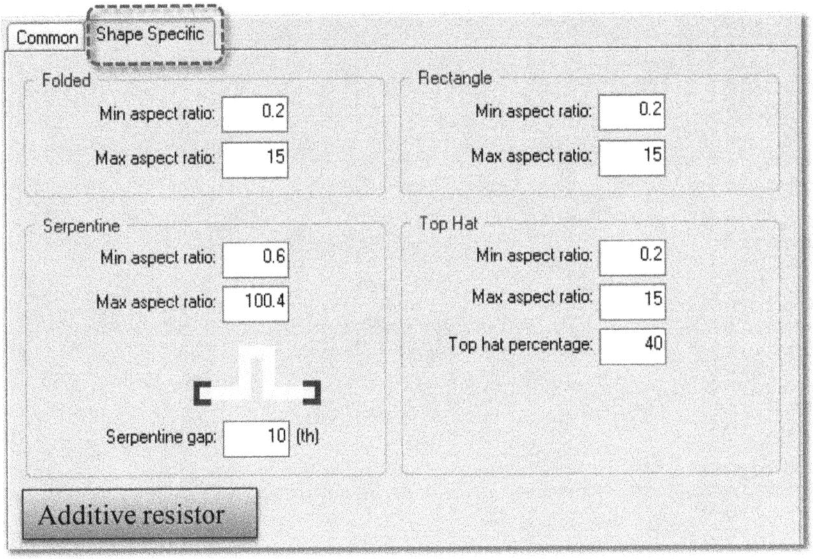

Figure 15.7 Shape Specific definition window for additive resistor.

- Pad extend, defines the pad's lateral epitaxial parameters (the relative extra distance of resistor material).

2) Shape Specific tab
   Defines parameters for folded, rectangle, serpentine and top hat, as shown in Figure 15.7.
   a) Min aspect ratio: defines the minimum length–width ratio of the resistor.
   b) Max aspect ratio: defines the maximum length–width ratio of the resistor.
   c) Serpentine gap: defines the internal spacing of serpentine resistor.
   d) Top hat percentage: defines the top hat shape proportion in the whole resistor.

15.2.1.2 Subtractive Resistors

Figure 15.8 shows the common parameters definition window for subtractive resistor.

1) Common tab
   a) Materials, used for materials selection; it is different from the additive resistor process, it includes mainly Ohmega and Ticer resistor materials. The designer can also add custom material, and, as for the additive resistor process, the material which the designer adds must be defined in materials in advance, and must support the subtractive resistor process.
   b) Allowed forms, defines allowed shapes of resistor; as with additive resistor, it mainly includes the following four shapes: folded, rectangle, serpentine and top hat.
   c) Production mask, defines the mask lateral expansion relative to the pad size.
2) Shape Specific tab
   This has the same shape parameters as for additive resistor, but the values are independently defined. Because the process is different, the parameter definition is not the same as for the additive resistor process, as shown in Figure 15.9.

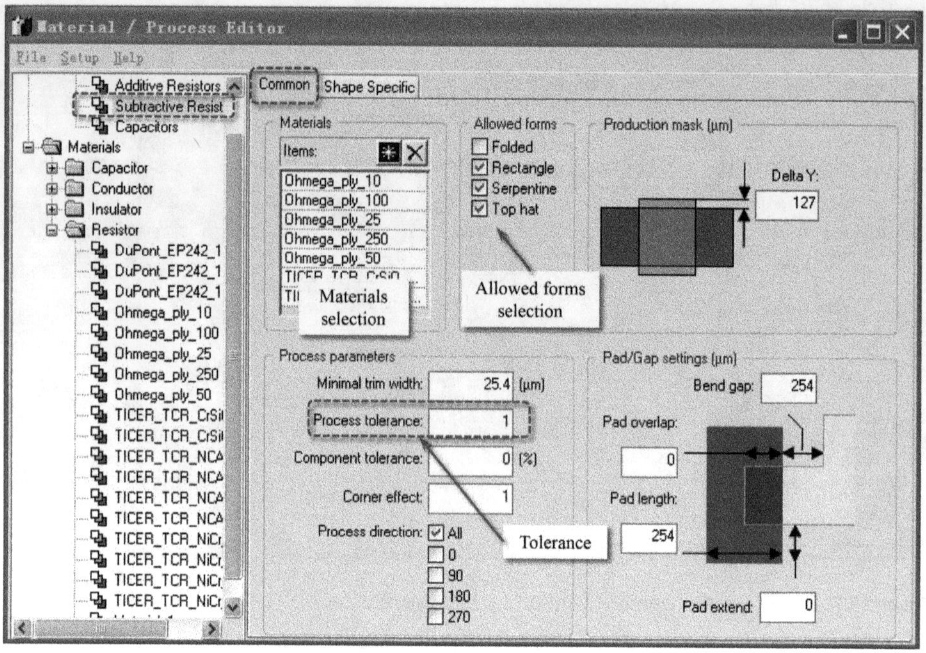

Figure 15.8 Common parameters definition window of subtractive resistor.

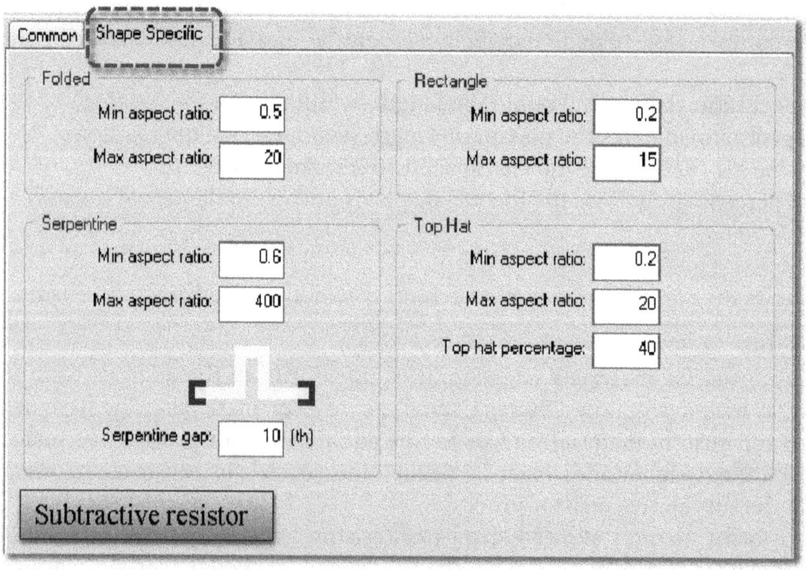

Figure 15.9 Shape Specific definition window of subtractive resistor.

15.2.1.3 Capacitors

Figure 15.10 shows the capacitor common parameters definition window.

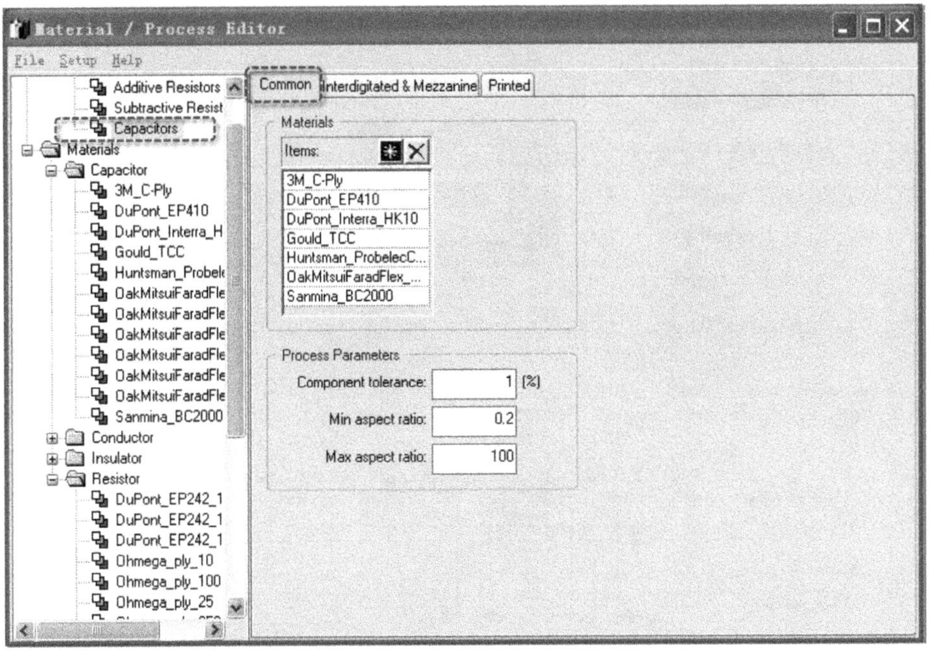

Figure 15.10 Capacitor common parameters definition window.

1) Common tab
   a) Materials, used for materials selection; lists materials options for capacitor, including 3M, DuPont, Gould, Huntsman, etc. The designer can also add customized material, and, as for the resistor process, the materials must be defined in capacitor materials in advance.
   b) Process parameters
      - Component tolerance, 1% stands for ±1% of the component value, the scope of the tolerance is 99%~101%.
      - Min aspect ratio, defines the minimum length–width ratio of the capacitor.
      - Max aspect ratio, defines the maximum length–width ratio of the capacitor.
2) Interdigitated & Mezzanine tab

   Figure 15.11 shows the capacitor Interdigitated & Mezzanine parameter-setting window.
   a) Interdigitated capacitor: this capacitor shape is like the crossed fingers of two hands; it is placed in the electrical layer as a whole element; internally it is filled with dielectric, as shown in Figure 15.12.
      - External finger width (A), refers to the width of metal A.
      - Internal finger width (B), refers to the width of metal B.
      - Slot width (C), the width of the gap between metals.
      - Term width (D), the width of the metal at the terminal of the capacitor.
      - Number of fingers.
      - IDC Dielectric Offset, refers to the amplification of dielectric relative to metal.

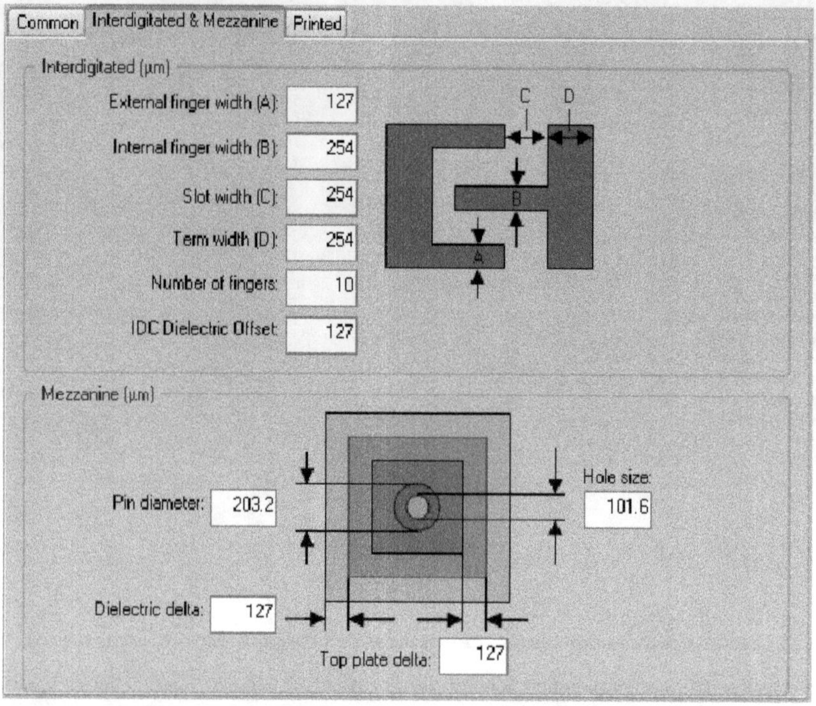

Figure 15.11 Capacitor Interdigitated & Mezzanine parameter-setup window.

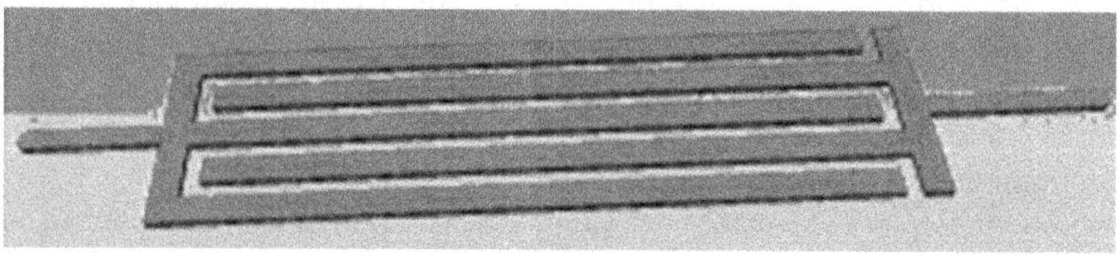

Figure 15.12 Interdigitated capacitor structure.

b) Mezzanine: this is a sandwich-type capacitor; its structure is more complex, and includes top metal, dielectric, bottom metal and a connect-via hole. Figure 15.13 shows the top view and side view of this type of capacitor's physical structure.
   - Pin diameter, the metal plate diameter of the hole.
   - Hole size.
   - Dielectric delta, the amplification of dielectric relative to the underlying metal.
   - Top plate delta, the reduction of the top metal relative to the bottom metal.

3) Printed tab

This relates to a printed capacitor; its structure includes two pieces of metal at the bottom of the capacitor, the two pins of the capacitor. One of the pins has a larger area; a dielectric layer is printed above this pin, then the conductor layer is printed above the dielectric layer, another end of the conductor is connected with the smaller metal pin, and the effective area is the overlapping area of the conductor and the big

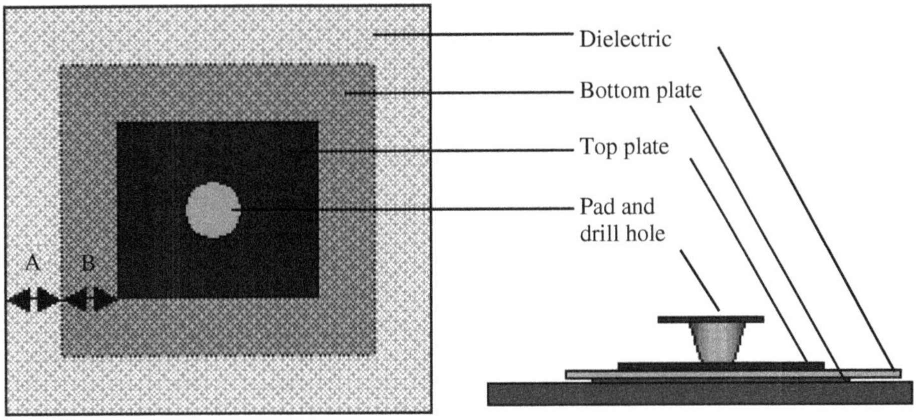

Figure 15.13 Mezzanine capacitor.

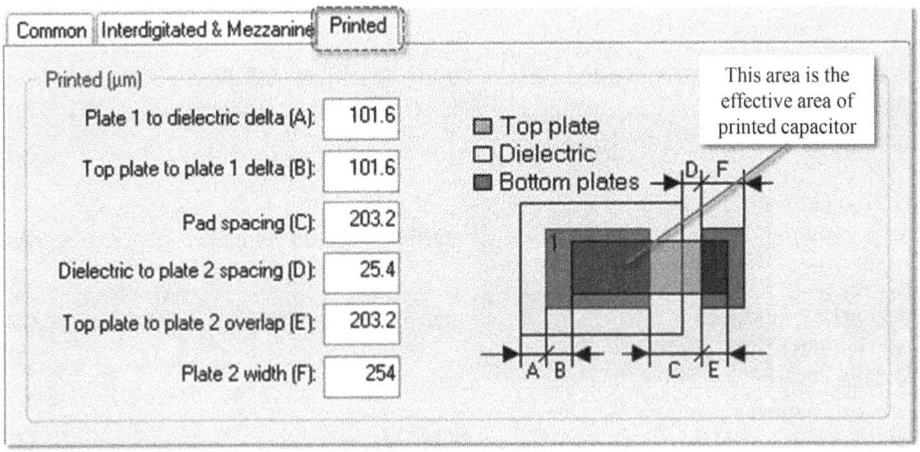

Figure 15.14 Capacitor Printed window.

pin separated by the dielectric layer. Figure 15.14 shows the capacitor Printed settings window.

- Plate 1 to dielectric delta (A), the amplification of the dielectric relative to the underlying metal (one pin of capacitor).
- Top plate to plate 1 delta (B), the reduction of the top conductor relative to the underlying metal.
- Pad spacing (C), the spacing of the two pieces of underlying metal (two pins of the capacitor).
- Dielectric to plate 2 spacing (D), the spacing between dielectric and the second pin of the capacitor.
- Top plate to plate 2 overlap (E), the overlap length of the top metal and the second pin of the capacitor.
- Plate 2 width (F), the length of the second capacitor pin.

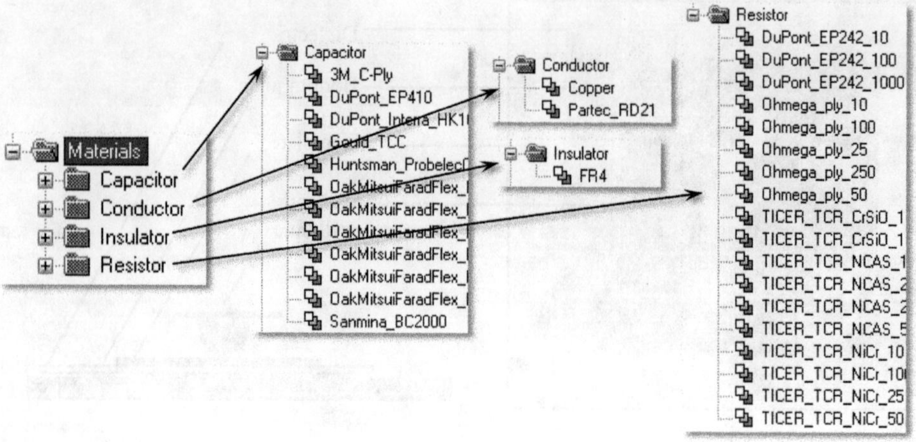

Figure 15.15 Materials definition list.

15.2.2 Embedding Materials

The materials used for embedded resistors and capacitors include the following four categories: capacitor materials, conductor materials, insulator materials and resistor materials, as shown in Figure 15.15.

15.2.2.1 Capacitor Materials

The parameters defined for capacitor materials include material properties and allowed forms.

1) Material properties
   - Manufacturers, such as DuPont, etc.
   - Description, describes the characteristics of the material.
   - Cost, material price.
   - Status, material status; this shows whether the material is verified or not; it contains verified and unverified options, and designers can adjust the status in different design stages. At the beginning of design, for example, the status would be unverified; after the material is used, its status can be changed to verified.
   - Weight, the weight of the material per unit area.
   - Thickness, the thickness of the material.
   - Dielectric constant, the dielectric constant of the material.
   - Loss tangent, the loss tangent of the material.

   For planar capacitors, material thickness and dielectric constant are the two key parameters; together they determine the capacitor value per unit area.
2) Allowed forms

   Allowed forms includes the three allowed capacitor shapes described above; checking all of them indicates all shapes are allowed.

   Figure 15.16 is the definition window for capacitor material parameters.

15.2.2.2 Conductor Materials

Conductor material parameters include material properties and valid conductor forms; the setup window is shown in Figure 15.17.

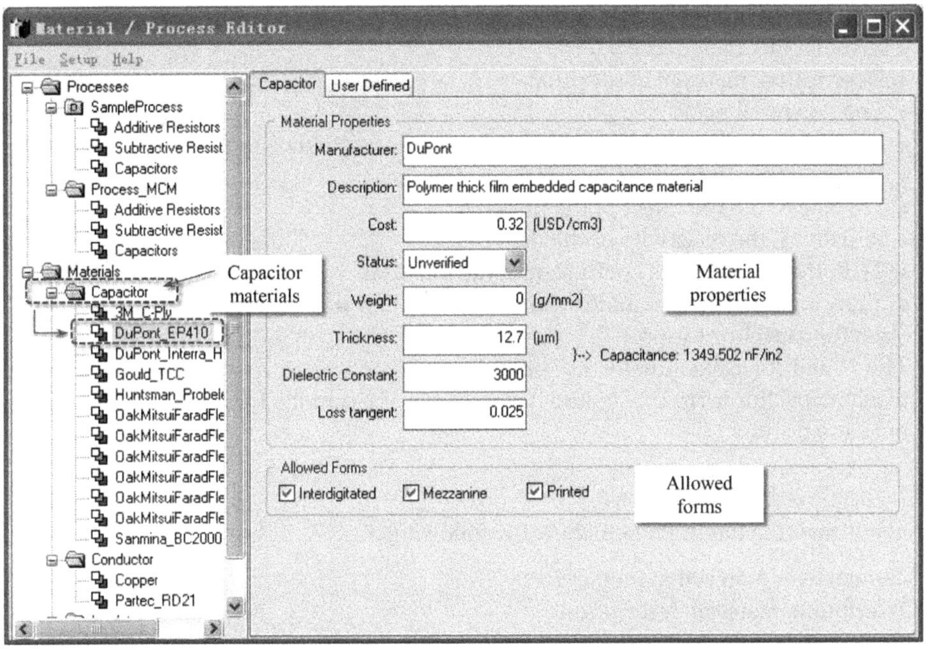

Figure 15.16 Capacitor material parameters window.

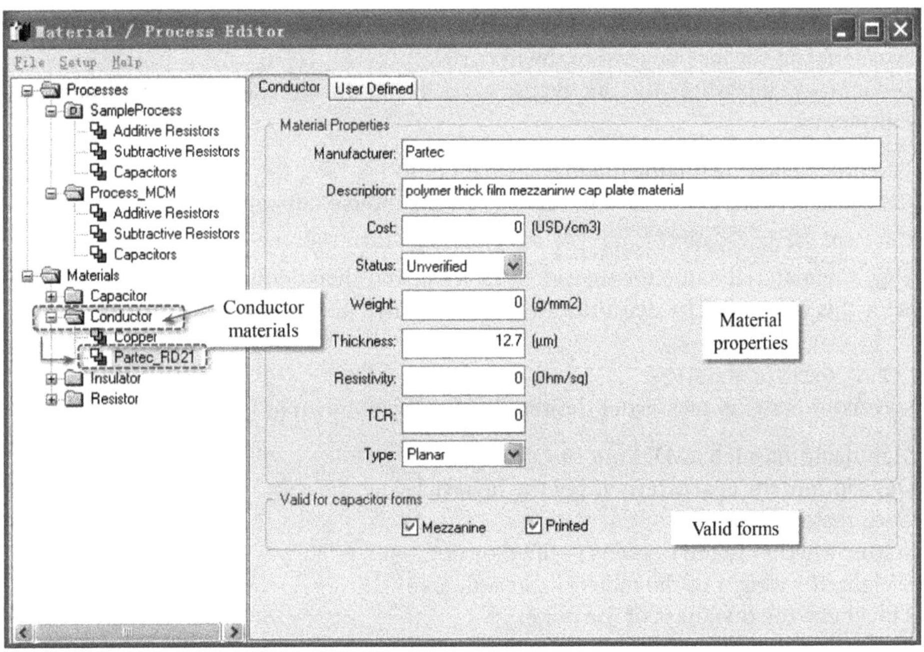

Figure 15.17 Conductor material parameters window.

1) Material properties
   - Manufacturer, such as Partec.
   - Description, material description.
   - Cost, material price.
   - Status, material status; whether the material is verified or not.
   - Weight, the weight of the material per unit area.
   - Thickness, the thickness of the material.
   - Resistivity, the resistivity of the material.
   - TCR, the temperature coefficient of the material.
   - Type, material type; includes planar material and wire material.
2) Valid for capacitor forms
   The "Valid for capacitor forms" options are mainly used to define which kind of planar capacitor form this material is valid for; the options for selection are mezzanine and printed.

15.2.2.3 Insulator Materials

Insulator material parameters include the following:.

- Manufacture, material manufacturer.
- Description, material description.
- Cost, material price.
- Status, material status, whether the material is verified or not.
- Weight, the weight of the material per unit area.
- Thickness, the thickness of the material.
- Dielectric constant, the dielectric constant of the material.
- Loss tangent, the loss tangent of the material.
- Technology, including the six types: core, substrate, prepreg, screened, additive, subtractive.

The only default insulator material in Expedition is FR4; the designer can add such materials according to actual use. Right-click the mouse button and select the pop-up menu item New Insulator Material to create new material; the designer only needs to change the material name for easy identification, and then define the relevant parameters for this material. The definition window is shown in Figure 15.18.

15.2.2.4 Resistor Materials

The resistor material properties definition window is shown in Figure 15.19.

- Manufacturer, such as DuPont or Ohmega.
- Description, the characteristics of the material.
- Cost, material price.
- Status, whether the material is verified or not.
- Weight, the weight of the material per unit area.
- Thickness, the thickness of the material.
- Power equation, the power consumption of the material per unit area.
- Voltage handling, the high-voltage resistance of the material.
- Technology, the applicable process, include additive process and subtractive process.

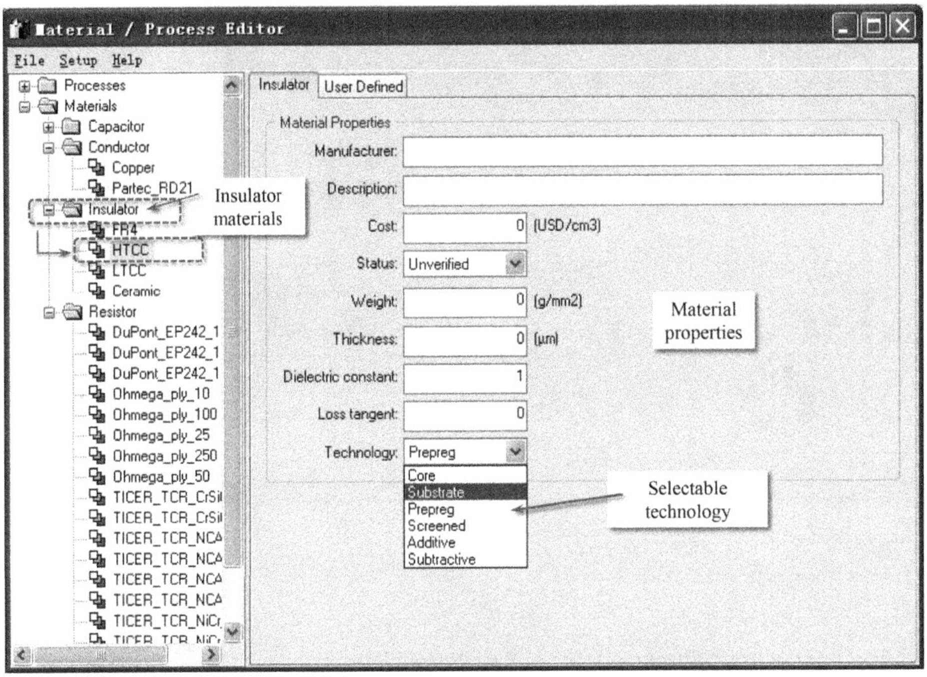

Figure 15.18 Insulator material properties window.

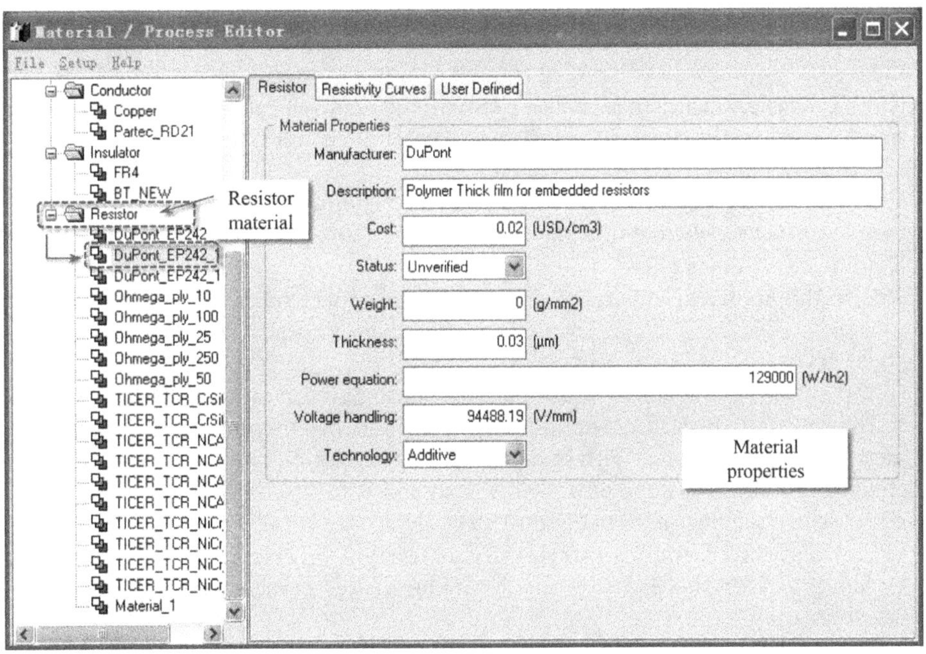

Figure 15.19 Resistor material properties window.

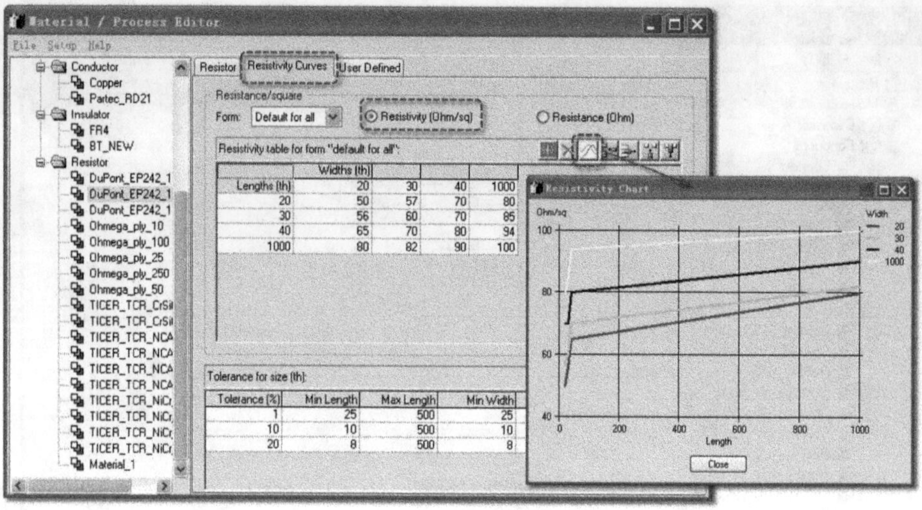

Figure 15.20 Nonlinear resistivity curve.

15.2.3 The Nonlinear Characteristics of Resistance Materials

The resistivity of some resistance material is not constant; as its shape or area changes, the resistivity will change accordingly, which shows its nonlinear characteristics.

1) Nonlinear resistivity curve
   When the length of the resistance material is constant, with an increase in width, the resistivity presents an upward trend; when the width of the resistance material width is constant, with an increase in length, the resistivity also presents an upward trend. So, when the area of the resistance material changes, the resistivity changes accordingly.

   The trend of change increases rapidly at the beginning (curve is relatively steep), then increases slowly (curve is slowing down), as shown in Figure 15.20.
2) Nonlinear resistance curve
   Due to the nonlinear resistivity, as the material aspect ratio or area changes, the resistance will also change. Toggle from resistivity to resistance to see the resistance curve.

   If the resistance material width remains constant, with an increase in length, its resistance value increases. For resistance materials of different widths, the rate of increase is not the same. With material width decrease, the increase rate is accelerated, from the curve slope of the 40th, 30th and 20th (th, thousandth inch, that is, mil). We can see this trend in Figure 15.21.

   For this kind of nonlinear resistivity material, if we calculate the resistor area by hand, besides the fact that there may be a large error, the workload will be very big.

   Use the automatic planner function software and the work will become simple. Expedition makes a comprehensive calculation, fully considering the nonlinear factors, so as to get an accurate result.

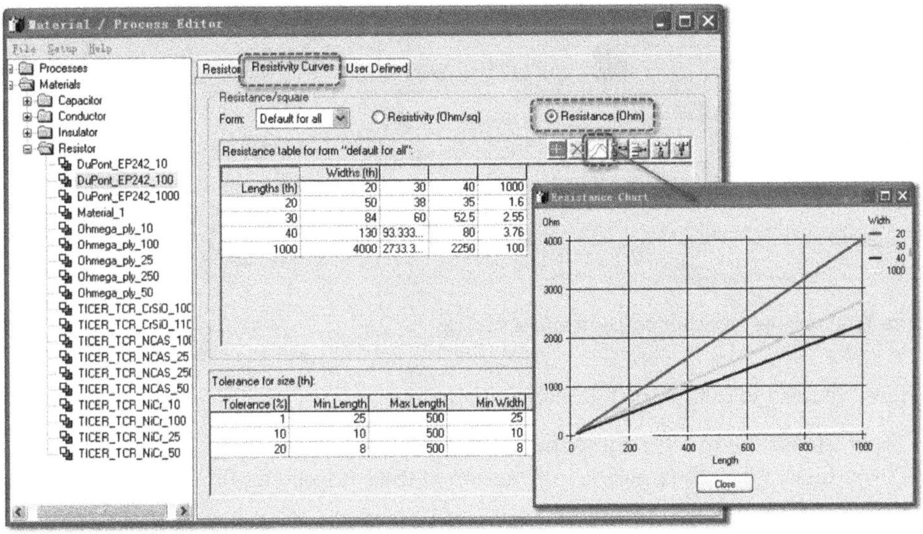

Figure 15.21 Nonlinear resistance curve.

15.3 Resistor and Capacitor Automatic Synthesis

15.3.1 Preparation for Automatic Synthesis

In SiP substrate design, after processes and materials have been defined in the central library, the designer can update the processes and materials from the central library to the local design library via forward annotation.

Of course, the designer can also start the function in the Expedition environment via Setup→Material/ Processes Editor; the setup method is the same as that in the central library. If this method is used, the setting is only effective in this design, while the central library setting will be effective for all designs that use the central library.

Before synthesis, the designer must input the property values for resistors or capacitors, such as component type, resistance, power consumption, etc. These parameters can be input into part properties in the create library, and can also be assigned in Part Editor in the current design.

Figure 15.22 shows the resistor properties defined in Part Editor.

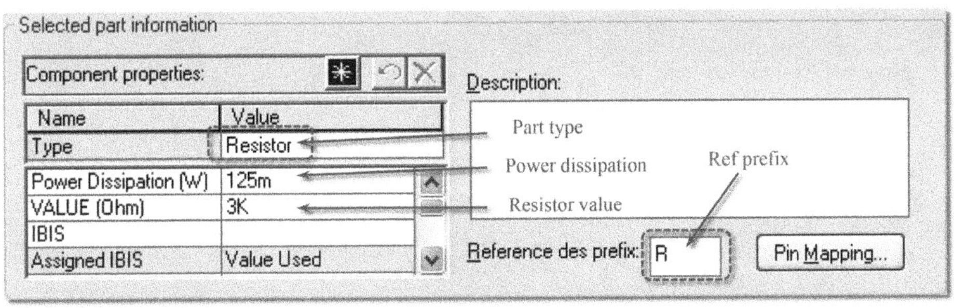

Figure 15.22 Define resistor properties in Part Editor.

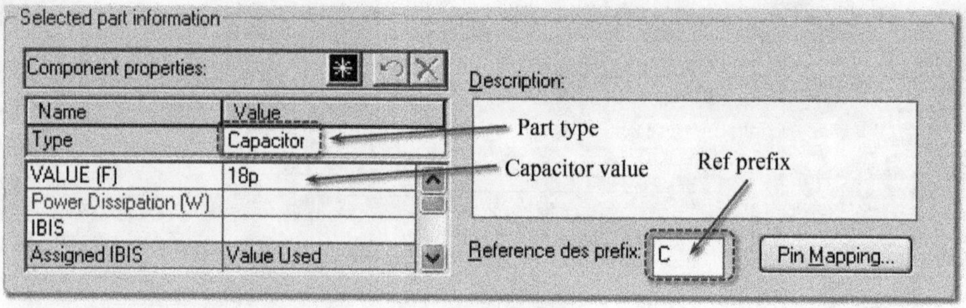

Figure 15.23 Define capacitor properties in Part Editor.

Here, we need to define the following parameters.

1) Type, choose the resistor component type in the drop-down list.
2) Power dissipation, the power consumption of the component; fill in the actual value of the resistor power consumption.
3) VALUE (Ohm), resistance value.
4) Reference des prefix, "R" is commonly used for resistors.

Figure 15.23 shows the capacitor properties in Part Editor; it is necessary to define the parameters below.

1) Type, component type, choose capacitor in the drop-down list.
2) VALUE (F), capacitance value.
3) Reference des prefix, "C" is commonly used for capacitors.

After defining the properties for resistors and capacitors in Part Editor, we need to enable the Enable Advanced Technology Pro (EP) license via Setup→Licensed Module→Advanced Technology Pro (EP), as shown in Figure 15.24.

Then select Setup→Embedded→Planner, invoke the Planner interface and select the Resistor Planner tab, as shown in Figure 15.25. The main settings are as follows:

1) Graph display mode, options for percentage or numerical display; in this case, select the percentage of resistance quantity.

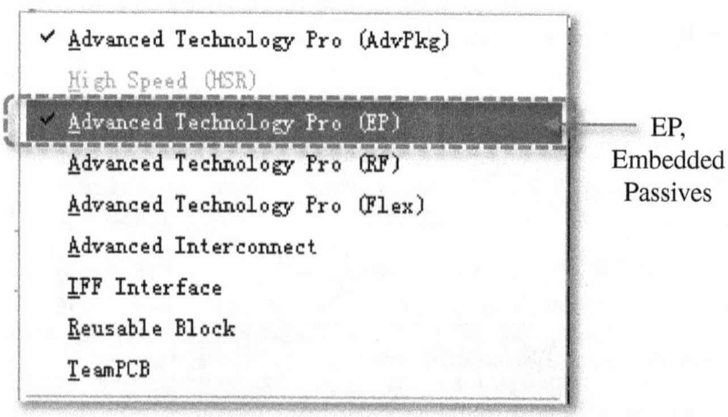

Figure 15.24 Enable Advanced Technology Pro (EP) license.

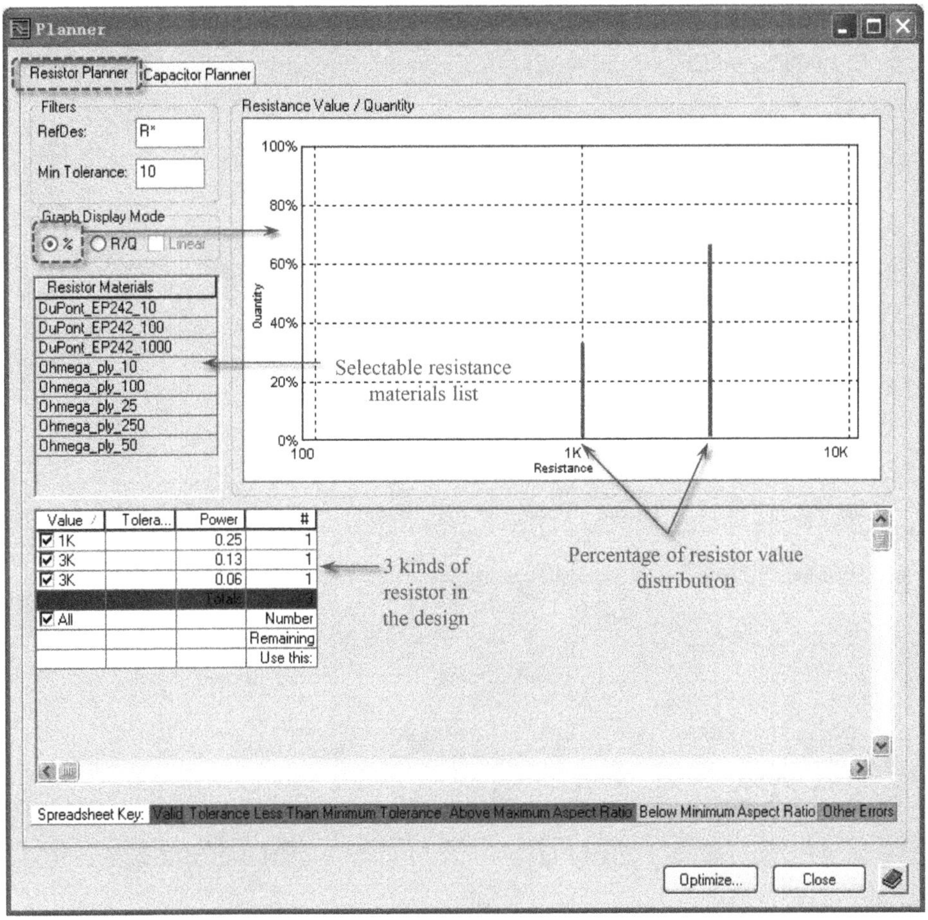

Figure 15.25 Resistor Planner interface.

2) Resistor materials, list all the resistance materials defined in materials definition.
3) Resistance value/quantity, resistance value (X coordinate) – quantity (Y coordinate) curve; it can be useful for seeing the percentage of different value resistors in the design.
4) At the bottom left of the Planner window, we can see the resistors needed for synthesis, in this case 1 K, 3 K, 3 K; the power consumption of the two 3 K resistors is different, respectively 1/8 w (0.13) and 1/16 w (0.06); the power consumption of the 1 K resistor is 1/4 w (0.25).

Switch to the Capacitor Planner tab, as shown in Figure 15.26; the main settings are as follows:

1) Graph display mode: options for percentage or numerical display; in this case, select the C/Q quantity and check the linear display option.
2) Capacitor materials: list all the capacitor materials defined in materials definition.

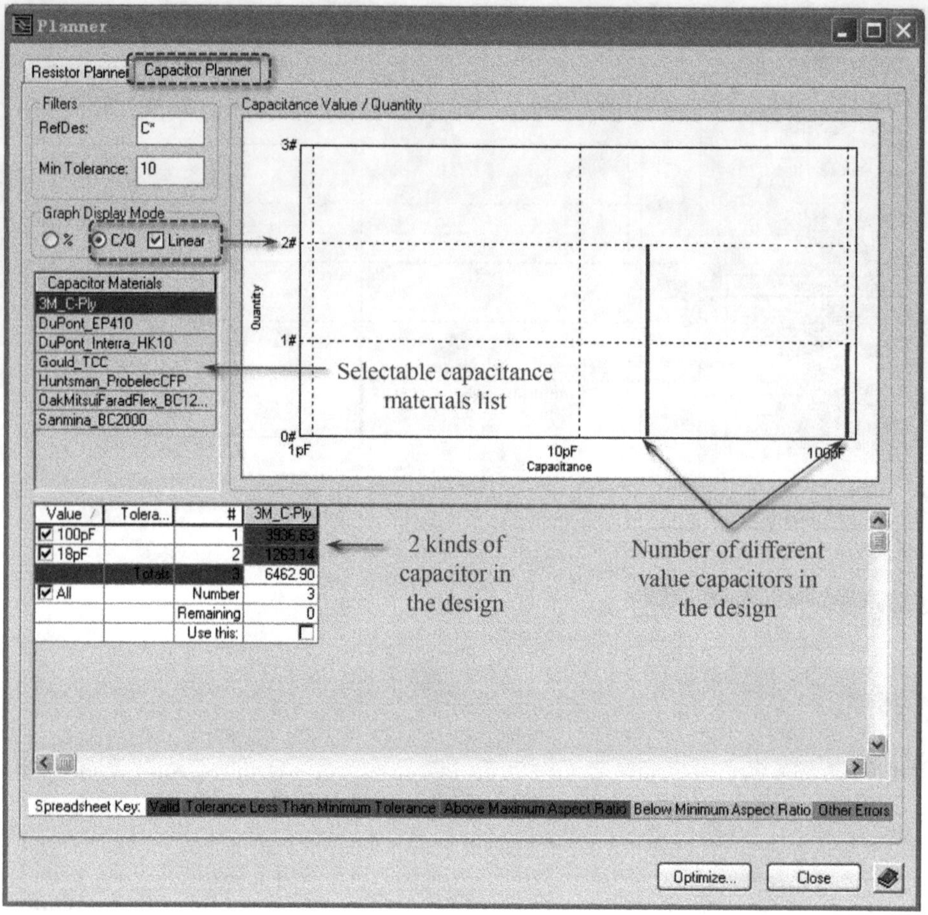

Figure 15.26 Capacitor Planner interface.

3) Capacitance value/quantity: capacitance (X coordinate) – quantity (Y coordinate) curve; it can be useful for seeing the percentage of different value capacitors in the design.
4) At the bottom left of the Planner window, we can see the capacitors needed for synthesis, in this case, one 100 pF and two 18 pF.

15.3.2 Resistor Automatic Synthesis

First, let's do the resistor automatic synthesis. Switch to the Resistor Planner tab and click the Optimize button at the bottom of the Planner window to start Optimizer, as shown in Figure 15.27.

In the resistor forms list, select all the forms; in the resistor materials list, select all the materials; for optimize method select By Area. The software will automatically optimize according to the principle of minimum area, choosing the smallest resistors from all the synthetic resistors.

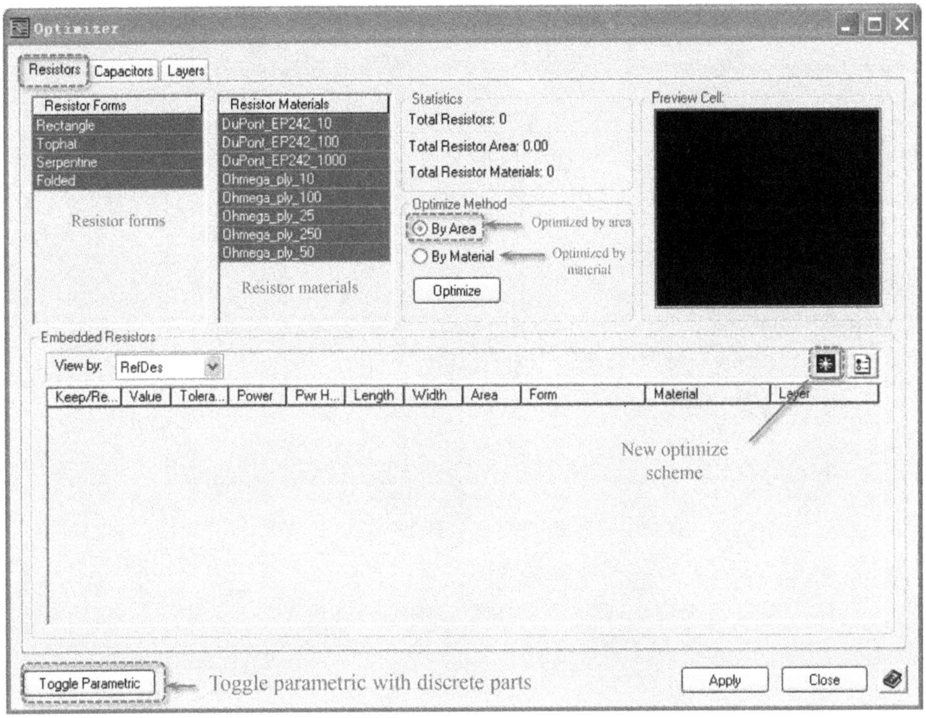

Figure 15.27 Resistor Optimizer interface.

Then click on the Layers tab and select the layers on which synthetic resistors can be placed; we can select different layers for different materials; the automatic synthetic tool supports placing embedded resistors or capacitors on any layer of the design.

In actual design, the designer needs to choose the allowed layers according to the technology and production capacity. For example, some processes only support surface resistors, meaning that we can only choose the surface layer for synthesis. In an actual design process, we must fully consider production and manufacturing issues, that is DFM (Design for Manufacture). In this case, we select layer 1 and layer 2 for the various materials, as shown in Figure 15.28.

Then click the Apply button to return to the Optimizer Resistor tab; in the Embedded Resistors subwindow, click the New button and create a new optimization scheme; the system will open the Toggle Optimizer window. In this window put the resistors for synthesizing to the right side of the window, as shown in Figure 15.29.

Click the OK button to return to Optimizer; at this time we can see that there are many resistors in the embedded resistors list. In the automatic synthesis process, the software will automatically synthesize multiple types of resistors and capacitors. We can select them for optimization By Area or By Material, and can also select manually (using the check items before the embedded resistors list), as shown in Figure 15.30.

In Optimize Method, select By Area, that is, prioritize according to area. Then click on the Optimize button, wait a moment, optimization will complete. We can see

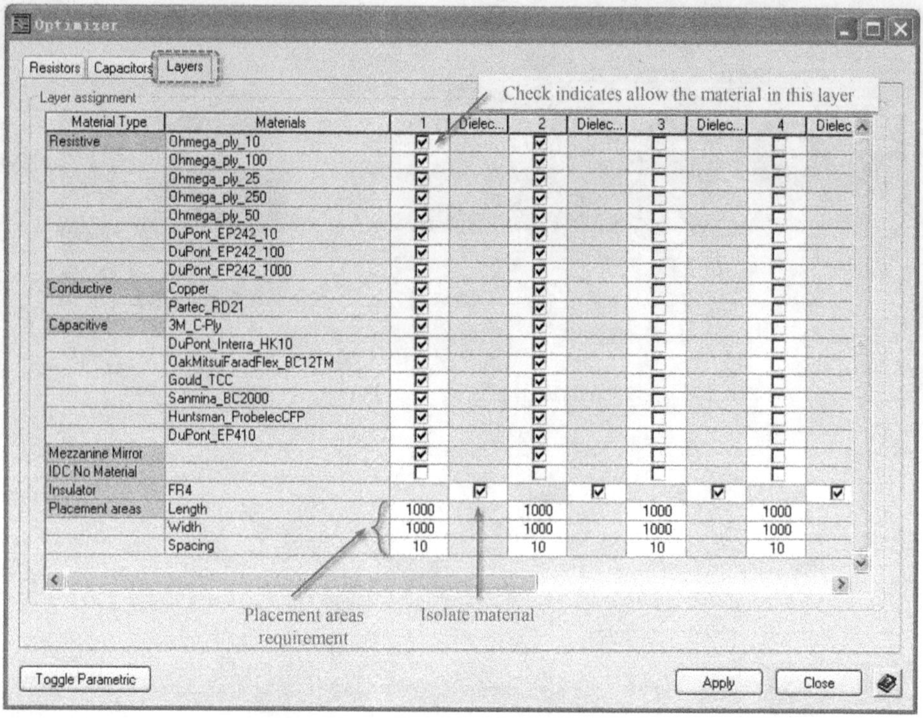

Figure 15.28 Define the layers allowed for embedded materials.

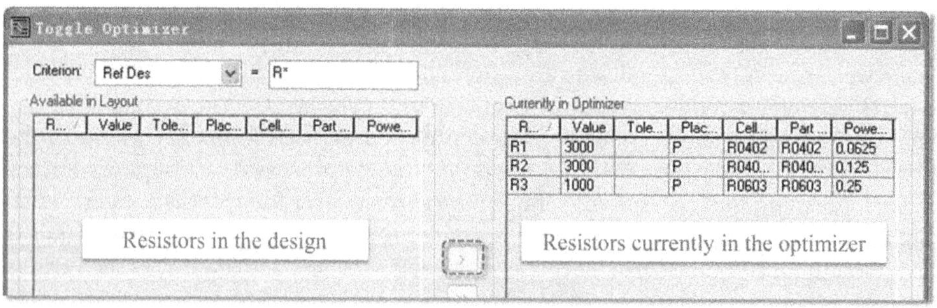

Figure 15.29 Put resistors for synthesizing to the right side of the window.

the optimized resistor cell appearing in the Preview Cell window, as shown in Figure 15.31.

Then click on the Apply button at the bottom of the Optimizer window; we can see the planar resistors as shown on the right side of Figure 15.32. This kind of resistor can be located in the substrate surface or embedded into any layer of the substrate.

We can see from Figure 15.32 that, although the value of R1 and R2 is the same, due to the difference in power consumption, the size of the synthesized resistor is different; basically, R2 is amplified in relation to R1, and the color of R3 is different from that of R1 and R2, showing that it uses different resistance material.

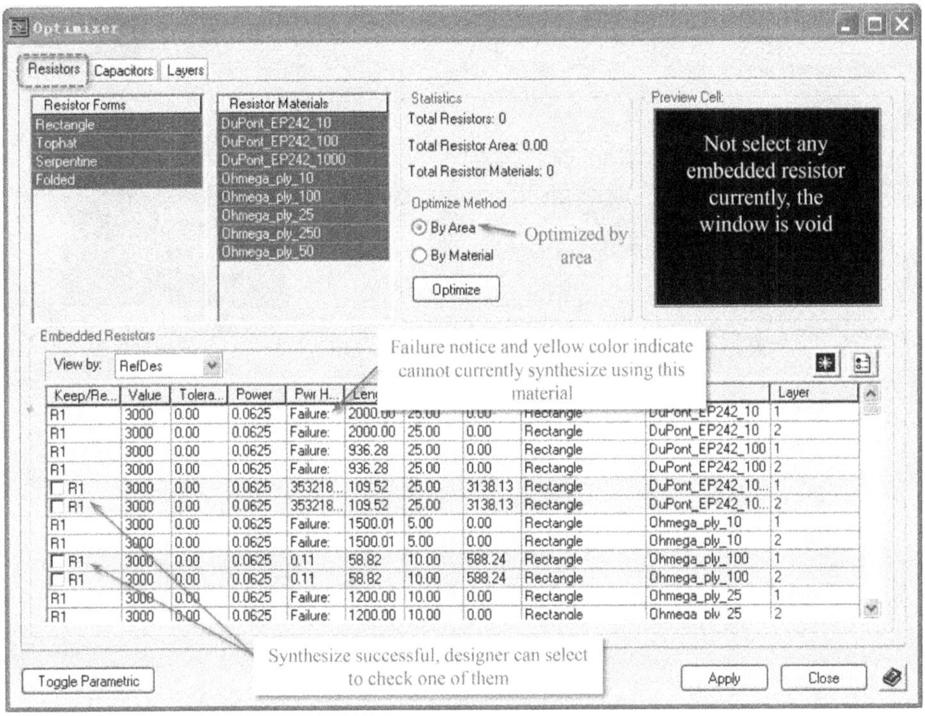

Figure 15.30 Return to Optimizer interface.

In actual design, there may be a situation after the completion of automatic synthesis where the designer wants to toggle planar resistors to return them to discrete resistors. We can click the button [Toggle Parametric] and in the pop-up window select the resistors to be returned to discrete resistors, as shown in Figure 15.33. Then click the Apply button below the window. In layout, we can see the selected resistors toggle back to the original discrete resistors.

15.3.3 Capacitor Automatic Synthesis

Capacitor automatic synthesis is also very simple; after synthesis the software can automatically optimize according to the By Area or By Material parameter, or as manually selected by the designer.

Planar capacitors can have the three forms described previously (see Figure 15.34 below); the choice of capacitor materials and conductive materials must also be considered; select them in the various columns.

In the Optimizer window, switch to the Capacitor tab, select all forms in the capacitor forms list, select all in the capacitor materials list and select all in the conductive materials list (in actual design we should select materials according to the project demand and permission). For optimize method select By Area, as shown in Figure 15.34.

Then click the New button [icon] to create a new optimize scheme; in the Toggle Optimizer window, put the capacitors for synthesizing to the right side of the window, and then click the OK button to return to the Optimizer interface.

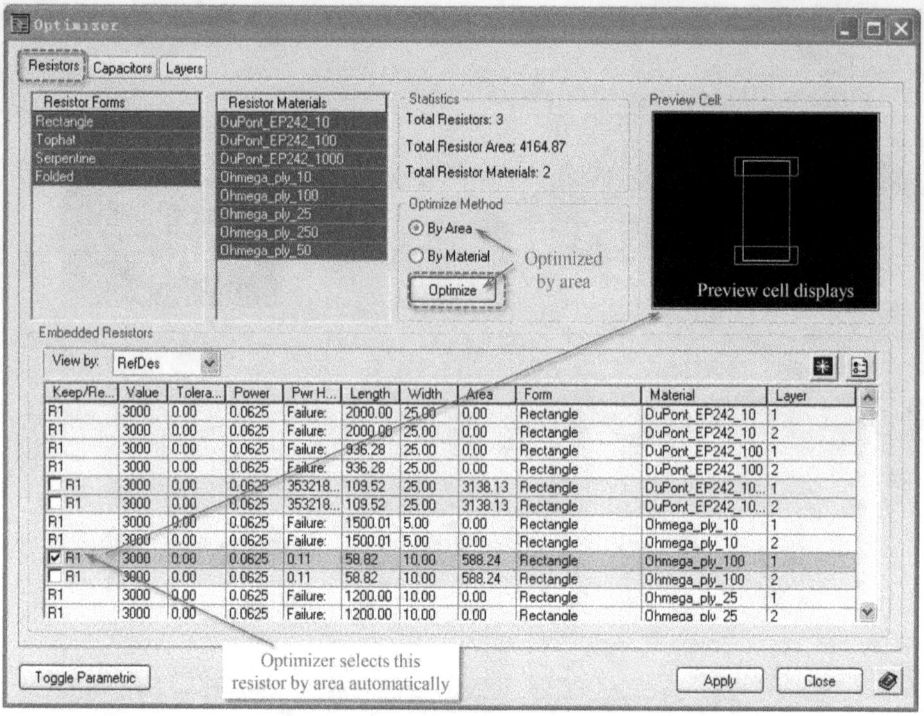

Figure 15.31 Resistors are synthesized and optimized.

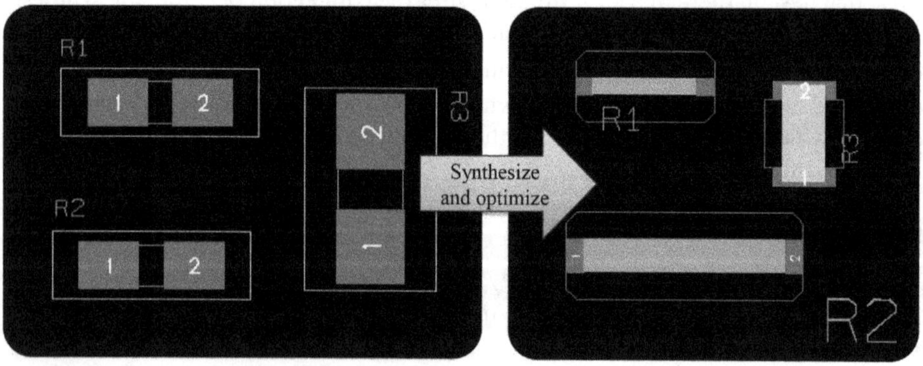

Figure 15.32 Discrete resistors are automatically synthesized to planar resistors.

At this time, in the embedded capacitors area, we can see many capacitors in the list; choose the optimize method By Area, click the Optimize button, and wait a moment. The optimization is complete, and we can see the synthesized capacitor cell in the Preview Cell window, as shown in Figure 15.35.

Then, click the Apply button; we see the capacitors as shown on the right side of Figure 15.36.

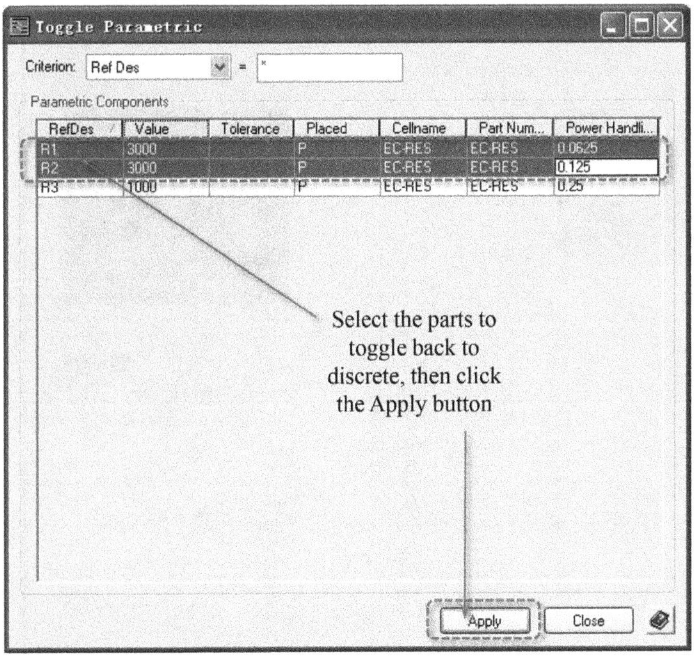

Figure 15.33 Toggle planar resistors to discrete resistors.

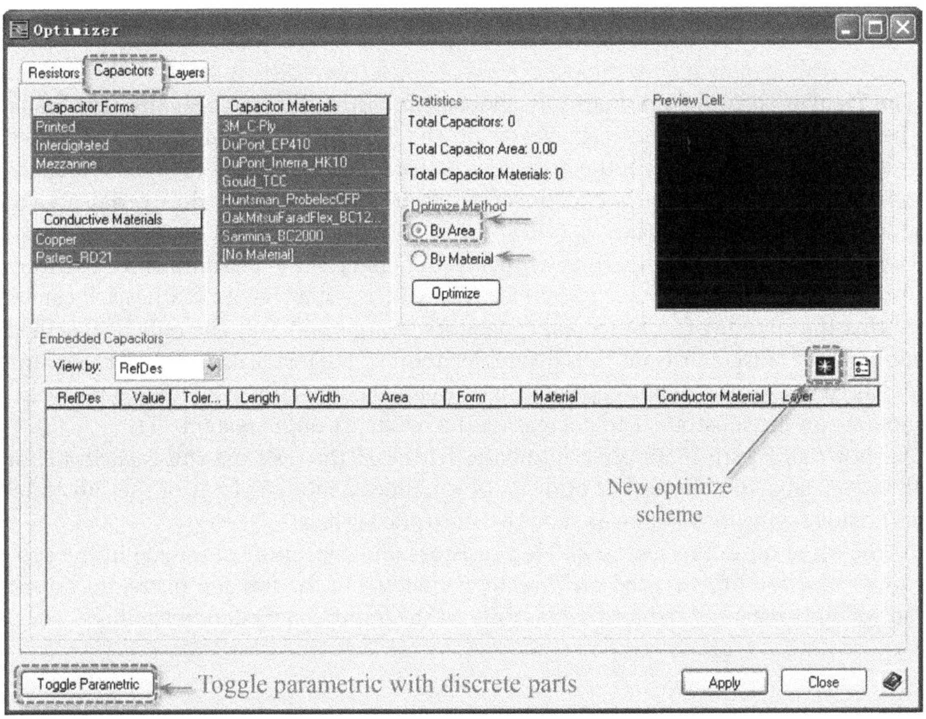

Figure 15.34 Capacitor Optimizer interface.

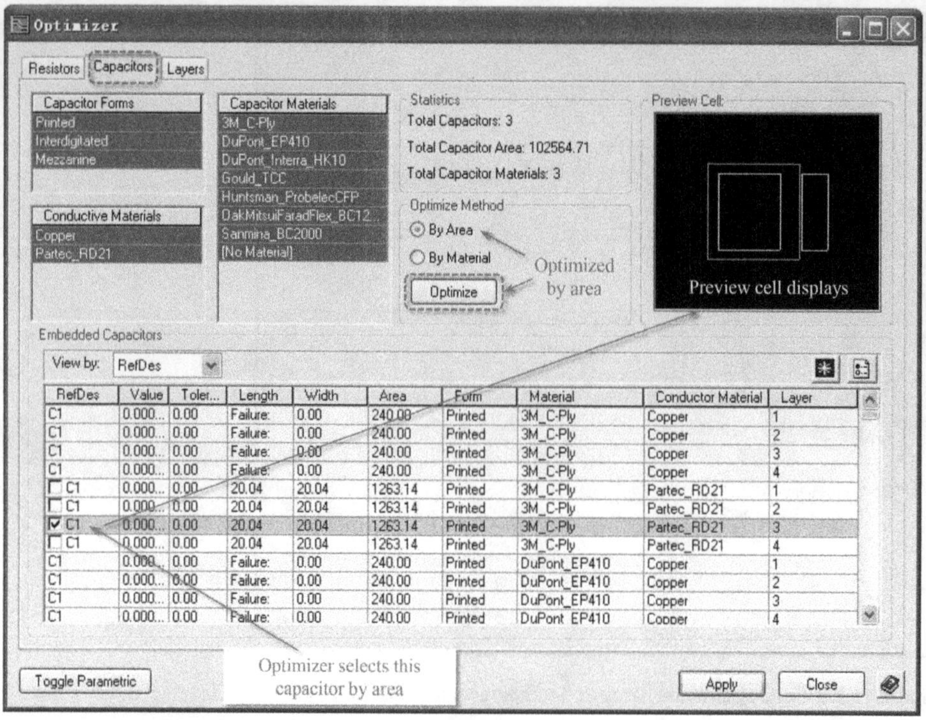

RefDes	Value	Toler...	Length	Width	Area	Form	Material	Conductor Material	Layer
C1	0.000...	0.00	Failure:	0.00	240.00	Printed	3M_C-Ply	Copper	1
C1	0.000...	0.00	Failure:	0.00	240.00	Printed	3M_C-Ply	Copper	2
C1	0.000...	0.00	Failure:	0.00	240.00	Printed	3M_C-Ply	Copper	3
C1	0.000...	0.00	Failure:	0.00	240.00	Printed	3M_C-Ply	Copper	4
☐ C1	0.000...	0.00	20.04	20.04	1263.14	Printed	3M_C-Ply	Partec_RD21	1
☐ C1	0.000...	0.00	20.04	20.04	1263.14	Printed	3M_C-Ply	Partec_RD21	2
☑ C1	0.000...	0.00	20.04	20.04	1263.14	Printed	3M_C-Ply	Partec_RD21	3
☐ C1	0.000...	0.00	20.04	20.04	1263.14	Printed	3M_C-Ply	Partec_RD21	4
C1	0.000...	0.00	Failure:	0.00	240.00	Printed	DuPont_EP410	Copper	1
C1	0.000...	0.00	Failure:	0.00	240.00	Printed	DuPont_EP410	Copper	2
C1	0.000...	0.00	Failure:	0.00	240.00	Printed	DuPont_EP410	Copper	3
C1	0.000...	0.00	Failure:	0.00	240.00	Printed	DuPont_EP410	Copper	4

Figure 15.35 Capacitors are synthesized and optimized.

The Planner software automatically chooses the printed form capacitor; it is easier to implement. In the figure, capacitors C1 and C2 are 18 P, and C3 is 100 P; see Figure 15.36 (b). The mesh in the graphic is dielectric material, the yellow rectangle is conductor, the red is capacitor pad; the effective area of capacitor is the overlap area of conductor, dielectric and pad.

Finally, let's look at a comparison of the overall effect of the resistors and capacitors before and after synthesis. As shown in Figure 15.37, the state before synthesis, it can be seen that the volume of resistors and capacitors is large, and they can only be installed on the surface of the substrate, thus occupying more of the installation area. At the same time, the discrete resistor and capacitor need welding connections, increasing the welding points on the substrate, and decreasing the reliability of the system.

As shown in Figure 15.38, after automatic synthesis, the resistors and capacitors not only have a smaller volume, but also can be embedded into any layer of the substrate, thus greatly saving installation area on the substrate surface.

At the same time, because embedded resistors and capacitors are made in the substrate production process, and are directly connected to the trace or plane, they don't need welding, meaning reduced solder joints, thus improving system reliability.

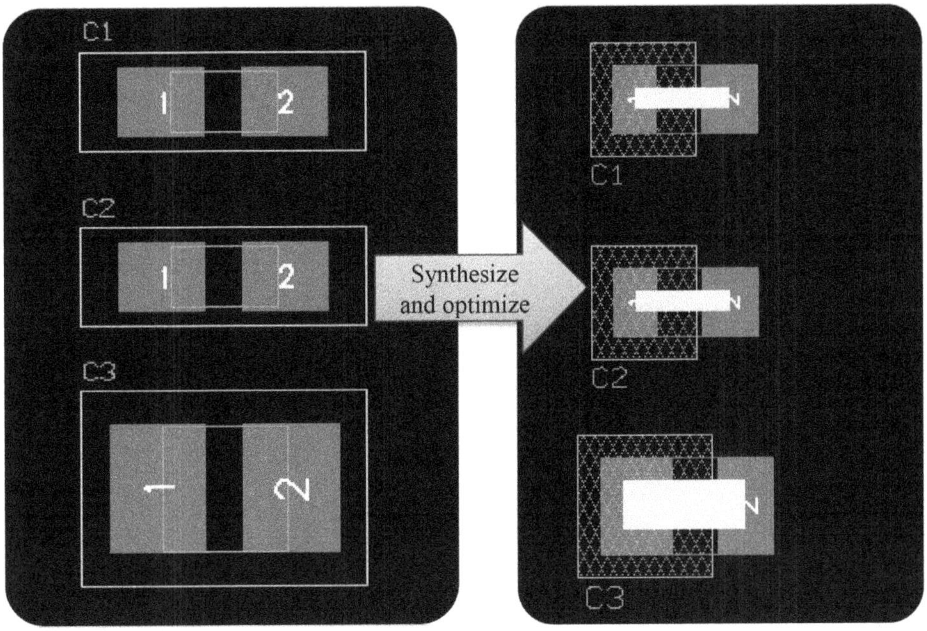

Figure 15.36 Discrete capacitors are automatically synthesized to planar capacitors.

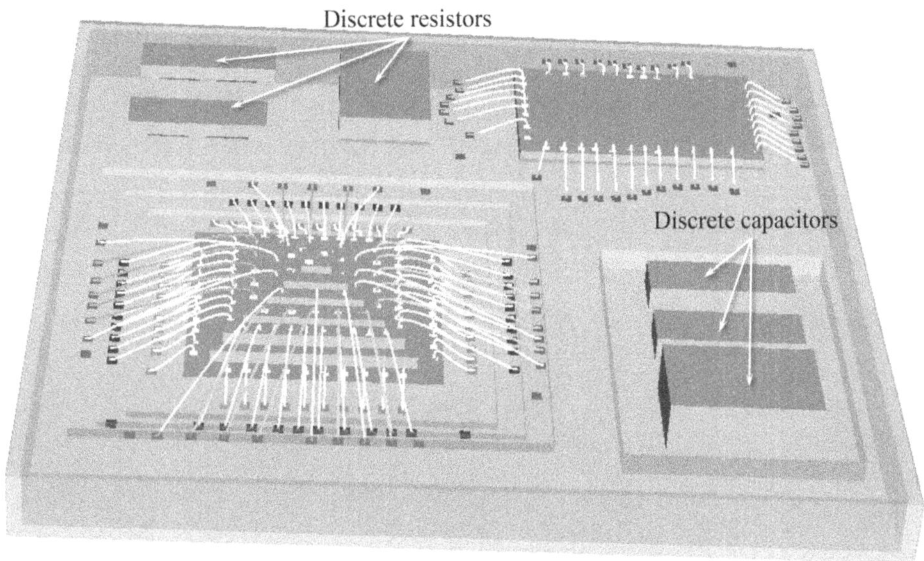

Figure 15.37 Resistors and capacitors before synthesis.

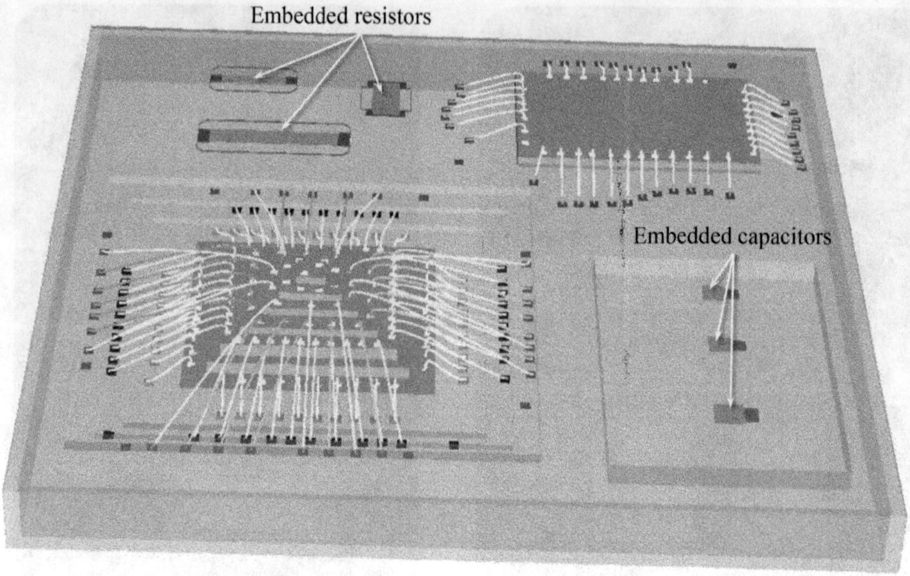

Figure 15.38 Resistors and capacitors after synthesis.

16

RF Circuit Design

16.1 RF SiP Technology

RF SiP technology, as the name implies, is a technology integrating RF and SiP.

RF SiP design technologies include microstrip line, stripline, loop inductance, cross finger capacitor, filter, mixer, etc., which are commonly used in RF circuit design, and bond wire, flip chip, chip stack, cavity, embedded resistor and capacitor, etc., which are commonly used in SiP design.

RF SiP technology is widely used in various fields of the electronic information industry. Currently, the most important research and application area for RF SiP is physical layer circuit for wireless communication.

For commercial RF chip, it is difficult to use the silicon planar process for implementation. The RF integration realized by SoC technology is relatively low, and such performance may not meet the RF requirements.

Due to the fact that the physical-layer-circuit working frequency is very high, and that it needs a variety of matching and filtering nets containing a large number of passive components, SiP has technical advantages in this area.

SiP has the advantages of shorter chip-interconnect, smaller volume and lower power consumption, and it is faster and more versatile. For example, full-featured RF SiP encapsulates RF chip, baseband chip and flash chip within a module, so, when it is necessary to change wireless front-end analog circuits, the system can quickly form and provide corresponding function circuits.

RF SiP technology has gradually become mainstream in wireless design.

RF circuit design can be integrated in different design phases, including IC design, SiP design and PCB design. In this book, RF SiP technology refers to integrating RF circuit into the SiP design process.

On the one hand, RF SiP technology can help to simplify RF IC chip design, shorten the IC product-development cycle and reduce cost; on the other hand, RF SiP technology can greatly reduce the number of RF components on the PCB, simplify PCB design, reduce the area of PCB and enhance PCB reliability.

RF engineers usually choose ADS or AWR (which has become a subsidiary of NI) for special RF tools for RF circuit design and simulation. Package or SiP designers usually choose SiP tools, such as Mentor EE Flow or Cadence APD or SiP.

SiP System-in-Package Design and Simulation: Mentor EE Flow Advanced Design Guide,
First Edition. Suny Li (Li Yang).

But if the designer wants to add RF circuits in SiP substrate layout, or to design bond wire, cavity, digital, analog and RF hybrid circuits in the same substrate, ADS and other RF tools don't meet the requirements. ADS does not support complex SiP or package design very well, or complex digital circuit design.

So, for RF SiP design, the designer needs to find a tool that can support a combination of RF technology and SiP technology; support RF, digital and analog mixed circuits in a uniform environment; and support RF shapes, bond wire, cavity, chip stack, flip chip, etc.

Currently, Mentor EE Flow is the only design platform that can support so many technologies in one unified environment.

16.2 Mentor RF Design Flow

Mentor EE Flow can support SiP, package and MCM design as well as digital, analog and RF mixed-circuit design simultaneously making it the best choice for RF SiP designers.

In EE Flow, the designer can realize flexible RF circuit design and simulation. Using the RF Connect interface, RF design data can be exported to Agilent ADS for simulation and verification; at the same time we can also import ADS design data or simulation results into DxDesigner and Expedition also using the RF Connect interface, so as to ensure RF design and simulation data synchronization and integrity.

Figure 16.1 shows the RF circuit design and simulation process in EE Flow.

In EE Flow we design the RF schematic in DxDesigner and the RF layout in Expedition, and, using RF Connect, we can connect EE Flow with RF Simulator ADS/AWR to transfer design and simulation data either way.

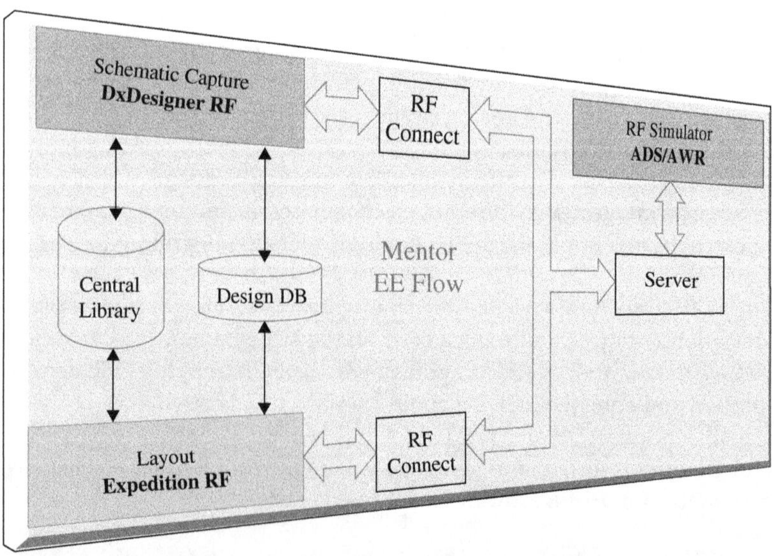

Figure 16.1 The RF circuit design and simulation process in EE Flow.

16.3 RF Schematic Design

16.3.1 RF Shapes Library Configuration

Before RF schematic design, the first task is to configure the RF shapes library by importing RF shapes from the software installation directory into the designer-specified central library. The most convenient method is batch commands.

First, write a library configuration batch file; its content is shown in Figure 16.2, and save it as CL.bat.

The function of this file is to copy the parameterized RF shapes library from the Mentor software installation directory to the central library the designer is currently using. In this case, the RF shapes path is as follows: D:\MentorGraphics\EE7.9\7.9 EE\SDD_HOME/standard/RF/ShapesLibrary.

This path is located in the EE installation directory. For example, if we install EE7.9.5 or VX on the E disk, the path should be changed to E disk EE7.9.5 or the corresponding VX directory.

The current design central library path is as follows: F:\SIP\MENT_LIB.

Please see Figure 16.3 for the directories of the two kinds of libraries.

After finishing the batch file, save and run the CL.bat file in CMD.exe window; when the file is running, we will see prompt information on the content and progress of the RF shapes library copying.

```
%SDD_HOME%\common\win32\bin\RFSymMergeUtility -ads -rfcl
D:\MentorGraphics\EE7.9
\7.9EE\SDD_HOME\standard\RF\ShapesLibrary\ShapesLibrary.lmc -all RF*
-update -userCL
F:\SIP\MENT_LIB\MENT_LIB.lmc
```

Figure 16.2 Library configuration batch file content.

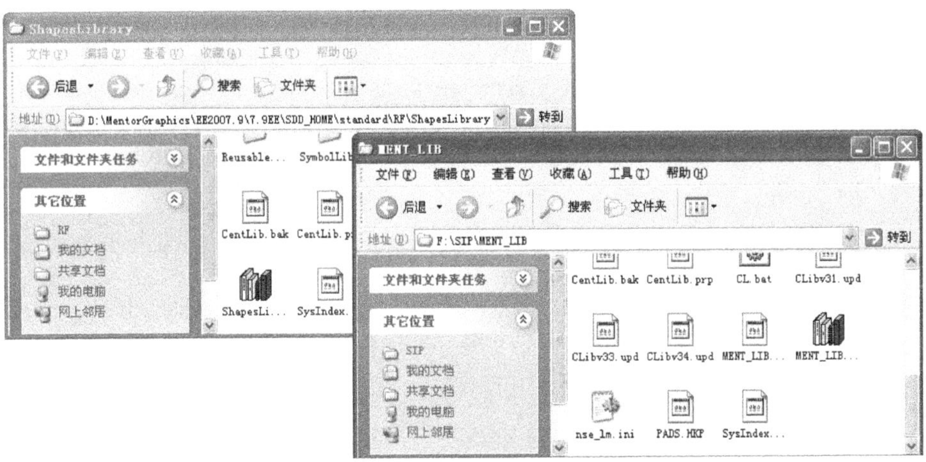

Figure 16.3 Parameterized RF shapes library path and current design central library path.

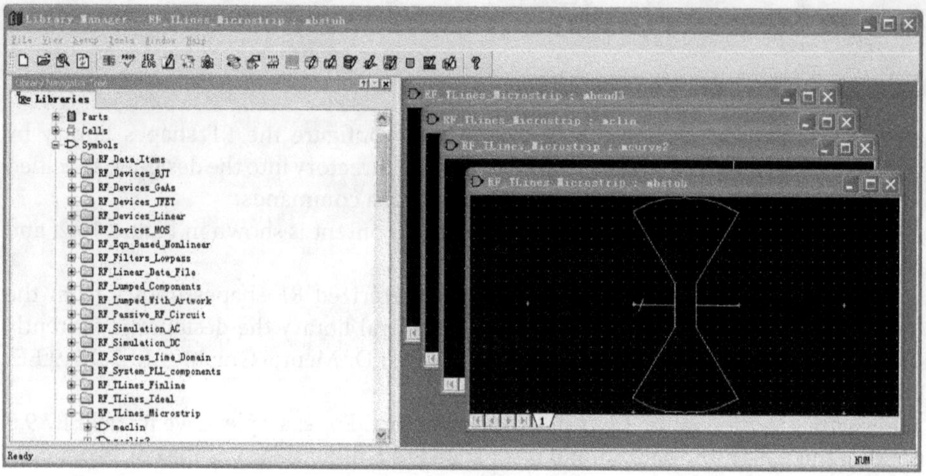

Figure 16.4 RF shapes imported to user central library.

When the copying process is finished, open the central library F:\SIP\MENT_LIB\MENT_LIB.lmc; under Symbols we can see that there are many RF symbol partitions – these are the parameterized RF shapes library copied from the installation directory. These RF symbols are fully compatible with ADS (developed jointly by Mentor and Agilent); see Figure 16.4.

16.3.2 RF Schematic Toolbar

After starting the schematic tool DxDesigner, we need to check the RF design license: DxRFEngineer. Select from the menu Setup→Settings→License, and check DxRFEngineer, as shown in Figure 16.5.

When the RF design license is enabled, the RF tools show in DxDesigner, as shown in Figure 16.6.

1) RF Connect

RF Connect is used to connect RF simulation tools such as ADS\AWR, and transfer RF schematic design data to ADS or AWR for simulation. In the process of simulation, we usually need to adjust the RF parameters so as to achieve the optimal result; the results of this kind of adjustment can also be transferred back to DxDesigner using RF Connect. Figure 16.7 shows RF Connect status before and after connection.

2) RF Group

RF Group is used for the creation and management of RF components; designers can form an RF group with similar function RF symbols and an RF group can also have a sub-group. When normal connection to RF Connect has been established, RF groups can be transmitted to RF simulation tools, such as ADS/AWR; see Figure 16.8.

3) RF DRC

RF DRC is used for RF design rule check; the check results show in the Output window. By clicking on the information text, the designer can find the corresponding RF shape symbol in the schematic, which is selected automatically, as shown in Figure 16.9.

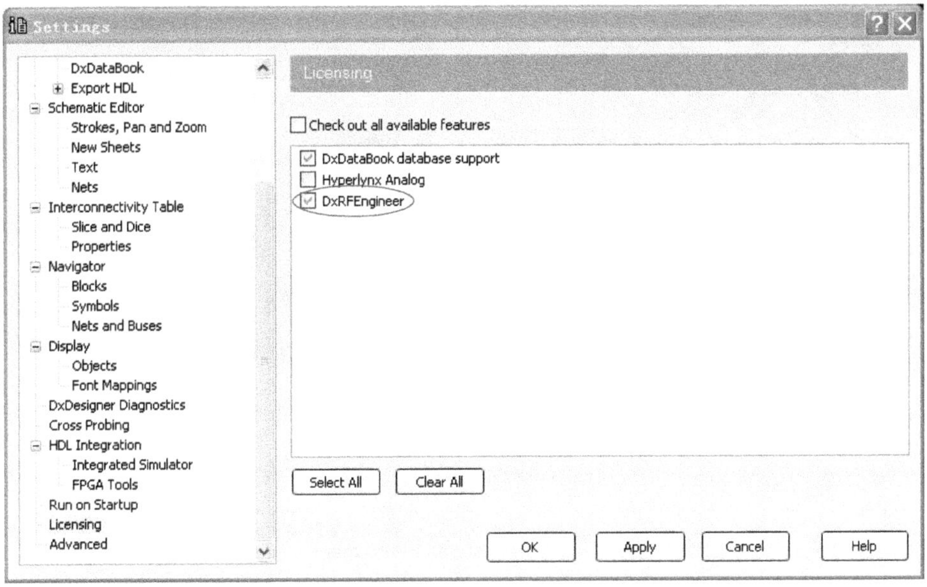

Figure 16.5 Enable DxRFEngineer license.

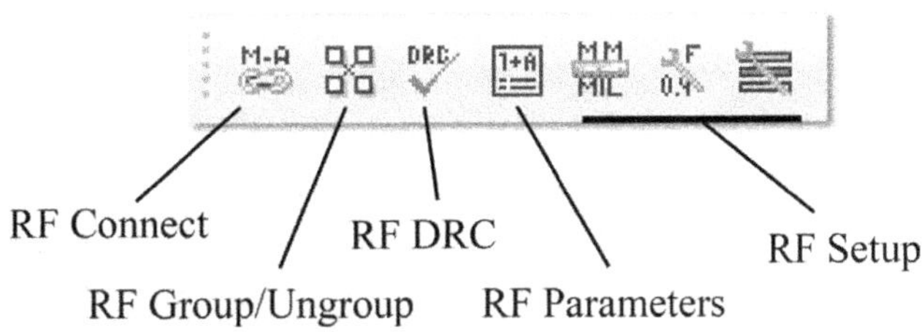

Figure 16.6 RF tools in DxDesigner.

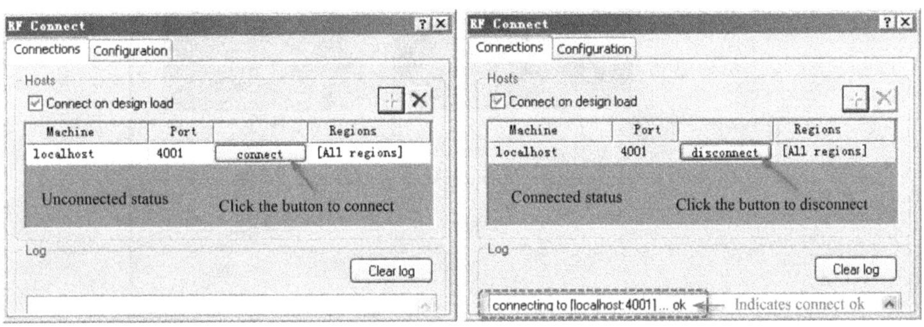

Figure 16.7 RF Connect status before and after connection.

4)  RF Parameters

RF Parameters is used to set up and adjust the parameters of RF shapes, as shown in Figure 16.10.

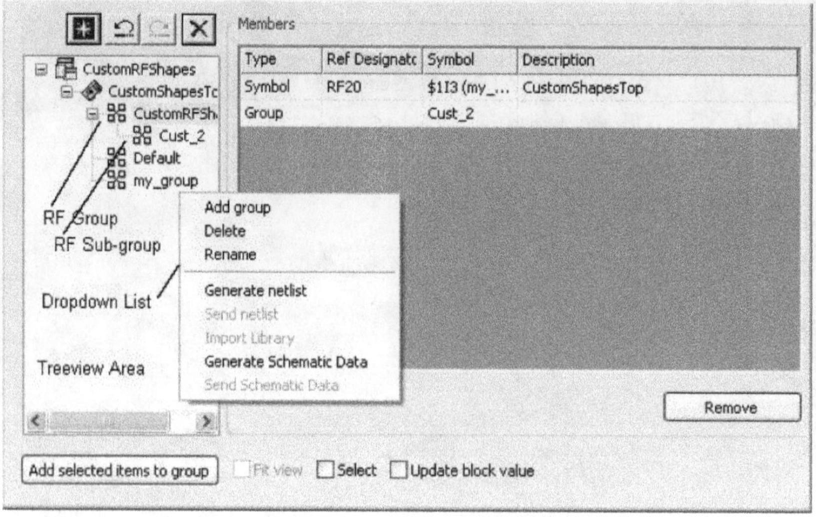

Figure 16.8 RF Group interface and RF data generate and send menu.

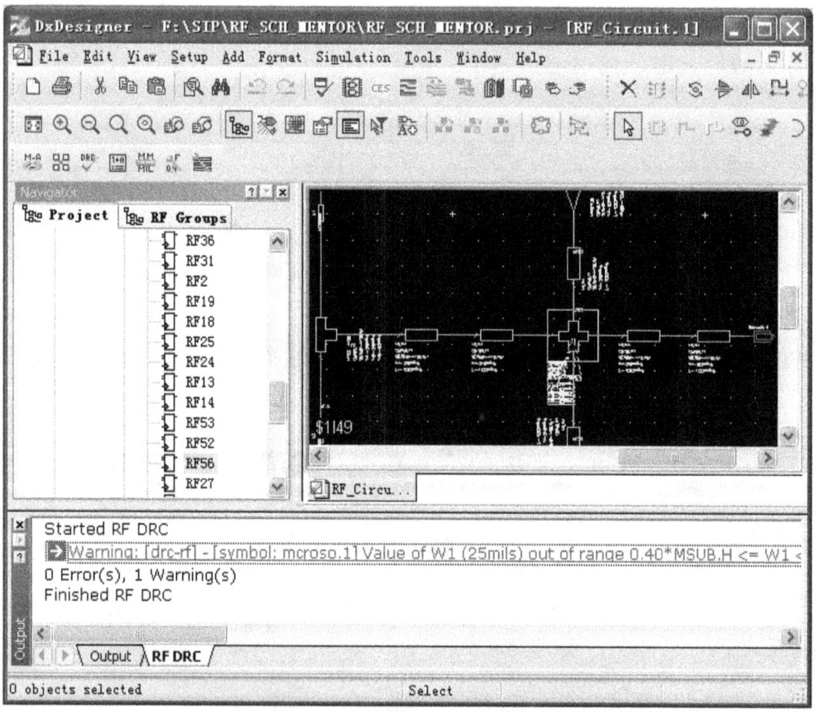

Figure 16.9 Find corresponding RF symbol by clicking the DRC message.

5) Default Units

Default Units allows the setting of various parameters for RF design units; the parameters of the unit can be selected in the drop-down list, as shown in Figure 16.11.

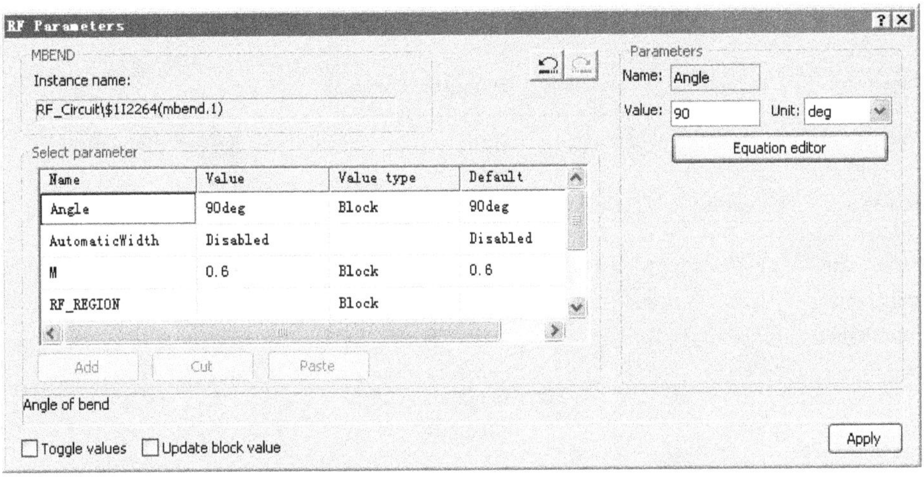

Figure 16.10 RF Parameters window.

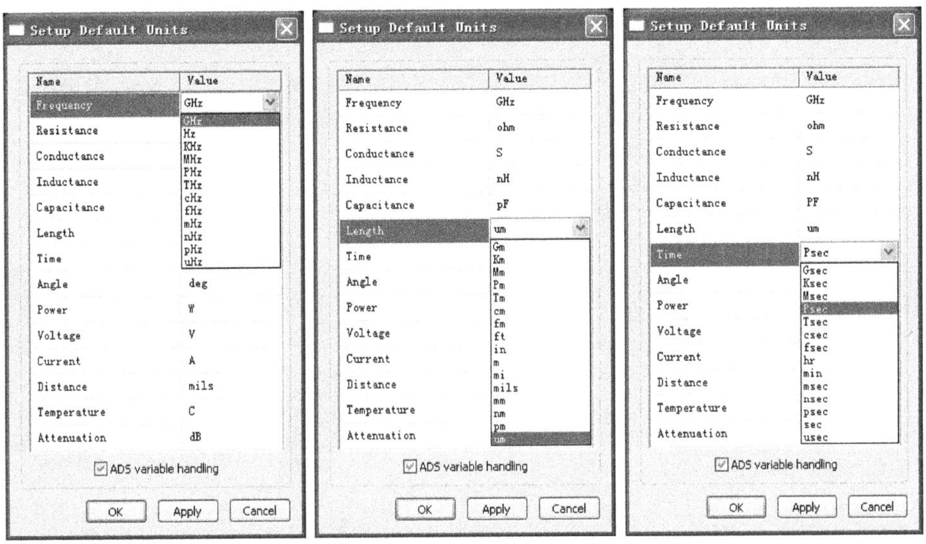

Figure 16.11 RF default-unit setting window.

Any change of RF unit in the schematic will be updated to layout design. Here, for example, the RF length unit is changed to um in DxDesigner; after synchronization, when the designer draws with RF Meander in the Expedition layout environment, the default RF length unit will be um.

In the RF default-unit setup window, if the ADS variable handling option is selected, the software will automatically check whether related processing methods for RF variables are compatible with ADS. When this option is checked, for example, if a parameter does not have a defined unit, the software automatically changes the base unit to the default unit; if this option is not selected, the software keeps the base unit unchanged.

The DxDesigner RF base units, default units and supported units are shown in Table 16.1.

Table 16.1 RF units.

Dimension	Base unit	Default unit	Supported units
Frequency	Hertz	GHz	GHz, Hz, MHz, PHz, THz, cHz, fHz, mHz, nHz, pHz, μHz
Resistance	Ω (Ohm)	Ω (Ohm)	Gohm, Kohm, Mohm, Pohm, Tohm, cohm, fohm, mohm, nohm, ohm, pohm, μohm
Conductance	Siemen	S	GS, KS, MS, PS, S, TS, cS, fS, mS, nS, pS, μS
Inductance	Henry	nH	GH, H, KH, MH, PH, TH, cH, fH, mH, nH, pH, μH
Capacitance	Farad	pF	F, GF, KF, MF, PF, TF, cF, fF, mF, nF, pF, μF
Length	Meter	mil	Gm, Km, Mm, Pm, Tm, cm, fm, ft, in, m, mi, mils, mm, nm, pm, μm
Time	Second	psec	Gsec, Ksec, Msec, Psec, Tsec, csec, fsec, hr, min, msec, nsec, psec, sec, μsec
Angle	Radian	Deg	deg, rad
Power	Watt	W	GW, KW, MW, PW, TW, W, cW, dBm, fW, mW, nW, pW, μW
Voltage	Volt	V	GV, KV, MV, PV, TV, V, cV, fV, mV, nV, pV, μV
Current	Ampere	A	A, GA, KA, MA, PA, TA, cA, fA, mA, nA, pA, μA
Distance	Meter	mil	Gm, Km, Mm, Pm, Tm, cm, fm, m, mm, nm, pm, μm
Temperature	Celsius	C	C
Attenuation	Decibel	dB	dB

6) Frequency Range

The Frequency Range tool allows the setting up of the specific RF circuit model frequency range when doing simulation in RF simulators such as ADS. The frequency range set in DxDesigner can be passed to RF Simulator. Open the Setup window, as shown in Figure 16.12, set up the minimum and maximum frequency values, and then click the OK button.

In DxDesigner's RF DRC tool, the RF components for which frequency is not included in the RF frequency range will be detected and displayed in the RF DRC report window.

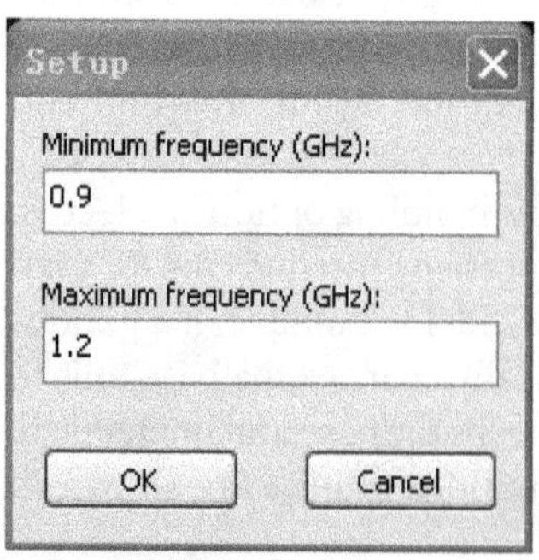

Figure 16.12 RF Frequency Range Setup window.

7) Substrate

This tool allows RF substrate layer setting; in DxDesigner, the default substrate layers are as shown in Table 16.2.

In the Set up substrates window, we can create a new substrate or edit substrate properties, as shown in Figure 16.13.

Use the New button to create a new substrate, and the Edit button to edit substrate properties. As shown in Figure 16.14, when a property in the RF Substrates properties edit window is clicked, the interpretation of this property can be seen in the status bar, and its value and unit can then be set according to the design demands.

Button is used to delete user-defined substrates. Note that the default substrate cannot be deleted. If the designer clicks the delete button for the system default substrate, the prompt shown in Figure 16.15 will appear.

Table 16.2 The default substrate layers.

CPWSUB	FSUB	MLSUBSTRATE2
MLSUBSTRATE3	MLSUBSTRATE4	MLSUBSTRATE5
MLSUBSTRATE6	MLSUBSTRATE7	MLSUBSTRATE8
MLSUBSTRATE9	MLSUBSTRATE10	MLSUBSTRATE12
MLSUBSTRATE14	MLSUBSTRATE16	MSUB
PCSUB1	PCSUB2	PCSUB3
PCSUB4	PCSUB5	PCSUB6
PCSUB7	SSSUB	SSUB
SSUBO		

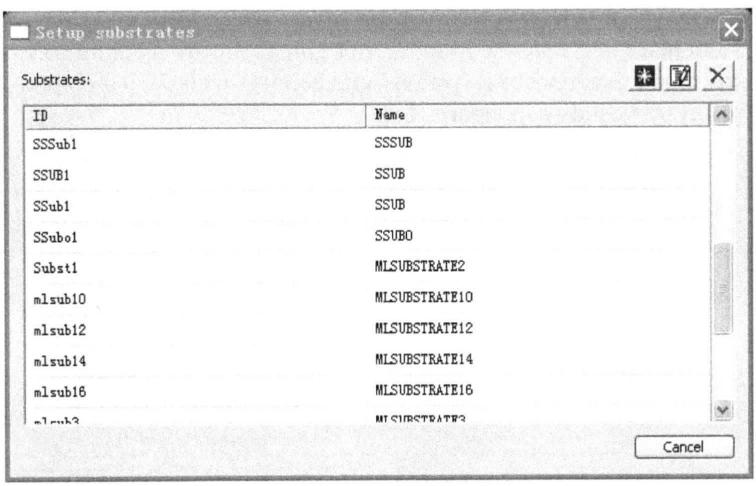

Figure 16.13 RF Set up substrates window.

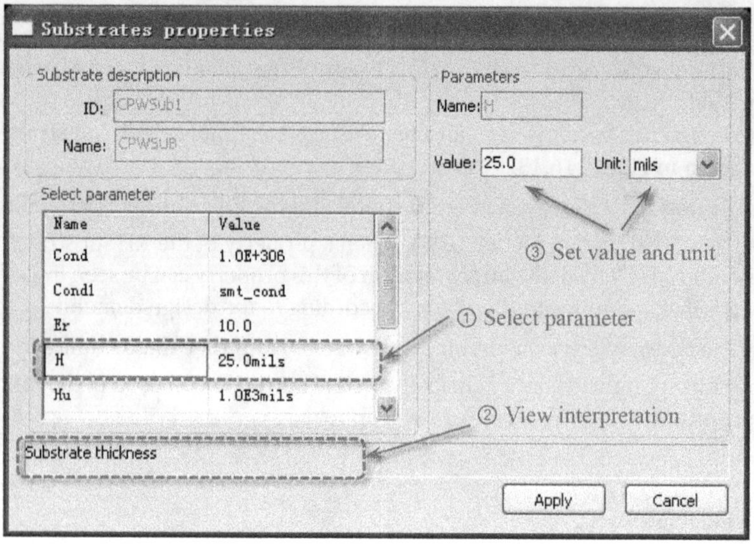

Figure 16.14 RF Substrates properties edit window.

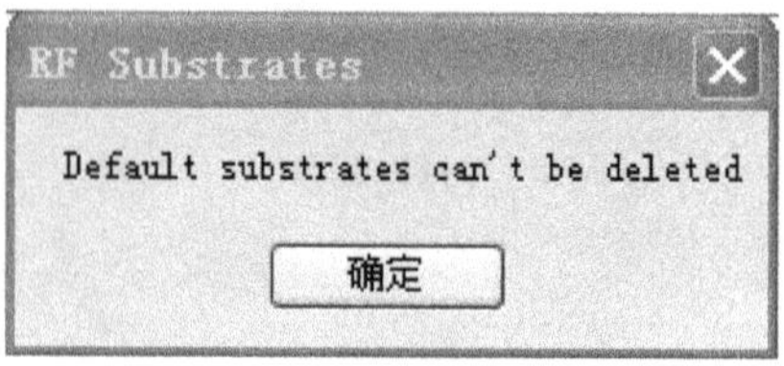

Figure 16.15 Default substrates can't be deleted.

16.3.3 RF Schematic Design

Here, we start to input RF schematic data in DxDesigner. Click button [icon] to invoke DxDataBook, select CL View tab at the bottom of the DxDataBook window, then switch to Symbol View; in RF_Tlines_Microstrip partition, select MBEND2, MLIN, MRIND, MRSTUB and place in schematic from left to right, as shown in Figure 16.16.

Add nets to the schematic to connect RF symbols, and connect with the left chip; we obtain a schematic diagram as shown in Figure 16.17.

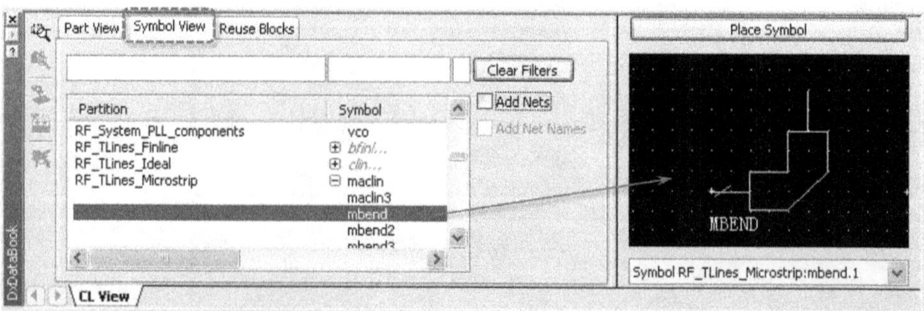

Figure 16.16 Select RF shapes in Symbol View.

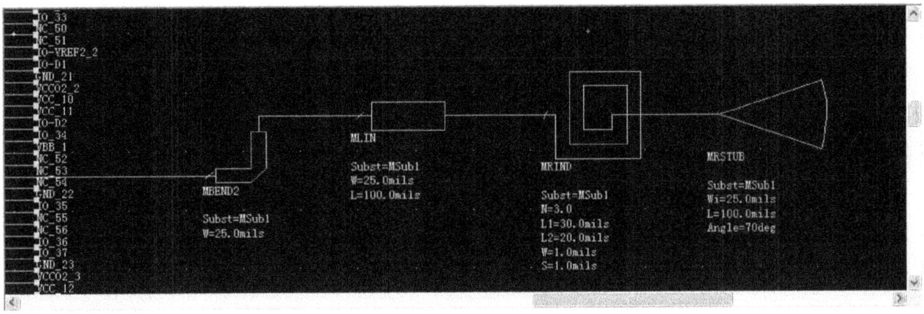

Figure 16.17 Input RF symbols in DxDesigner and connect with nets.

From Figure 16.17, we can see that each RF symbol has corresponding parameters. Select one RF symbol, and then click the toolbar RF button to open the RF Parameters window; here we can easily set or adjust various parameters of RF shapes.

Because the schematic contains RF shapes and common bare die chips, we need to run RF DRC and common schematic Verify ; when RF DRC and Verify have passed, package the design, then click the Expedition button to go into the layout design environment.

In Expedition, open Setup→Project integration window; we can see the forward annotation status lamp has become amber, the text behind also suggests "forward annotation required". Mouse-click the amber status lamp to forward annotate. After the completion of forward annotation, all status lamps turn green, and the prompt text becomes "No connectivity changes to be forward annotated", indicating that the schematic data and layout data are synchronized, as shown in Figure 16.18.

After data synchronization, select from the menu Setup→Licensed Modules→Advanced Technology Pro (RF), enable the RF layout license, then switch to RF mode (the RF mode icon is located on the right of the three traditional operation modes), and then click the Place RF Shapes soft key at the bottom of the layout window. In the pop-up Place Parts and Cells window, select the parameterized RF shapes and place into layout, as shown in Figure 16.19.

For the situation after the completion of RF shape placement, see Figure 16.20.

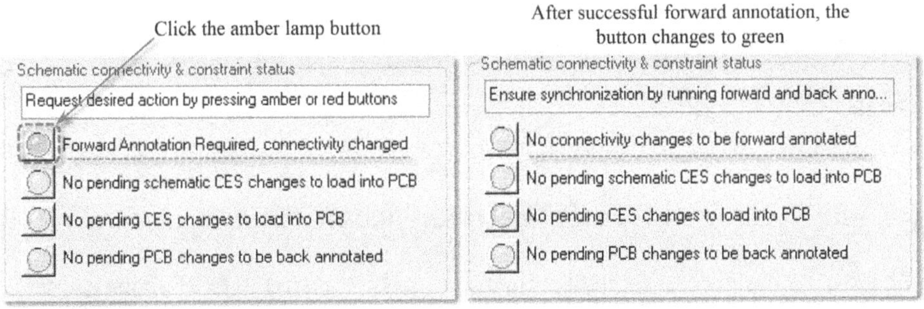

Figure 16.18 The status lamps and prompt texts change before and after forward annotation.

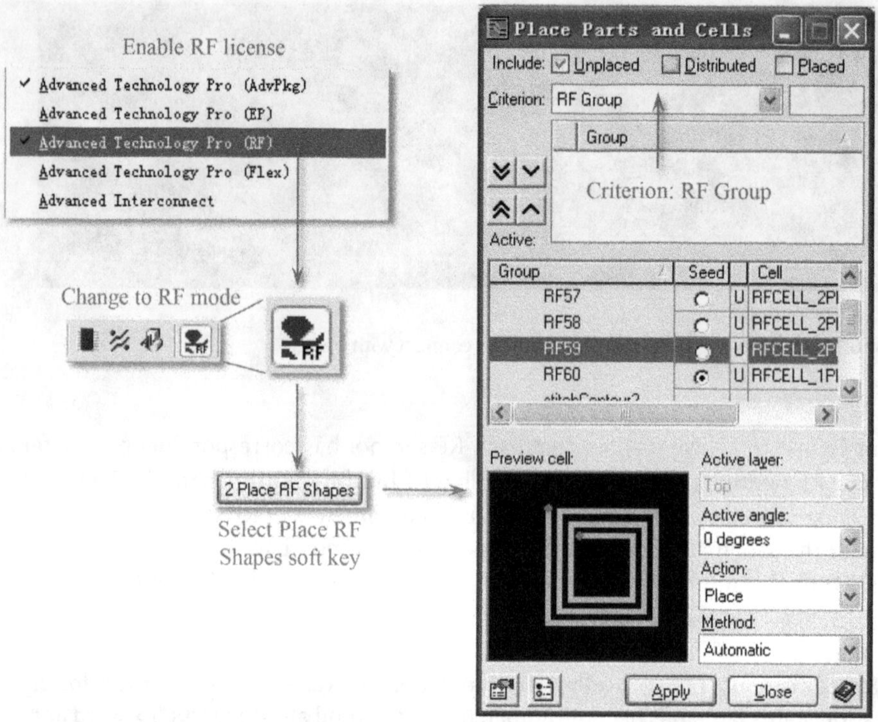

Figure 16.19 Placing parameterized RF shapes in layout.

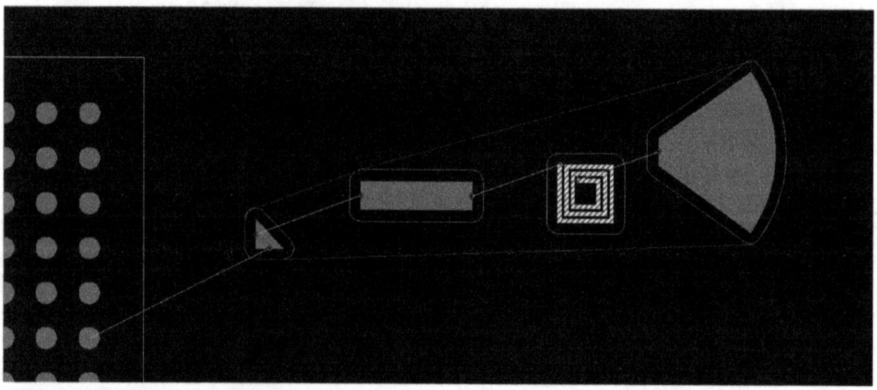

Figure 16.20 RF shapes layout.

16.4 RF Parameter Transfer Between Schematic and Layout

In EE Flow, RF parameters set in the schematic can be transferred to RF shape elements in layout. RF shapes are often defined as RF cells and stored in the local library by Expedition after automatic synthesis, and their names will be the same as the RF symbols in the schematic.

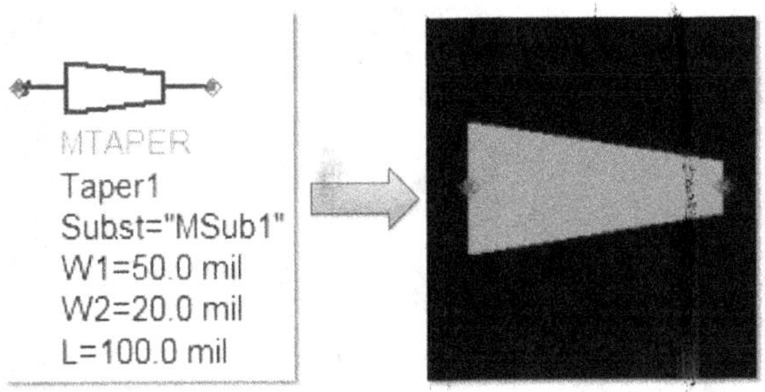

Figure 16.21 Transfer RF parameters from RF symbol to RF cell.

In package design and the forward annotation process, depending on the parameters of RF symbols in the schematic, the RF synthetic engine automatically synthesizes RF cells whose shape and size conform to the parameters defined in the RF symbol, as shown in Figure 16.21.

Using the specific example below, let's realize the process of interactive transfer of RF parameters.

First, check MRIND in the schematic; the parameters L1 and L2 are 20 mils and 20 mils respectively; after package and forward annotation, we can see that the RF cell is transferred to layout and inherits the parameters of the RF symbol in the schematic, as shown in Figure 16.22.

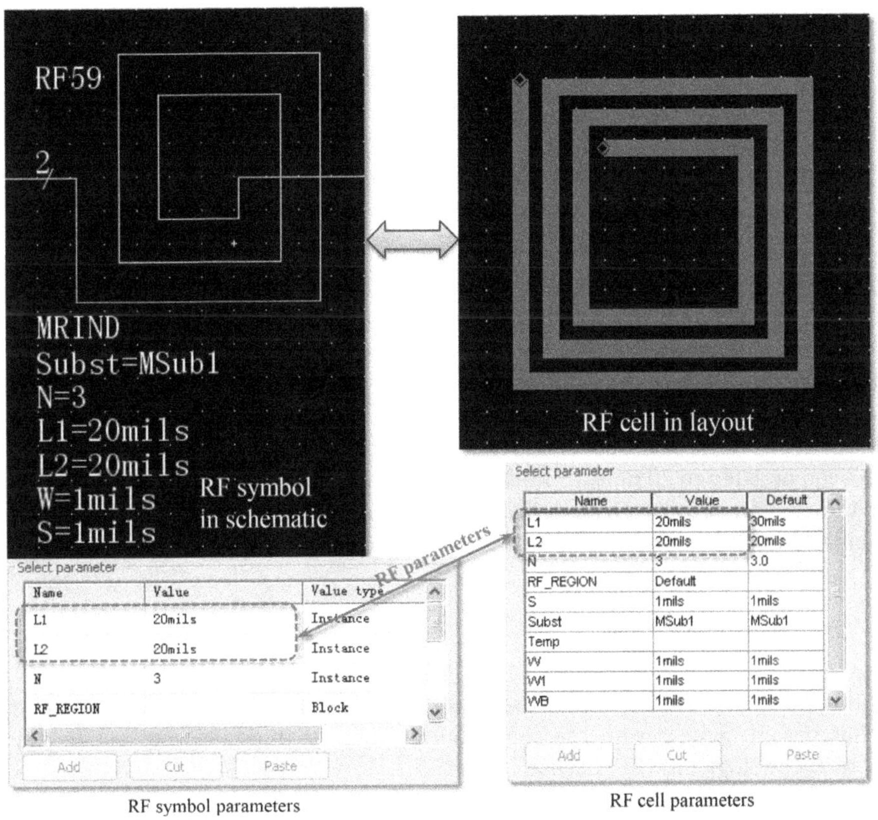

Figure 16.22 RF parameters can be transferred between schematic and layout.

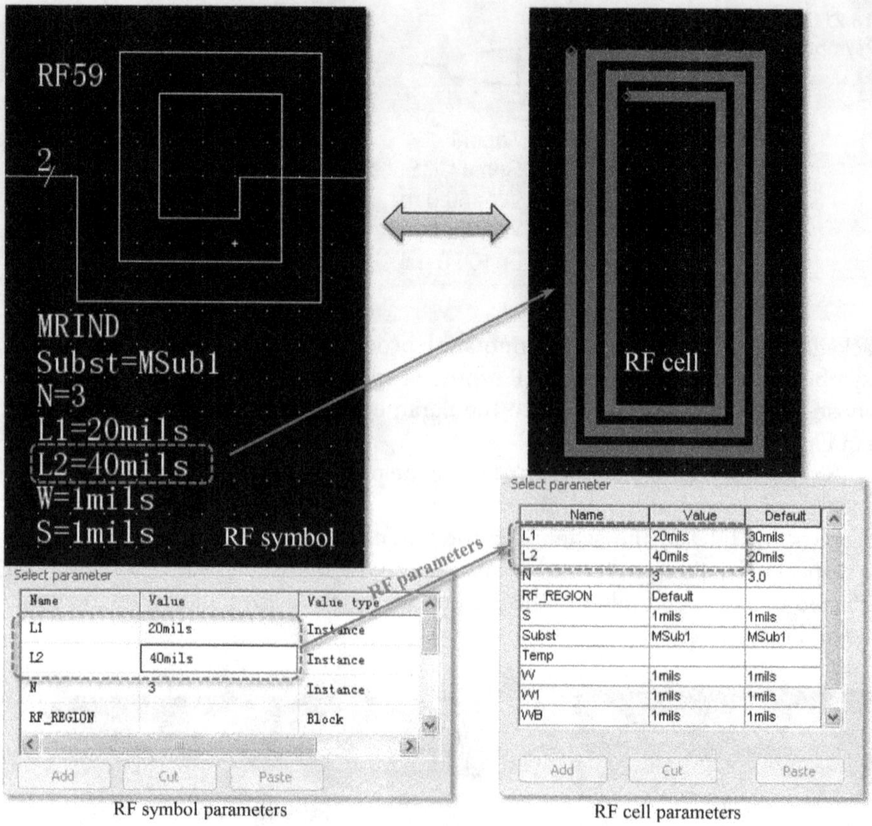

Figure 16.23 RF parameters are transferred from schematic to layout.

Next, change MRIND parameters L1, L2 to 20 mils, 40 mils in the schematic; after package and forward annotation, we can see that in layout the RF cell parameters L1 and L2 change to 20 mils, 40 mils. The cell appearance also changes accordingly, as shown in Figure 16.23.

Then, change MRIND RF cell parameters L1, L2 to 20 mils, 15 mils in layout and run back annotation; checking the schematic we can see that the RF symbol parameters L1, L2 have been updated to 20 mils, 15 mils. See Figure 16.24.

From the example above, we can see that the RF parameters mutually transfer between the schematic and layout, a very convenient feature of EE Flow.

16.5 RF Layout Design

As mentioned in the previous section, in Expedition, a specific RF license is required for RF circuit design. Select from the menu Setup→License modules and enable the Advanced Technology Pro (RF) license.

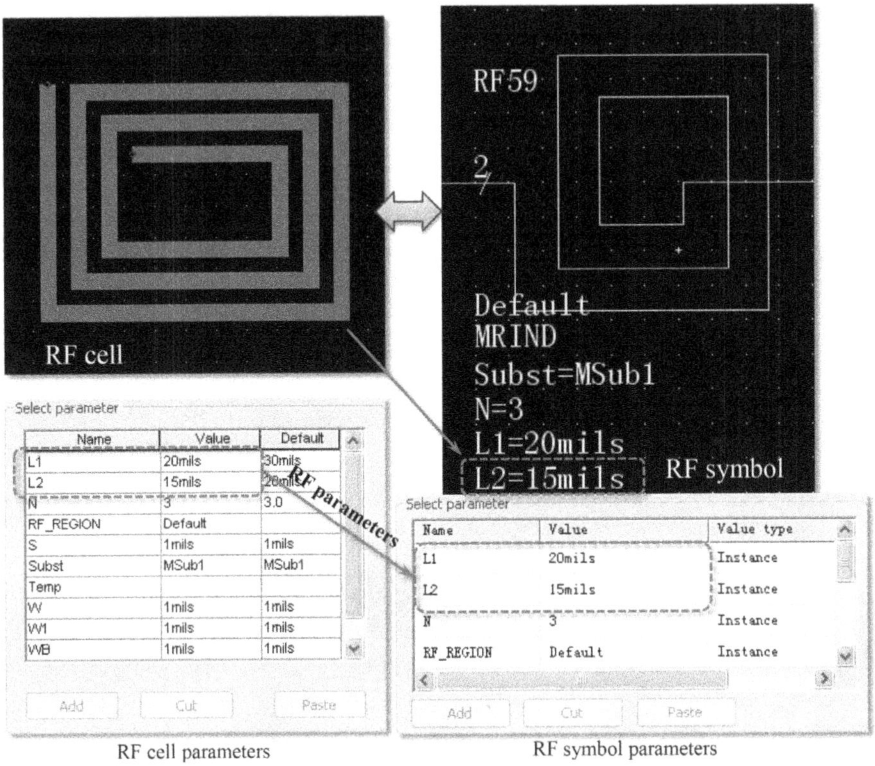

Figure 16.24 RF parameters are transferred from layout back to schematic.

Figure 16.25 RF toolkit.

16.5.1 RF Toolkit in Layout

After the RF license has been enabled, the RF Toolkit icon changes from gray to colored . Click RF Toolkit, Expedition will open the RF toolbox automatically, as shown in Figure 16.25.

We can also use menu Setup→RF to call the RF tool, as shown in Figure 16.26.

Another way to open the RF tools is to use the right-mouse-click menu from RF, as shown in Figure 16.27.

It is important to note that these three methods for invoking RF tools are valid only in RF mode; in the other three traditional operating modes (place, route and draw), we cannot call RF tools.

RF toolkit can be divided into three sections depending on the function: RF Shape toolbar, RF Node and Segment toolbar, RF General toolbar, explained as follows.

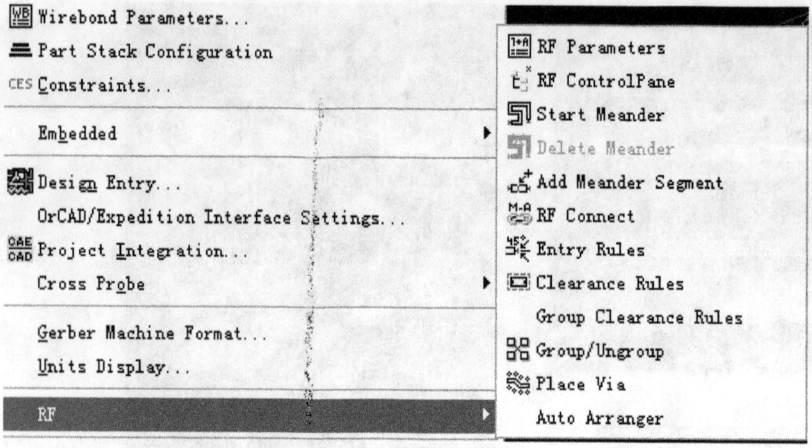

Figure 16.26 Select from the menu Setup→RF.

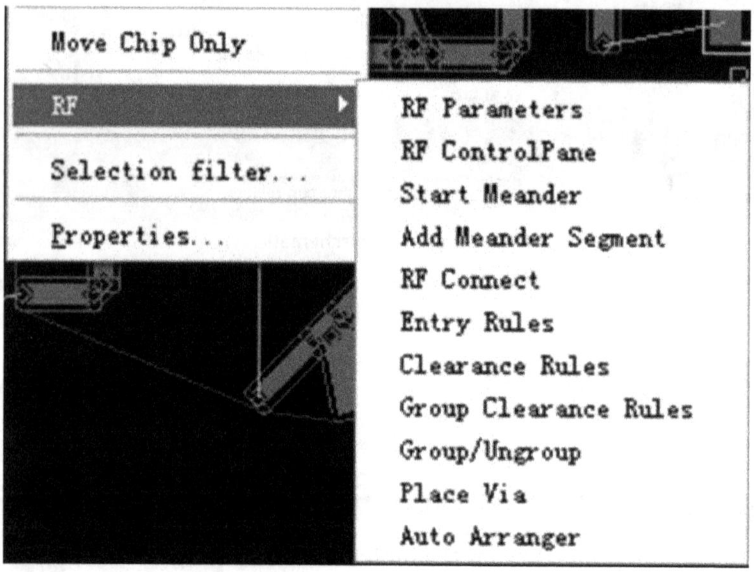

Figure 16.27 Right-click menu to invoke RF tools.

1) RF Shape toolbar 

   The RF shape toolbar is mainly used for RF shape moving, rotating, copying, etc.; the operational objects include library shapes (RF shapes loaded from the central library) and meander (RF trace drawn directly in layout); see Table 16.3.

2) RF Node and Segment toolbar 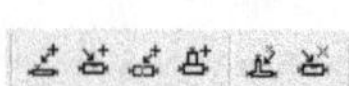

   This toolbar is mainly used for RF shape node placement and meander editing; operational objects include library shapes and meander; see Table 16.4.

Table 16.3 RF Shape toolbar description.

RF tools	Operation	Library shape	Meander
Move shape	Connecting RF shapes by moving	Yes	Yes
Angle shape	Rotating an RF shape	Yes	Yes
Mirror about X	Mirroring an RF shape about X axis	Yes	Yes
Mirror about Y	Mirroring an RF shape about Y axis	Yes	Yes
Copy	Copying a meander	No	Yes

Table 16.4 RF Node and Segment toolbar description.

RF tools	Operation	Library shape	Meander
Add edge node	Adding an edge node	Yes	No
Add floating node	Adding a floating node	Yes	No
Add segment	Adding a segment	No	Yes
Add stub	Adding a stub	No	Yes
Delete node stub	Deleting the selected node stub	No	Yes
Delete floating node	Deleting the selected floating node	No	Yes

3) RF General toolbar

This toolbar is mainly used to draw and edit RF meander in layout, for RF parameter setting and for RF group operations; see detailed description in Table 16.5.

Table 16.5 RF General toolbar description.

RF Toolkit	Description
Add meander	Add RF meander, opens the Meander Properties dialog box. Use the Meander Properties dialog box to define the meander properties.
Delete meander	Deletes the selected meander.
Edit meander	Alters the angle of the selected meander and segment position.
Convert RF shape	Converts a conductive polygon to an RF shape.
Parametric properties	Opens the RF Shape Parameters dialog box in which the designer can alter RF shapes and define block variable parameters.

(Continued)

Table 16.5 (Continued)

RF Toolkit	Description
Group/Ungroup	Opens the Group/Ungroup dialog box in which the designer can alter RF groups containing RF shapes and non-RF shapes.
Control window	Opens the RF Control dialog box in which the designer can alter RF design hierarchy. The designer can also access RF group, segment and node property values with the RF Design tab and view RF library shapes with the RF Library Shapes tab.
Clearance rules	Opens the Clearance Editor dialog box in which the designer can define RF shape and segment clearances.
Entry rules	Opens the RF Entry Rules dialog box in which the designer can define the node-entry angles for node-to-node and node-to-trace connections.
Place vias	Opens the Place Via dialog box in which the designer can place stitching vias.
RF Connect	Opens the RF Connect dialog box (layout) in which the designer can connect the layout tool to an RF simulator for bidirectional data transfer.
Selection Filter	Opens the Selection Filter dialog box in which the designer can define selectable objects in the layout tool.

In addition to the RF toolbar, RF soft keys are arranged at the bottom of the design window; their functions correspond to F1 to F12 on the keyboard. The functions of the soft keys will change with the operations carried out by the designer. Figure 16.28 displays soft-key functions when an RF unit is not selected, mainly showing some common RF operation commands.

Figure 16.29 shows the soft-key functions when an RF unit is selected; these include operation commands specifically for RF units.

When the designer clicks an RF soft key, the key will automatically change into a sub-soft key, allowing the designer to choose further functions so as to complete the corresponding operation.

RF soft-key general functions and the sub-soft-key main functions are described in Table 16.6.

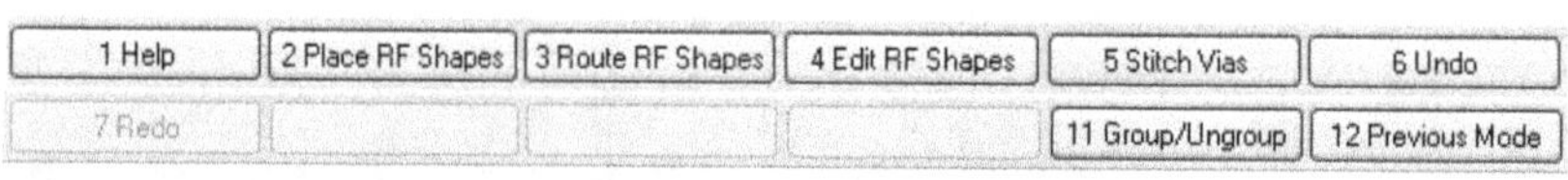

Figure 16.28 RF soft-key functions when an RF unit is not selected.

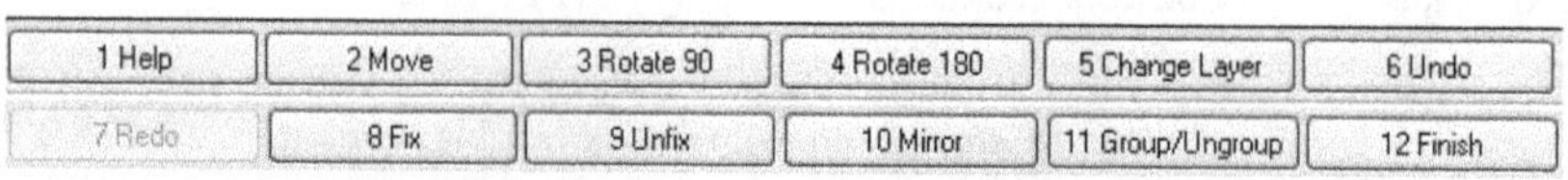

Figure 16.29 RF soft-key functions when an RF unit is selected.

Table 16.6 RF soft-key and sub-soft-key function descriptions.

The name of the soft key	Soft-key or corresponding sub-soft-key functions
F2→Place RF shapes	Opens the Place Parts and Cells dialog box, with which the designer can place RF groups, RF shapes and parts in the design.
F3→Route RF shapes	Enables the designer to route RF shapes. The Action keys change as follows: F2→Add Meander – opens the Meander Properties dialog box.
F4→Edit RF shapes	Enables the designer to edit RF shapes. F2→Add a node to RF shape. F3→Add a floating node to RF shape. F4→Delete RF node.
F5→Stitch vias	Opens the Place Via dialog box, with which the designer can place stitching vias.
F11→Group/Ungroup	Opens the Group/Ungroup dialog box, with which the designer can group design objects into RF groups.

16.5.2 Three Kinds of RF Unit

In RF layout design, RF units are divided into three types: RF standard shape, meander and user-defined RF shape, described below.

1) RF standard shape

   RF standard shape is re-defined RF shape. RF standard shape is available from the central library, and can be added to the schematic in the design flow using package and to layout from the schematic using forward annotation. RF standard shape cannot be added directly to layout.

   RF standard shape is called from the central library and added to the schematic, and its parameters can then be accurately passed to RF layout. From the previous section we have learned that the RF-standard-shape parameters can be adjusted in the schematic and passed to layout; they can also be adjusted in layout and passed to the schematic. RF parameter transmission is bidirectional, as shown in Figure 16.30.

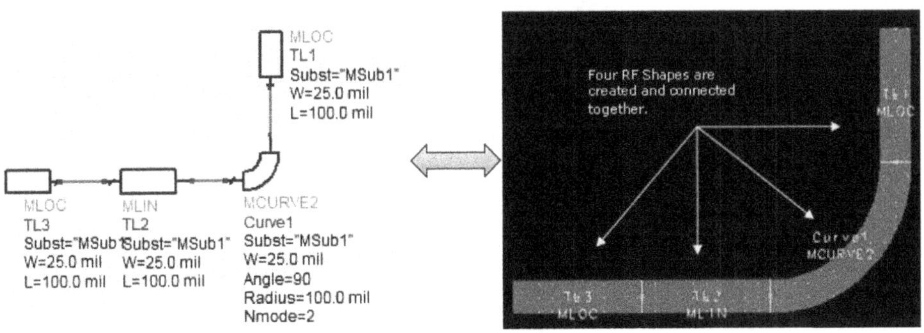

Figure 16.30 RF parameter two-way transfer between schematic and layout.

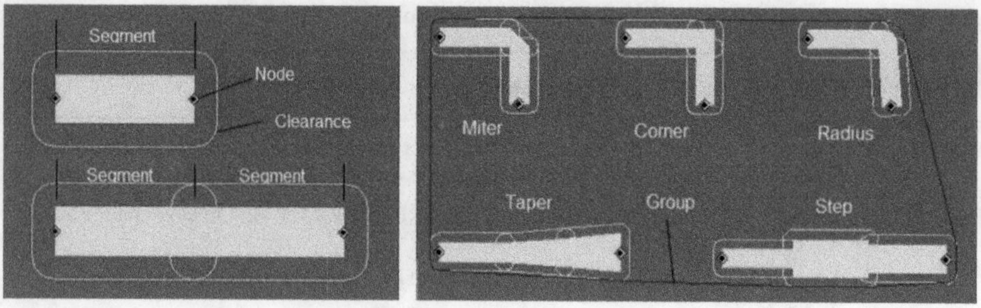

Figure 16.31 Segments and nodes of RF shapes.

2) Meander
   Meander is a segment or a series of segments of conductive RF shapes. Add meander to connect parts or to build a complex RF shape in layout. Meander is directly drawn in layout, independent of the schematic. The parameters of meanders are not transferred to the schematic. Each segment of meander is a polygon metal conductor.
3) User-defined RF shape
   User-defined RF shape is the RF shape defined by the user in layout, and is usually composed of polygon metal. Generally, first, the designer draws an ordinary shape, and then changes this to an RF shape by command. We can add custom RF shapes to the library, which we can then place in the schematic.

   The above three kinds of RF shapes are all composed of nodes and segments; the node's role is as the pin of a common component; the segment is conductor or copper, just like the body of a common component. Segments are connected by nodes, as shown in Figure 16.31.

16.5.3 Add and Edit Meander

Next, we learn how to add and edit meander in layout.

1) Add Meander

   Click the Add Meander button; the Meander properties window will pop up and the mouse cursor will change to a full-screen cross automatically. Then, we can start drawing meander in layout; during the process of drawing we can change meander properties at any time to meet different requirements, as shown in Figure 16.32.

   The Meander properties window is divided into three categories: General, Shape specific and Serpentine specific; each category contains multiple properties, as explained in Table 16.7.

   When drawing a meander, the first task is to set the current layer, then switch layer using the up and down arrows on the keyboard; the designer can look at the lower-right corner of the status bar to view the current layer and paired layer. For example, 1H, 8V suggests that the current layer is 1 and the paired layer is 8, 2V, 7H suggests that current layer is 2 and its paired layer is 7; V represents vertical, H represents horizontal, respectively indicating the default routing direction of the layer.

   Designers can also use the Layer tab in display control to switch current layer; click the layer in display control to set the current layer directly. In the process of

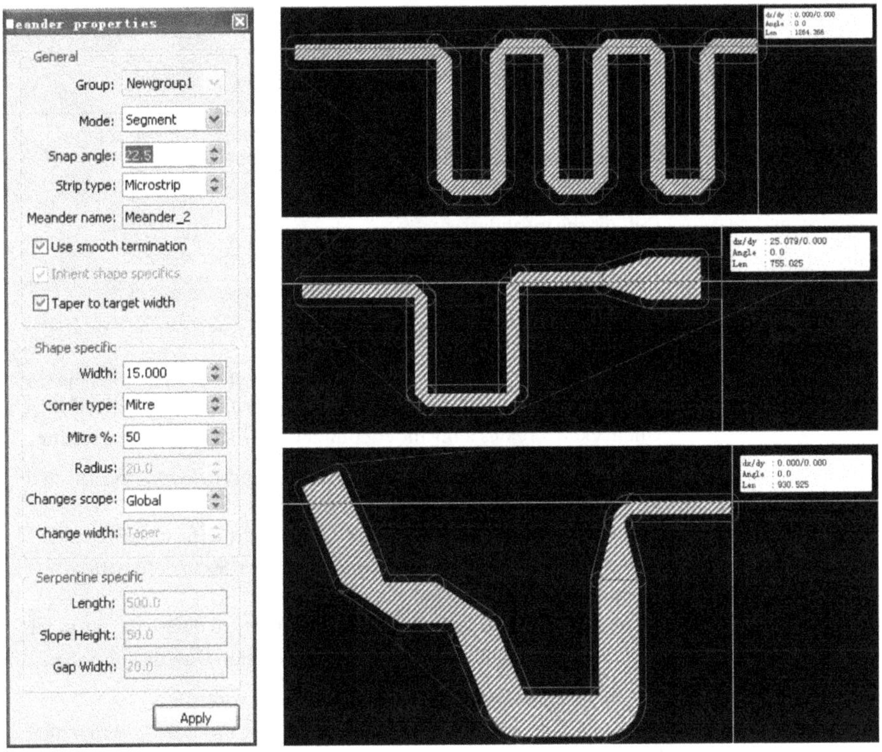

Figure 16.32 Draw meander in layout.

Table 16.7 Meander properties description.

Meander properties name		Properties description
General	Group	Defines all RF groups that are not fixed, locked or frozen. • <Inherit> – Applies the RF group name of the selected object to the meander and uses the node of the selected object as a starting point (RF group contents). • <Mouse select> – Applies the RF group name of the selected object to the meander. The meander does not need to be physically connected to the object. • <RF group name> – Applies the active RF group name to the meander. The meander does not need to be physically connected to the RF group.
	Mode	Defines the meander. • Segment – Defines a portion (segment) of a meander (the portion between mouse-clicks). • Serpentine – Defines a curved shape resembling the letter “s” (refer to the Serpentine field at the end of this table).

(Continued)

Table 16.7 (Continued)

Meander properties name		Properties description
	Snap angle	Defines the meander placement angle. For example, if you set the snap angle to 45°, the placement angles are 0°, 45°, 90°, 135° and so forth.
	Strip type	Defines the meander as a: • Stripline – A trace between ground planes. • Microstrip – A trace on the top or bottom layer of the PCB.
	Meander name	Defines the meander name.
	Use smooth termination	Checked, automatically creates an RF shape connection with the least amount of signal reflection (smooth termination connection). Unchecked, creates a meander connection with the current settings.
General	Inherit shape specifics	Checked, enables the meander to inherit properties from the selected RF shape. Selected shapes are: • Meander – Uses all RF shape-specific properties from the meander. • RF shape – Uses only width and layer properties to define the meander. • Non-RF shape – Uses only the layer property to define the meander. Unchecked, populates the shape-specific section with default values.
	Tape to target width	Checked, tapers meander width to the target width. Unchecked, enables the designer to attach the meander to different target widths.
Shape specific	Width	Defines the meander width.
	Corner type	Defines the meander corner type as: • Free Radius – The largest possible radius for arbitrary meanders. • Corner – A snap angle. • Mitre – A mitre angle. • Radius – An arc. Free Radius Corner Mitre Radius 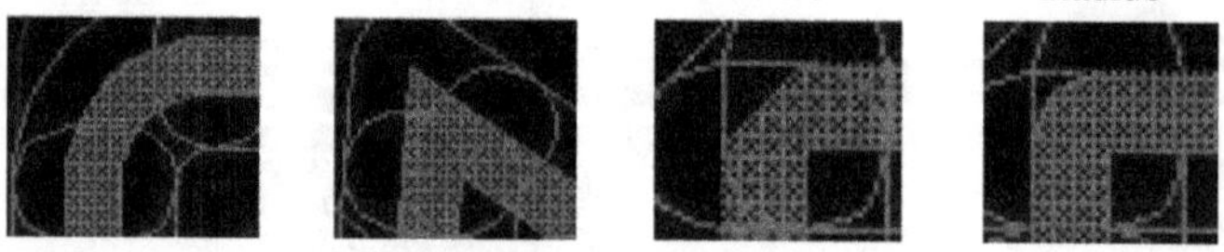
Shape specific	Mitre %	(Mitre corner-type only) Defines the percentage of the outer corner to be removed.
	Radius	Defines the radius of the arc applied to the outer corner.
	Change scope	Defines the meander scope. • Local – Applies changes only to selected meander segments. • Global – Applies changes to the entire meander.
	Change width	Defines how the width changes between segments: • Taper – Gradual width change. • Step – Abrupt width change.

Table 16.7 (Continued)

Meander properties name		Properties description	
Serpentine specific	Length	Serpentine meander length	Serpentine specific section – Defines the serpentine characteristics. Serpentine attributes are only available in the serpentine mode (Mode field). 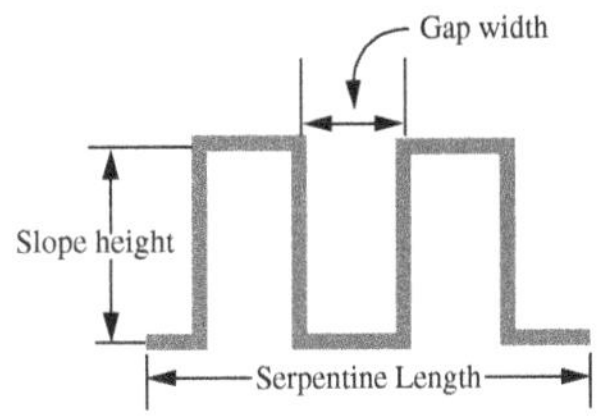
	Slope height	Serpentine meander slope height	
	Gap width	Serpentine meander gap width	

drawing a meander, if the designer wants to switch the layer, this can be done by adding a via. The most convenient method for adding a via is to click the layer in display control with the mouse button, as shown in Figure 16.33.

2) Edit meander

If it is necessary to edit a meander, we can use the Edit meander command button; when editing, drag the mouse according to the design needs. Figure 16.34 shows a graphical comparison of before and after meander edit.

16.5.4 RF Control Window

The RF Control window is mainly used to view and edit RF groups, segments, nodes, etc., and the value of RF properties.

The RF Control window contains two tabs: the RF Library Shape tab and the RF Design tab.

The RF Library Shape tab lists the RF units loaded from the central library; these properties are read-only in the RF Control window.

The RF Design tab lists the RF units in this design; their properties can be modified in the RF Control window. The designer can select the RF unit (including shape and

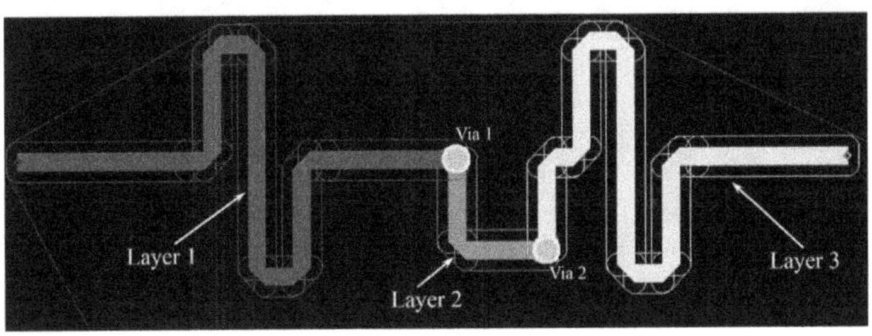

Figure 16.33 Switch meander layer through via.

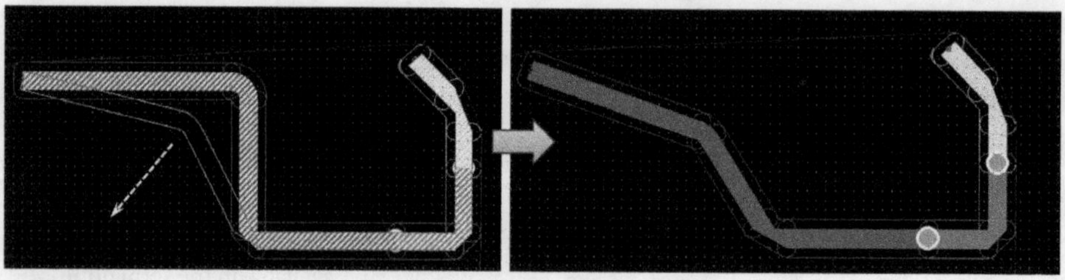

Figure 16.34 Graphical comparison of before and after meander edit.

meander) in the list, modify the properties directly, then click the Apply button; the RF unit changes can be seen in the preview window.

For example, to change a certain segment start point width from 15 to 30, after applying the change we can immediately see the changed starting point width, as shown on the right side of Figure 16.35.

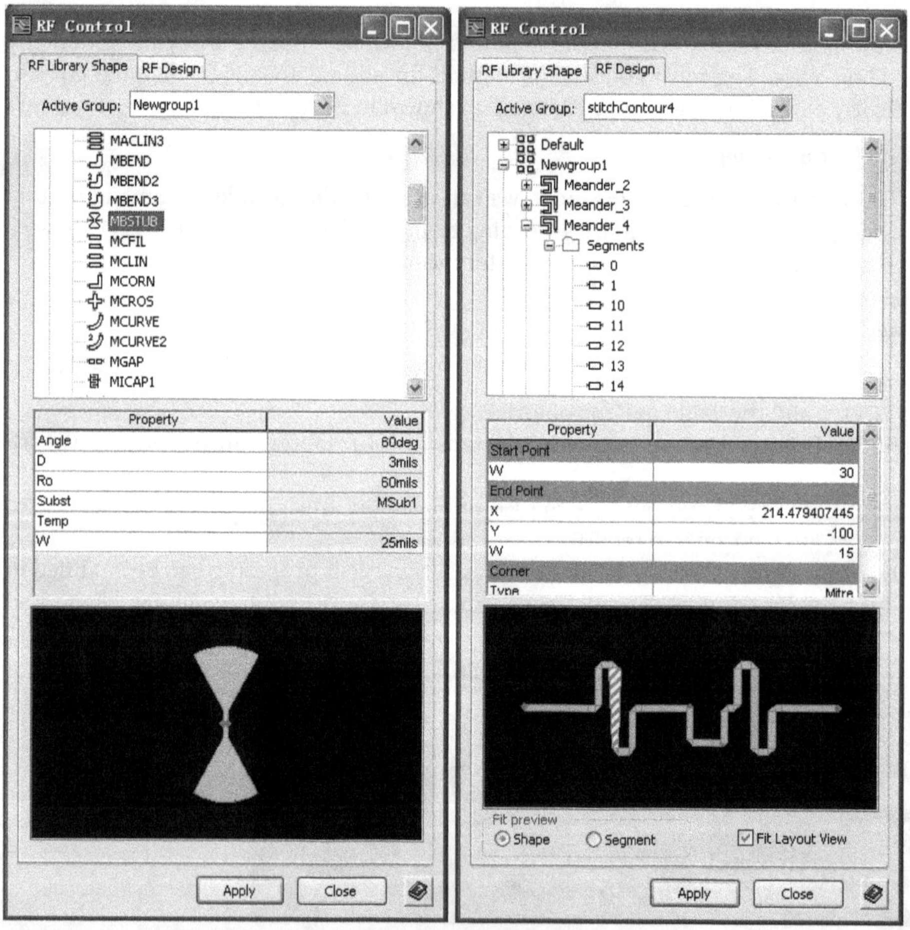

Figure 16.35 RF Library Shapes tab and RF Design tab.

16.5.5 Create Custom RF Shape

In RF circuit design, if we want to create a special RF shape, we can first draw a conductive shape in draw mode, and then click the Convert RF shape button ; the conductive shape we have drawn will be converted into an RF shape, as shown in Figure 16.36.

The converted RF shape is placed in the RF default group; if we want to move this RF shape to a given group, we can do this in the RF Group window; see section 16.5.7 below for details.

16.5.6 RF Via

In RF circuit design, we often need to add various types of vias to connect different layers of RF meander or different conductor layers, or to add vias to RF shapes, etc., as shown in Figure 16.37.

In the RF design toolkit, there is a specific tool for placing via. Click on the Place Vias button , and the Place Via window will pop up. This window contains five tabs: Interactive, Stitch Contour, Stitch Shape, Radial, Array.

1) Interactive
   Interactive place via refers to manual placement of via; in this tab we can choose via padstack, net name, via span layer, lock status and DRC options. After setting these,

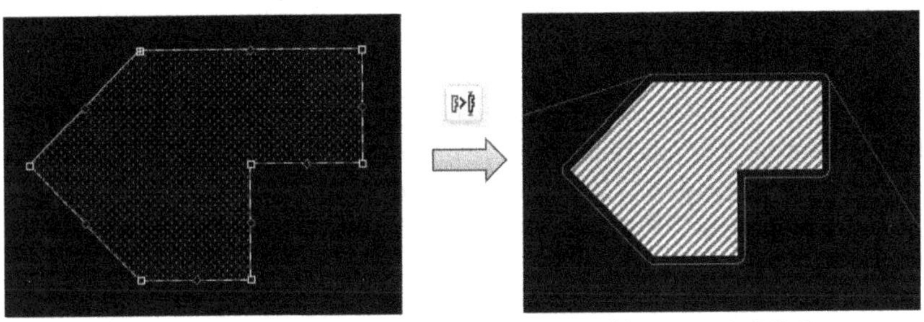

(a) Draw conductive shape (b) Change to RF shape

Figure 16.36 Create custom RF shape.

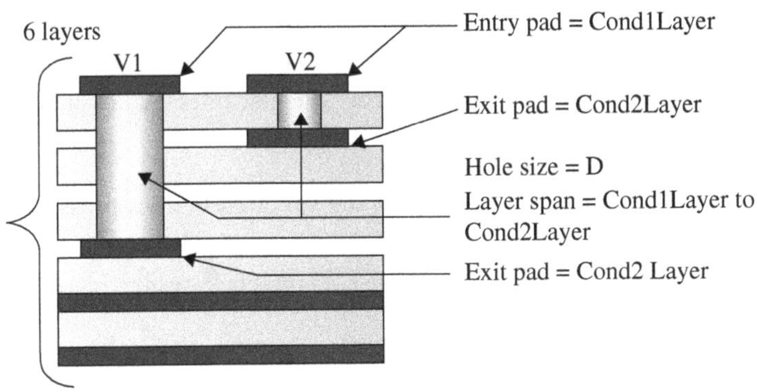

Figure 16.37 Using RF via to connect different layers.

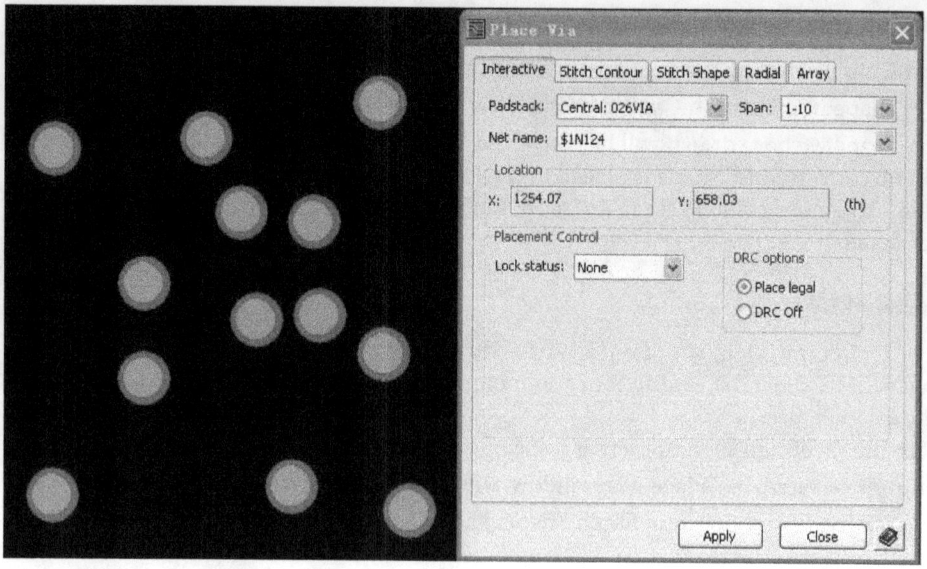

Figure 16.38 Interactive place via.

click the Apply button and then use the left-mouse button to place via on the layout. When a via is placed, its X and Y coordinates will be displayed in the Location bar, as shown in Figure 16.38.

2) Stitch contour
Stitch contour means automatically placing via along a design element edge. This design element can be RF meander, trace, plane, cavity, etc. Comparing this with interactive place via, in addition to setting via padstack, net name, etc., it is also necessary to set algorithm, distance, rows, via spacing and row spacing, etc. in the Placement Control area.

After setup, click the design element on which to place the stitch via, such as an RF meander, then click the Apply button. Expedition will place the stitch via automatically according to the rule settings, as shown in Figure 16.39.

3) Stitch shape
We can also use this tool to place via in plane shape or RF shape; the button is used to select the shape where the via will be placed, as shown in Figure 16.40. After pressing this button, select the shape, such as a conductive shape; this is then automatically added to the shape bar. Then set the via placement rules in the Placement Control area, and click Apply. See Figure 16.40.

4) Radial
For radial place via, the button is used to select the via circle center; press this button, use the mouse to click in the layout to set the center of the via circle, then set the via placement rules in the Placement Control area, and click Apply; the result is as shown in Figure 16.41.

5) Array
For Array place via, the button is used to select the first via position of the array. After pressing this button, use the mouse to click the first via position; the location

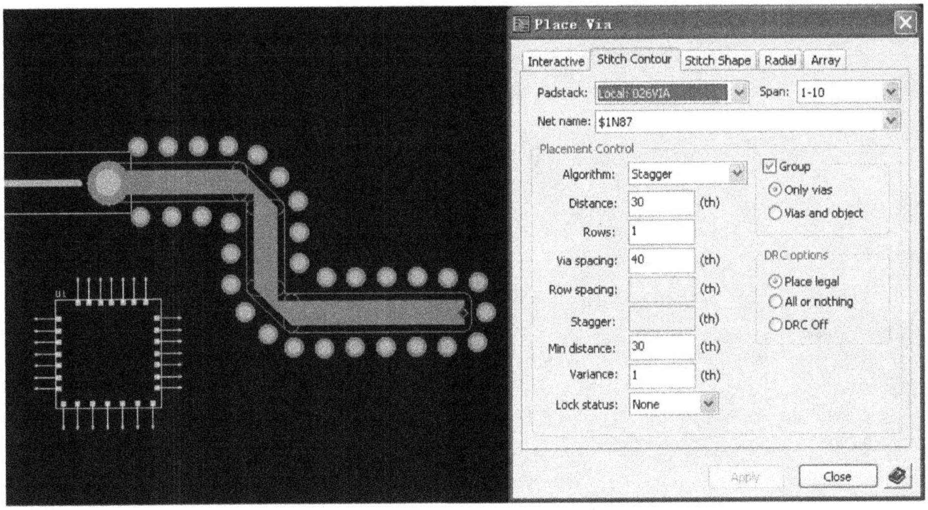

Figure 16.39 Stitch contour.

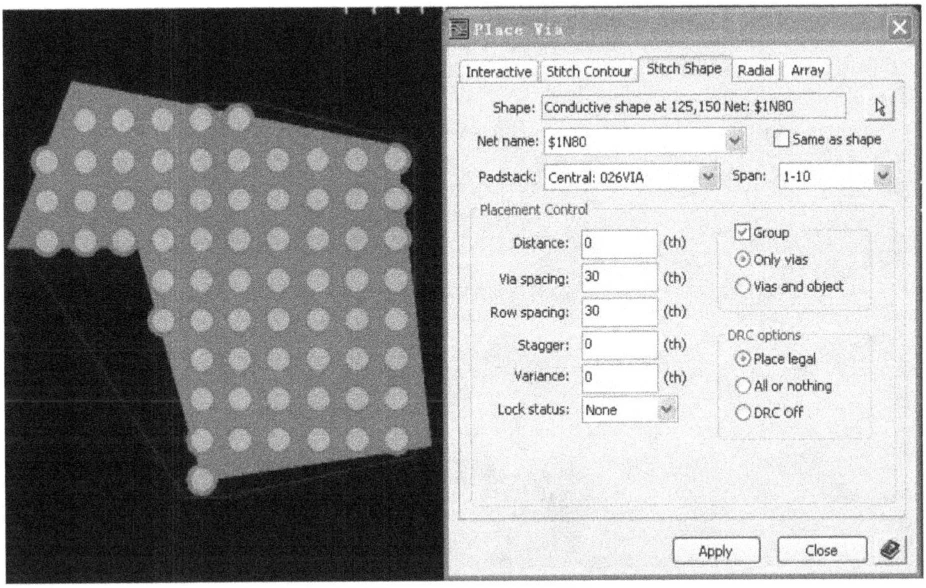

Figure 16.40 Stitch shape.

bar will automatically be updated with the mouse-click point coordinates; then set the via placement rules in the Placement Control area, and click Apply; the result is as shown in Figure 16.42.

16.5.7 RF Group

The RF Group tool is composed of the same or similar functions as apply to RF units. All RF design units belong to a particular RF group. In the Expedition environment,

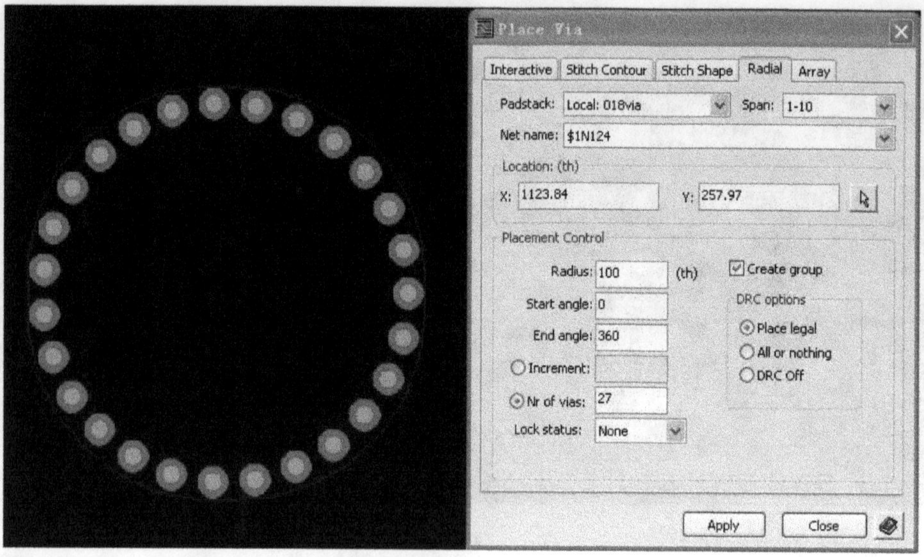

Figure 16.41 Radial place via.

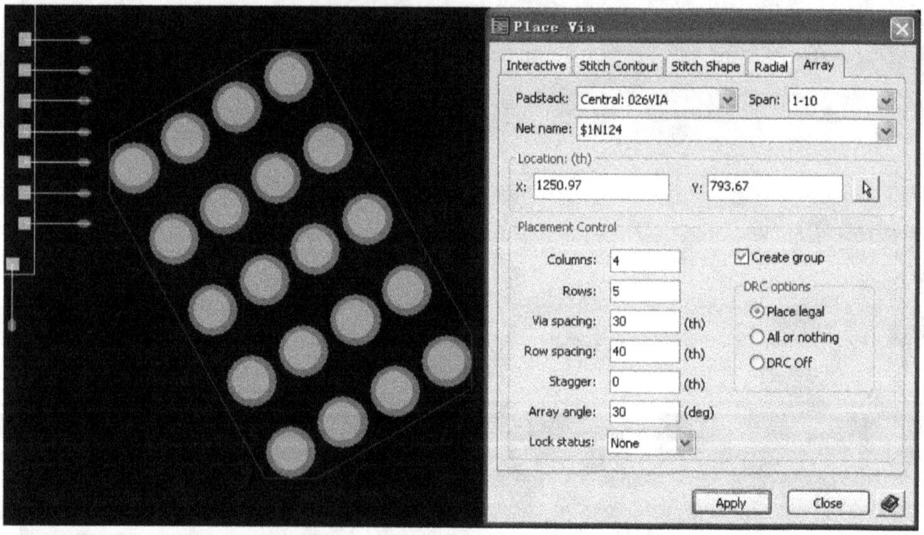

Figure 16.42 Array place via.

many RF commands use RF group as the unit, such as generate RF netlist, generate RF layout data, etc.

Click the button in the RF toolbar to open the RF Group editor window. In this window, we can do a variety of operations on RF groups.

If all the design elements in an RF group have been placed into layout, we can use the freeze command to freeze the RF group. After being frozen, the RF group cannot be modified. If any element of the RF group is selected, the whole frozen group will be

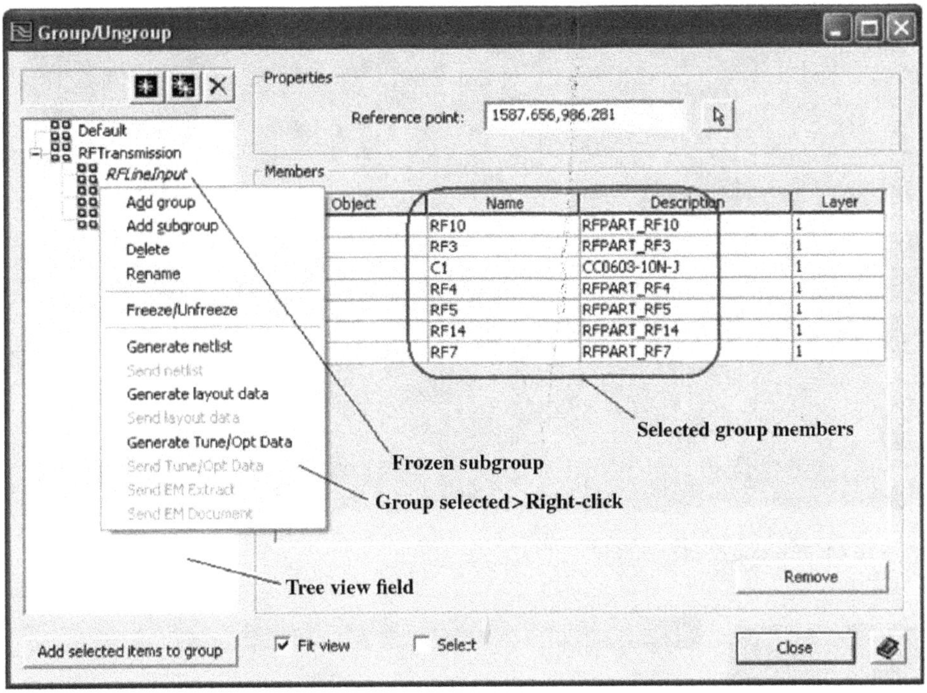

Figure 16.43 RF group operations.

selected, and the designer can move or rotate the whole group. For frozen RF groups, the layer cannot be changed, and the designer cannot add or remove elements or change the parameters; the group name is displayed in italics. Note that the default RF group can't be frozen.

Figure 16.43 shows RF group operations. In the RF Group/Ungroup window, we can add RF groups or sub-groups and delete RF groups with the buttons .

All RF units belonging to the same RF group are usually surrounded by a line frame; the line frame represents the RF group as a whole. The designer can control the display and color of RF groups in the display control Groups tab. Figure 16.44 shows RF group display control.

16.5.8 Other RF Edit Functions

1) Parametric properties

   Using the button, we can open the RF Shape Parameters window. In this window, the designer can view and change RF parameters; the parameter changes will be reflected in the RF shapes in real-time, as shown in Figure 16.45. The parameter changes can also be passed to the schematic using back annotation.

2) RF shape clearance rules

   Select an RF shape, then click the Clearance Rules button to open the Clearance Editor window. In this window, first create a new clearance rule, and then click the

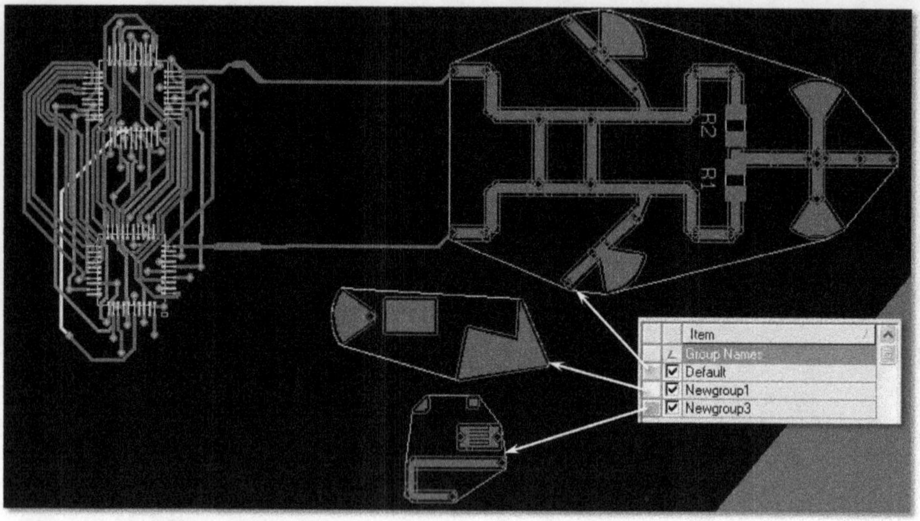

Figure 16.44 RF group display control.

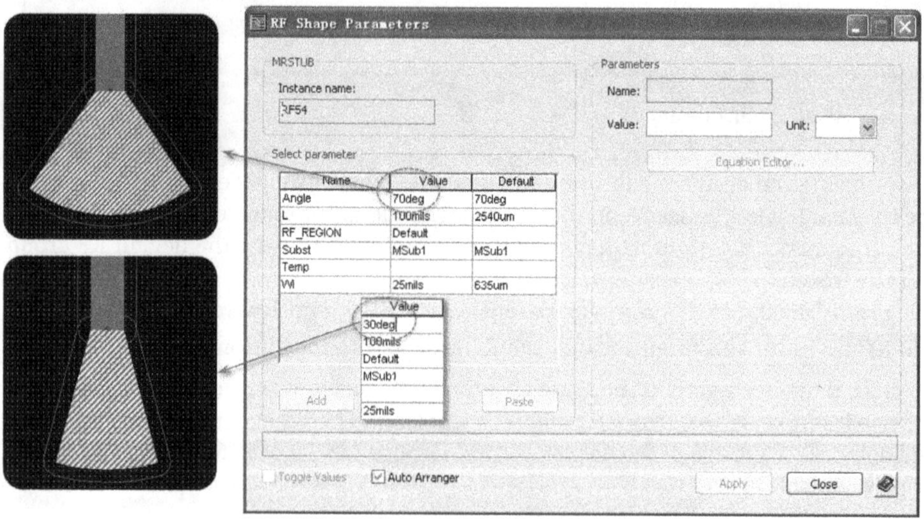

Figure 16.45 RF parameter changes are reflected in RF shape change.

white line frame in the window. The selected part turns red, and its operating point (square endpoint) is amplified accordingly; drag the operating point to realize the spacing rules modification. The designer can also enter a specific value in the Value bar to change the clearance rule, as shown in Figure 16.46. After editing is complete, click the Apply button to apply the change to layout design.

3) RF entry rules

In the RF toolbar, click the Entry Rules button to open the RF Entry Rules window. Here, the designer can define the node-to-node-entry rules and the node-to-trace entry rules, as shown in Figure 16.47.

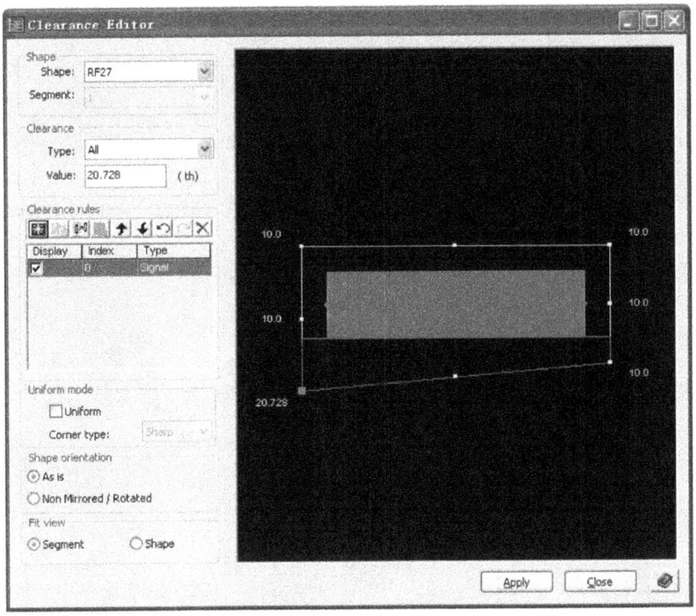

Figure 16.46 Edit RF shape clearance.

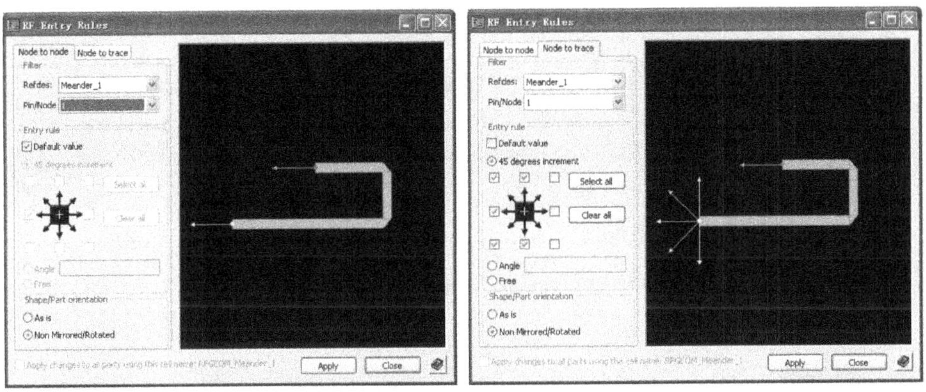

Figure 16.47 Define RF entry rules.

Under normal circumstances, the default value is selected, but if the designer wants to define these rules himself or herself, he or she should first uncheck the default value, and then choose from the options beneath. The 45 degrees increment field indicates an increase by 45° within the 360° scope.

The designer can also input the required angles directly into the Angle bar, with a semicolon as separator, for example, "15; 33; 44; 55", etc. "Free" indicates that the entry rules can be any angle.

4)  RF selection filter

RF filter is used to select different types of elements; it is divided into an Object tab and a Layer tab. The Object tab is used to choose different RF units in the design, the Layer tab is used for the selection of layer filters.

RF filter is different from the filter in Circuit Move and Copy, as shown in Figure 16.48.

5) RF Connect

RF Connect is mainly used to link RF simulation tools for data transfer. The RF Connect window is divided into three tabs: the Connections tab, mainly used for simulator connection and connection status display; Configuration tab, mainly used for the configuration of port and timeout settings, etc.; and the Extended Configuration tab, mainly used for the configuration of transport layer and data type options. Figure 16.49 shows the three tabs of RF Connect.

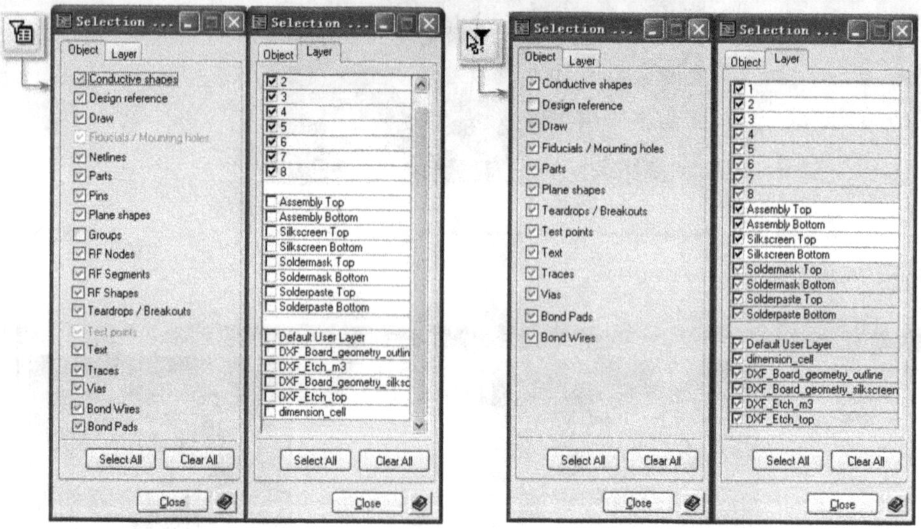

(a) RF selection filter

(b) Circuit selection filter

Figure 16.48 RF filter and Circuit Move and Copy filter.

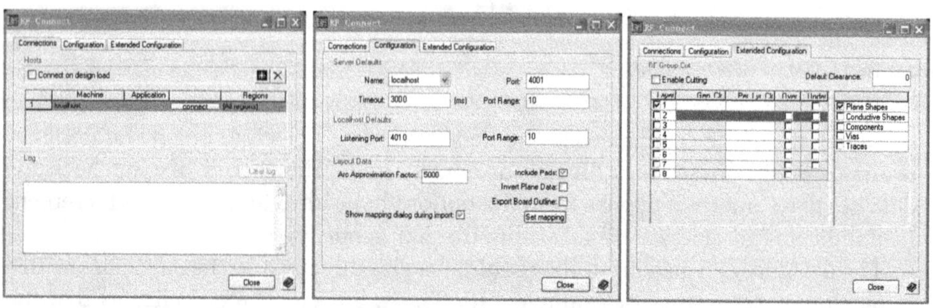

(a) Connections tab

(b) Configuration tab

(c) Extended Configuration tab

Figure 16.49 Three tabs of RF Connect.

16.6 Connect RF Simulation Tools and Transfer Data

16.6.1 Connect RF Simulation Tools

Using RF Connect, RF schematic data and RF layout data can be transferred to RF simulation tools. At present the supported simulation tools include ADS and AWR; in this book we use ADS as an example.

First, install MentorDA_DK (Mentor Dynamic Access_Design Kit) in ADS; its path is the ADS installation directory MentorDA_DK folder; for example: ...\ ADS2009U1 \ pcb_connect \ MentorDA_DK, as shown in Figure 16.50.

After installation, the Mentor DA menu will appear in ADS; select from the menu Start MentorDA Server; keeping the default options, click the OK button. After the completion of the connection, a prompt window stating "MentorDA Server started" will pop up, indicating that the connection is successful, as shown in Figure 16.51.

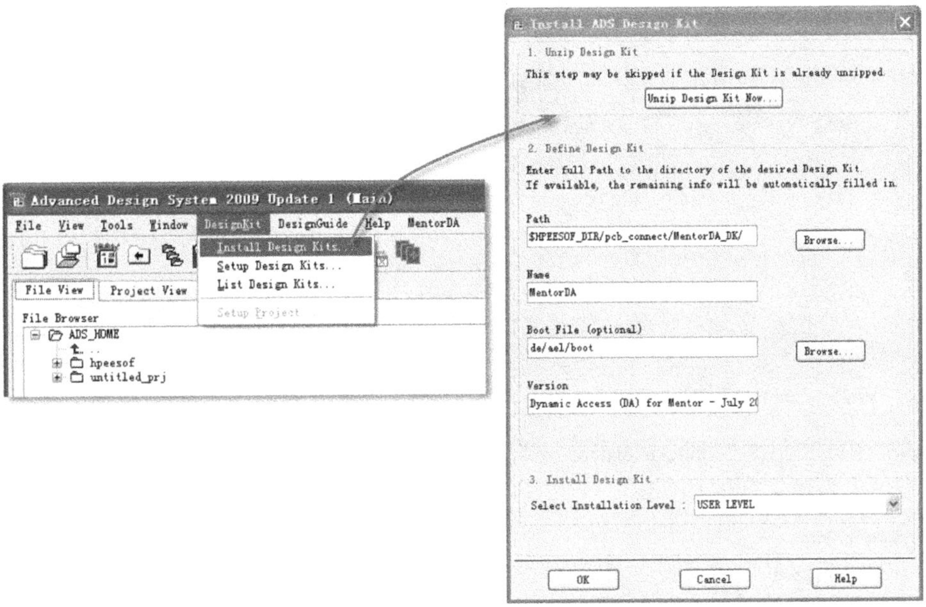

Figure 16.50 Mentor Design Kit installed in ADS.

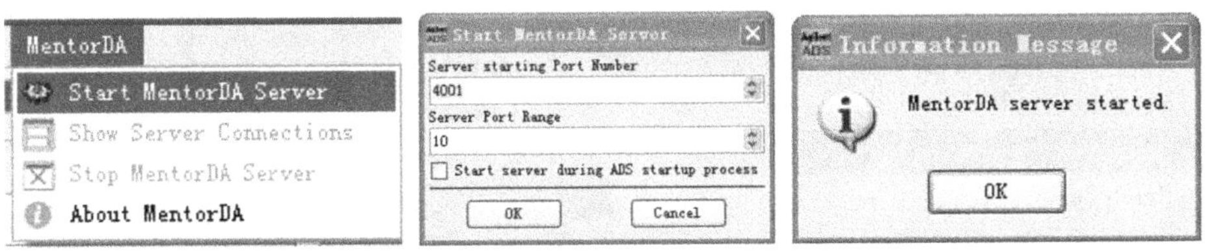

Figure 16.51 Start Mentor Server connection in ADS.

After successful connection in ADS, in the Expedition environment start RF Connect; click the Connect button and the Application bar will display "ADS"; in the Log area the message "Connecting to (localhost: 4001) ... ok" will show, indicating that the connection is ok, as shown in Figure 16.52.

16.6.2 Layout RF Data Transmission

After RF Connect is successful, the next step is to transfer RF design data to ADS. In the RF Group window, select the RF group we want to transfer; using the right-click menu, select Generate layout data, and a viewer log window will be displayed. Check the log window, if there is no error report, then select the Send layout data option to transmit the RF layout data to ADS, as shown in Figure 16.53.

After Send layout data is selected, data will be transmitted to ADS. Open ADS; we can see the RF layout data has been transferred to ADS, as shown in Figure 16.54.

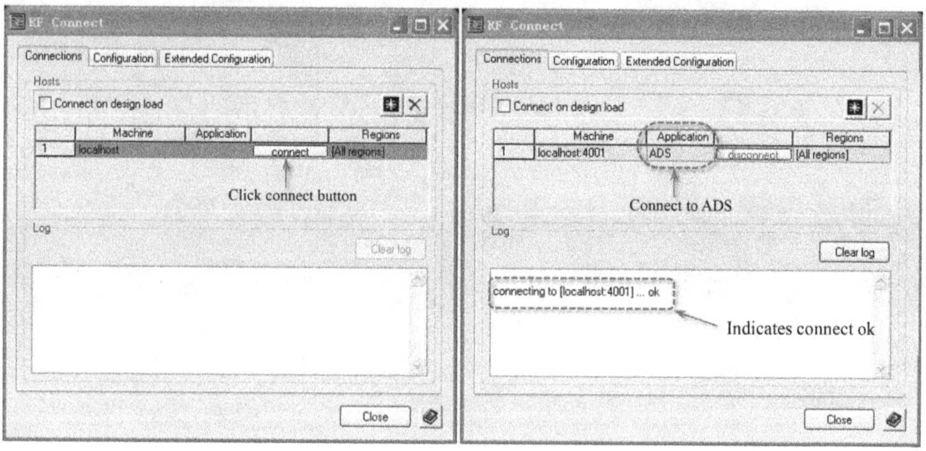

Figure 16.52 Start RF Connect in Expedition.

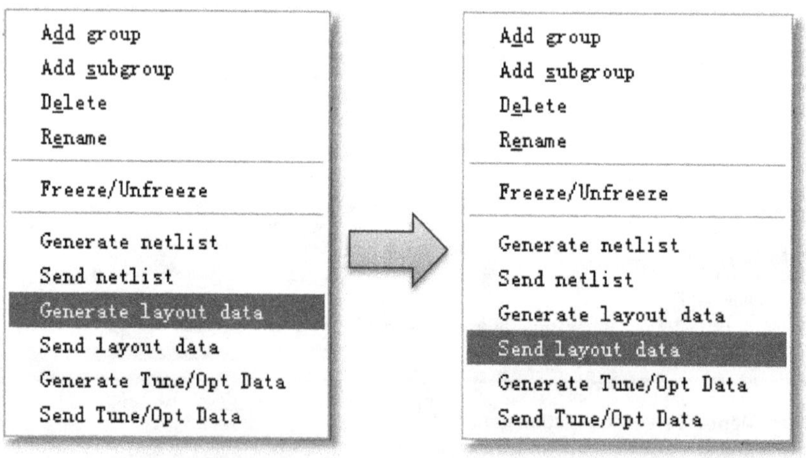

Figure 16.53 Transfer layout design data to ADS.

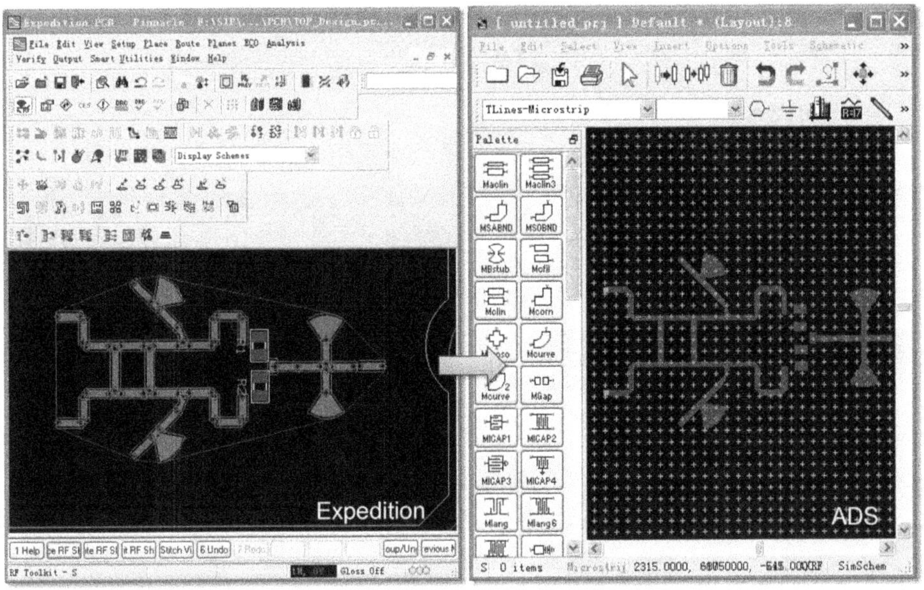

Figure 16.54 Transmit RF layout data from Expedition to ADS. *(See color plate section for the color representation of this figure.)*

Usually Generate layout data and Send layout data are mainly used for one-way data transfer, namely to transfer design data from Expedition to ADS for simulation. If it is necessary to transfer design data changes in ADS back to Expedition, we will need to use Generate Tune/Opt Data and Send Tune/Opt Data. Due to space limitations, this is not discussed in detail; the reader should refer to the relevant Mentor help documents.

16.6.3 Schematic RF Data Transmission

In the schematic tool DxDesigner, we can also connect with ADS using RF Connect. After successful connection, in the RF Group window of DxDesigner, using the right-click menu, choose Generate Schematic Data and Send Schematic Data to ADS, as shown in Figure 16.55.

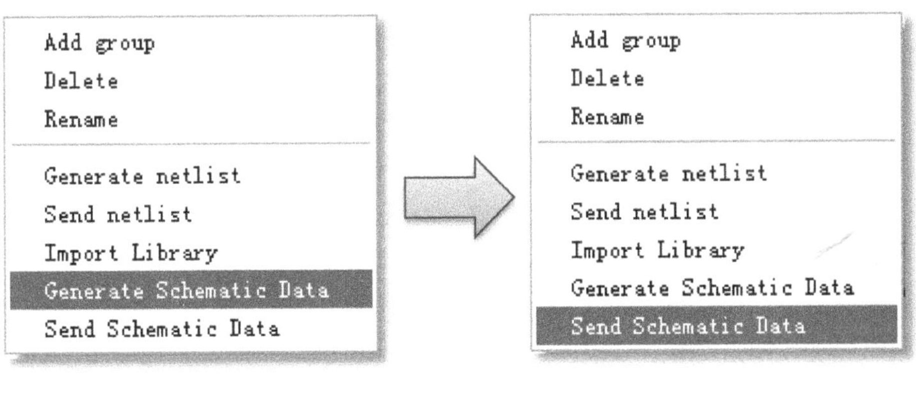

Figure 16.55 Transfer schematic RF data to ADS.

After command execution, schematic RF design data will be completely transferred from DxDesigner to ADS, as shown in Figure 16.56.

At the same time, RF parameters defined in DxDesigner will be accurately transferred to ADS, as shown in Figure 16.57.

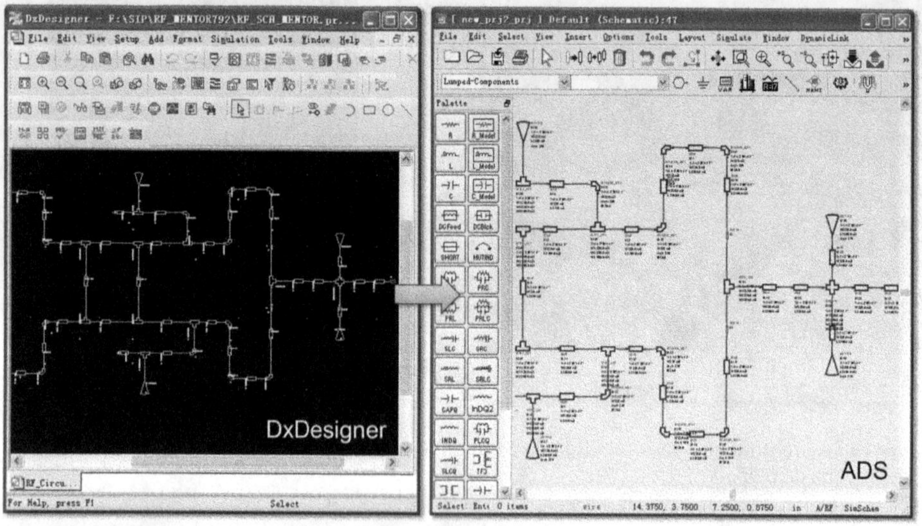

Figure 16.56 Transmit Schematic RF data from DxDesigner to ADS.

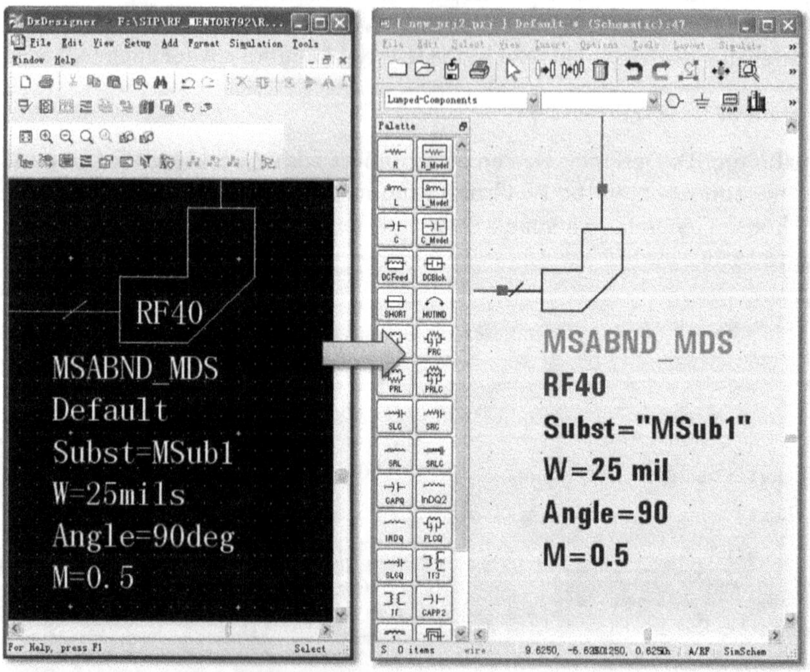

Figure 16.57 RF parameters in DxDesigner are transferred to ADS.

17

Concurrent Layout Design

17.1 Concurrent Layout Design Technology – Xtreme

Concurrent layout design technology refers to the possibility of multiple designers participating in one layout design at the same time.

In the concurrent design process, each designer can perform global layout design, and at the same time, they can set a temporary protection area to avoid operations by other designers impacting completed design sections.

In Mentor EE Flow, concurrent layout design has been popular for many years; it is a very mature technology.

As shown in Figure 17.1, multiple players design a project in real-time. In the concurrent design process, designers, in addition to completing their own edit operations, must also communicate with others; this kind of communication can be realized by meeting or telephone, or via the built-in information window that allows designers to convey messages to other designers.

In Expedition, the concurrent layout design tool is Xtreme, which has real-time concurrent design ability, supporting multiple designers operating on one layout design at the same time. The Xtreme concurrent design process is completely based on dynamic network access operations in real-time, without any human intervention, which greatly improves the design efficiency and quality.

Xtreme has the following features.

1) Real-time concurrent design, real-time dynamic collaborative routing. The real-time operations include component placement, design rule setting, automatic routing, plane shape handling, silkscreen adjusting, etc. Real-time collaborative design can run throughout the whole process of layout design.
2) Real-time concurrent design is based on network access and share technology, for the whole design process, without any design splitting and merging, thus guaranteeing the consistency of the design data.
3) It supports multiple designers simultaneously working on online collaborative layout; the number of designers can be adjusted dynamically according to the project progress and difficulty.
4) Design data can be located on a server or on one of the clients; any one of the clients can be designated the data server and can share data with other clients involved in the concurrent design.

SiP System-in-Package Design and Simulation: Mentor EE Flow Advanced Design Guide,
First Edition. Suny Li (Li Yang).

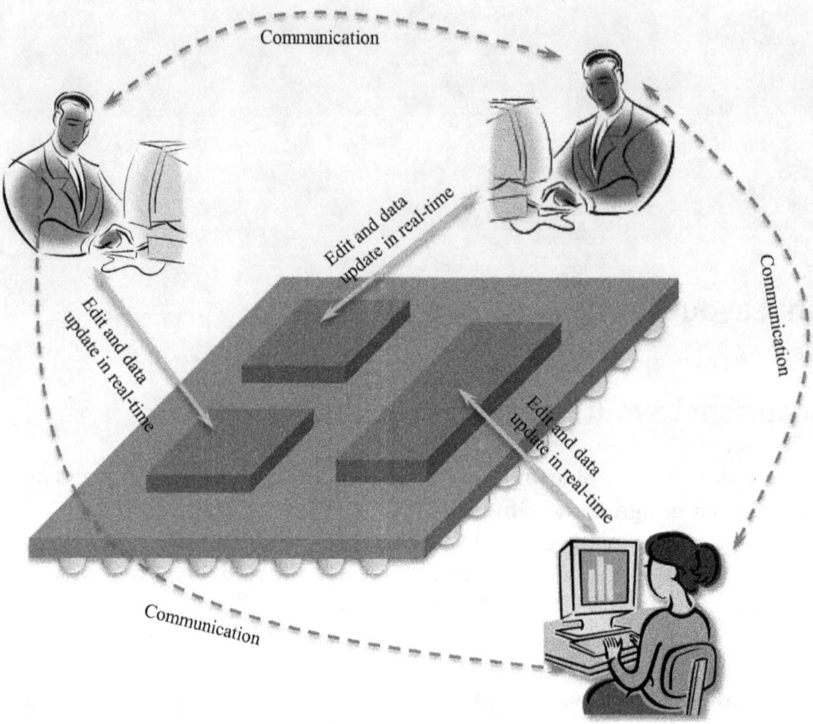

Figure 17.1 Multiplayer concurrent layout design sketch map.

In traditional collaborative layout design, artificial "splitting" must be carried out on design data to let more designers participate in design; the interface relationship definition of each segment is multifarious, and after the completion of split design, it is also necessary to "merge" the design. If the design changes frequently, the time required for splitting and merging can be even longer than the time required for design itself.

For practical design, design change is inevitable, and a design probably will change many times. In the split/merge method of task allocation, there is no dynamic adjustment of design tasks to match the project schedule. Because one designer can only see completed design for his or her own part, if they want to see the status of the whole project, they must download and merge the design data. This type of operation meant that it was not possible to realize data updating in real-time and dynamically. This old collaborative technology has been gradually abandoned by Mentor.

Xtreme is the latest Mentor concurrent design technology; it is real-time and dynamic, and there is no need to split design and define interfaces, or to have a data-merge procedure. Designers can enter and exit the layout design dynamically, that is to say, if two people are working on the design together, and another person becomes involved, three people can now design at the same time without the need for any change and preparation by the original two designers. When, late in design, only one person is needed, the other two just exit the design, without the necessity for any design merge process.

Xtreme automatic synchronization technology is exciting; each designer can see the operations of all the others in real-time.

The process server uses priority selection and DRC inspection to prevent user editing conflicts. In addition, members can also create protected areas or can open a protected area around the cursor to show the scope of their work. In Xtreme, when one designer needs to make design changes which may affect other designers, he or she can issue a notice to all users and let them make the choice to accept or reject the changes.

Xtreme concurrent design technology has been developed and applied in layout design for many years, its technology is very mature, and many companies and research institutes worldwide have applied Xtreme in their PCB layout design, with great social and economic benefits.

According to industry statistics, Xtreme can improve design efficiency by as much as 40%~70%, thus greatly shortening the project development cycle, improving the competitiveness of products, and providing the best solution for shortening design time for sudden design tasks, thus giving the company or institute a competitive advantage.

The author of this book began to use Xtreme concurrent design technology in 2005; from the perspective of actual projects, Xtreme can typically save more than 50% of the time required for layout design, and can greatly alleviate pressure on the designer, thus improving the quality of the design.

17.2 Xtreme Configuration

Xtreme concurrent design technology includes the elements shown in Table 17.1.

The data transmission relationships in Xtreme concurrent design are shown in Figure 17.2; under the unified management of server tool XDS, every designer's design data is transmitted to the server in real-time. After the integration of all design data, the server transmits the data to each designer's client computer. From the perspective of the designer, his or her own operations and other designers' operations are carried out at the same time.

In various software/hardware configurations, a variety of modes can be used in Xtreme; we introduce here the two most common modes.

The first mode has an independent file server; the design data are put on the file server, and at the same time RSCM (Remote Server Configuration Manager provided by Mentor) is installed on the server and used to manage the design data. All designers involved in the design are in client mode and have write access to the data on the file server.

Table 17.1 Xtreme concurrent design configuration table.

Xtreme Design Session (XDS)	Server management tool for concurrent design, manages design data, allows people to access design data at the same time. The maximum allowable designers for concurrent design is 15. Multiple XDS can be launched from one central server to manage multiple concurrent design projects simultaneously.
Xtreme Design Client (XDC)	Client management tool for concurrent design that allows users to join in a concurrent design environment.
Server	File server, can be a public computer or one of the designers' computers.
Handle	Identification of the designer, it is used to allow designers involved in concurrent design to identify themselves and other designers involved in the same design; the default name is the designer's computer's name, the designer can modify the name on entering the concurrent design; the name should be less than 16 characters, and the designer should avoid leaving blank spaces in the handle name.

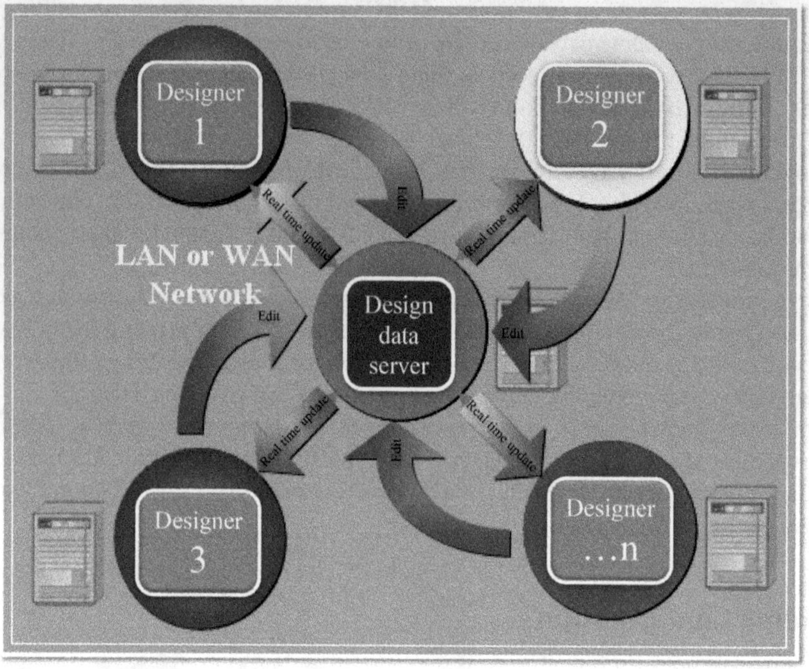

Figure 17.2 Data transmission in Xtreme concurrent design.

This mode is suitable for a company or institute regularly using concurrent design; it requires configuration of a dedicated server, and the use of access rights management to ensure each design is only open to the relevant people and that others cannot access the design or can only access it in read-only mode. See Figure 17.3.

The second mode is client folder-sharing mode. One designer's computer serves as file server, the design data is put on this computer, and RSCM is also installed on this computer to manage the design data. Using the folder-sharing function, all designers involved in the design have write access rights in this folder.

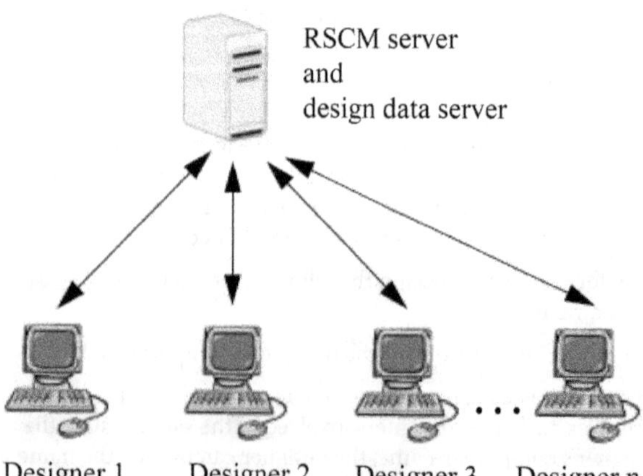

Figure 17.3 Independent server mode.

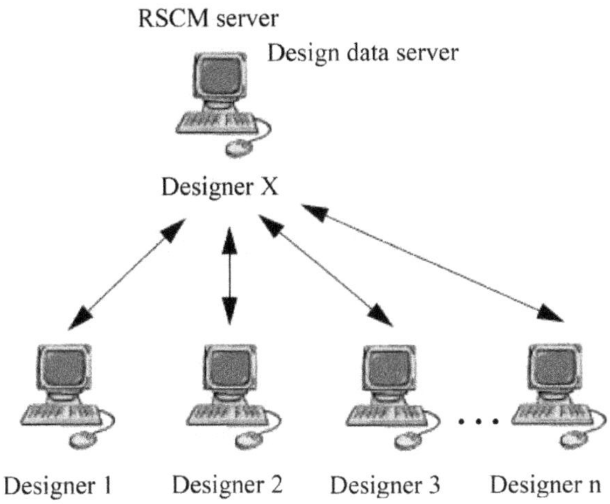

Figure 17.4 The client folder-sharing mode.

Configuration of this mode is more flexible, and more suitable for a company or institute occasionally using concurrent design. This mode does not need configuration of a dedicated server; for urgent projects, a designer can invite colleagues to help him or her with the main layout; in actual design, this mode plays an important role in urgent projects. See Figure 17.4.

17.3 Start Xtreme Concurrent Design

Click the Xtreme Design Session icon in Dashboard (the icon must be dragged out in advance from the Layout Toolboxes folder to Dashboard Shortcuts), start concurrent design and select the desired data source and parameters; the general data source is a *.pcb file, as shown in Figure 17.5.

Then, in the pop-up window, select the desired Xtreme Design Client licenses, such as AdvPkg license for SiP or package design, EP license for embedded resistor and

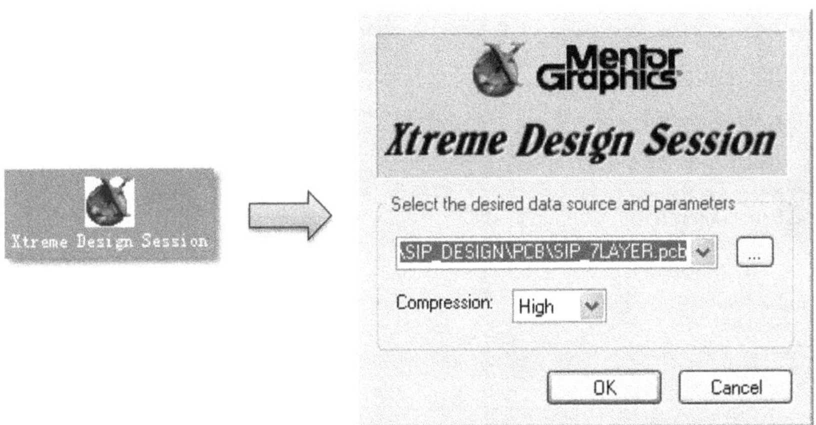

Figure 17.5 Start Xtreme Design Session.

capacitor design, RF Layout Design Option license for RF circuit design, etc. Here it is important to note that only the Pinnacle license supports Xtreme concurrent design; Ascent and Ascent LX do not support Xtreme.

The Xtreme window will pop up and XDS will start normally, as we can see from Figure 17.6; at this time no designers have joined the concurrent design.

Then, different designers open the design with Xtreme Design Client, as shown in Figure 17.7.

After the designers join the design, we can see those designers in the XDS window, as shown in Figure 17.8.

At this time, we can proceed to concurrent layout design. For designers there is no essential difference between concurrent design and independent design; all operating procedures are exactly the same, and the designer just needs to operate according to their own preferences.

In the concurrent design process, each designer can see other designers' operations in his or her design window in real-time. They can define display color and pattern according to their own preferences; Display Control is relatively independent for each designer.

Figure 17.9 shows the design window of different designers in concurrent design.

Each designer can see all the other designers' operations in real-time. The designer can also create protected areas temporarily or show a protected area around the cursor of their current work.

As shown in Figure 17.10, designer A and designer B have protected circular areas around the cursor; in these protected areas other designers are not allowed to enter and carry out edit operations.

We can also see from Figure 17.10 that the protected area handle identification of designer A is not shown, while the protected area handle identification of designer B is shown, which indicates that the screenshot comes from the computer of designer

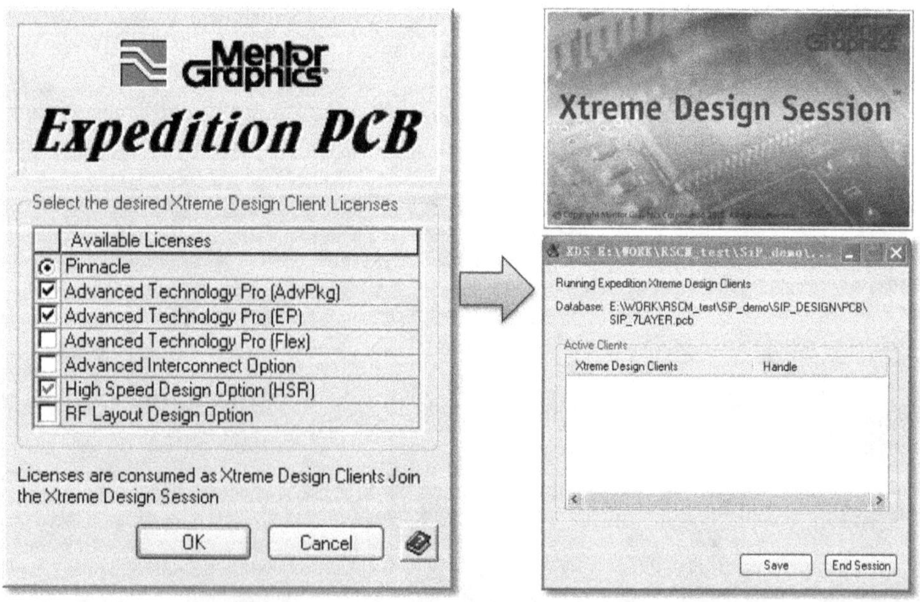

Figure 17.6 XDS starts normally.

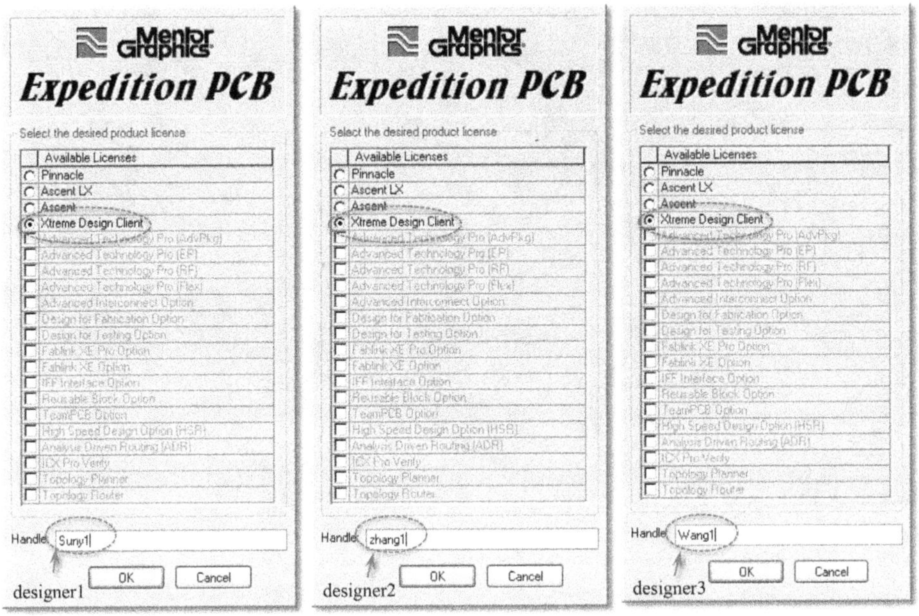

Figure 17.7 Each designer opens the same design with Xtreme Design Client (XDC).

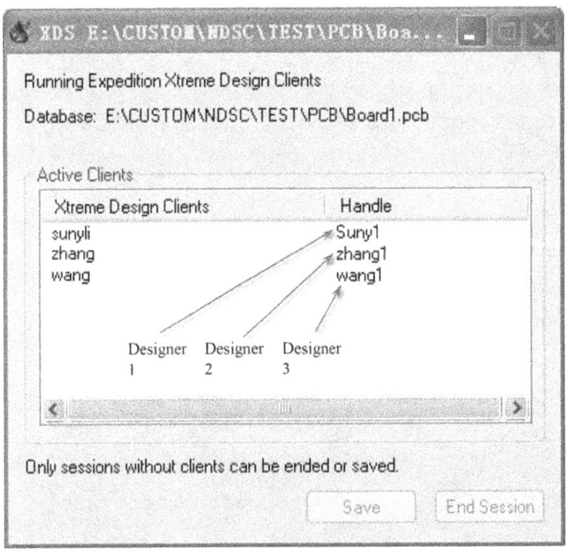

Figure 17.8 The participating designers are shown in the XDS window.

A. By default, a local handle identification does not show, but other designers' handles show automatically, to help the local designer identify edit operations by other participants.

When Xtreme concurrent design is complete, or in the process of concurrent design, participant designers can exit the design environment. When exiting Xtreme, the designer will be prompted to save the design; select Yes for ok.

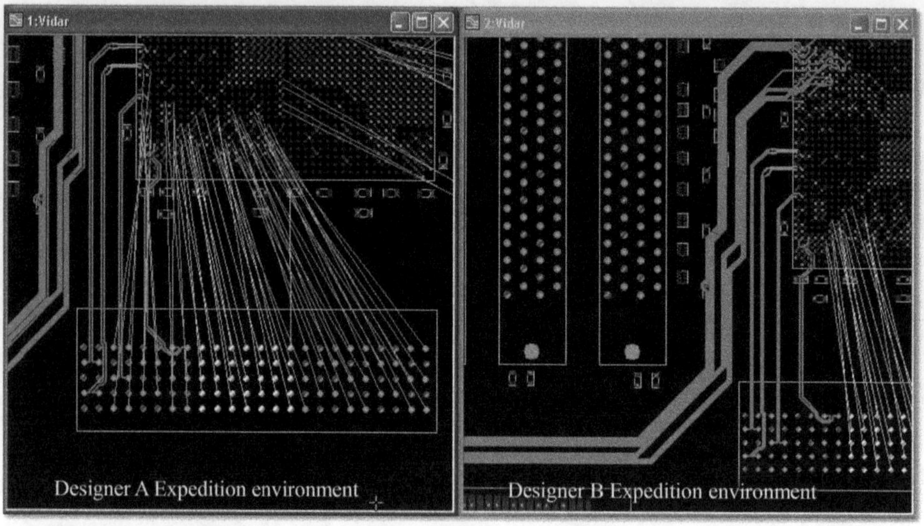

Figure 17.9 Different designers' design windows in concurrent design.

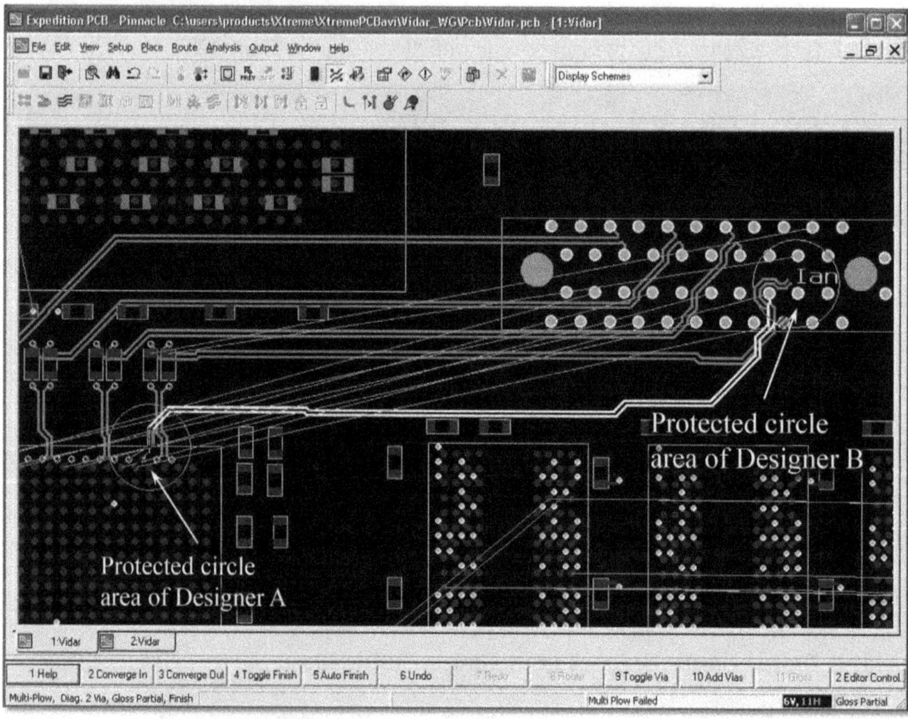

Figure 17.10 The edit operation shown in the interface of designer A.

When the last designer exits Xtreme concurrent design, the system will notify him or her that they are the last participant in this design session; click End on the prompt window; the system displays another prompt window, click Yes, as shown in Figure 17.11.

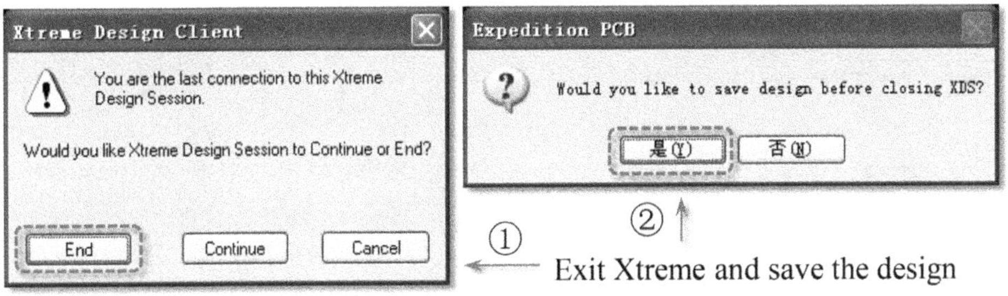

Figure 17.11 Notification to the last designer exiting the design session.

After the design has been saved, Xtreme concurrent design is completed.

In Xtreme the design data format is exactly the same as for general layout design; in subsequent design processes, the design can be continued by a single designer or Xtreme can be restarted and the design can be handled by many designers using concurrent design.

17.4 Matters to Note in Xtreme

There are several matters that should be noted regarding Xtreme concurrent design.

1) The software version of Expedition should be the same on each participant's computer.
2) RSCM should be installed and its service should run properly.
3) The running RSCM version should be the same as the Expedition version.
4) Sometimes the Windows firewall should be shut down.
5) The project should not run in single user mode; SingleUserMode value should be auto or no; see Figure 17.12.

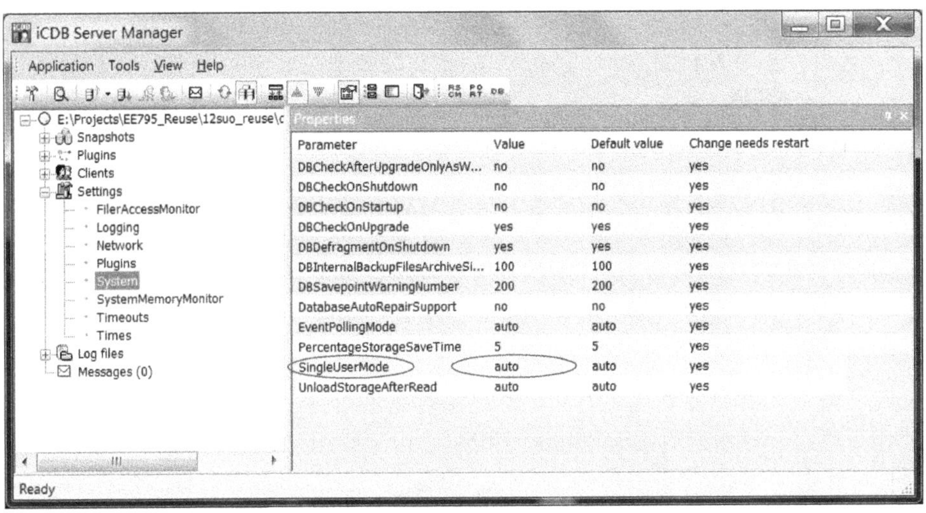

Figure 17.12 SingleUserMode value should be auto or no.

18

3D Real-time DRC

In the Expedition layout design environment, there are two 3D display and check environments; one is Wire Model Editor, located on the Wire Bond toolbar, the other is 3D Viewer, located in the Window menu, as shown in Figure 18.1. These two tools are introduced here.

18.1 Wire Model Editor 3D Display and DRC

18.1.1 Wire Model Editor 3D display

Wire Model Editor is mainly used for bond wire 3D model checking and adjusting. At the time of inspection and adjustment of 3D bond wire, we need to first select the bond wire (usually by selecting the corresponding bond pad or die pin), and then click the Wire Model Editor icon. In the window, the bond wire and its surrounding elements are displayed in 3D graphics, as shown in Figure 18.2.

In Figure 18.2, the left side of the window is the bond wire model list and related parameters, the right side of the window comprises four sub-windows which display a front, left, top and 3D view of the bond wire. In the 3D view window, the selected bond wire displays green, to distinguish it from other bond wires.

The Wire Model Editor window is mainly used to adjust the selected bond wire profile and parameters in real-time. If no bond pad or die pin is selected, the window will show as blank. For example, click the blank space in the design window, or any elements not related to bond wire, and the window will display blank, as shown in Figure 18.3.

At this time, if the designer selects a bond pad or die pin connected to the bond wire, the Wire Model Editor window will immediately display the bond wire from the front, left and top, the 3D view and the related parameters, as shown in Figure 18.4.

In the three view windows (front, left and top), the designer can adjust the operation point of the bond wire. During the adjustment process, the 3D window will display bond wire profile changes in real-time, comparing it with other bond wires and 3D elements. The designer can see the result of the adjustment and can also look at the parameter changes, to judge the amount of bond wire adjustment.

At the same time, the designer can also directly modify the value of related parameters in the Points or Variables areas of the window to change the bond wire profile.

SiP System-in-Package Design and Simulation: Mentor EE Flow Advanced Design Guide,
First Edition. Suny Li (Li Yang).

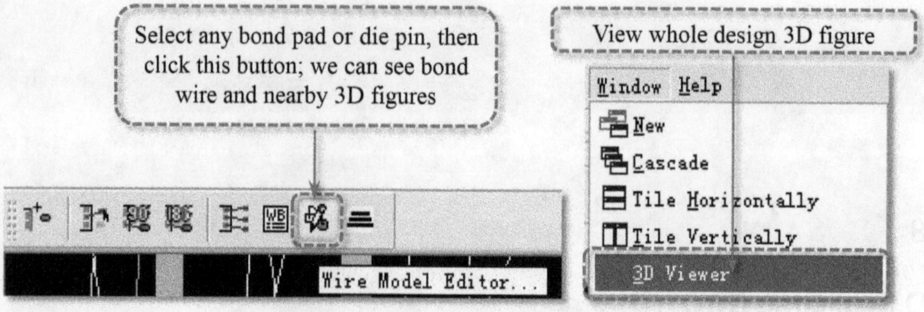

Figure 18.1 Wire Model Editor and 3D Viewer.

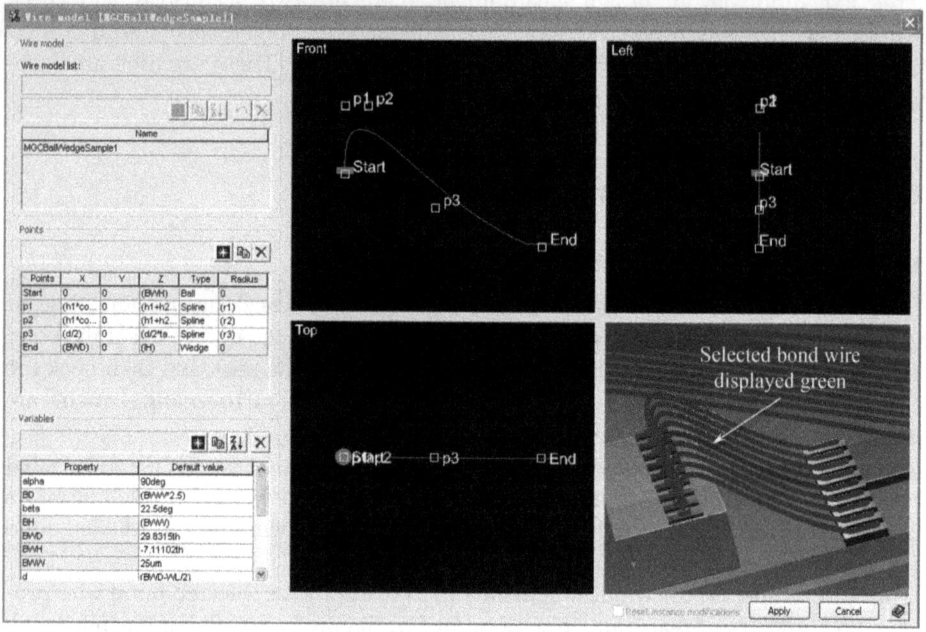

Figure 18.2 Bond wire model in Wire Model Editor.

In Figure 18.5, the designer drags point P3 upward in the front-view window, adjusts the height of the bond wire, and the adjustment effect is seen in the 3D window in real-time.

18.1.2 Wire Model Editor 3D DRC

In the process of adjustment, if the distance between the bond wire and other elements does not meet the DRC rule settings, DRC errors will be notified automatically. In Figure 18.6, for example, two bond wires are too close; this has violated the bond wire to the bond wire distance rule.

Currently, this is a 3D rule, that is, the distance between the bond wires is checked in 3D space. Some bond wires may overlap in 2D space (top view), but because their heights are different, they meet the DRC rules in 3D space.

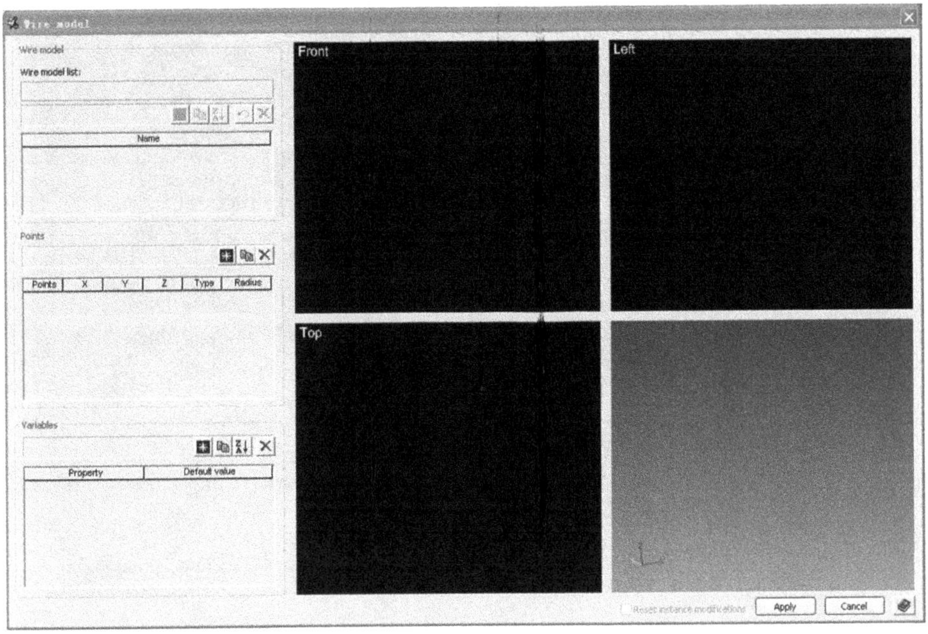

Figure 18.3 The window shows blank when elements not related to bond wire are selected.

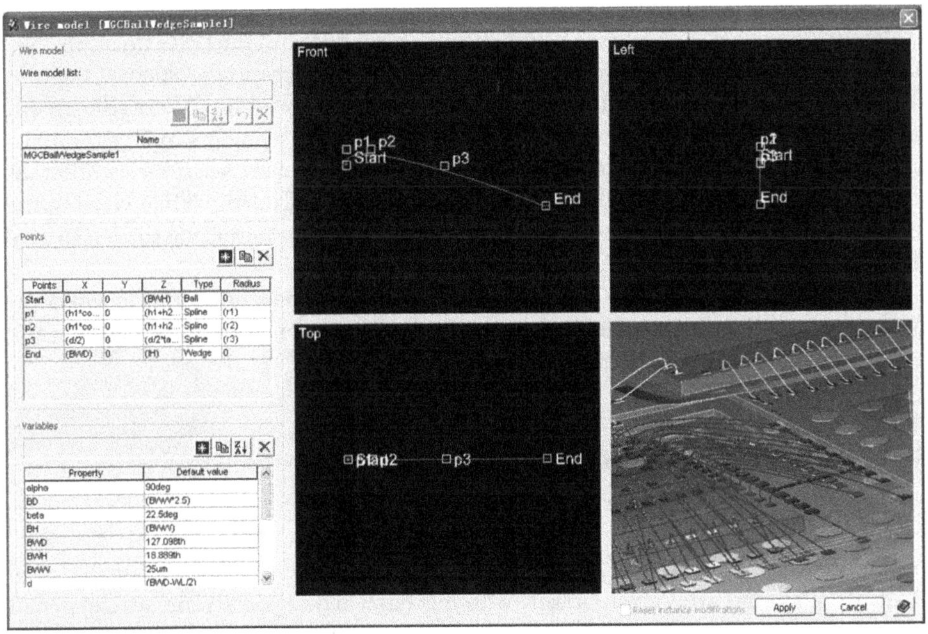

Figure 18.4 The display state when the bond wire is reselected.

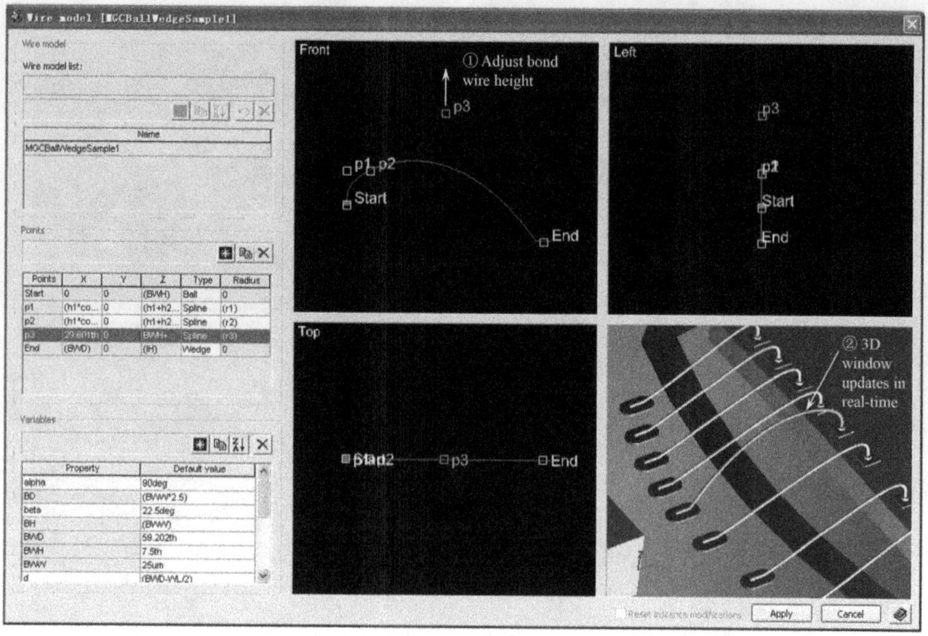

Figure 18.5 The adjustment result of bond wire displays in real-time.

In the case of a violation of the rules, click the Apply button below the window; the alert window will automatically pop up notifying the designer of the DRC conflict, and ask the designer whether he or she wants to make a change with the DRC violation. 'No' should be chosen in order to avoid later production and processing problems.

Although the Wire Model Editor 3D window is used mainly for bond wire adjustment and 3D DRC check, the window can also show other design elements, such as components, pads, traces, vias, layers, cavities, planes, etc. Designers can see from six different direction in the 3D view: top, bottom, east, west, south and north. While viewing the elements, the designer can rotate the 3D graphics in any direction using the left mouse button, as shown in Figure 18.7.

To allow easier viewing of the design elements, the 3D window has a translucent components option; Figure 18.8 shows the two display effects of opening and closing the translucent components option.

18.2 3D Viewer Display and DRC

18.2.1 3D Viewer Introduction

To view more elements in 3D display and to check the whole layout design in 3D view, sometimes also effectively collaborating with mechanical design software, we can install the 3D Viewer plugin or EDMD Collaborator for Expedition.

Because of space limitations, we only touch on simple data interaction with mechanical design software, so only 3D Viewer is discussed here. For more about EDMD Collaborator, see Mentor's related help documents.

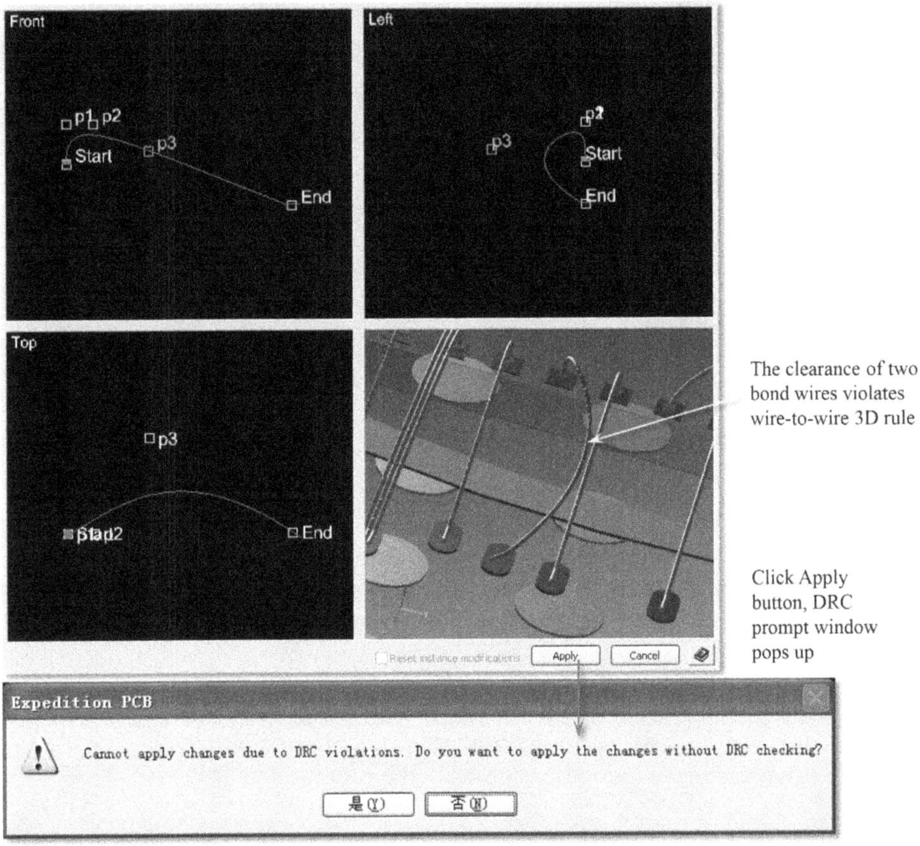

Figure 18.6 Alert window pops up automatically when 3D DRC rules are violated.

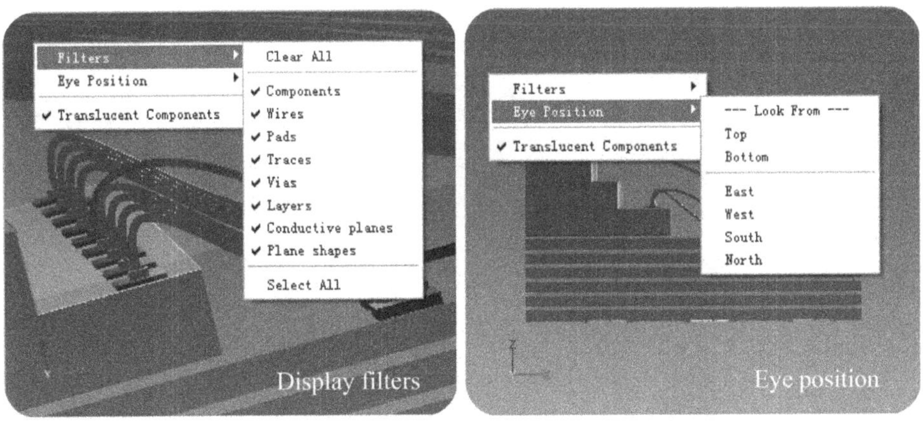

Figure 18.7 Display filters and eye position options.

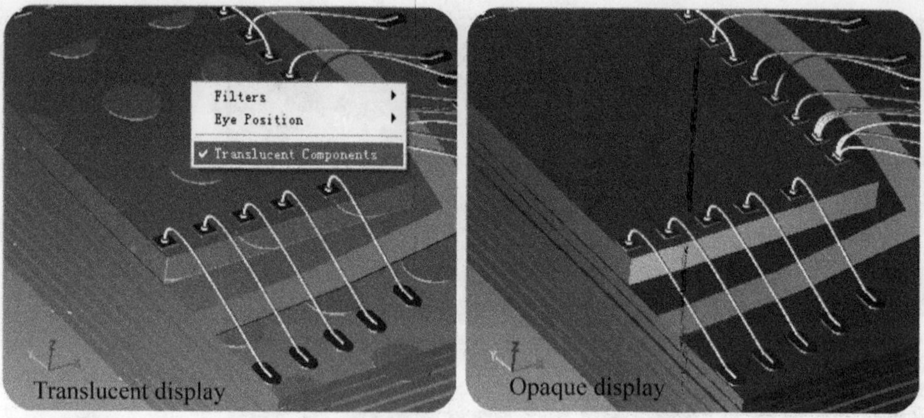

Figure 18.8 Display effects of translucent and opaque modes.

In 3D Viewer, the designer can see all elements on the substrate, and 3D Viewer can synchronously update with the 2D window in real-time. For example, if components, trace, via, etc. are moved in the 2D window, 3D Viewer will update in real-time. At the same time, 3D Viewer can also interactively check the design structure with mechanical software.

Using 3D Viewer, the designer can check all the design elements in Expedition – substrate, die, bond wire, cavity, trace, via, etc. – and can also check whether these design elements have interference with the package shell, etc. Figure 18.9 shows the display interface of 3D Viewer.

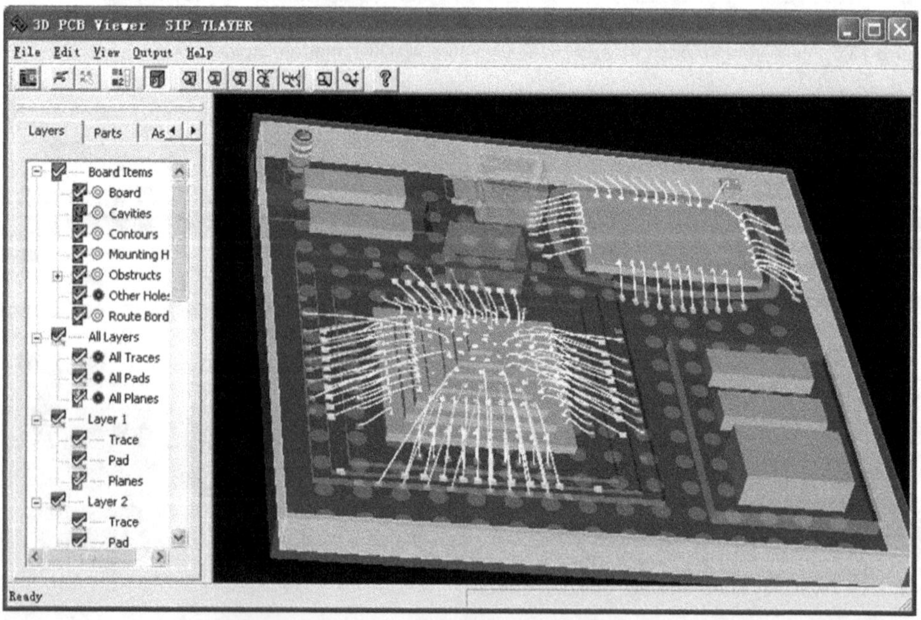

Figure 18.9 The display interface of 3D Viewer.

The display control is located on the left side of the 3D Viewer window; it is divided into three tabs, Layers, Parts and Assembly, which each control different displayed content, as shown in Figure 18.10.

In the 3D Viewer, the designer can use toolbar button to open or close the Display Control window.

1) Layers tab, mainly controls the display of board and layer elements. Board items include board, cavity, contours, counting holes, via, routing board, etc.; layer items include all layers and individual control of each layer; layer elements include trace, pads and planes.
2) Parts tab, mainly controls the display of components, die pins, bond wires, embedded resistors and capacitors, etc.
3) Assembly tab, mainly controls the display of PCB components; at the same time it can also display external elements that have been imported, such as package shell, for example, SiP_Mold1 and SiP_Ball1 as shown in Figure 18.10.

18.2.2 Real-Time Check in 3D Viewer

Using 3D Viewer, the designer can check for trace, through-hole via, blind and buried via, plane data, etc. Figure 18.11 shows a comparison of opening and closing of plane data. On the left side of the figure we can clearly see the trace, via and layer stackup; on the right we can view the plane data of different layers.

In 3D Viewer we can set the display mode; right-click on each display item to switch between two kinds of display mode, transparent and solid. From the small circle in front of the display items, we can clearly identify whether the current mode is transparent or solid; when the small circle is solid this indicates that the current display is in solid

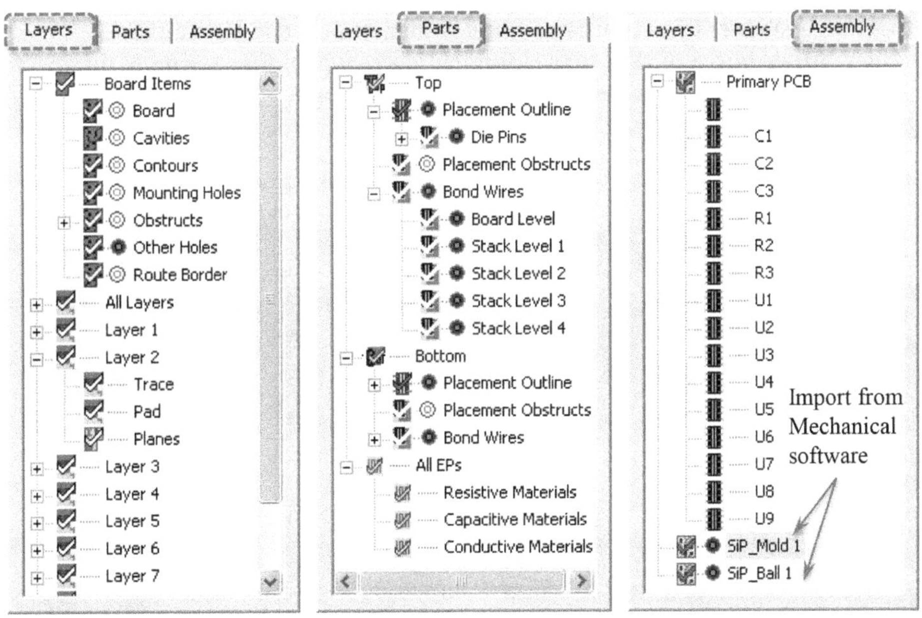

Figure 18.10 3 Display control tabs in 3D Viewer.

Figure 18.11 Substrate 3D view when plane data is opened and closed.

mode, when the small circle is hollow this shows that the current display is in transparent mode, as shown in Figure 18.12.

18.2.3 3D Simulation of SiP Production Processing

In 3D Viewer, using a set of 3D figure displays, we can simulate the whole process of SiP production.

1) First, when layout design has been completed, and the trace, plane, via and other data are generated and sent to the substrate factories, turn off all in 3D Viewer and then open pads, trace, planes and other holes, as shown in Figure 18.13.
2) After obtaining the production data, the manufacturer carries out substrate production; the 3D figure below shows the situation after completion of substrate production. In 3D Viewer open board and route border, as shown in Figure 18.14.
3) The substrate board is taken to the assembly workshop for chip assembly, die attach, etc.; continue by opening the component placement outline in 3D Viewer, as shown in Figure 18.15.

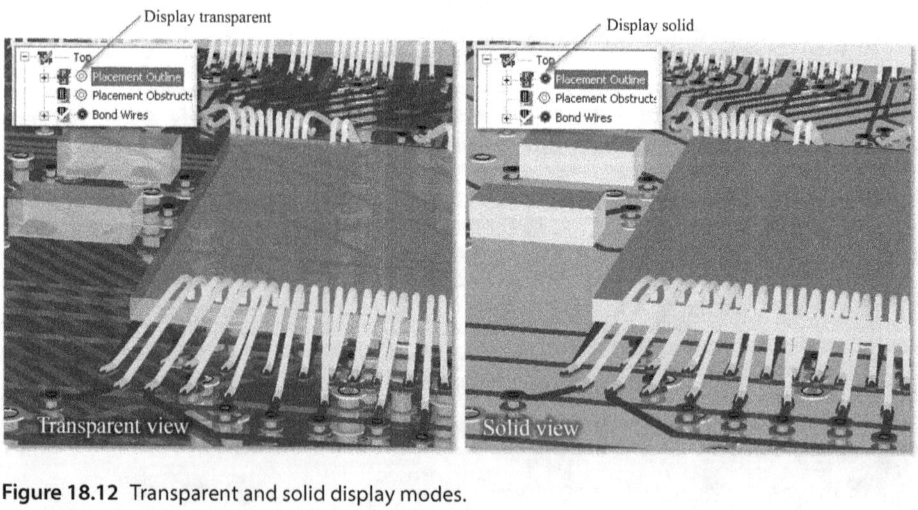

Figure 18.12 Transparent and solid display modes.

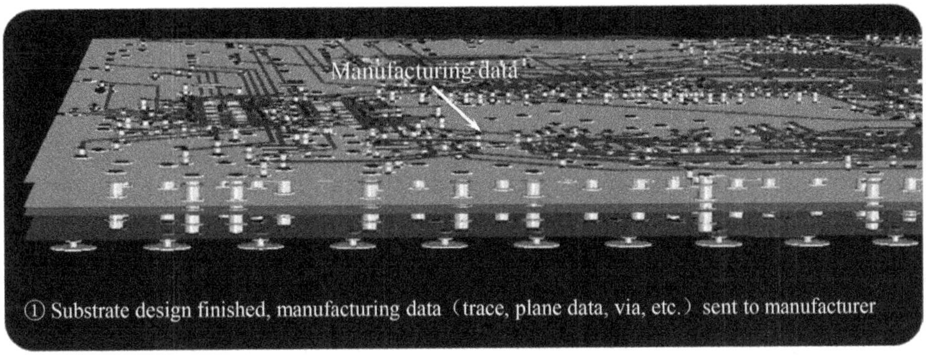

Figure 18.13 3D display of trace, plane data and via in substrate.

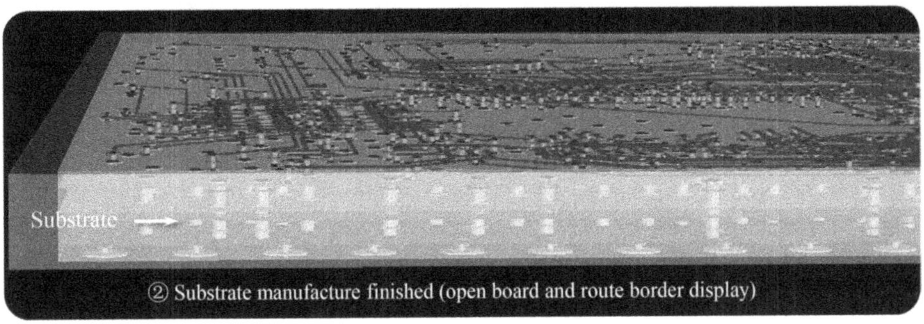

Figure 18.14 3D display after the completion of substrate production.

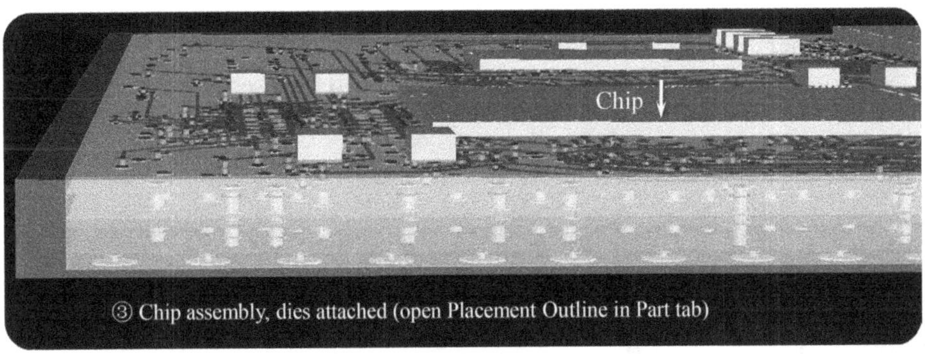

Figure 18.15 3D display after chip assembly and die attach.

4) After completion of assembly, the substrate is sent to the bonding workshop for bare die wire bonding; continue by opening the bond wire display in 3D Viewer, as shown in Figure 18.16.
5) After completion of assembly and wire bonding, usually the die molding is sealed. In the plastic package process, a special mold is used for sealing plastic glue to fix and protect the components on the substrate. For ceramic or metal package, the corresponding processes are vacuum pumping or nitrogen inflation, following which the package is covered

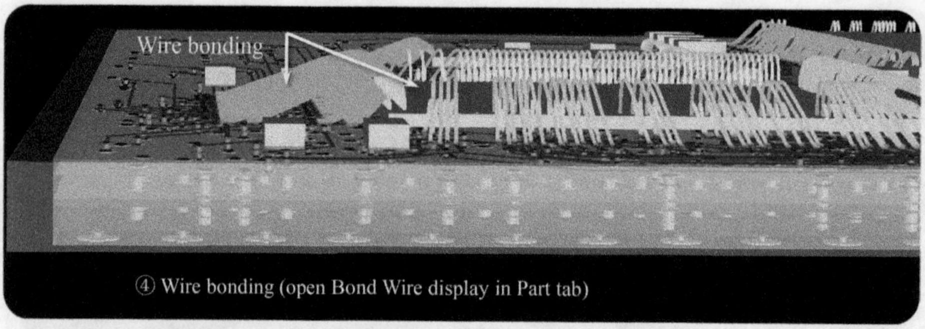

Figure 18.16 3D display of finished wire bonding.

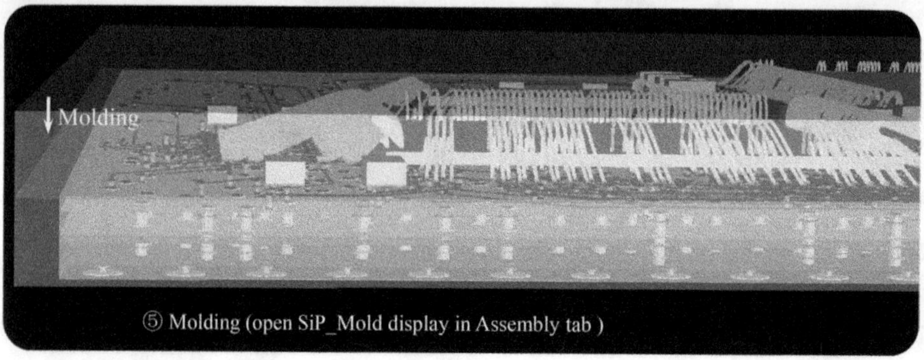

Figure 18.17 3D display after molding.

and sealed. To continue, open the SiP_mold imported in advance in 3D Viewer, as shown in Figure 18.17. For more on mechanical data import, please see section 18.2.4.

6) After the completion of molding, the next operation is usually ball bumping. Choose appropriate tin ball, and by welding or other processes, fix on the BGA substrate pad; the balls are the signal pathway for SiP package and PCB. To continue, open the SiP_ball imported beforehand in 3D Viewer, as shown in Figure 18.18.

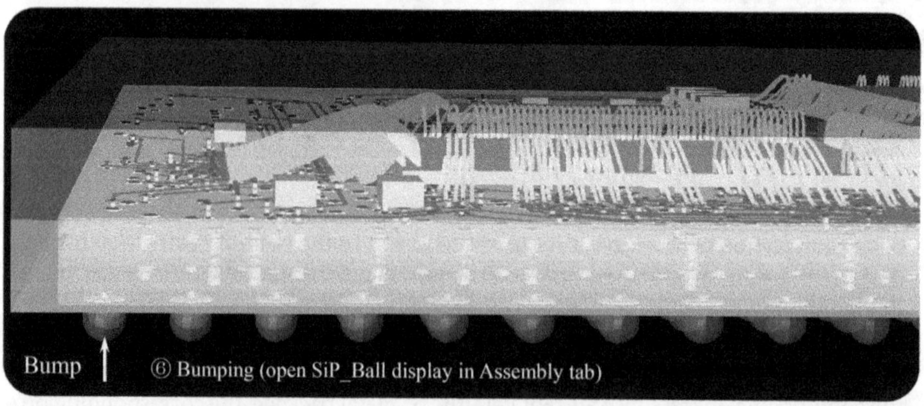

Figure 18.18 3D display after bumping.

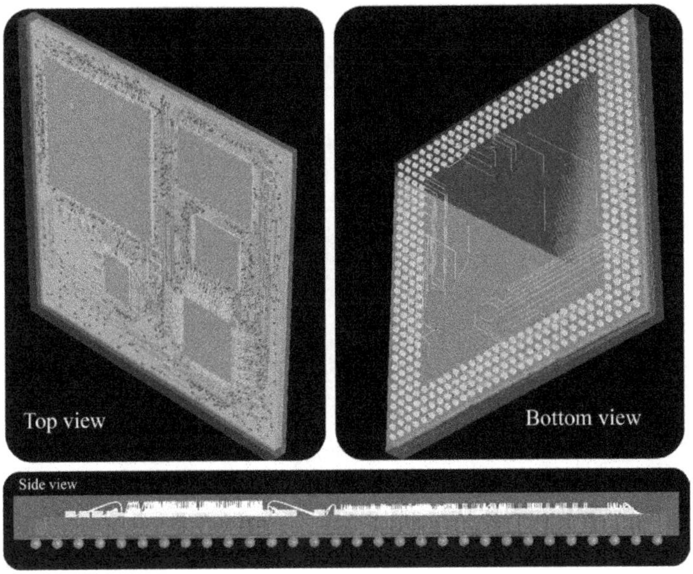

Figure 18.19 View SiP product from three directions in 3D Viewer. *(See color plate section for the color representation of this figure.)*

7) After the completion of the ball bumping process, SiP package production is almost finished; all that remains is to check and test. Figure 18.19 shows the SiP package product from three different directions.
   From the above SiP production process simulation, we can see that in Expedition's 3D Viewer, we can simulate the full process of production and processing by time sequence; it can also be seen as a kind of virtual production process, because this process is driven by real data, and thus has important reference meaning.
   It is intended to generate awareness of possible production problems in the design phase, and can also effectively solve SiP manufacturability problems in the design process, thus realizing DFM (Design for Manufacture).

18.2.4 Import 3D Mechanical Data

1) Import mold and bump data
   By importing the mechanical design data, such as package shell and bumps, we can implement effective data interaction and checks. In 3D Viewer, the designer can easily import mechanical design data, including 3D component models and mechanical shell models.
   Although it depends on the actual project design demands in relation to the creation of mechanical data, usually the data is generated by the mechanical engineer. Of course the SiP designer can also create mechanical data using ProE, SolidWorks or other mechanical tools.
   The author adopted the open-source lightweight 3D Blender software; the specific method is not dealt with in this book.
   The mold and bumps created in Blender are shown in Figure 18.20 and Figure 18.21.

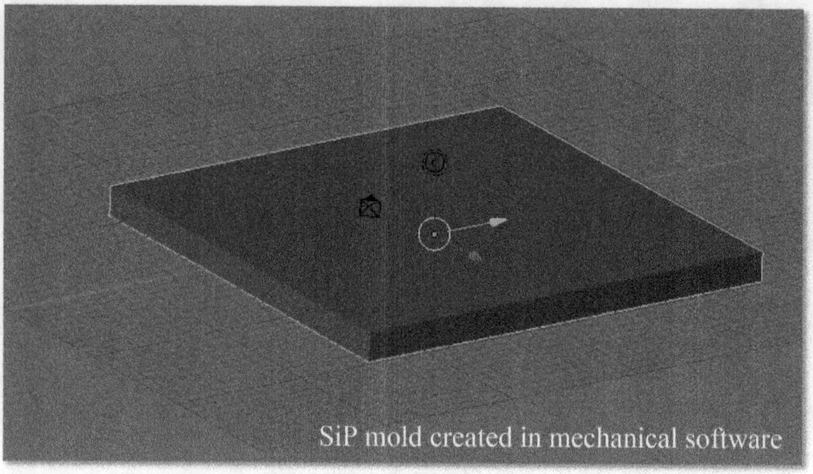

Figure 18.20 SiP mold designed in Blender software.

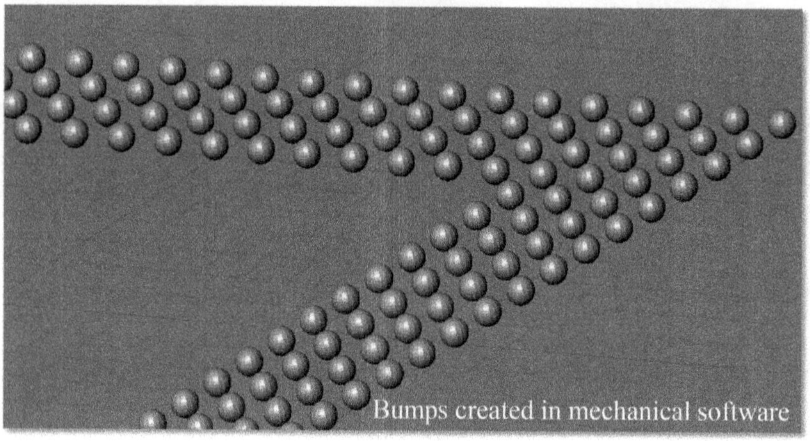

Figure 18.21 SiP bumps designed in Blender software.

After designing SiP mold and bumps in mechanical software, we need to export the data in a format identifiable by 3D Viewer. In EE7.9.x, these formats include *.e3d, *.pcb, *.wrl, *.obj and *.stl. The author adopted the *.wrl format, and output the mechanical design data as SiP_Mold.wrl and SiP_Bump.wrl respectively.

Then, in the 3D Viewer menu, select File→Auxiliary File Import, import the mold and bump data; in the pop-up window select SiP_Mold.wrl and SiP_Bump.wrl, and select units, scale factor, rotate and offset parameters, so as to ensure that the imported data accurately coordinates with the design data, as shown in Figure 18.22 and Figure 18.23.

The display switch for imported mold and bump is located in the Assembly tab; the designer can select display on or off, and at the same time can switch between transparent and solid mode.

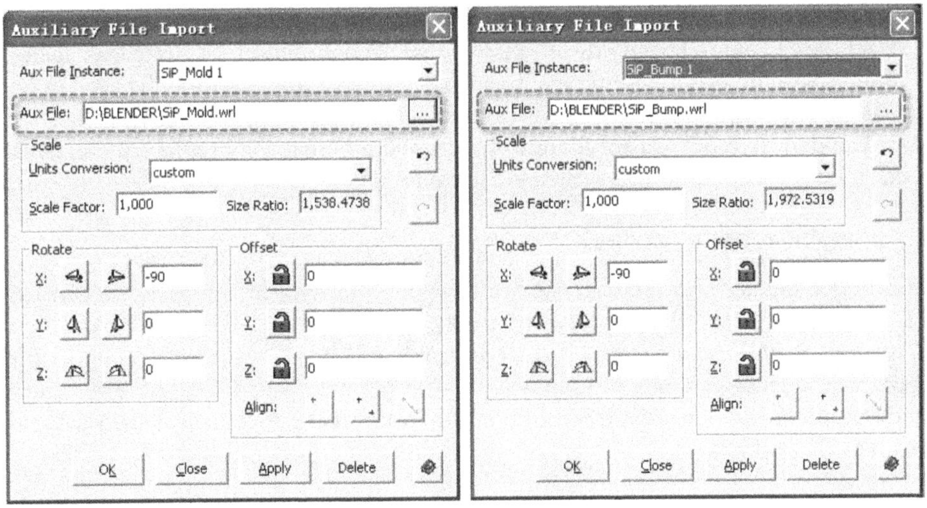

Figure 18.22 Import mold and bumps in 3D Viewer.

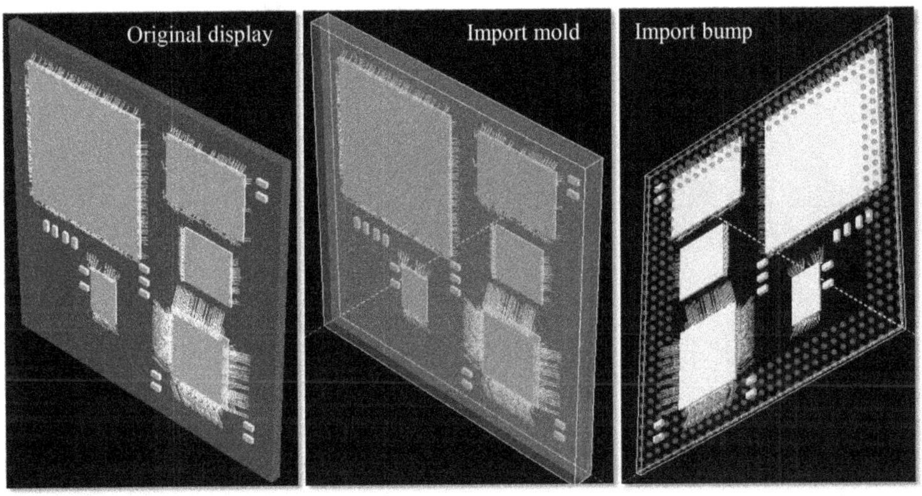

Figure 18.23 Imported mold and bump display in 3D Viewer.

2) Import 3D cell data

Cells designed in Expedition usually have only length, width and height information, and the 3D cell model is usually characterized by a combination of cuboid, cylinder and multilateral structures, which cannot be accurately reflected in a 3D model. If the designer wants to accurately reflect a component in a 3D model, he or she needs to precisely design the 3D model in mechanical software, and then import into 3D Viewer using the 3D cell import function for assembly check and simulation.

The 3D cell-data formats supported by 3D Viewer include *.e3d, *.wrl, *.obj, *.stl, etc.; we simply need to output a corresponding file from the mechanical software.

In 3D Viewer, select from the menu File→3D Cell Import, in the pop-up window, select the cell name, specify the corresponding 3D file, and set the unit, scale factor, rotate angle and offset parameters, so as to ensure that the imported data can accurately locate and cooperate with the original design, as shown in Figure 18.24.

The imported 3D cell mechanical model will automatically replace the original cell model in Expedition, as shown in Figure 18.25.

18.2.5 Real-time DRC in 3D Viewer

After importing accurate 3D cell models from the mechanical software, we can use real-time 3D DRC, and realize virtual assembly.

To run real-time 3D DRC and virtual assembly effectively, we need to import the mechanical shell data. Using the same method as outlined above, in 3D Viewer select from the menu File→Auxiliary File Import, and import the mechanical shell data, as shown in Figure 18.26.

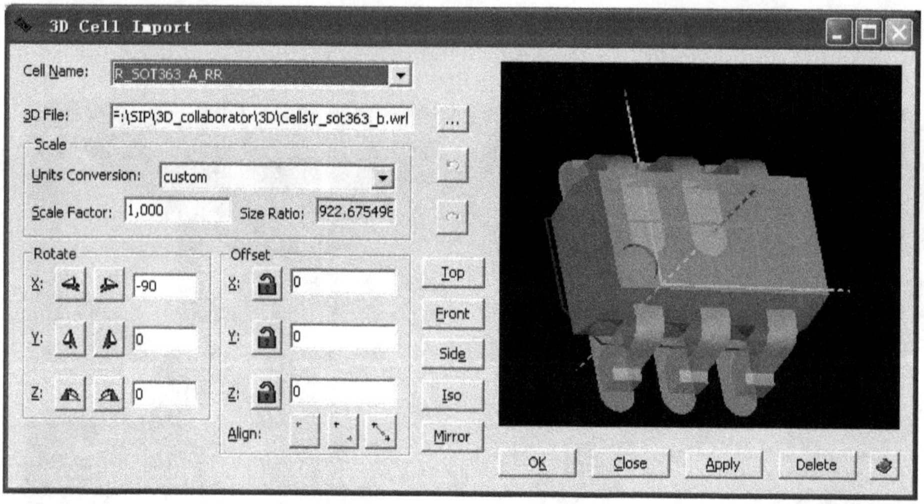

Figure 18.24 3D Cell Import window.

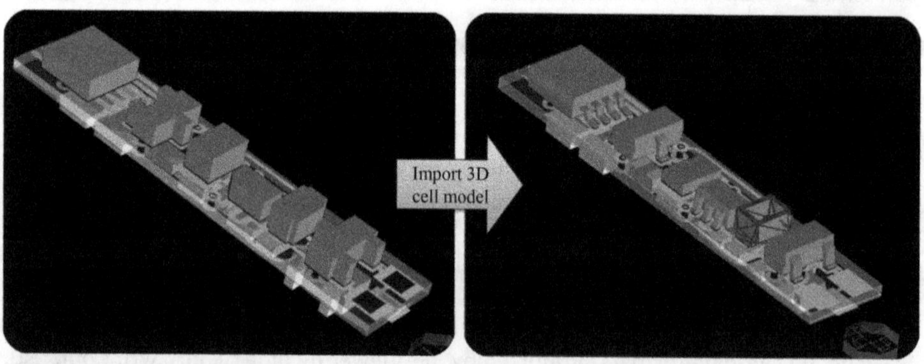

Figure 18.25 Comparison of before and after import of mechanical 3D cell.

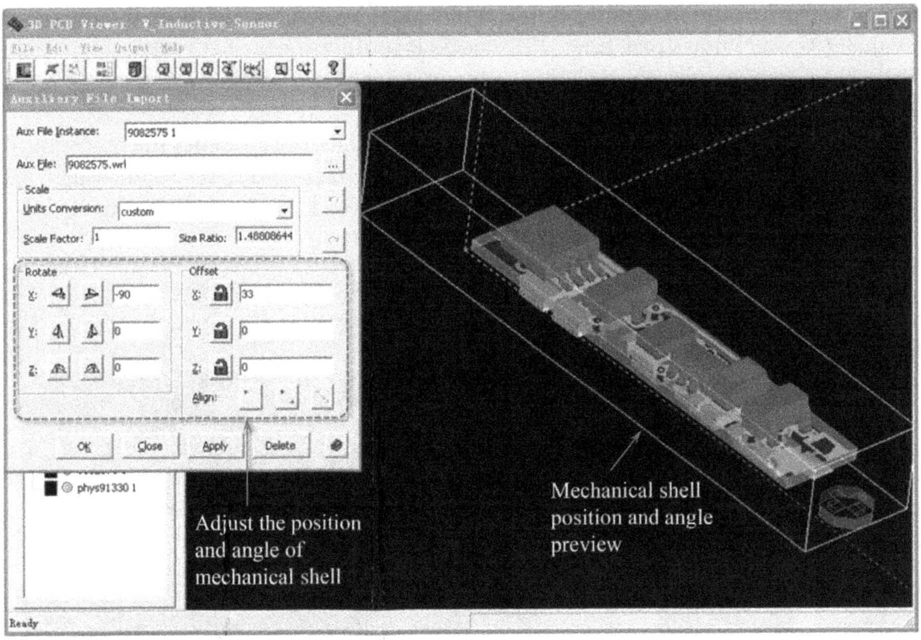

Figure 18.26 Import mechanical shell data.

Set unit, scale factor, rotation angle and offset parameters, then click Apply or OK to complete mechanical shell import, as shown in Figure 18.27.

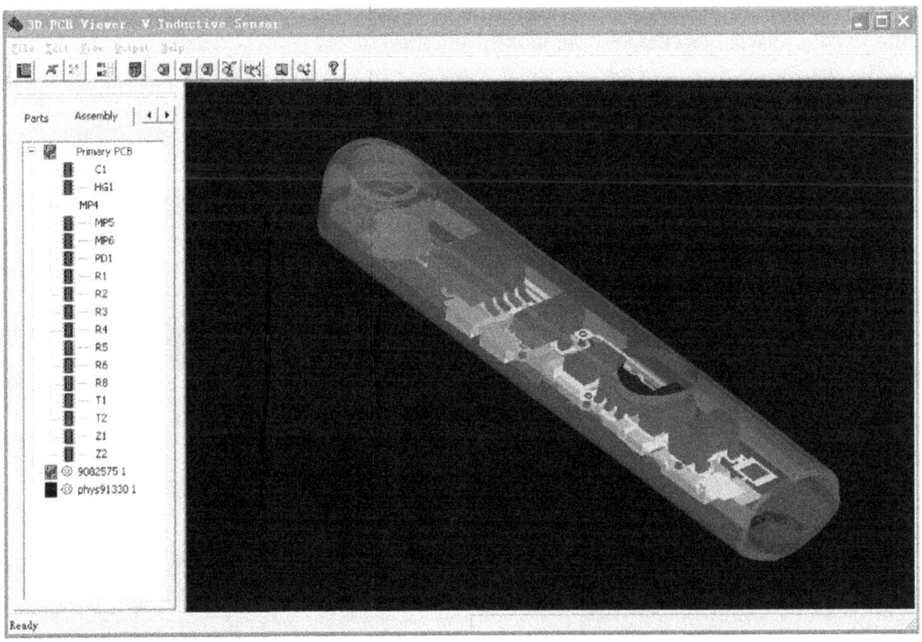

Figure 18.27 After mechanical shell import.

When import of the accurate 3D cell model and the mechanical shell data is completed, 3D Viewer allows real-time DRC check; if conflicts occur between components or between components and shell, this is shown in real-time, with warning information displayed in the status bar.

If a component is moved, for example, if the component body touches the mechanical shell, a conflict occurs; the status bar information will be shown in red, indicating a 3D DRC conflict, and such component placing will be banned, as shown in Figure 18.28.

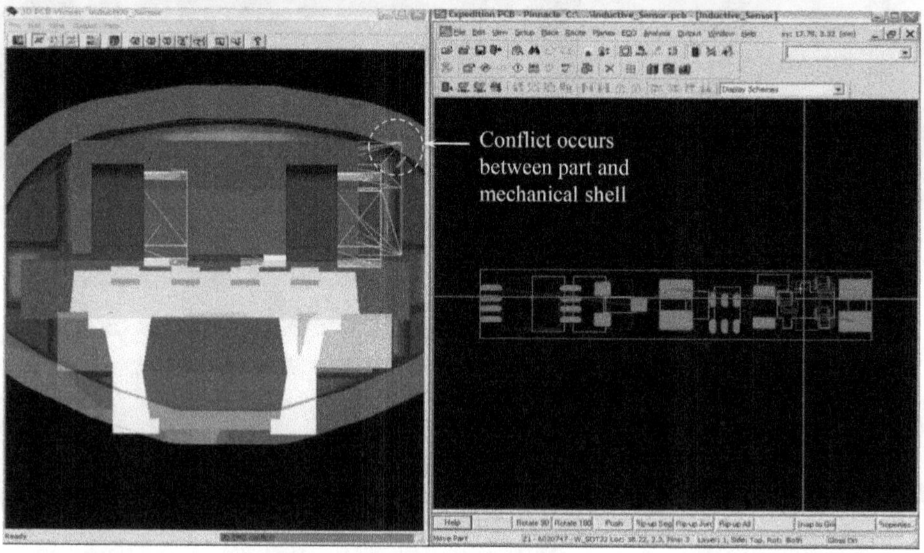

Figure 18.28 Conflict occurs between component and shell.

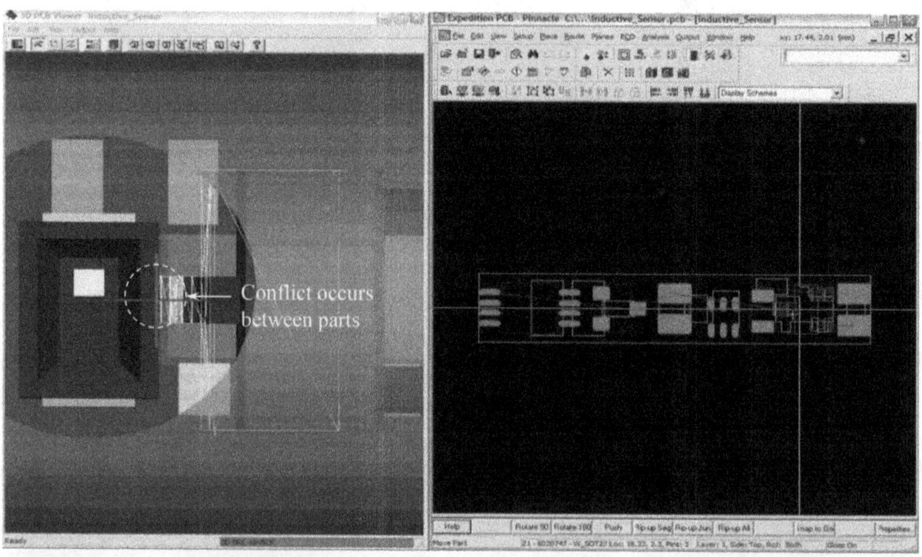

Figure 18.29 Conflict occurs between components.

Similarly, if a component is moved and touches any other component, a conflict occurs and the status bar information will show in red, indicating a 3D DRC conflict, and such component placing will be banned, as shown in Figure 18.29.

Using 3D real-time DRC, we can effectively avoid possible conflicts between components and between components and mechanical shell in the design process; this ensures that design is correct, and carrying out virtual assembly can solve possible DFM problems in SiP design.

19

Design Review

19.1 Online DRC

In Expedition layout design, real-time interactive DRC is turned on by default.

Generally, with DRC turned on, the designer can be sure that design is correct and that there is no DRC error. Sometimes, however, due to special design needs, DRC is intentionally turned off by the designer.

This kind of real-time interactive DRC is often called online DRC.

The online DRC option is located in the Common Settings area of Editor Control; by unchecking the Interactive Place/Route DRC item, we can turn off online DRC. A prompt box will notify the designer that DRC is turned off. Some features, such as auto route, gloss, tuning, etc. will be disabled, and the designer will be asked to confirm.

When "Yes" is chosen, online DRC is turned off, and the system automatically pops up a box notification that online DRC has been disabled; this box is always displayed in the upper-left corner of the workspace until online DRC has been re-enabled.

Figure 19.1 shows the notification when online DRC is turned off, and the box prompt.

When online DRC is turned off, designers can execute any operation; Expedition does not check them. Such operations are sometimes necessary in special design cases. For example, the designer may need to place one or two resistor pads overlapping, or at the time of soldering choose a certain resistor for assembly. This method is often used in RF circuit design for mobile phones and other wireless devices.

Whether online DRC is turned on or off, when a design is completed the designer usually needs to run a comprehensive DRC for the whole design; this is called batch DRC.

19.2 Batch DRC

Batch DRC plays the role of a detailed project review, to ensure that all elements of the design are within the scope of design rule definitions. Batch DRC can be carried out at any stage of the design process, and is often run after the design is finished.

Batch DRC has similar review items to online DRC, but batch DRC is more comprehensive, and can be configured by the designer.

Select from the menu Analysis→Batch DRC to start batch DRC, or click the tool icon  in the toolbar; after start-up, the Batch DRC interface appears; it is divided into two tabs, as shown in Figure 19.2.

SiP System-in-Package Design and Simulation: Mentor EE Flow Advanced Design Guide,
First Edition. Suny Li (Li Yang).

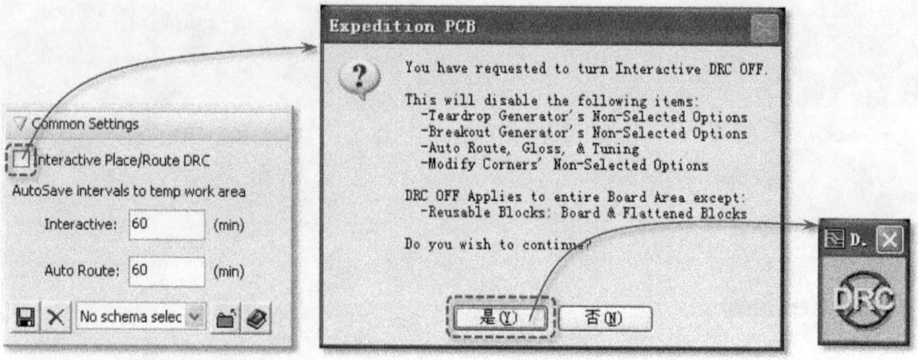

Figure 19.1 Online DRC turned off and the notification window and box.

19.2.1 DRC Settings

1) Proximity area
In the DRC Settings tab, Proximity area is the first item; here we select the entire design, as shown in Figure 19.3. If you want to check part of the design, you need to draw a DRC window in design. A DRC window can be drawn in draw mode, it must be a polygon, and there can only be one DRC window in a design; if one already exists, a new DRC window will automatically replace the existing DRC window.
If the DRC window is selected, batch DRC only runs design-rule checking in the scope of the DRC window, which can save a lot of time in complex designs. It is usually used to check different regions in the design process. If there is no DRC window in the design, this option is grayed out and cannot be selected.
Due to space limitations, DRC windows are not discussed in detail here.
2) Rules to check
If some nets are selected in the design, the Check selected nets only option will be selected automatically, and the number of selected nets displayed; if no nets are selected, this option is grayed out and can't be selected, as shown in Figure 19.4.
Please note, if there are no design changes in design review, but a second check gives different results, establish whether the Check selected nets only option was checked both times (so that only the selected nets were checked) or the other time this option is accidentally not checked (so that the entire design was checked).
Connectivity and special rules are usually selected by default; if not selected, the Connectivity and special rules tab (introduced in the next section) shows grayed out and not enabled.
In the Proximity options, Net class clearances and rules, Plane clearances and rules and Bond Wire clearances and rules are selected by default. If no test point is present in the design, the Test Probe clearances and rules option shows gray and cannot be selected.
If the General and element to element rules option is not selected, the Advanced Element to Element Rules button is gray and cannot be selected, as shown on the right side of Figure 19.5.

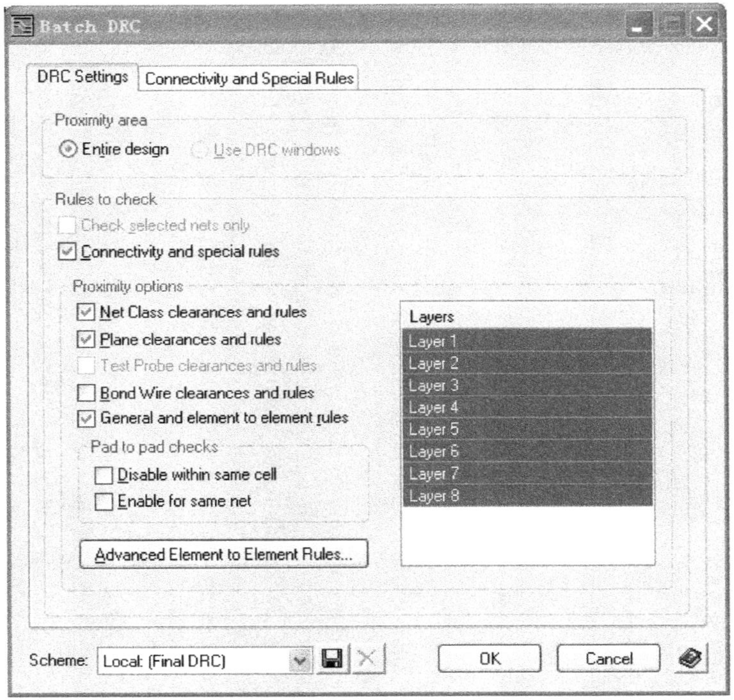

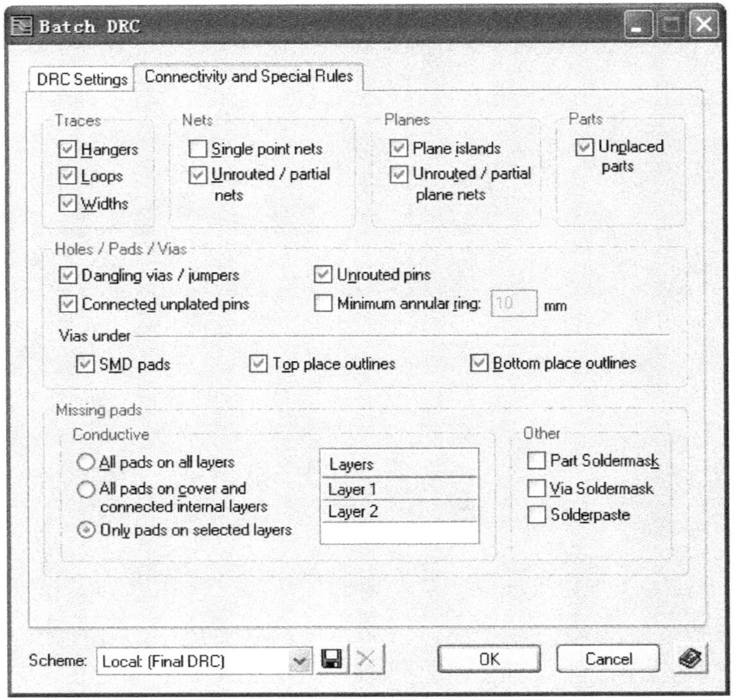

Figure 19.2 The two tabs of the Batch DRC interface.

Figure 19.3 Check entire design.

(a) Check selected nets only (b) Check all nets in design

Figure 19.4 Check selected nets only and check all nets of the design.

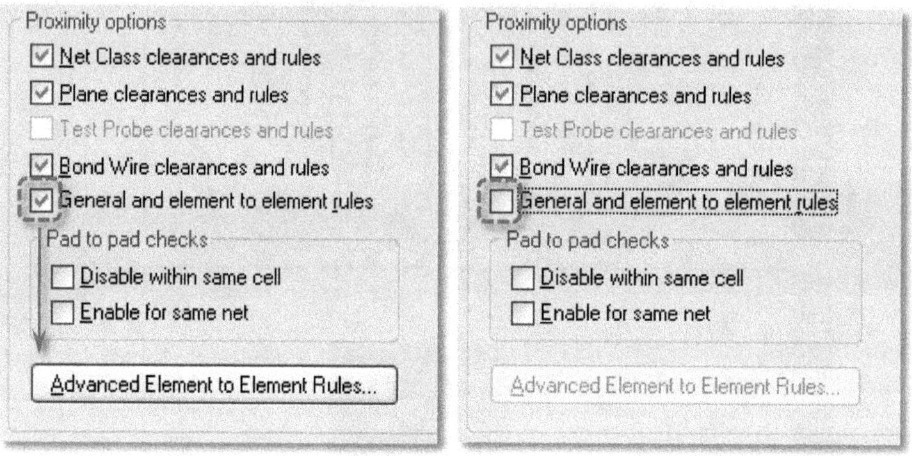

Figure 19.5 General and element to element rules.

After selecting General and element to element rules, click the Advanced Element to Element Rules button, and the Element to Element rules window pops up, as shown in Figure 19.6.

By default, the contents of this window are inherited from CES clearance rules. Here, the user can change the rules or add new rules. The font color of the default rules is black; when the default rules are changed, the font color will become red; when new rules are added, the font color is blue. Normally, if there are no special design requirements, keep the default rules, and the same applies to CES definitions.

If the designer has changed the default rules and then wants to restore the default rules from CES settings, this can be done by clicking the Load Default Rules button at the lower-left corner of the window; a warning message will pop up as shown in Figure 19.7, notifying the designer that the customization of the rules will be discarded. If the designer clicks "Yes", the system will automatically load the default CES rules, and replace the customization changes.

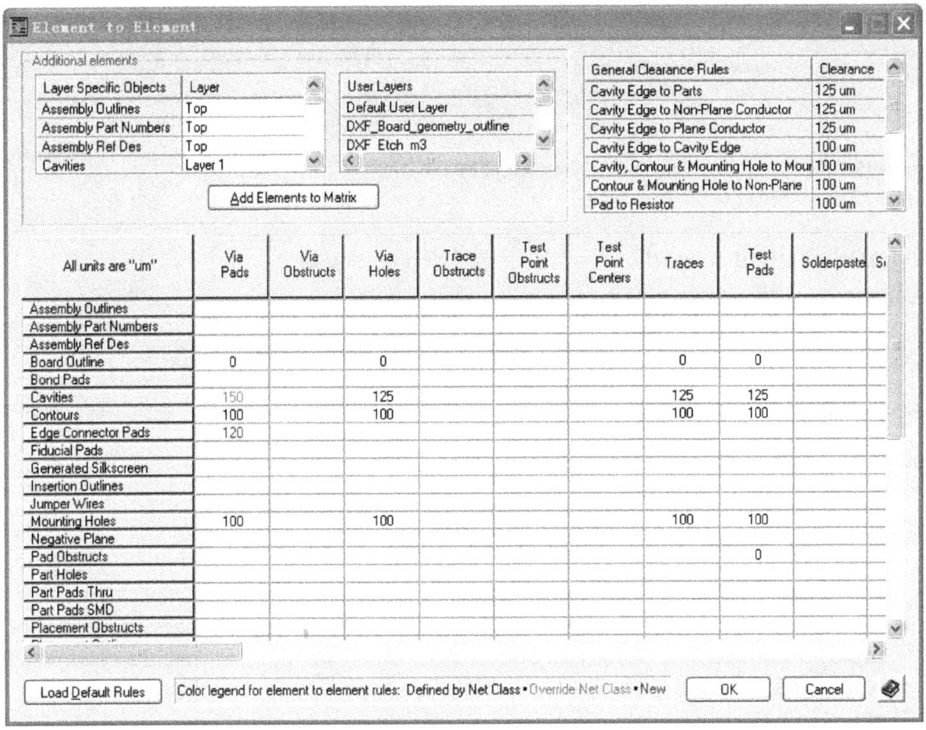

Figure 19.6 Element to Element rules window.

The Pad to pad checks option is mainly used to ignore or check pad-to-pad spacing within the same component cell or between the same nets, as shown in Figure 19.8.

If Disable within same cell is checked, batch DRC will not check the pad-to-pad spacing inside components; if Enable for same net is checked, batch DRC will check pad-to-pad, pad-to-via and via-to-via spacing within the same net.

19.2.2 Connectivity and Special Rules

Switch to the Connectivity and Special Rules tab; this deals with setting checks for traces, nets, plane islands, unplaced parts, holes/pads/vias, etc. These are presented one by one in the following:

1) Trace
   - Hangers: dangling traces that didn't finish, one end of the trace is not connected to any pad or via.
   - Loops: ring route, that is, the trace itself forms a loop, the loop may be on one layer or across multiple layers.
   - Widths: trace width, checks the width of trace, so as to ensure it is the same as that defined in CES.
2) Net
   - Single point net: checks and reports whether there are single point nets in the design.
   - Un-routed or partial net: checks connectivity of all non-plane-layer nets.

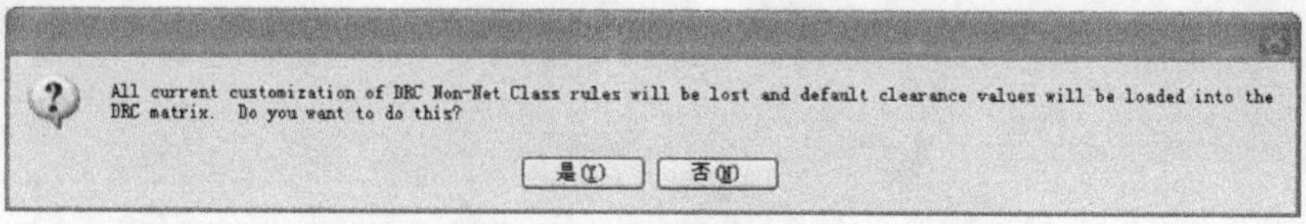

Figure 19.7 Restore default clearance rules from CES.

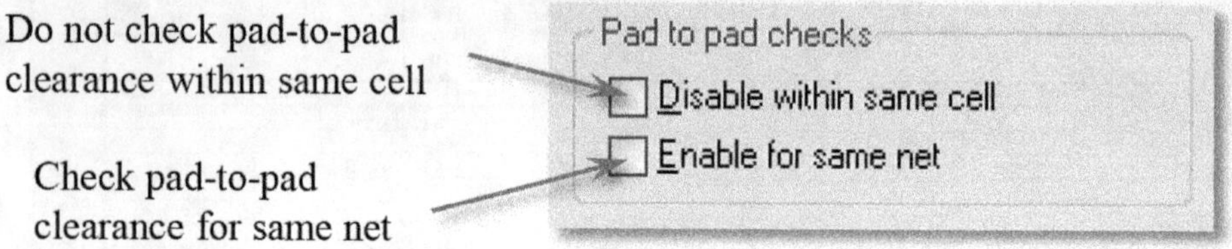

Figure 19.8 Pad to pad checks option.

3) Plane
   - Plane islands: checks if there are any islands in plane data. Note that shielding net and negative plane data are not checked.
   - Un-routed or partial plane nets: checks connectivity of all plane nets, and whether there are plane nets that have not been connected by copper or that are only partly connected.
4) Parts
   Unplaced parts: checks if there are components in the layout database that have not been placed in layout.
5) Holes/pads/vias
   - Dangling vias/jumpers: checks vias and jumpers that are not connected by trace or plane.
   - Un-routed pins: checks whether the component pins are connected by trace or plane.
   - Connected un-plated pins: checks whether the nets are connected to un-plated pins.
   - Minimum annular ring: checks the smallest annular ring width of vias or through-hole pins; if minimum annular ring width is less than the value set, this will be reported in DRC. Minimum annular ring width is usually related to process technology, since it is easy to break small annular rings in the drilling process, resulting in disconnection of nets.
6) Vias under
   - SMD pads: checks vias on SMD pads; usually not allowed in through-hole design, and often used in blind or laser-hole design.
   - Top place outline: checks if there are vias inside the top components' outline; this is usually allowed and the designer can ignore this inspection.
   - Bottom place outline: checks if there are vias inside the bottom components' outline; as above, it is usually allowed and the designer can ignore this inspection.
7) Missing Pads
   - All pads on all layers: checks all the through-hole pins to ensure there are pads in each layer.

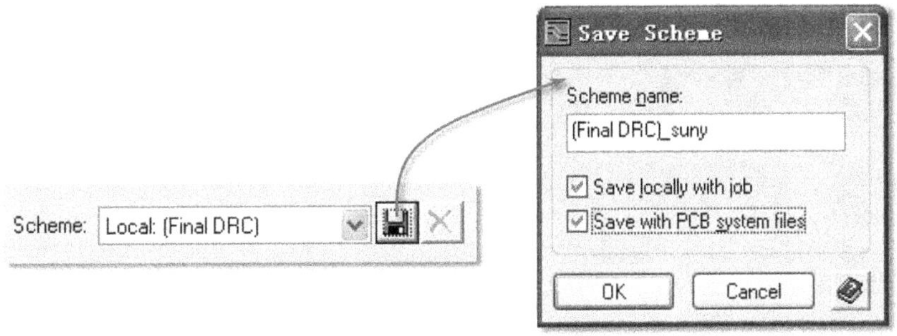

Figure 19.9 Save DRC scheme.

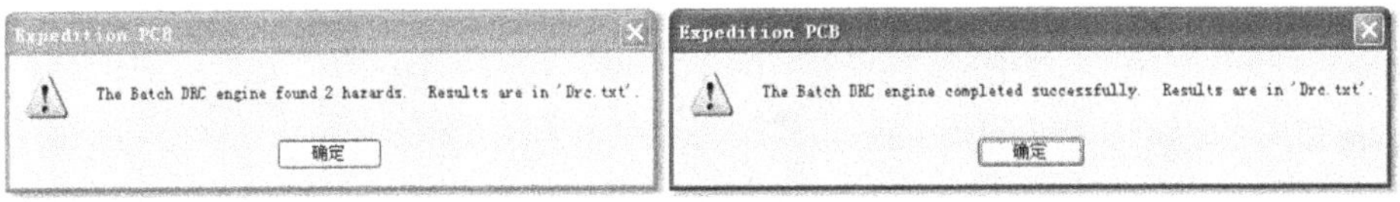

(a) Found 2 DRC errors (b) DRC check passed

Figure 19.10 Batch DRC prompt interface (the left showing hazards found, the right no hazards).

- All pads on cover and connected internal layers: checks all the top and bottom pins, as well as the inner pins connected to trace to ensure there are pads for connection.
- Only pads on selected layers: checks the pads on user-selected layers; the user can select the layers in the list boxes on the right.

8) Others
   - Part Soldermask: checks whether all components have soldermask or not.
   - Via Soldermask: checks whether all vias have soldermask or not.
   - Solderpaste: checks whether all SMD pins have solderpaste or not.

19.2.3 Batch DRC Scheme

After completion of the batch DRC settings, they can be saved into a scheme and reused or edited. The scheme can be saved locally with the job or saved with the PCB system files, as shown in Figure 19.9.

After the completion of all settings, the designer can run the batch DRC check; when the check is completed, a prompt window will pop up automatically showing how many hazards were found in the design. If there are no hazards, the prompt window shows that DRC completed successfully, as shown in Figure 19.10. The check results are saved in the Drc.txt file.

Designers can use the Expedition File View tool to review the Drc.txt report; the file gives a detailed report for each DRC item. Figure 19.11 shows partial contents of the file.

19.3 Review Hazards

In addition to the Drc.txt file report, Expedition provides a more convenient and intelligent way to review DRC checks – Review Hazard.

```
Non-Net Class Element To Element Clearances And Rules
-----------------------------------------------------

    Route Border TO Via Pads Layer 1     Clearance: 0um
        No Hazards.

    Route Border TO Traces Layer 1     Clearance: 0um
        No Hazards.

    Part Pads SMD Layer 1 TO Route Border     Clearance: 0um
        No Hazards.

    Bond Pads Layer 1 TO Route Border     Clearance: 0um
        No Hazards.

    Board Outline TO Via Holes Layer 1     Clearance: 0um
        No Hazards.

    Board Outline TO Via Pads Layer 1     Clearance: 0um
        No Hazards.

    Board Outline TO Placement Outlines Layer 1     Clearance: 254um
        No Hazards.
```

Figure 19.11 Partial content of Drc.txt file.

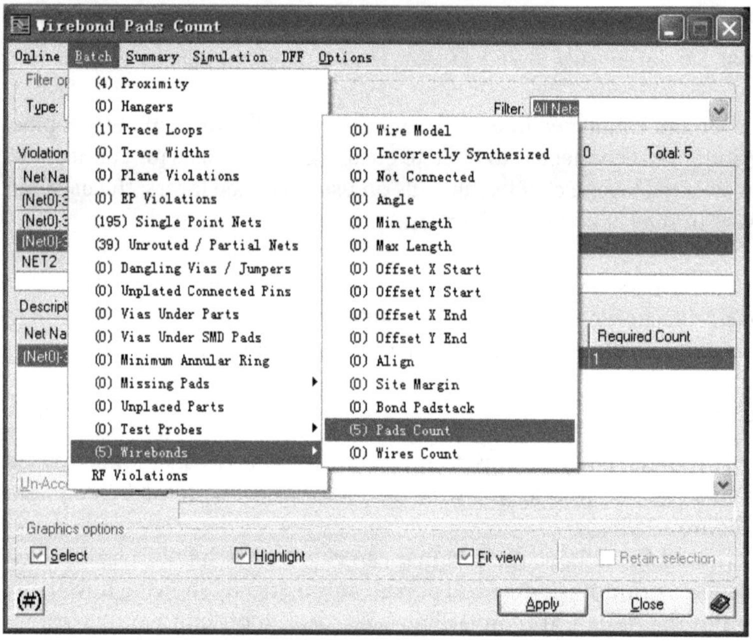

Figure 19.12 Review Hazard window.

Select from the menu Analysis→Review Hazard, or choose toolbar icon to open the DRC review window. First, click on the update button (#) in the lower-left corner of the window, and review the batch DRC results, as shown in Figure 19.12.

The word "hazards" is a proper noun expressing DRC errors or problems in Expedition. There are two kinds of hazard list: batch hazards and online hazards, corresponding to batch DRC errors and online DRC errors.

Batch hazards are irregularity results from batch DRC checking, and comprise 18 results, including proximity, hanger, trace loop, EP violations, wire bonds, RF, etc.

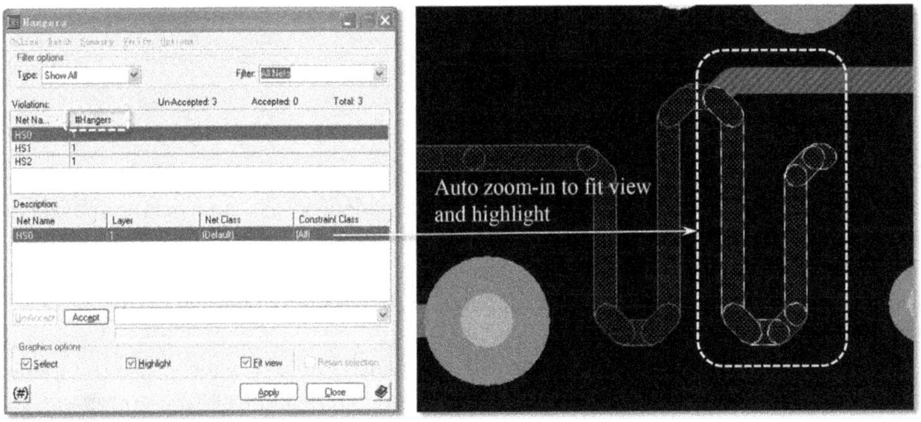

Figure 19.13 Fit view for the hanger hazard.

Each DRC result can be located to a specific place in layout design, as shown in Figure 19.13, the hanger hazards. Choose the first hazard, then check the Graphics options Select, Highlight and Fit view, and click Apply; the software will automatically jump to the hanger-hazard location; the hazard is selected and highlighted.

On finding the specific hazard location, the designer can correct the hazard, and re-run the batch DRC to confirm that the hazard has been successfully corrected. Now, the designer can use the Check selected nets only option to save time.

Designers can find and correct all hazards according to the method described above, and then re-run batch DRC to check the entire design to confirm whether all the hazards have been successfully corrected.

Some hazards might be deliberately made for special design needs. Some connectors, for example, will be placed outside or across the board outline; this kind of hazard just needs to be confirmed and accepted by the designer.

When all hazards are corrected or accepted, re-run batch DRC and check the entire design until DRC passes or all hazards are accepted by the designer.

Online hazards are online DRC check results; they will be updated automatically in real-time as the design changes, and there is no need to run special DRC check commands. Online hazards report violations of physical and electrical design rules in real-time. Online hazards are divided into eight categories comprising 24 items; the violation number is shown in front of each item.

Online hazards are updated in real-time; when the designer corrects a certain hazard, the hazard will disappear immediately from the online hazard list. As with batch hazards, all online hazards should be checked and corrected until DRC passes or all hazards are accepted by the designer.

19.4 Verify Design Library

Select from the menu Analysis→Verify the local to central library command; this compares the parts, cells, padstacks and vias in the current local project library with those in the central library, finding any differences.

Generally, the results are as follows:

1) Older than central library.
2) Newer than central library.
3) Only in local library.

The inspection report file VerifyLocal2CentralLibrary.txt is provided after verification, and will open automatically, reporting pad, padstack, cell and part comparison results, as shown in Figure 19.14, which shows part of the contents of the file.

The reasons why the local library might not completely match the central library include: (1) the central library data was updated in the process of design; (2) the designer edited the local library data in design.

To synchronize the local library data with the central library, we can use the menu command ECO →Update Cells & Padstacks or the ECO→Replace Cell command, as shown in Figure 19.15.

Newer than central lib — Older than central lib — Only in local lib

Summary:

Library Object	# Newer	# Older	# Only in Local Library
Pad Names	0	0	4
Hole Names	0	0	0
Padstack Names	1	0	6
Cell Names	1	0	16
Part Numbers	0	1	0
Reusable Blocks	0	0	0

Figure 19.14 Partial contents of VerifyLocal2CentralLibrary.txt file.

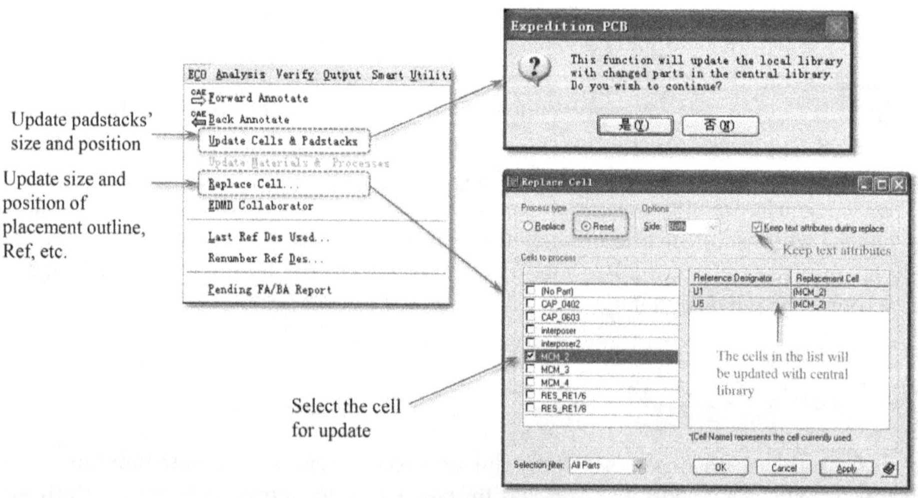

Figure 19.15 Local library and central library data synchronization.

In general, the update cells & padstacks command is mainly used for position and shape updates for cells and padstacks. The replace cell command is mainly used to update the text and outline of the cell, such as reference designator position and size updates. If the designer needs to keep the change of text attributes in layout, the "Keep text" attribute must be checked during the replace; the text attributes will then be preserved in the update process.

After the whole design has been reviewed, it will be ready for SiP design data output for manufacture.

20
Manufacturing Data Output

SiP manufacturing data is mainly divided into three parts:

1) Used for SiP substrate manufacturing – Gerber and drill data.
2) Used for SiP component assembly (including adhesive, bonding, welding, etc.) – the drawing file and component coordinate data.
3) Used for SiP package manufacturing – mold, ceramic and metal encapsulation shell structure design data.

Because mold tooling and encapsulation shell structure are not dealt with in this book, this chapter mainly focuses on how to output the first two kinds of manufacturing data in Expedition.

20.1 Drill and Gerber Data Output

20.1.1 Drill Data Output

In Expedition, select from the menu Output→ NC Drill, open the NC Drill Generation window.

1) Drill Options tab
   The NC Drill Generation tab is shown in Figure 20.1.
   a) NC drill machine format file: select the drill file format; there are two kinds of file format to choose from: DrillEnglish.dff for the English format, DrillMetric.dff for the metric format.
   b) NC drill output directory: the output path of drill files, the default output directory is the PCB\Output\NC Drill subdirectory.
   c) Drill generation: sweep axis option, scanning direction, includes horizontal and vertical; bandwidth, scanning bandwidth, the default unit is 100th or 2500 μm. The designer can make reasonable adjustments according to the density of the substrate board; if the substrate density is high, choose a smaller bandwidth to avoid back-and-forth drilling movement that reduces drilling efficiency; if the substrate density is loose, choose a larger bandwidth in order to improve the scanning speed.

SiP System-in-Package Design and Simulation: Mentor EE Flow Advanced Design Guide,
First Edition. Suny Li (Li Yang).

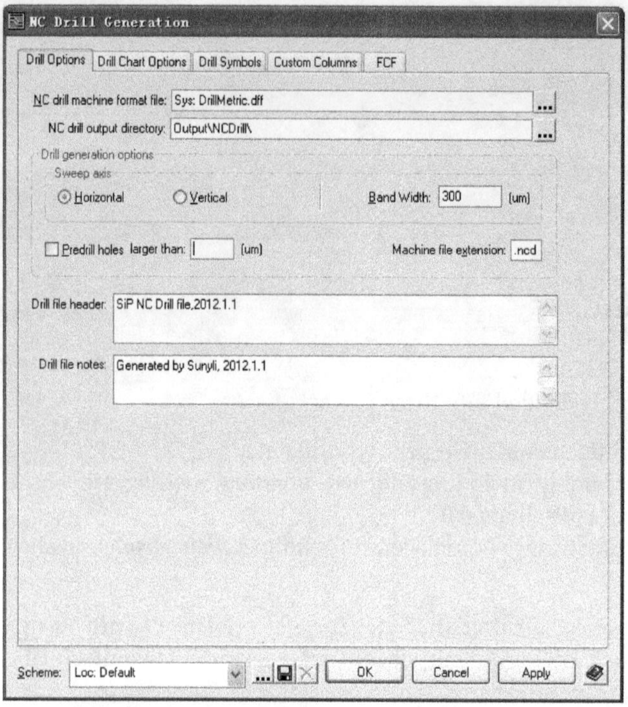

Figure 20.1 Drill Options tab.

d) Predrill holes larger than: if a hole diameter is large, we can choose to drill a small location hole in advance, and then drill accurately; this is advantageous for accurate positioning of the large hole.
e) Drill file header and drill file notes: for adding comment information by the designer, the words will be added at the beginning and end of the drilling file.

2) Drill Chart Options tab

The Drill Chart Options tab is shown in Figure 20.2.

a) Columns area: mainly used to set the drill chart items and text arrangement, the designer can create new items or delete existing items.
b) Text settings: mainly used to set the font name, font size, line spacing, pen width, display precision, units and default tolerance.
c) Auto assign drill symbols: the system automatically selects drill symbols (i.e., the user does not specify). "Character" indicates automatically displayed text, "Symbols" indicates automatically displayed symbols.
d) All spans on single chart: if checked this indicates that all drill chart data is located in the same form; if not checked, then vias passing through different layers, such as blind or buried vias, are shown in a different form.
e) Drill symbols on separate layers: if checked, drill chart is output to different user-defined layers, convenient for separate viewing and printing.
f) Include contours: if checked, include contours in drill chart.
g) Line separators: select to add line separators to drill chart, including horizontal line and vertical line.
h) Text format: the text format of the drill chart, includes title case or uppercase options.

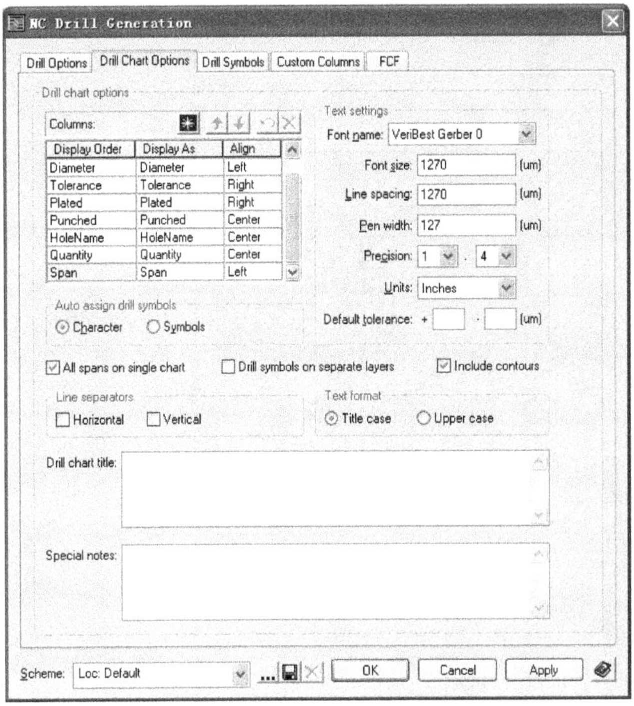

Figure 20.2 Drill Chart Options tab.

   i) Drill chart title: add drill chart title, located at the top of the drill chart generally.
   j) Special notes: add special annotation, located at the bottom of the drill chart generally.
3) Drill Symbols tab
   The Drill Symbols tab is shown in Figure 20.3; it is mainly used for drill symbol definition at design level; for different holes, corresponding symbols or characters can be defined, the designer can also let the system generate drill symbols automatically, or have no drill symbol for some holes.
   a) Automatic assign during output: the system automatically selects drilling symbols during output.
   b) None: don't generate drilling symbols.
   c) Use character as drill symbol: the options are uppercase and lowercase of the 26 English-language letters.
   d) Use drill symbol from list: use graphics as drill symbols; user can select from the list.

   The Custom Columns tab and FCF tab usually do not change; keep the default settings.

   After the completion of settings, click the OK button; the system will generate the drill file and put it into the Output\NC Drill sub-folder. Depending on the layer-stackup setting, the number of files will be different. Figure 20.4 shows the drill files for layer stackup 3+4+3 generated by Expedition.

   In display control, open Drill Drawing-related Display options; the drill drawing and drill chart are as shown in Figure 20.5.

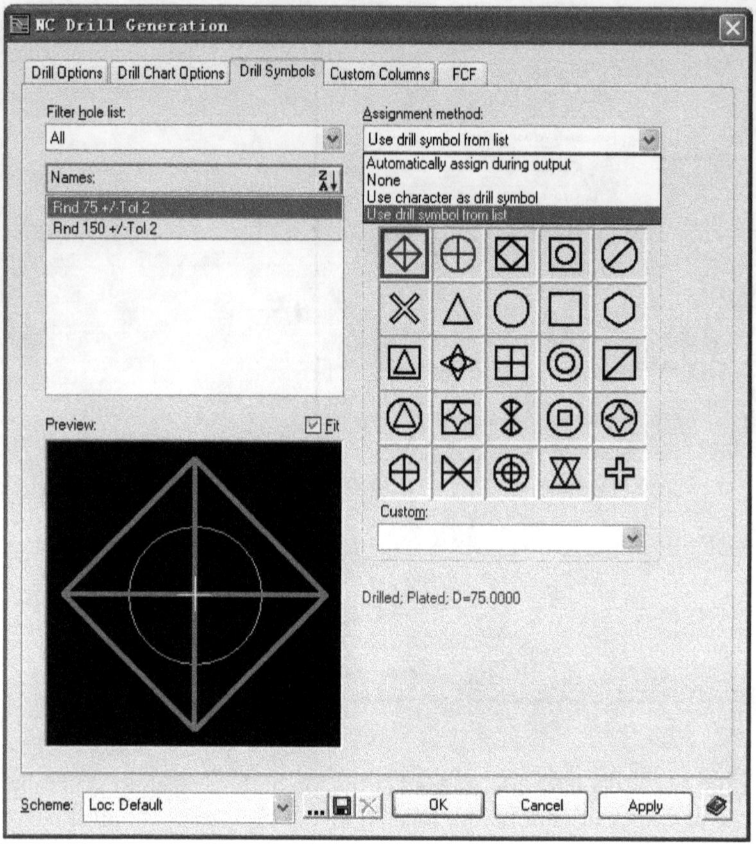

Figure 20.3 Drill Symbols tab.

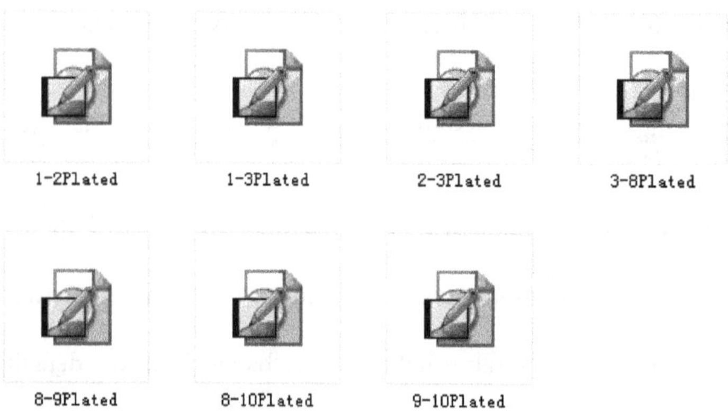

Figure 20.4 Drill files for layer stackup 3+4+3.

After drill files are generated, we need to inspect the drilling data in order to avoid mistakes; the designer can check the drill data with third-party tools, such as CAM350, and can also import drill data back into Expedition for inspection.

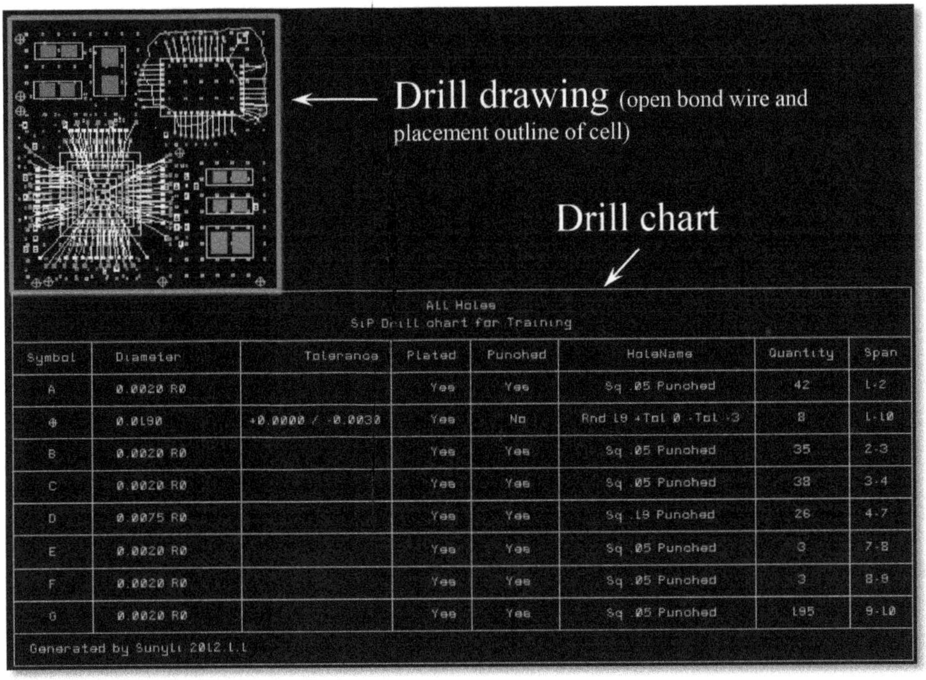

Symbol	Diameter	Tolerance	Plated	Punched	HoleName	Quantity	Span
A	0.0020 R0		Yes	Yes	Sq .05 Punched	42	1-2
⊕	0.0190	+0.0000 / -0.0030	Yes	No	Rnd 19 +Tol 0 -Tol -3	8	1-10
B	0.0020 R0		Yes	Yes	Sq .05 Punched	35	2-3
C	0.0020 R0		Yes	Yes	Sq .05 Punched	38	3-4
D	0.0075 R0		Yes	Yes	Sq .19 Punched	26	4-7
E	0.0020 R0		Yes	Yes	Sq .05 Punched	3	7-8
F	0.0020 R0		Yes	Yes	Sq .05 Punched	3	8-9
G	0.0020 R0		Yes	Yes	Sq .05 Punched	195	9-10

Figure 20.5 Drill drawing and drill chart.

To import drill data into Expedition, we first need to enable the FabLink XE or FabLink XE Pro license; select from the menu Setup→License Modules→Acquire FabLink XE or Acquire FabLink XE Pro.

Then, select from the menu File→Import→Drill to import drill data, open the corresponding layer in display control, and review the drill data, as shown in Figure 20.6.

20.1.2 Gerber Machine Format

In the Gerber output process, the first task is to set the Gerber machine format. Select from the menu Setup→Gerber Machine Format; the window shown in Figure 20.7 pops up.

Usually, we keep the default options; specific options are explained in Table 20.1.

20.1.3 Gerber Data Output

SiP Gerber data are different from ordinary PCB Gerber data; usually silkscreen is not needed. The main reasons for this are that the silkscreen is too small to print, or that compared with PCB design, the quantity of components in SiP is less, making identification easier. Therefore, Gerber data for SiP substrate generally does not include the silkscreen layer.

Typical SiP Gerber data is composed of signal layer, plane layer, soldermask layer, solderpaste layer and drill-drawing layer, etc.; the designer can also add or delete layers according to the requirements of the actual project. In SiP Gerber output, for example, the designer often increases the bond wire layer.

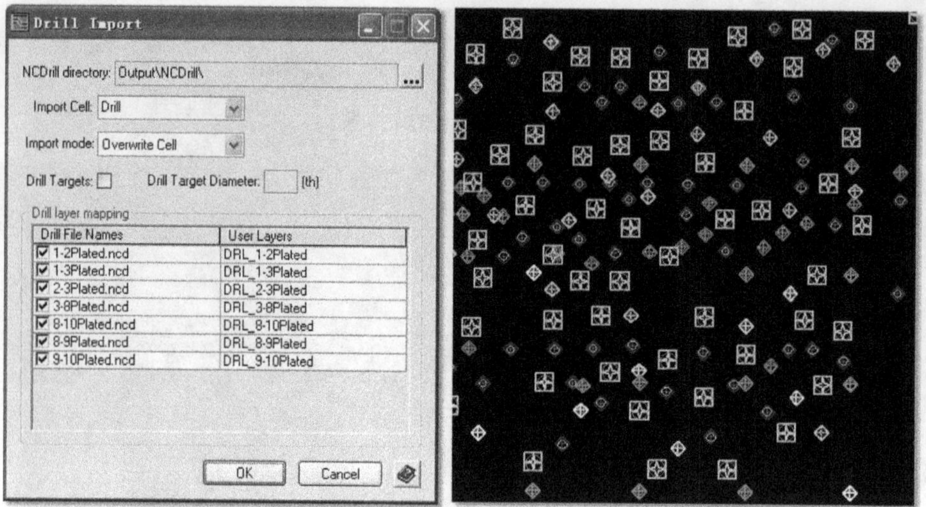

① Import drill data ② Review drill data

Figure 20.6 Import drill data for review (including drill symbols).

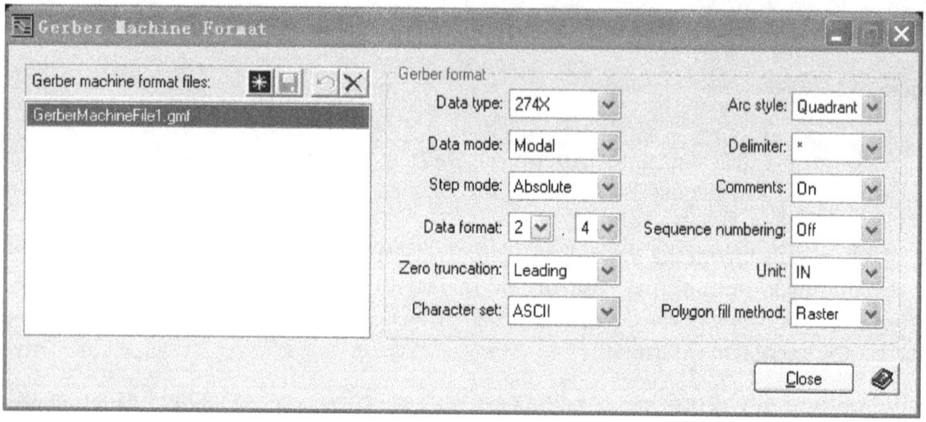

Figure 20.7 Gerber machine format setting.

In Expedition, select from the menu Output→Gerber to open the window shown in Figure 20.8.

In this window, the Gerber Plot Setup file is used to save the user's settings, and the Gerber output directory is used to set the Gerber file output path.

1) Parameters tab
   a) Output files area: mainly used for creating and renaming the Gerber output file according to the requirements of the project. In Expedition the Gerber file name usually uses .gdo as a suffix; Figure 20.8 shows bonding_finger.gdo, bongding-wire.gdo, etc. For D-Code Mapping File, select Automatic.
   b) Header text and Trailer text: used to add annotations at the beginning and end of the file.

Table 20.1 Gerber file format options description.

Parameter name	Options	Description
Data type	274X/274D	274X – (Default) EIA standard RS-274X (Extended Gerber format) embeds aperture information into the file. 274D – EIA standard RS-274D requires aperture information to be provided in a separate D-Code mapping file.
Data mode	Modal/ Non-modal	Modal (default) – Includes only words that have changed since the previous control statement. For example, if the X coordinate did not change since the last control statement, the X coordinate word is omitted from the current statement. Non-Modal – Includes all words in the control statement, including Preparatory Functions (G-codes), Coordinates (X, Y, I, J), Draft Functions (D-codes) and Miscellaneous Functions (M-codes).
Step mode	Absolute/ Incremental	Absolute (default) – Gives absolute coordinates in relation to the board origin. Incremental – Gives relative coordinates.
Data format	0~5,0~5	Defines the number of coordinate digits to the left and right of the decimal point. Default is 2.4.
Zero truncation	Leading/ Training/None	Leading (default) – Eliminates zeros before whole numbers (if any) to the left of the decimal point. Trailing – Eliminates zeros after whole numbers (if any) to the right of the decimal point. None – Does not truncate zeros.
Character set	ASCII/ EBCDIC	Defines the character set as ASCII (default) or EBDCDIC. ASCII – American Standard Code for Information Interchange. EBCDIC – Extended Binary-Coded Decimal Interchange Code.
Arc style	Quadrant/Full Circle/None	Defines the Arc style: Quadrant – Converts all circles into four arcs (90-degree sweep axis). Full circle – Creates full-circle hardware arcs (360-degree sweep axis). None – Does not allow the use of hardware arcs. Arcs and circles are approximated with line strings.
Delimiter	*/#/@	Enables selection of a delimiter (*, #, or @) that is recognized by the plot device. The delimiter indicates the completion of a command in the output file. Default is *.
Comments	On/Off	Includes (On) or does not include (Off) comments in the Gerber output file. Default is On.
Sequence number	On/Off	Applies (On) or does not apply (Off) sequence numbering to the output file. Default is Off.
Unit	Inch/mm	Defines the unit's value in inches (default) or millimeters. Note: Gerber data is written in the indicated units regardless of the units used in the design file.

(Continued)

Table 20.1 (Continued)

Parameter name	Options	Description
Polygon fill method	Raster/Draw	Defines the polygon fill method: Raster – Uses raster fill to create the data with "sharp" corners. Draw – Uses vector fill to create the data with drawn or "rounded" corners. Note: This option is only available for the 274X data type.

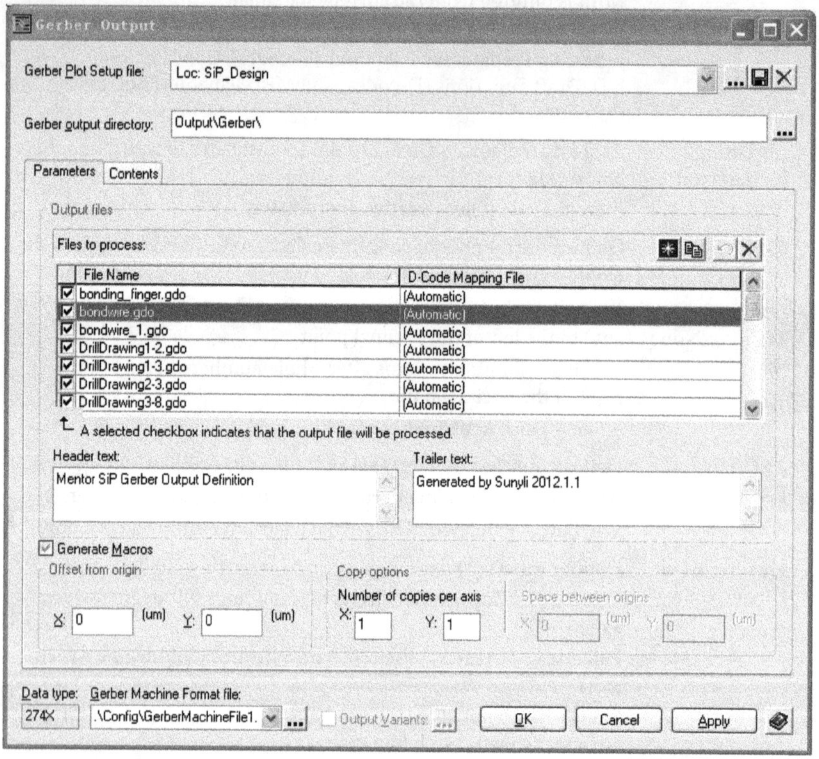

Figure 20.8 Gerber Output tab.

c) Generate Macros: used to generate customer apertures with shapes and lines.
d) Offset from origin: used to define the Gerber data offset relative to the origin, usually set to (0, 0).
e) Copy options: used to generate duplicate files; usually we keep the default.
f) Space between origins: when choosing to generate more copy documents, set the space between them.
g) Data type, data format: the same as defined for the Gerber machine file format; the designer only needs to select the defined file.

2) Contents tab

The Contents tab is mainly used to select the content for the output files defined by the designer; there are five options: conductors, board items, user-defined layers, cell types and cell items.

a) Conductors area: contains the metal conductor-related items in the substrate, usually conductive layer-related elements such as bond pad, parts pad, trace, via, etc.
b) Board items: substrate-related items, mainly board outline, cavity, mounting hole, silkscreen, soldermask, solderpaste, etc.
c) User-defined layers: mainly contains the user-defined layers and the layers imported from external files.
d) Cell Types: includes all the cell types in the design for designers to select from.
e) Cell Items: contains items related to cell, such as bond wire, die pin, placement outline, assembly outline, etc. First select the layer, and then select the elements in this layer.

In Expedition, the content definition of the Gerber file is very flexible; the designer can put different options together to form different Gerber files.

For example, if the designer needs to output the file named "bondwire.gdo" as bonding graphics, he or she can select items as shown in Figure 20.9.

According to the above settings, the output bond wire Gerber file graphics would be as shown in Figure 20.10.

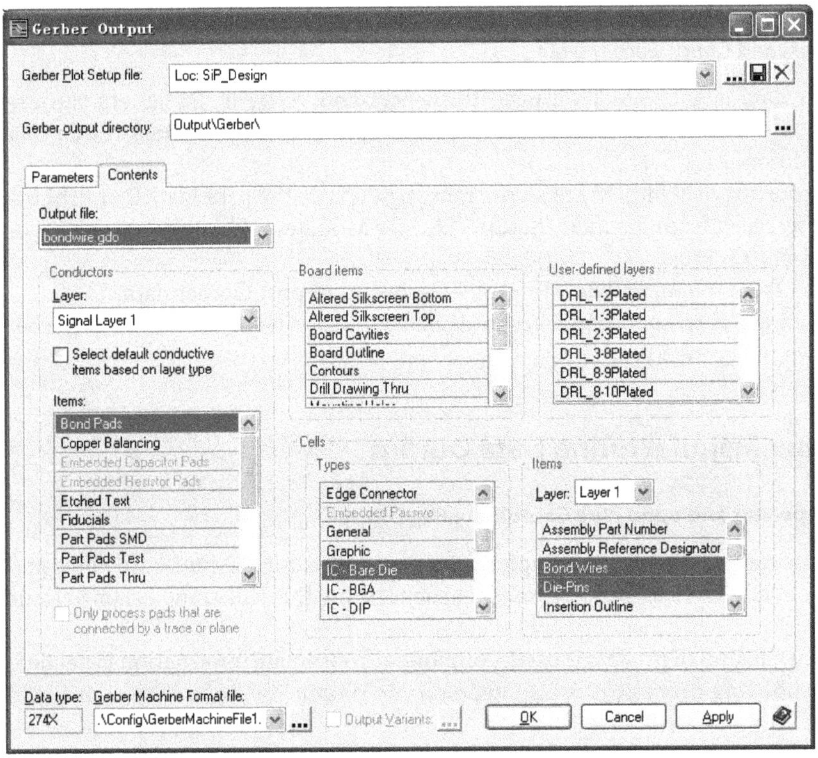

Figure 20.9 Gerber file content selection.

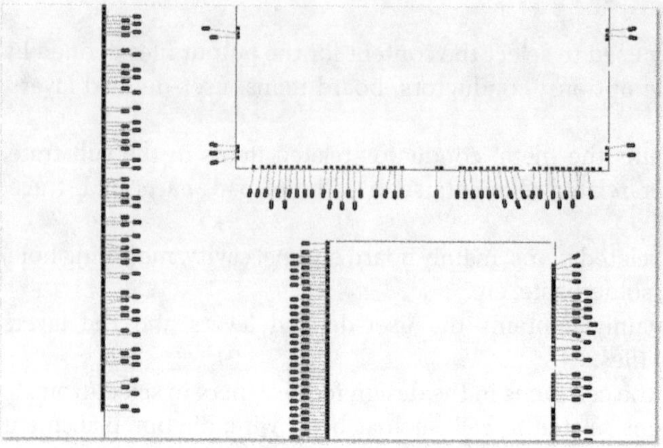

Figure 20.10 Output bond wire graphics (partial).

For details of Gerber file output configuration, the designer can refer to Table 20.2. User-defined layers are not included in all designs, so these are not present in this table. If necessary for the design, the designer can define a user-defined layer and add it to the output file.

When Gerber file configuration is finished, click the OK button to generate Gerber files. The generated Gerber files are put in the PCB\Output\Gerber subdirectory, each layer is a separated file.

20.1.4 Import and Check Gerber Data

After Gerber data is generated, we need to inspect and verify it. Designers can use third-party tools, such as CAM350, to check Gerber data, and can also check the Gerber data in Expedition.

As with importing drill files, the designer must first enable the FabLink XE or FabLink XE Pro license; select from the menu Setup→License Modules→Acquire FabLink XE or Acquire FabLink XE Pro.

Then, select from the menu File→Import→Gerber, to import Gerber data.

Finally, open the corresponding layer in display control, and we can review Gerber data, as shown in Figure 20.11.

20.2 Other Manufacturing Data Output

20.2.1 Component and Bond Wire Coordinate File Output

When SiP substrate processing is complete, the next step is to assemble bare chips and other components such as resistors and capacitors on the substrate using various processes.

No matter what kind of process is used, component coordinate information is needed. For the wire bonding process, it is also necessary to output the bond wire coordinate information.

Select from the menu File→Export→General Interface, in the pop-up window select Generic AIS, and then click OK, as shown in Figure 20.12.

Table 20.2 Output Gerber file configuration.

Output file	Conductor layer	Conductor items	Board items	Cell types	Cell items
Bondwire.gdo	Signal Layer 1	Bond pads	Board Outline	IC-Bare Die	Layer 1: bond wires die-pins
Positive layer	Corresponding signal or plane layer	Select all (exclude via holes)	Board Outline	Select all	None
Negative layer	Corresponding plane layer	Plane data	Board Outline	Select all	None
Soldermask Top	None	None	Board Outline Soldermask Top	Select all	None
Soldermask Bottom	None	None	Board Outline Soldermask Bottom	Select all	None
Solderpaste Top	None	None	Board Outline Solderpaste Top	Select all	None
Solderpaste Bottom	None	None	Board Outline Solderpaste Bottom	Select all	None

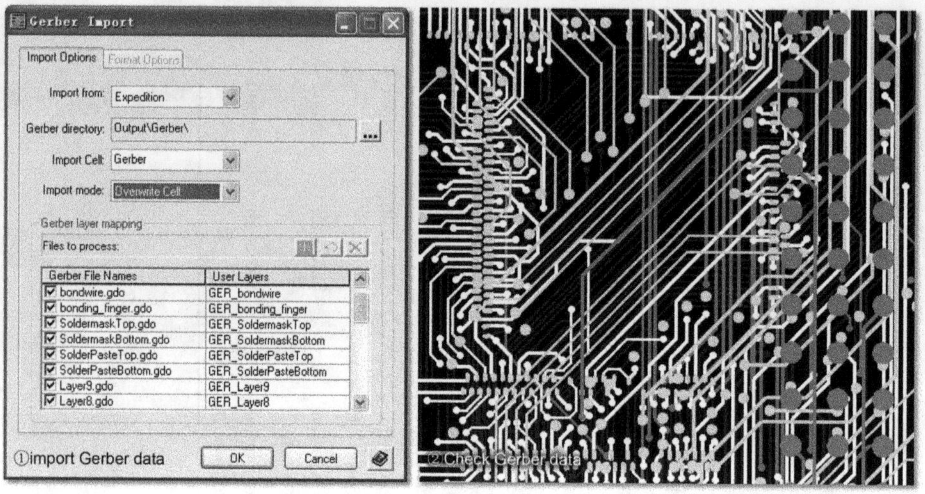

Figure 20.11 Import and review Gerber data.

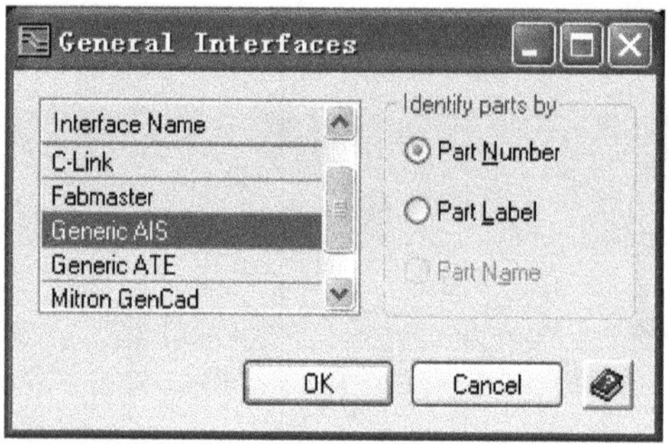

Figure 20.12 Output Generic AIS file.

After the file has been generated, the prompt window will pop up showing that the file has been generated successfully, and in the PCB\Output subdirectory, the designer will find vb_ais.txt file.

Open this file to view content as shown in Figure 20.13.

The file content is divided into seven columns, described in the list below.

1) Component reference designator, or bond wire name.
2) Component part number, for bond wire this is NULL.
3) Component rotation angle, for bond wire this is 0°.
4) Component (X, Y) coordinates, for bond wire these are the coordinates of the chip connecting to this bond wire. If the bond wire connects to the chip via die to die, the value is the start point chip coordinate.
5) The layer of the component or bond wire, at the top or bottom of the substrate.
6) Component or bond wire mirror placed or not, usually this is NO.

According to the project production needs, the designer can generate an accurate report file with bond wire start and end point coordinates. The file can be generated by running WireReport.vbs which is located in the Mentor installation directory ..\

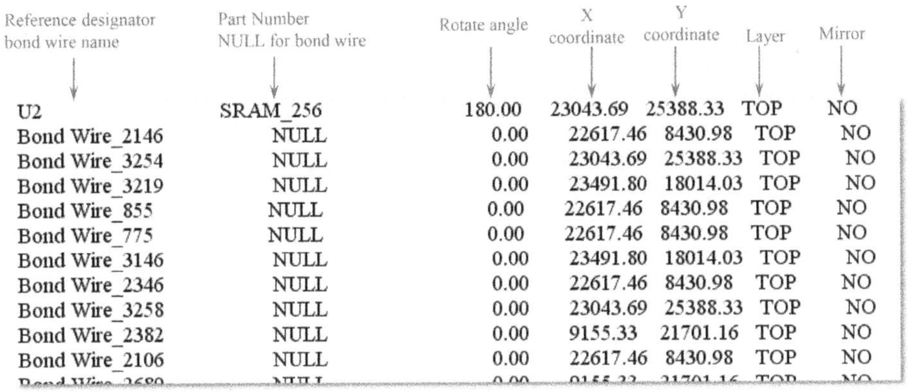

U2	SRAM_256	180.00	23043.69	25388.33	TOP	NO
Bond Wire_2146	NULL	0.00	22617.46	8430.98	TOP	NO
Bond Wire_3254	NULL	0.00	23043.69	25388.33	TOP	NO
Bond Wire_3219	NULL	0.00	23491.80	18014.03	TOP	NO
Bond Wire_855	NULL	0.00	22617.46	8430.98	TOP	NO
Bond Wire_775	NULL	0.00	22617.46	8430.98	TOP	NO
Bond Wire_3146	NULL	0.00	23491.80	18014.03	TOP	NO
Bond Wire_2346	NULL	0.00	22617.46	8430.98	TOP	NO
Bond Wire_3258	NULL	0.00	23043.69	25388.33	TOP	NO
Bond Wire_2382	NULL	0.00	9155.33	21701.16	TOP	NO
Bond Wire_2106	NULL	0.00	22617.46	8430.98	TOP	NO

Figure 20.13 The vb_ais.txt file format.

SDD_HOME\standard\examples\PCB\Automation\Scripts folder; he or she only needs to drag the WireReport.vbs file to the Expedition design window. After the file is generated, a prompt window automatically pops up showing the generated report file path and name, as shown in Figure 20.14.

Open the Wirebonds1 file; its format is shown in Figure 20.15.

The file content is divided into ten columns, explained below.

1 The name of the bond wire.
2, 3, 4 The bond wire start point (X, Y and Z) coordinates.
5 The start point layer.
6, 7, 8 The bond wire end point (X, Y and Z) coordinates.
9 The end point layer.
10 The bond wire model.

Figure 20.14 Wirebond report file was created.

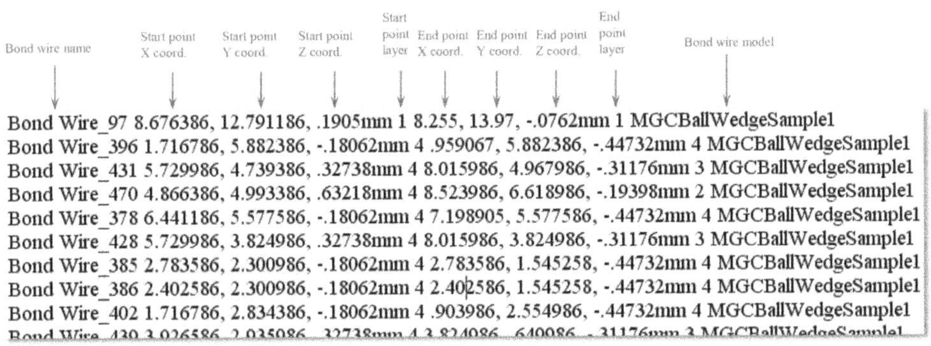


```
Bond Wire_97 8.676386, 12.791186, .1905mm 1 8.255, 13.97, -.0762mm 1 MGCBallWedgeSample1
Bond Wire_396 1.716786, 5.882386, -.18062mm 4 .959067, 5.882386, -.44732mm 4 MGCBallWedgeSample1
Bond Wire_431 5.729986, 4.739386, .32738mm 4 8.015986, 4.967986, -.31176mm 3 MGCBallWedgeSample1
Bond Wire_470 4.866386, 4.993386, .63218mm 4 8.523986, 6.618986, -.19398mm 2 MGCBallWedgeSample1
Bond Wire_378 6.441186, 5.577586, -.18062mm 4 7.198905, 5.577586, -.44732mm 4 MGCBallWedgeSample1
Bond Wire_428 5.729986, 3.824986, .32738mm 4 8.015986, 3.824986, -.31176mm 3 MGCBallWedgeSample1
Bond Wire_385 2.783586, 2.300986, -.18062mm 4 2.783586, 1.545258, -.44732mm 4 MGCBallWedgeSample1
Bond Wire_386 2.402586, 2.300986, -.18062mm 4 2.402586, 1.545258, -.44732mm 4 MGCBallWedgeSample1
Bond Wire_402 1.716786, 2.834386, -.18062mm 4 .903986, 2.554986, -.44732mm 4 MGCBallWedgeSample1
```

Figure 20.15 Bond wire report file format.

20.2.2 DXF File Export

DXF means Drawing Exchange Format; it was developed by Autodesk for data exchange between AutoCAD and other software.

Because AutoCAD is the most popular CAD software, the DXF format has been widely used and has become the de facto standard. DXF is an open-vector-data format, and the majority of CAD software can import or export DXF files.

In Expedition, to export a DXF file, select from the menu File→Export→DXF; the window shown in Figure 20.16 pops up. Select the output file path and file name, set appropriate units, check the required layer, and click the OK button to export DXF.

If the designer creates design schemes such as SiP chip placement, cavity, etc. in AutoCAD, he or she can also use DXF to transfer the design scheme to Expedition for use as reference for chip layout or cavity design. In Expedition, select from the menu File→Import→DXF to import the DXF file.

20.2.3 Layout Design Status Output

In order to gain a comprehensive understanding of current layout design status, we can use Menu→Design Status to output the design status file. This file contains basic information on the board size, layer stackup, via setting, nets, components, bond wires, etc., as shown in Figure 20.17.

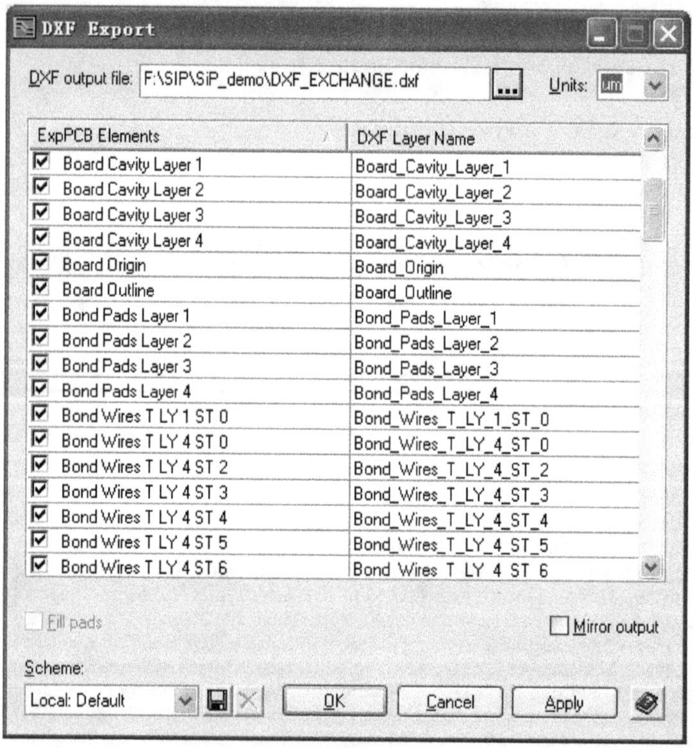

Figure 20.16 DXF Export window options.

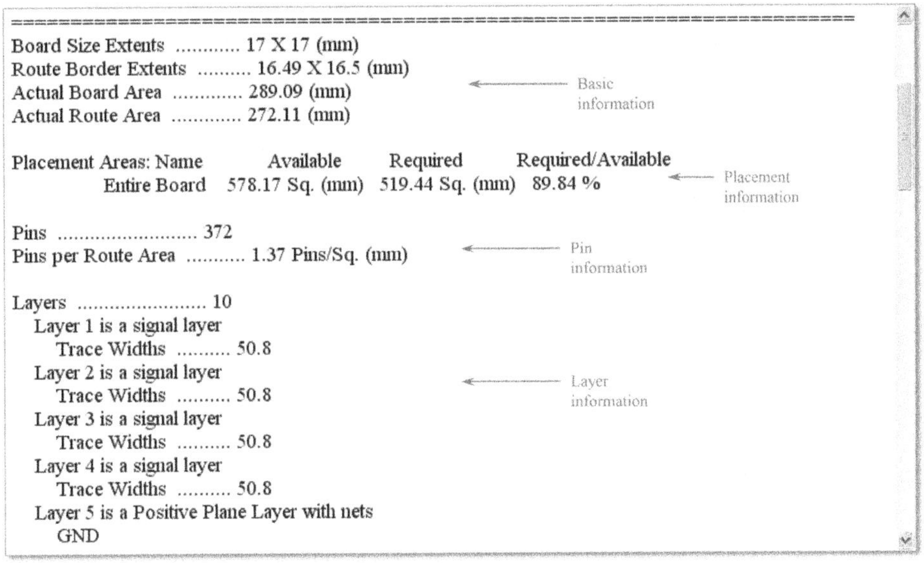

```
================================================================================
Board Size Extents ............ 17 X 17 (mm)
Route Border Extents .......... 16.49 X 16.5 (mm)
Actual Board Area ............. 289.09 (mm)
Actual Route Area ............. 272.11 (mm)

Placement Areas: Name          Available         Required          Required/Available
                 Entire Board  578.17 Sq. (mm)   519.44 Sq. (mm)   89.84 %

Pins .......................... 372
Pins per Route Area ........... 1.37 Pins/Sq. (mm)

Layers ........................ 10
  Layer 1 is a signal layer
    Trace Widths .......... 50.8
  Layer 2 is a signal layer
    Trace Widths .......... 50.8
  Layer 3 is a signal layer
    Trace Widths .......... 50.8
  Layer 4 is a signal layer
    Trace Widths .......... 50.8
  Layer 5 is a Positive Plane Layer with nets
    GND
```


Figure 20.17 Design Status file format.

20.2.4 BOM Output

At schematic stage, the designer can output the partlist, a list of components for production and procurement.

Corresponding to this function, after the completion of layout design, the designer can also export Bills Of Material (BOM) for components, containing the final state of production data, since the design is fixed at this time and the components won't change any more.

Select from the menu Output→Bill Of Material, and the BOM window pops up, as shown in Figure 20.18.

First, in the Bill of Material file bar, set the path of the output file, and then set the BOM configuration in the Settings area.

1) The check item after part number indicates that part number is a classification rule in BOM.
2) The check item after reference designator indicates that the reference designator is a classification rule in BOM. At least one of these two items must be checked, and both can also both checked.
3) The up and down arrows control entry order in BOM. The check item in front of each entry shows whether the entry is included in the BOM; the entry title can be edited, the designer can change the name according to his or her preference.
4) By checking the Include user properties option, the designer can include user-defined properties by selecting the required output items from the user properties area located on the right side of the window.
5) Checking the Include mechanical cells option means that mechanical cells will be included in the BOM, such as interposers used in chip stack.

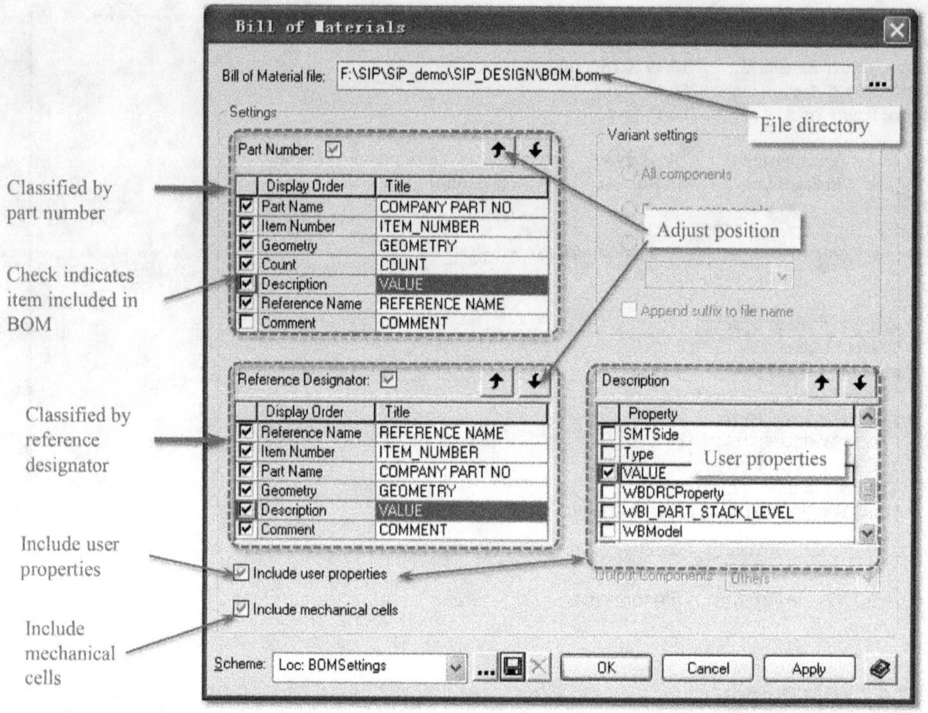

Figure 20.18 BOM export interface.

```
# Bill of Materials - Other Components
# date: Sat Jan 07 23:38:39 2012


COMPANY PART NO   ITEM_NUMBER   GEOMETRY     COUNT   VALUE              REFERENCE NAME

BGA100P           1             BGA100P      1                          U4
C0402             2             C0402        2       0.000000000018     C1 C2
C0504             3             C0504        1       0.0000000001       C3
interposer1       4             interposer   1                          interposer
R0402             5             R0402        1       3000               R1
R0402_2           6             R0402_2      1       3000               R2
R0603             7             R0603        1       1000               R3
SIP_1             8             SIP_1        1                          U1
SIP_2             9             SIP_2        1                          U2
SIP_3             10            SIP_3        1                          U3
SIP_4             11            SIP_4        1                          U5
SIP_5             12            SIP_5        1                          U6
SIP_6             13            SIP_6        1                          U7
SIP_7             14            SIP_7        2                          U8 U9

REFERENCE NAME   ITEM_NUMBER   COMPANY PART NO   GEOMETRY     VALUE             COMMENT

interposer       4             interposer1       interposer                     ---
C1               2             C0402             C0402        0.000000000018    ---
C2               2             C0402             C0402        0.000000000018    ---
C3               3             C0504             C0504        0.0000000001      ---
R1               5             R0402             R0402        3000              ---
R2               6             R0402_2           R0402_2      3000              ---
R3               7             R0603             R0603        1000              ---
U1               8             SIP_1             SIP_1                          ---
U2               9             SIP_2             SIP_2                          ---
U3               10            SIP_3             SIP_3                          ---
U4               1             BGA100P           BGA100P                        ---
U5               11            SIP_4             SIP_4                          ---
U6               12            SIP_5             SIP_5                          ---
```

Classified by part number

Classified by reference designator

Figure 20.19 BOM file (includes two kinds of classification rules).

After completion of the settings, click Save to save the scheme, and then click OK or Apply to output the BOM from layout.

Find the BOM file in the settings path and open with text editor; we see the BOM format as shown in Figure 20.19.

In the first part of the file, part number is a classification rule; in the second part of the file, reference designator is a classification rule.

In addition to the bare dies, resistors, capacitors and BGA encapsulation, the BOM file also includes interposers used in the chip stack.

21

SiP Simulation Technology

21.1 SiP Simulation Technology Overview

With the rapid development of SiP technology, SiP design complexity is constantly increasing; this is illustrated by the fact that, as substrate layer technology improves rapidly, more and more bare dies and passive components are installed on the substrate.

The bare IC chip itself is also becoming more and more complex – pin numbers are increasing rapidly, working frequency is increasing, signal change rates are accelerating. At the same time, the density of substrate routing is constantly improving, with signal frequency increasing rapidly.

Chip performance is also improving rapidly, which makes the circuit problem increasingly prominent, and directly affects the performance and working reliability of the SiP system. Using design without simulation, systems usually cannot meet the requirements of the design. For example, if the SiP system can only work at a lower clock-frequency than the design specifies, strict screening of the chip may be needed for the system to work normally, and sometimes many reworkings may be needed.

In order to solve these problems, we need to adopt an advanced SiP design platform, draft strict SiP design processes and specifications, and rely on the experience of the engineers for circuit design and problem screening.

At present, engineers' experience with SiP design cannot compare with that in the PCB design field, because PCB design technology has been mature for many years, while SiP design has only begun to be popular in recent years. Lack of design experience in this area requires the designer to rely on simulation tools to ensure the success of SiP projects.

Therefore, after SiP design is completed, to ensure the success of the design in a version, simulation is indispensable. Using simulation we can ensure that the product has good signal integrity and power integrity, good heat dissipation, and can meet the requirements of various indicators of electromagnetic compatibility.

Depending on the needs of SiP simulation, Mentor provides HyperLynx Signal Integrity (SI) specifically for signal-integrity analysis, HyperLynx PI for power-integrity analysis, HyperLynx Thermal for thermal simulation, and Quiet Expert (now called HyperLynx DRC) for EMI/EMC check.

For the schematic design phase, Mentor provides HyperLynx Analog for digital–analog mixed-circuit simulation; it can be directly simulated on the DxDesigner schematic.

SiP System-in-Package Design and Simulation: Mentor EE Flow Advanced Design Guide,
First Edition. Suny Li (Li Yang).

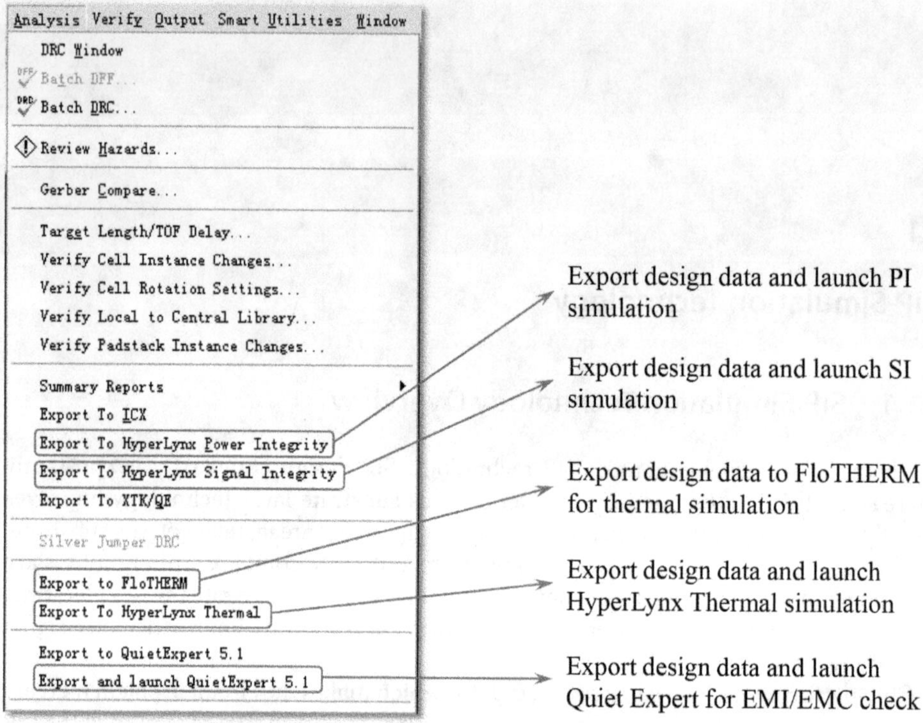

Figure 21.1 From Expedition export design data to simulation tools.

In Mentor EE Flow, using the analysis menu in Expedition the designer can directly output design data to simulation tools, and invoke the corresponding simulation tools automatically.

As shown in Figure 21.1, using the Analysis menu we can directly start SI – signal integrity simulation, PI – power integrity simulation, thermal analysis, EMC QuietExpert, etc.

21.2 Signal Integrity Simulation

Signal integrity refers to the signal waveform quality at the driver, receiver and transmission path; aspects include delay, reflection, crosstalk, timing, oscillation, etc.

With SiP system demands and IC device numbers increasing, signal-integrity problems are increasingly prominent. Serious signal-integrity problems will lead to system instability, where the system won't work normally.

Signal integrity has become something of concern in high-speed SiP design. Bare-chip working frequency and characteristics, the physical parameters of the SiP substrate, the placement of chips on substrate, high-speed signal routing etc., are factors that will cause signal integrity problems, and lead to system instability and eventually to system design failure.

Knowing how to consider signal integrity fully in the SiP design process and take effective control measures will greatly improve signal-integrity performance in SiP design.

21.2.1 HyperLynx SI Simulation Tool Introduction

Mentor HyperLynx SI is the industry's most popular high-speed-circuit signal integrity simulation tool. It contains a pre-simulation environment – LineSim, post-simulation environment –BoardSim, and multi-board analysis function. It can help the designer to simulate from low-frequency to thousands of MHZ (GHz) signals, and to simulate crosstalk and EMI, to eliminate hidden design risks, and to improve the success of SiP design.

HyperLynx SI is compatible with Mentor/Cadence, etc. layout design files, and provides a simple operation flow for high-speed SiP substrate simulation, such as the digital oscilloscope and spectrum analyzer in lab. From network topology planning, impedance design, high-speed rule definition and optimization in the early stages of design, to the final board-level validation, etc., all can be done in HyperLynx SI, effectively avoiding the necessity for repeated design.

Figure 21.2 shows screenshots from HyperLynx SI simulation.

1) HyperLynx LineSim
   The HyperLynx pre-simulation function LineSim can do "what-if" analysis for high-speed signaling in the schematic before layout; it can inspect signal transmission

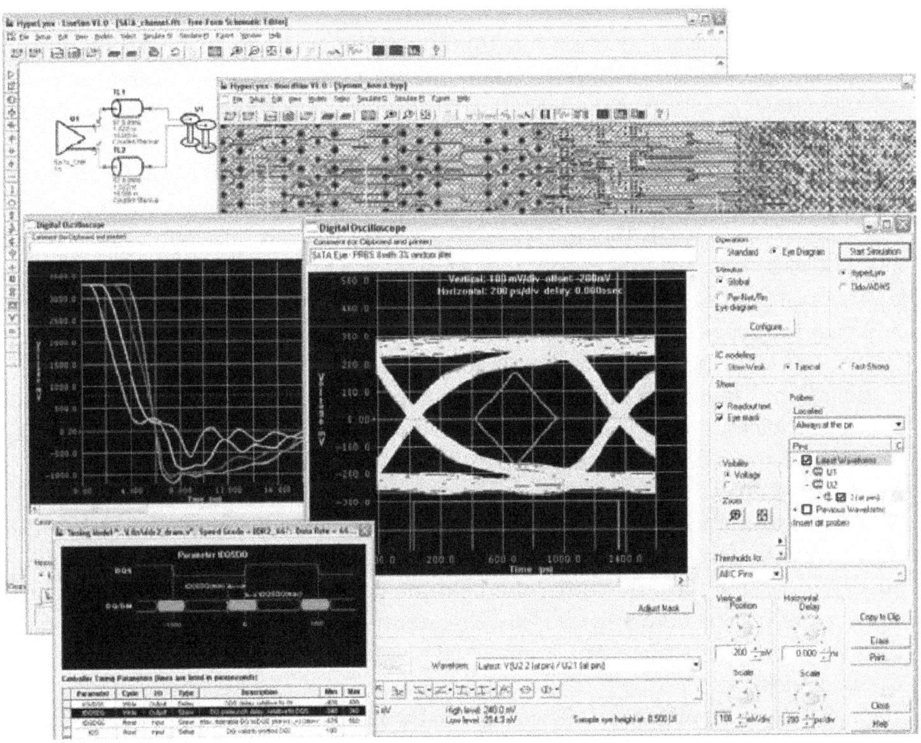

Figure 21.2 HyperLynx SI simulation screenshots.

effect under the virtual layer stackup and layout parameters; and help the designer to optimize a set of suitable layer-stackup, trace-impedance and high-speed design rules (trace width, trace length, spacing, etc.).

HyperLynx SI provides a visual window for layer stackup design, which helps engineers to quickly arrange the signal/power layer structure, edit the parameters of the substrate layer, and watch layer stackup and material parameter influence on trace impedance and loss tangent.

LineSim can easily link with the schematic design tool DxDesigner, import the key net topology from DxDesigner, and specify simulation models for the components connected to the nets.

Net connections can be assigned microstrip line, stripline, cable, connector and other transmission line models and can also assume the net parameters (width, length, spacing, etc.). The simulation can then be run to test the signal quality and electromagnetic radiation in the virtual environment. Investigation of overshoot, delay, crosstalk, EMI spectrum, eye-diagram, etc. indicators can be used to verify the SiP conditions assumed in the design process.

After the simulation has been debugged, the result will be a set of suitable layer stackups, net topologies, routing parameters and constraints for key nets; these constraints can be transferred from LineSim back to DxDesigner and passed to the Expedition layout design environment, where the constraints driver routing will be implemented.

2) HyperLynx BoardSim

The HyperLynx post-simulation function BoardSim can import an Expedition layout design file, extract the layer stackup and physical parameters of the substrate, calculate the characteristic impedance of the transmission line, and carry out signal-integrity, crosstalk and EMI analysis.

BoardSim can undertake interactive simulation for critical nets and output accurate signal transmission waveform, eye diagram or EMI radiation spectrum.

Designers can modify the trace parameters and carry out simulation again to find and improve unreasonable parameters. Designers can also directly modify various passive component parameters in BoardSim, update the schematic in DxDesigner, and pass the schematic to layout to quickly realize data synchronization and avoid artificial modification errors and omissions.

BoardSim has a batch simulation function to quickly scan all nets in a board; find overshoot, delay, crosstalk and EMI problems that go outside the design requirements; and give a detailed analysis report.

3) Multi-board analysis function

The multi-board analysis function can analyze the signal integrity of a high-speed digital system composed of multiple boards; examine the key nets' transmission effects in different layout boards; and help the designer to quantify the impacts of signal transmission across the boards. The multi-board analysis function facilitates the user in placing package or SiP on a PCB board for co-simulation.

4) Crosstalk analysis

HyperLynx SI supports crosstalk analysis; in the pre-simulation stage running crosstalk analysis can help the designer to optimize layout spacing and set coupling length, etc. routing rules to solve coupling problems between signals. The crosstalk analysis function can also be used to design differential impedance; depending on the differential impedance requirements, the designer can calculate suitable parameters such as differential spacing and line width.

In the post-simulation stage, the designer can do batch simulation for all nets in a board to summarize the nets that go beyond the safety index. HyperLynx SI supports the crosstalk positioning function; when any net is selected, the software will identify other nets with a crosstalk relationship with the selected net, and highlighting the coupling area. It is very easy to modify layout design according to the simulation results, thus improving design efficiency.

5) EMI/EMC analysis
HyperLynx SI supports the EMI/EMC analysis function. It can help the designer to investigate the effect of layer stackup, layout parameters, termination, shielding, etc. on signal electromagnetic radiation.

The designer can use the virtual current probe and standard (3 m/10 m/30 m) antenna probe to map electromagnetic radiation analysis of strong interference sources, to calculate the radiation value of interference sources in each frequency band.

HyperLynx SI provides a spectrum analyzer to show the analysis results of EMC. The spectrum analyzer integrates international EMI standards such as FCC, CISPR and VCCI, and designers can also build custom EMI standards in the spectrum analyzer.

Relative to high EMI test fees, HyperLynx SI software can provide a cost-effective EMI solution by finding EMI/EMC problems using simulation in the product design phase and putting forward suggestions for improvement. This can save much research and development cost and help to shorten the product development cycle.

6) Eye-diagram analysis
HyperLynx SI supports the eye-diagram analysis function; it can be used to check the working state of high-speed data signals and clock signals, and help the designer to test the signal transmission effect when the system is subject to abnormal phenomena such as clock offset, jitter, etc. This helps to confirm the stability of the system.

HyperLynx SI supports PRBS, 8b/10b, Toggling and USB2.0 compliance stimulus and customized stimulus, and USB2.0, PCI-E, SATA, SERDES, etc. eye mask and user-customized eye mask, saving much simulation time.

For layout design above GHz frequencies, the transmission line skin effect and dielectric loss cannot be ignored for signal waveform. Using analysis of transmission line loss, we can examine the relationship of layer stackup, routing and operating frequency with the two kinds of loss, provide guidance for layer stackup and dielectric definition of high-frequency routing rules, and give the designer more confidence about the high-frequency circuit design.

7) Model output function
HyperLynx SI supports S-Parameter-model and SPICE-model output functions. It can export the nets, containing vias, transmission lines, IC pins and matching circuits, to S-Parameter or SPICE formats.

HyperLynx SI is compatible with a variety of simulation model formats such as IBIS, SPICE, S-Parameter, etc. HyperLynx SI provides a powerful IBIS-model library for high-speed simulation, including CMOS/TTL standard process IC models such as CPU/DSP and FPGA/CPLD. It freely provides as many as 18000 proven IBIS models.

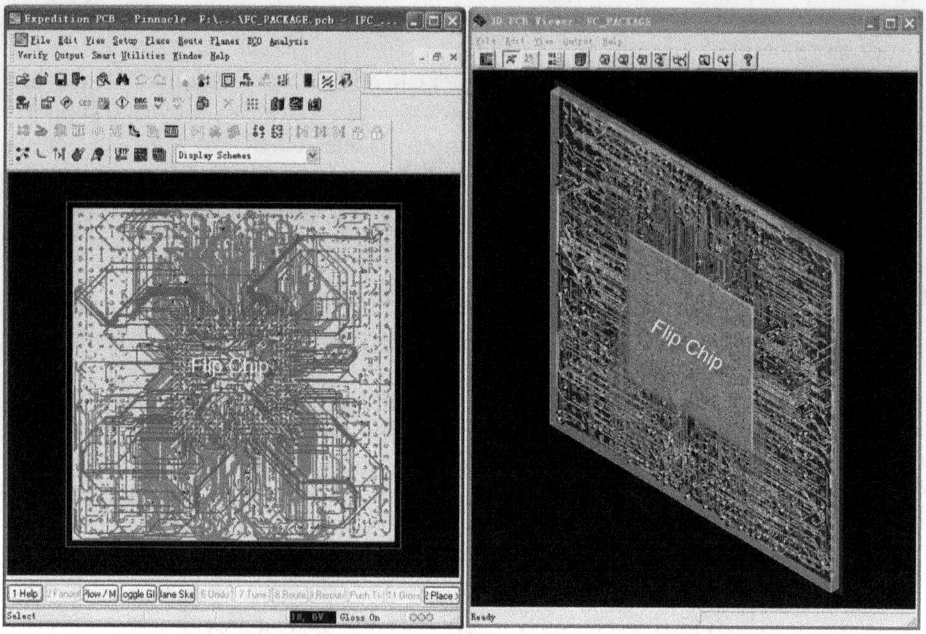

(a) 2D design view (b) 3D design view

Figure 21.3 Flip chip layout design in Expedition.

In addition to the powerful model library, HyperLynx SI provides a model creation wizard; the designer only needs to input component parameters such as switch time, parasitic value and process type, and the wizard can create a simple IBIS model for simulation.

HyperLynx SI include pre-simulation planning, design space exploration, design constraints generation and post-simulation verification and multi-board analysis functions; all these functions are tightly integrated in the EE design flow, helping to realize the seamless connection of design and simulation.

21.2.2 HyperLynx SI Simulation Example

The following is an introduction to the HyperLynx SI simulation process using the example of a flip chip design.

1) Design data transfer
   Figure 21.3 shows a screenshot of a flip chip layout design in Expedition; select from the menu Analysis→Export to HyperLynx Signal Integrity; Expedition will transfer design data to HyperLynx and open the HyperLynx SI simulation tool automatically.

   In the HyperLynx SI interface, we can see the flip chip design has been imported, as shown in Figure 21.4.
2) The key signal SI simulation
   First, we look at key signal simulation. In the HyperLynx SI window toolbar, click the NET button; on the Select Net by Name window, select a key signal in the design, such as net GCK0 as shown in Figure 21.5.

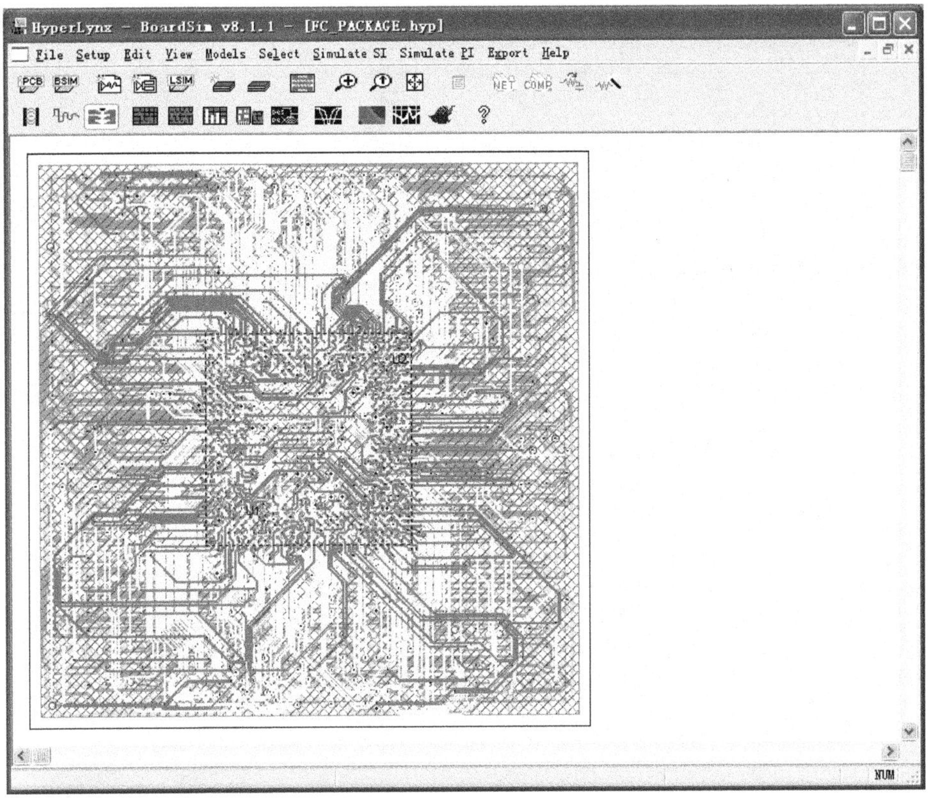

Figure 21.4 Flip chip design is imported in HyperLynx.

The selected GCK0 net will be highlighted while the other nets appear dimmed, as shown in Figure 21.6.

Then click the COMP button; in the pop-up Assign Models window click the Select button, in the Select IC Model window select model 105.ibs, and assign signal CLK_OUT to U1 and signal A0 to U2, as shown in Figure 21.7.

Now set Output to U1 and Input to U2 in Buffer Settings, as shown in Figure 21.8.

After the completion of model assignment, click the Run Interactive Simulation (SI Oscilloscope) button to start the digital oscilloscope, and set as shown in Figure 21.9. Then click the button Start Simulation to start SI simulation; the waveform soon shows in the oscilloscope window; from the simulation waveform as shown in Figure 21.9 we can see that the signal quality is not very good, there is obvious oscillation and overshoot.

To improve the signal quality, we can use a variety of methods, such as changing the signal driver ability, changing the layer stackup, re-routing, using terminal matching, etc. Here, we use the terminal matching method.

HyperLynx has an intelligent terminal matching tool, Terminator Wizard. Click the Run Terminator Wizard button , start the window as shown in Figure 21.10. Terminator Wizard will analyze the net automatically, give the results and recommend matching method and related values.

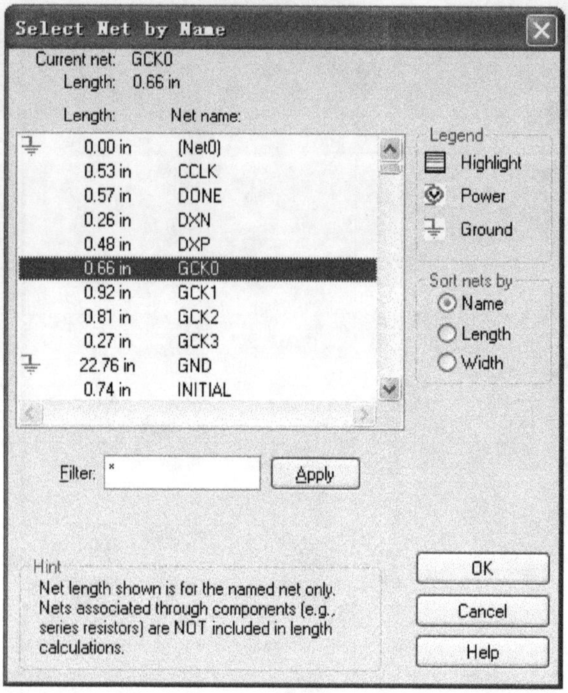

Figure 21.5 Select key net.

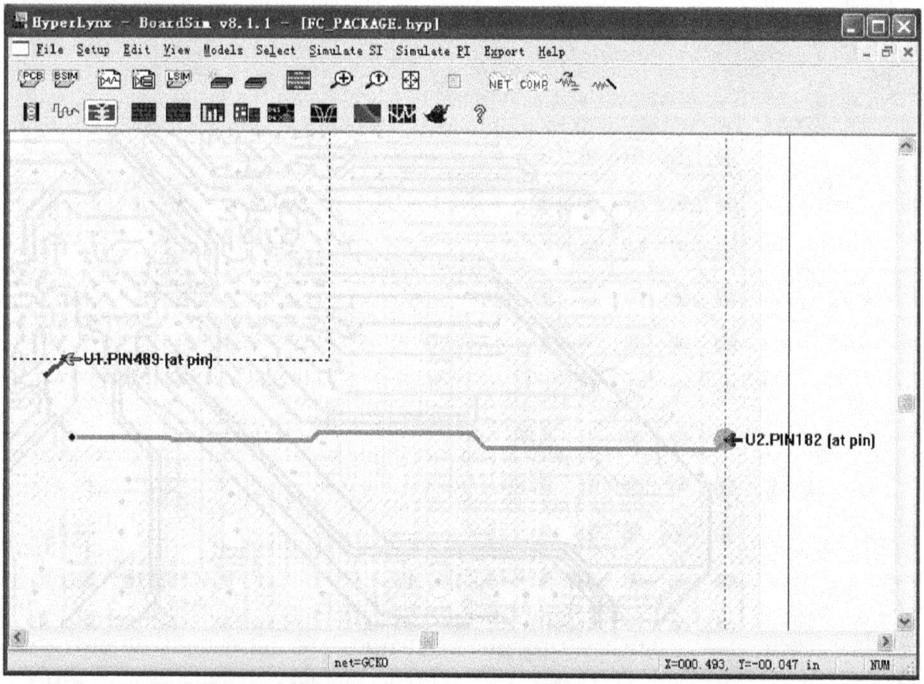

Figure 21.6 The key signal GCK0 net is selected.

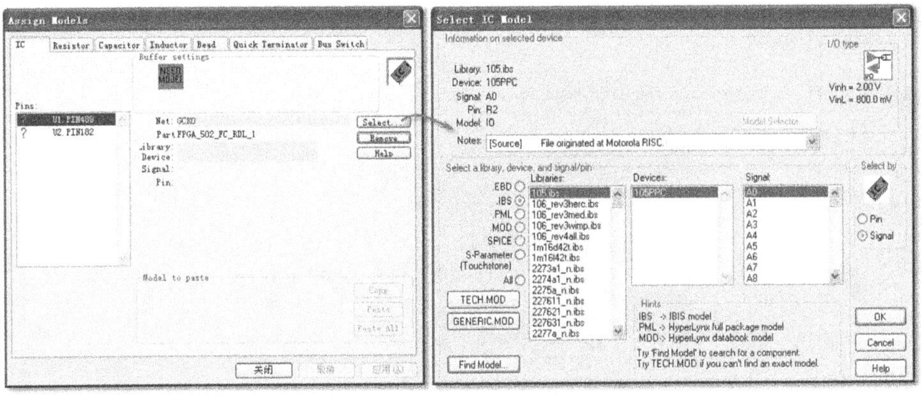

Figure 21.7 Specify simulation model.

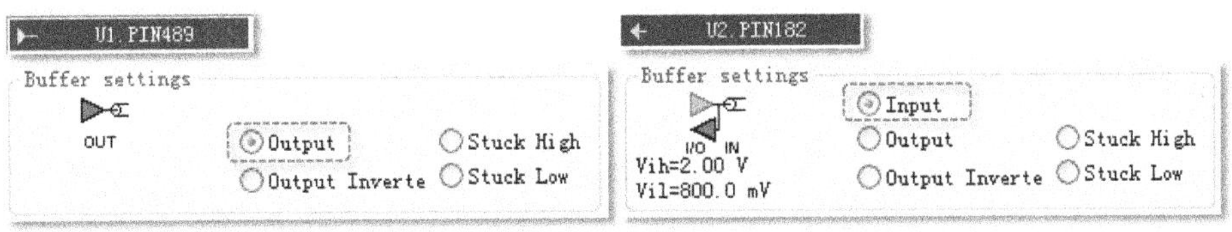

Figure 21.8 Specify the input and output buffers of the simulation model.

As shown in Figure 21.10, the termination suggested is series resistor and the value is 49.5 Ω.

Click the button [Apply Values] in the top-right corner of the Terminator Wizard, and then click the OK button to exit the window.

Re-start the simulation; now we get the results shown in Figure 21.11. Comparing this signal with the signal in Figure 21.9, we see that the signal quality has very much improved.

Compare the two simulation results, check the Previous Results box, and, close to the probe at the driver, open the probe at the receiving end and set to a different color, as shown in Figure 21.12. From the figure, the signal quality improvement after termination is obvious.

3) Transmission line S-parameter extraction

For SiP or package it is very important for the designer to extract the signal transmission path model; the model shows the impact that the SiP or package substrate routing has on the signal.

"Model after package" = "bare IC chip model" + "package's S-Parameter model"; with reference to this formula, in HyperLynx SI we can easily obtain the model after package; this model can be provided to the downstream PCB board-level system simulation, as shown in Figure 21.13.

In HyperLynx, select from the menu Export→Net To→S-Parameter Model, on the Extract S-Parameter Model window, click [Map Auto]; S-Ports will automatically map with component pins. Input a frequency range such as 0.1~10000 MHZ in the

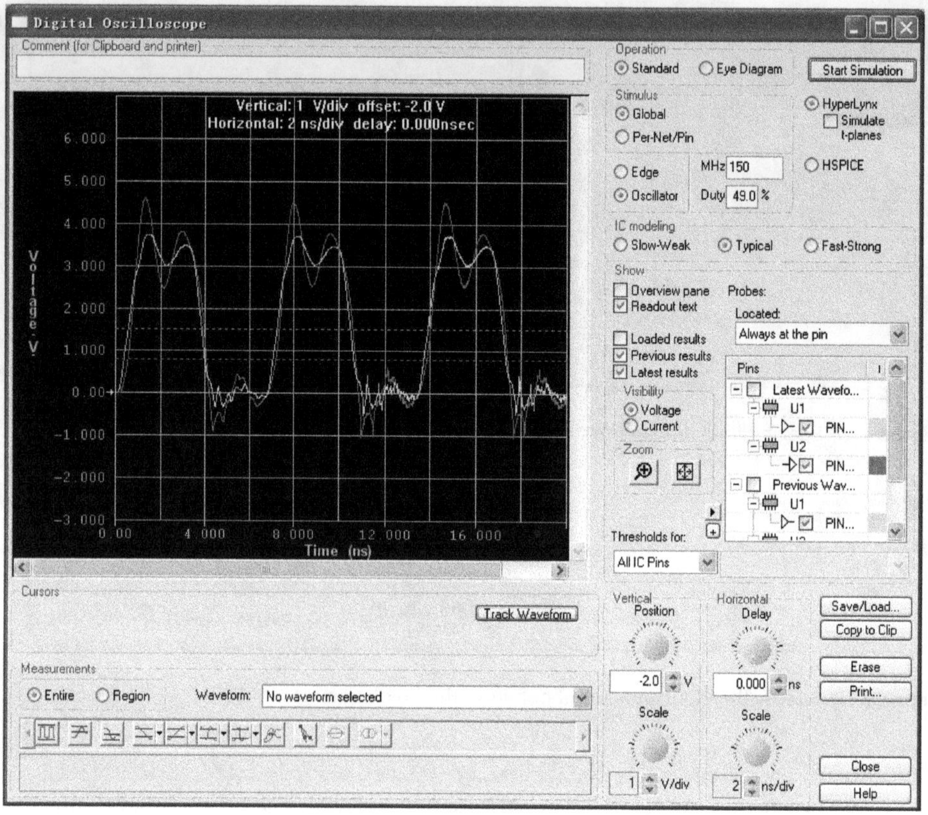

Figure 21.9 Digital oscilloscope displays simulation results.

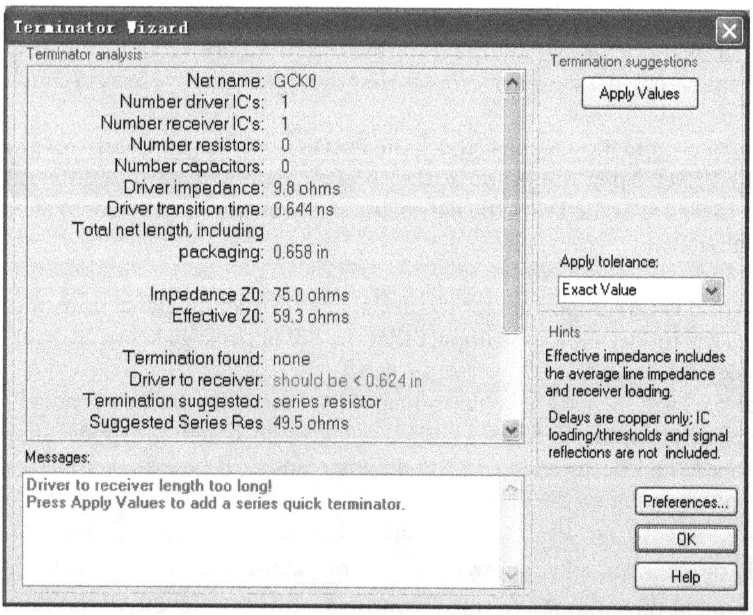

Figure 21.10 Terminator Wizard.

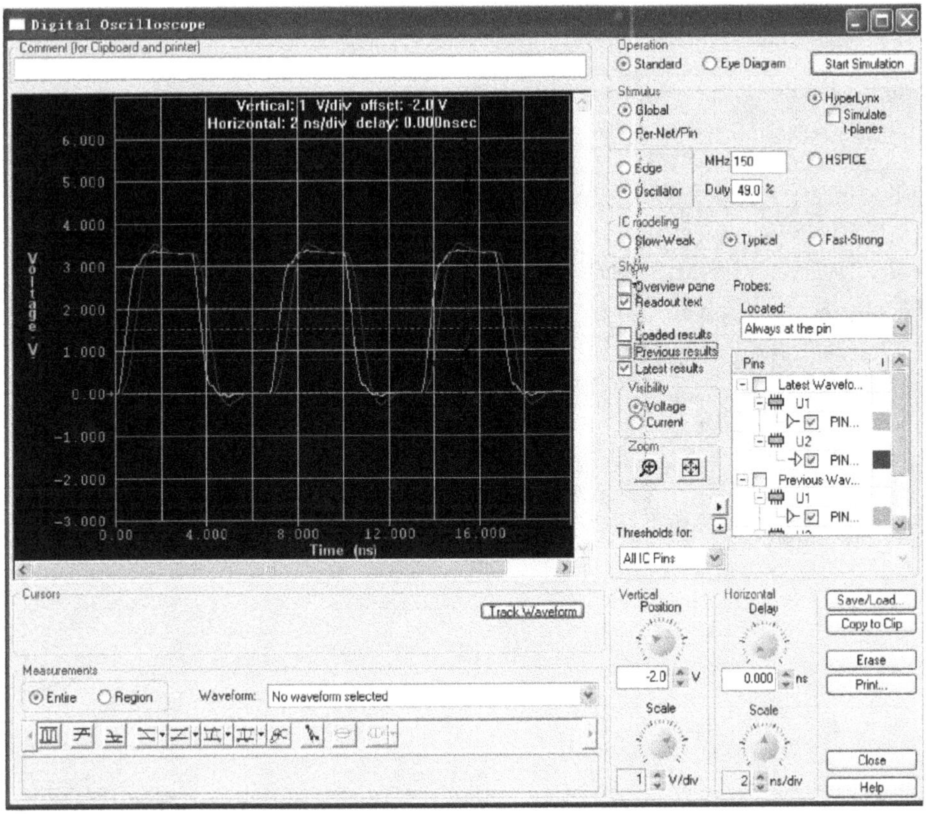

Figure 21.11 The simulation results after termination.

Modeling parameters area; in the Frequency-sweeping parameters area, select adaptive type; see Figure 21.14.

After the completion of settings, click the Create Model button at the lower-left corner of the window, input the model name to the pop-up window, for example, GCK0.s2p, then click Save; the S-Parameter model is generated.

Applying the S-Parameter model is very easily done in HyperLynx; create a new project, select from the menu File→New Free Form Schematic, click the Add IC to Schematic button ▷ to add the input and output symbols of IC components, and assign the corresponding model to the IC symbols, such as CLK_OUT signal for driver, A0 for receiver. Then click the Add Package to Schematic button, and add the package model. In the pop-up window select GCK0.s2p, map the port and set the pin side, as shown in Figure 21.15.

After assigning the S-Parameter model, we can see the driver and receiver connected by the S-Parameter model, as shown in Figure 21.16.

Then run a signal integrity simulation setting the parameters as before; we get the waveform shown in Figure 21.17. Comparing this with Figure 21.9, it can be seen that the two simulation waveforms are basically identical, thus verifying the accuracy of HyperLynx SI S-Parameter model extraction.

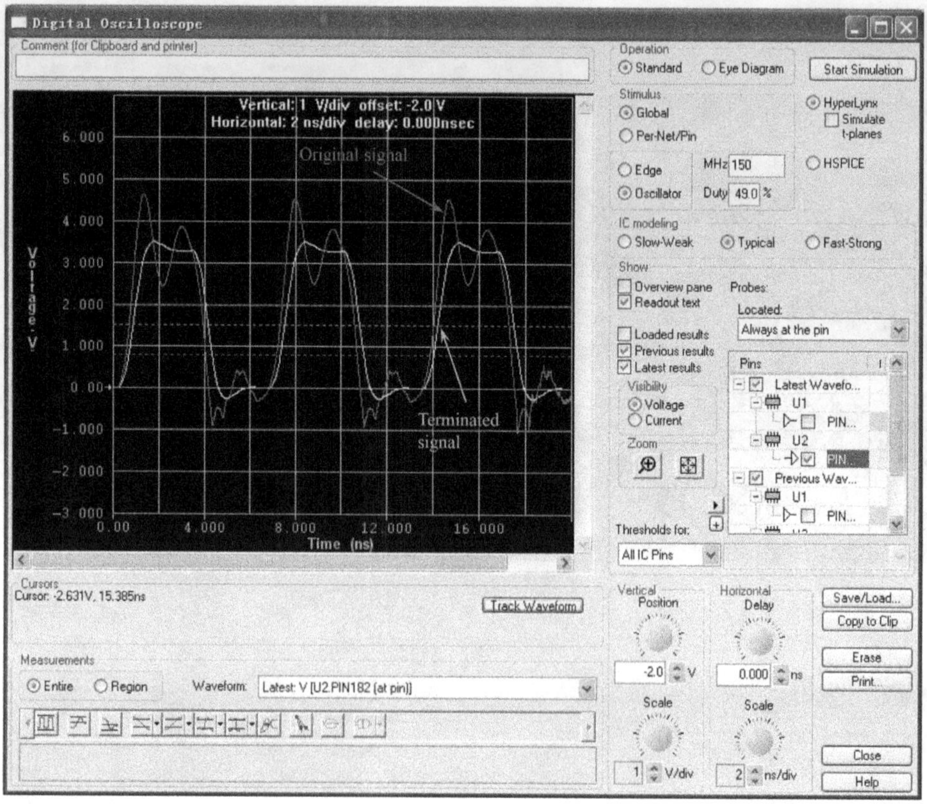

Figure 21.12 Comparison of original signal and terminated signal. *(See color plate section for the color representation of this figure.)*

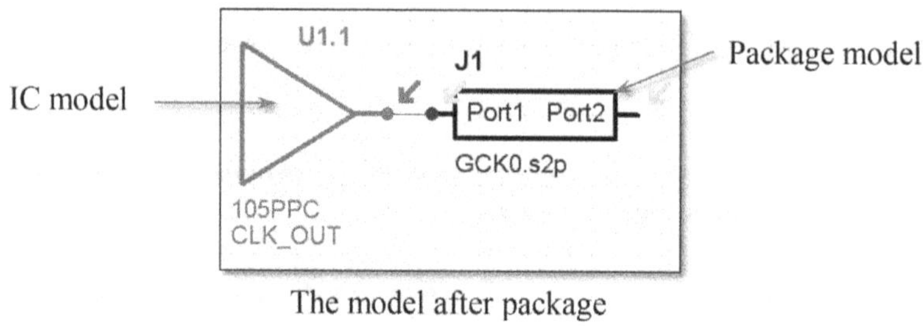

Figure 21.13 The chip model after package.

21.3 Power Integrity Simulation

With the pursuit of higher-performance systems, design complexity is gradually increasing and signal working frequency is becoming higher and higher.

SiP designers, in addition to paying attention to signal integrity analysis, such as reflection, crosstalk and EMI, must also focus on stable and reliable power supply, which is a focus of high-speed SiP design research.

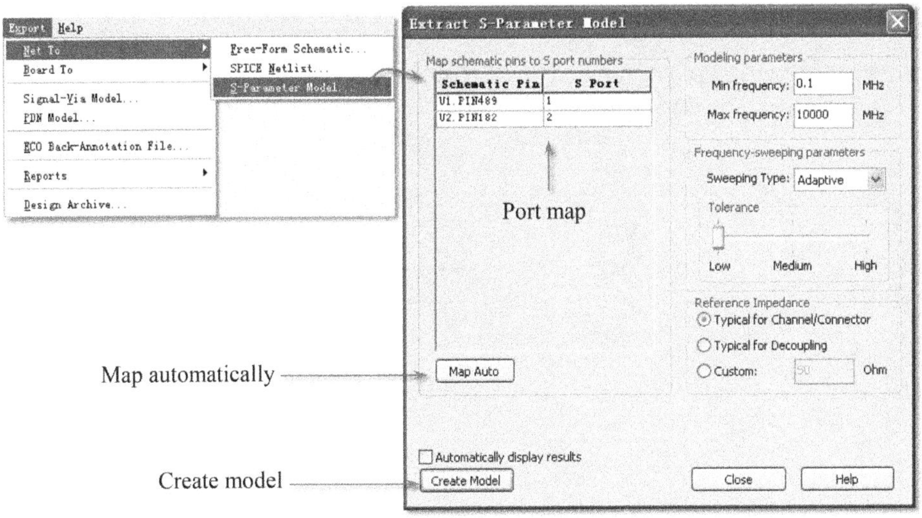

Figure 21.14 Extract S-Parameter model.

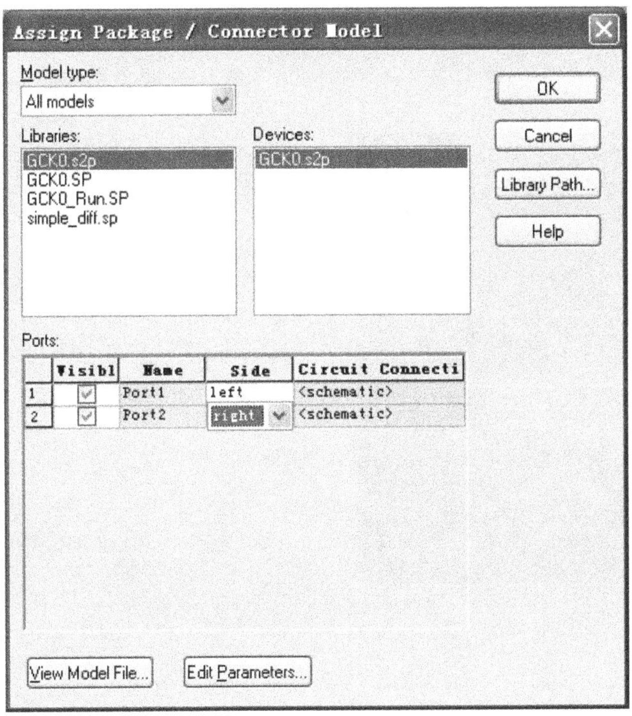

Figure 21.15 Assign S-parameter model.

With the number of chips increasing, the core voltage decreasing, and the types of power supply also increasing, we often design with multiple power supplies sharing the same plane layer. However, power supply volatility has a severe effect on the system, and suggests a new term: power integrity, hereinafter referred to as PI.

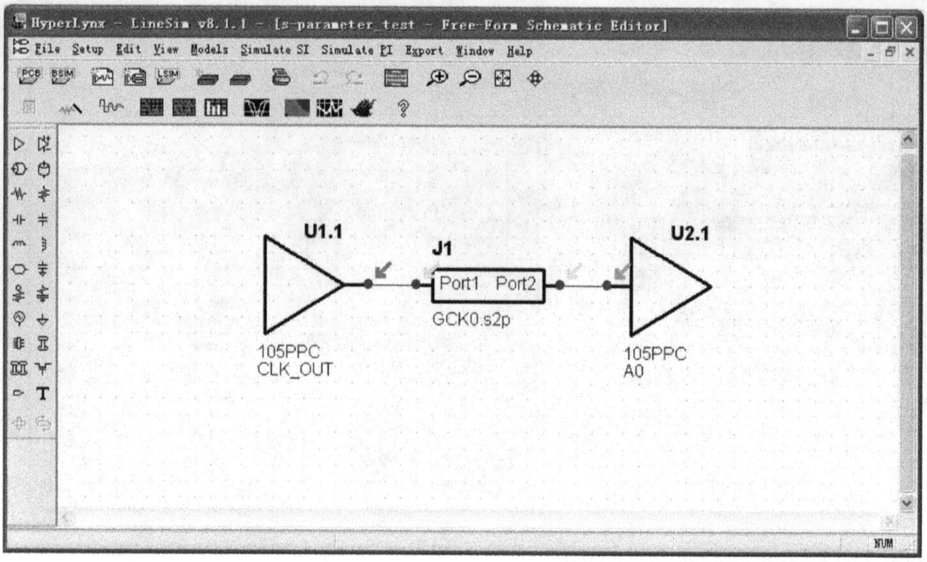

Figure 21.16 Connect driver and receiver through S-Parameter.

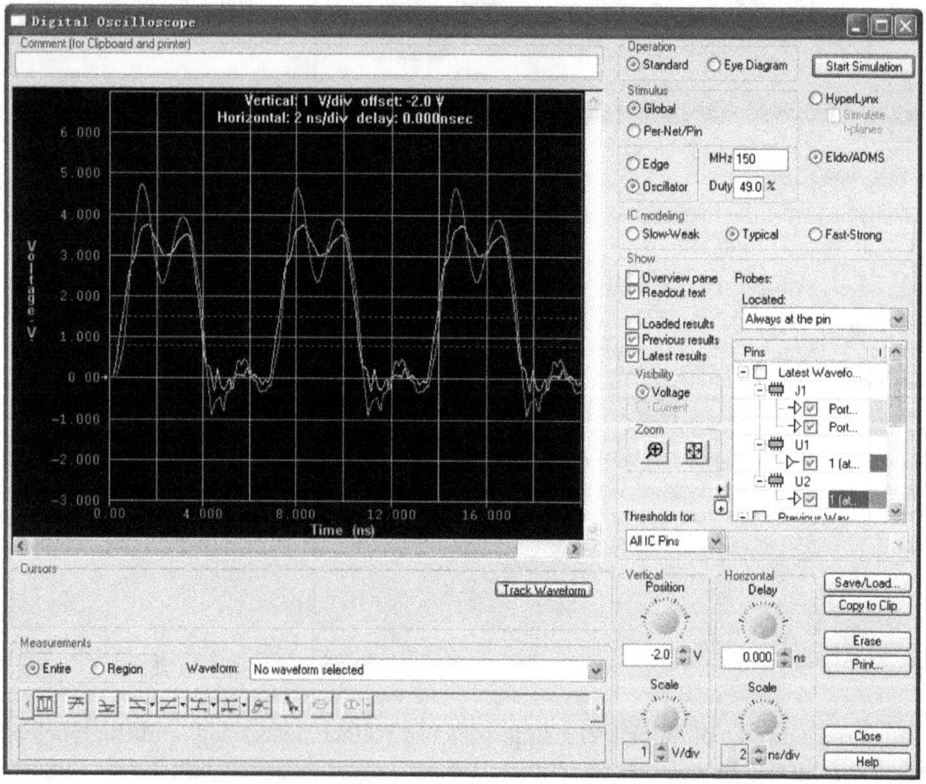

Figure 21.17 The simulation result with S-Parameter model.

In fact, PI and SI are closely linked; formerly, when simulation tools were used to carry out signal integrity analysis, it was generally assumed that the power supply was absolutely stable, but as the requirements of simulation precision have increased, this assumption obviously can no longer be accepted. PI analysis arises to meet this challenge.

In a sense, PI belongs to the SI research scope, and the new generation of signal integrity simulation must be based on reliable power integrity.

Although power integrity mainly involves discussion of the stability of power supply, because the ground plane and power plane are always inseparable in the actual system, ways to reduce the noise of the ground plane is also discussed as an important part of power integrity.

21.3.1 HyperLynx PI Simulation Tool Introduction

Power integrity (PI) analysis has become an indispensable part of modern electronic design. With increased IC power-supply types, power consumption and working frequency, and reduced noise margin, power supply system design is becoming more and more difficult.

If the power supply is insufficient, the signals on the circuit board will experience problems, and the whole design will fail due to logic errors.

Hardware engineers, circuit-design engineers and SI-analysis engineers can use the powerful features and benefits of HyperLynx PI for high-quality circuit design. For the newcomer to power integrity verification, a short study will enable easy use of the tool.

Using HyperLynx PI, even in the early design stage before layout, the designer can identify power allocation problems and at the same time can find a solution.

In the HyperLynx PI environment, the designer can use the "what-if" method to find the best solution. Once the layout is completed, the design can be verified using post-simulation, to ensure that each index of the design requirements is met.

HyperLynx PI will help to improve the reliability of products, reduce repeated design cycles, and shorten product development, speeding up the product's time to market.

Figure 21.18 shows a screenshot of HyperLynx PI simulation.

1) DC-drop and current-density analysis
   HyperLynx PI can identify potential DC power-allocation problems. Too much DC drop, for example, will result in the IC not working normally due to insufficiency of power-supply voltage; high-density current or excessive via current is likely to cause too much heat, interrupting the signal connection or damaging the substrate.

   The results of PI simulation can be viewed graphically and can also be generated in a report, helping designers to quickly find DC power-allocation problems.

   Figure 21.19 shows the plane-layer analysis result of DC drop.
2) Optimize power distribution net (PDN)
   HyperLynx PI can help designers to optimize the impedance of power distribution nets (PDN), and to determine the optimal distribution of decoupling capacitors, including a capacitor's number, position and installation mode, and get the impedance curve of power supply in the case of different capacitor distributions, to determine the optimal capacitor allocation. Figure 21.20 shows the impedance curve of the plane layer with different capacitor distributions.

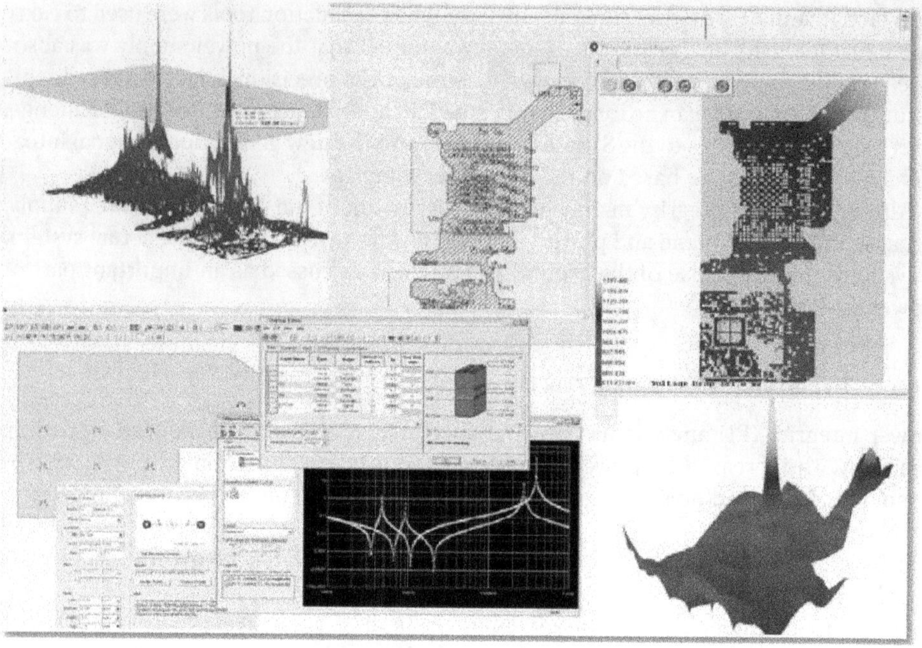

Figure 21.18 HyperLynx PI simulation screenshots.

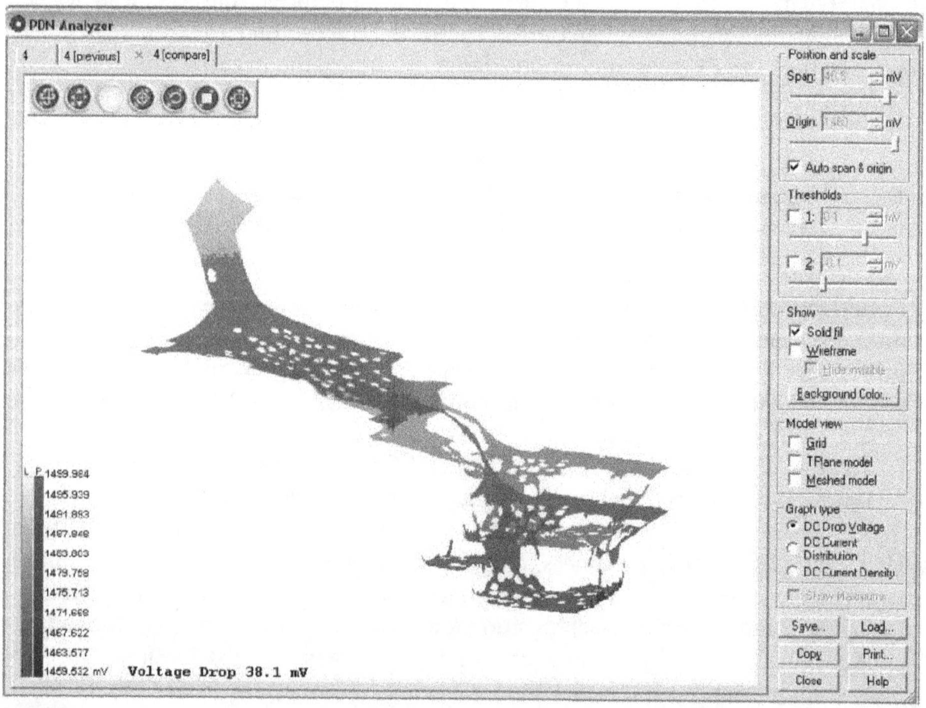

Figure 21.19 The results of DC drop analysis.

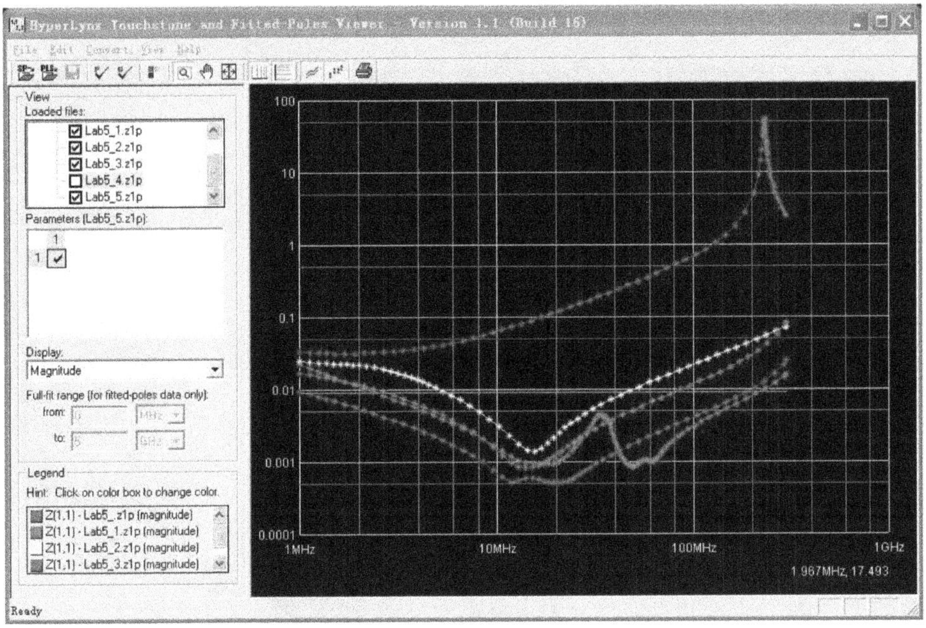

Figure 21.20 The impedance curve of the plane layer with different capacitor distributions.

Based on this function, HyperLynx PI can help designers to analyze the plane-layer impedance effect on noise propagation in the power supply layer; 3D waveform shows the spread of noise on the plane layer, as shown in Figure 21.21.

3) Model extraction

In the GHz field, reasonable parameterization of via is very important for SERDES design. HyperLynx PI can produce high-precision via models, including substrate decoupling nets, capacitors, vias and the influence of plane resonance.

HyperLynx PI supports PDN model extraction, and models can be extracted to S, Z or Y parameters, and can be easily applied in later simulations.

HyperLynx PI can help designers to detect power-supply system problems at an early stage, even before layout design, so as to reduce design iteration, shorten the design cycle and improve the reliability of products.

In HyperLynx PI, DC analysis can be used to find excessive DC drop and current density, so as to determine whether there are enough copper connections or vias; AC analysis can be used to determine the impedance of PDN and the size, number, installation method and location of decoupling capacitors; to test power plane spacing and location; and to check the power-supply noise transmission in the system.

Like HyperLynx SI, HyperLynx PI is also divided into pre-simulation (LineSim) and post-simulation (BoardSim); both simulations support DC-drop analysis, current-density analysis and decoupling and noise analysis.

21.3.2 HyperLynx PI Simulation Example

We introduce the HyperLynx PI simulation process simply here, using the example of flip chip design.

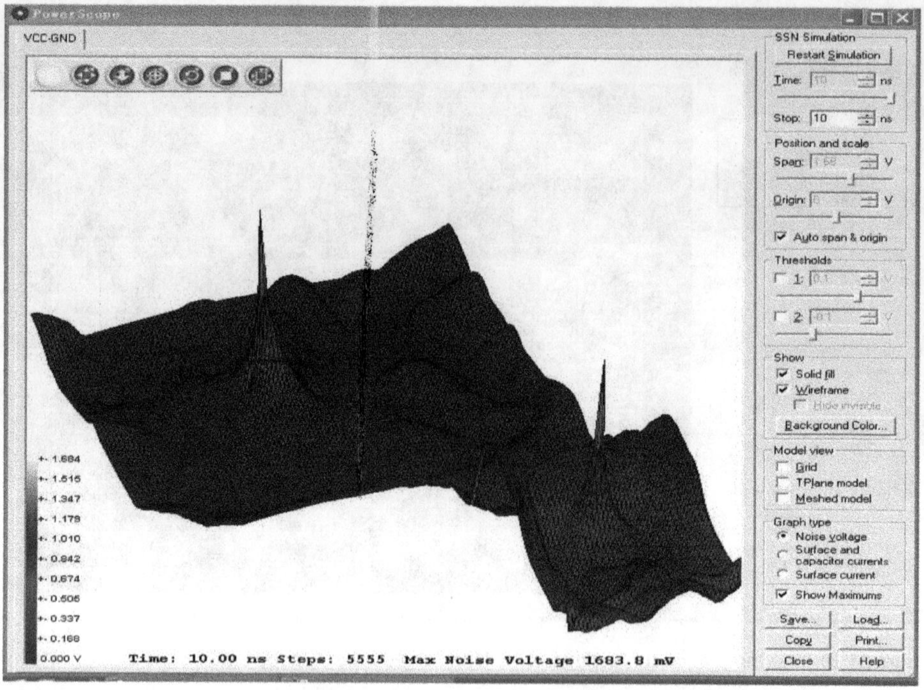

Figure 21.21 Plane-layer noise analysis.

In Expedition, select from the menu Analysis→Export to HyperLynx Power Integrity; the system automatically transfers data to the simulation tool and opens HyperLynx PI, as shown in Figure 21.22.

HyperLynx PI and HyperLynx SI share the same interface, and it is important to note that if analyzing PI, it is better to select Export to HyperLynx Power Integrity; similarly, if analyzing SI, it is better to select Export to HyperLynx Signal Integrity, as otherwise there will be a warning message showing that the data transmission mode is not correct.

In the HyperLynx PI interface, click the Run DC Drop Simulation button , and the DC Drop Analysis window pops up. In the Power/Ground Net to Analyze area, select VCC net. The VCC net distribution layer shows in the Display Area; clicking on each layer, we can see the VCC net distribution in this layer. Figure 21.23 shows the analysis interface of DC drop.

Click the Assign button Assign... and the window shown in Figure 21.24 pops up; we can assign DC parameters in this window. In this project, U2 is a BGA package, device VRM acts as power supply and bare IC chip U1 acts as DC sink.

Hold down the < Shift > key, select all the VCC net pins of U2 and click the Assign button below VRM Model; a window pops up with input parameters as shown in Figure 21.25. Click OK, and we can see the model has been assigned for U2 pins.

Hold down the < Shift > key, select all the VCC net pins of U1 and click the Assign button below DC Sink; a window pops up with input parameters as shown in Figure 21.26. Click OK, and we can see the model has been assigned for U1 pins.

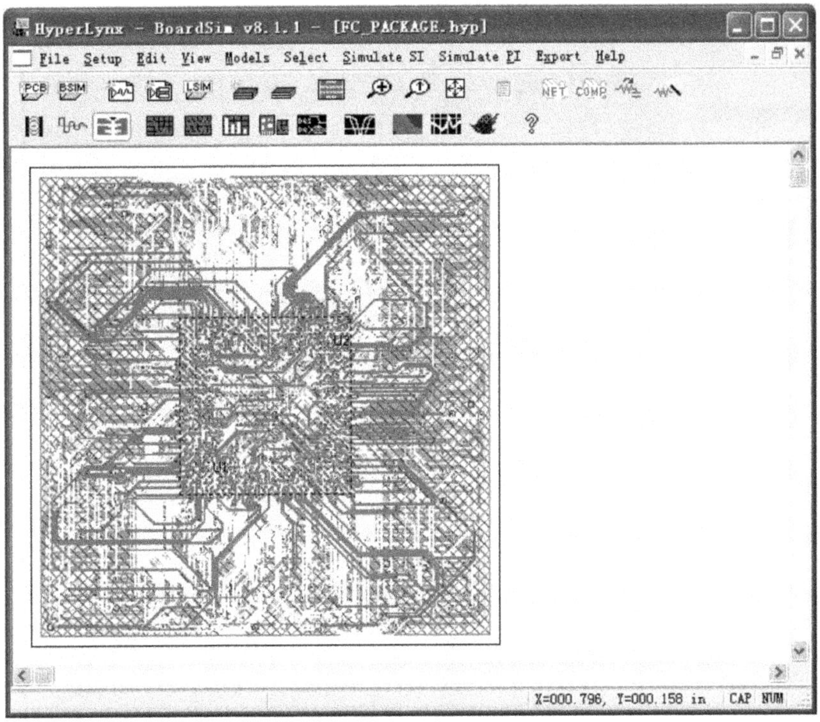

Figure 21.22 Transfer design data to HyperLynx PI.

After PI model assignment, click OK to exit the model assignment window. In the DC Drop Analysis window, click the Simulate button to start simulation. Depending on the complexity of the design, the time needed for simulation will be different.

After the completion of the simulation, two windows can be obtained, a 3D graphics display window as shown in Figure 21.27, and a report window as shown in Figure 21.28.

From the 3D graphics window, we can see that the power supply voltage of some IC pins has a larger decline, mainly due to long traces, and it is necessary to re-route these traces to improve the DC voltage drop.

The report window lists the voltage and current value of all pins and vias, and all the points have dynamic links with the DC drop analysis window; use the mouse to click the pins in the report window and the mouse cursor will automatically jump to the corresponding point on the graphics window – so smart!

In the 3D graphics window, we can check the DC Current Distribution or DC Current Density options to view the relevant current graphics, as shown in Figure 21.29.

21.4 Thermal Analysis

Because of SiP's small volume and greater number and concentration of IC chips, SiP consumes more power than ordinary package or PCB in the same unit area, so the power density in SiP is higher than in ordinary package or PCB; therefore, thermal analysis is particularly important for SiP.

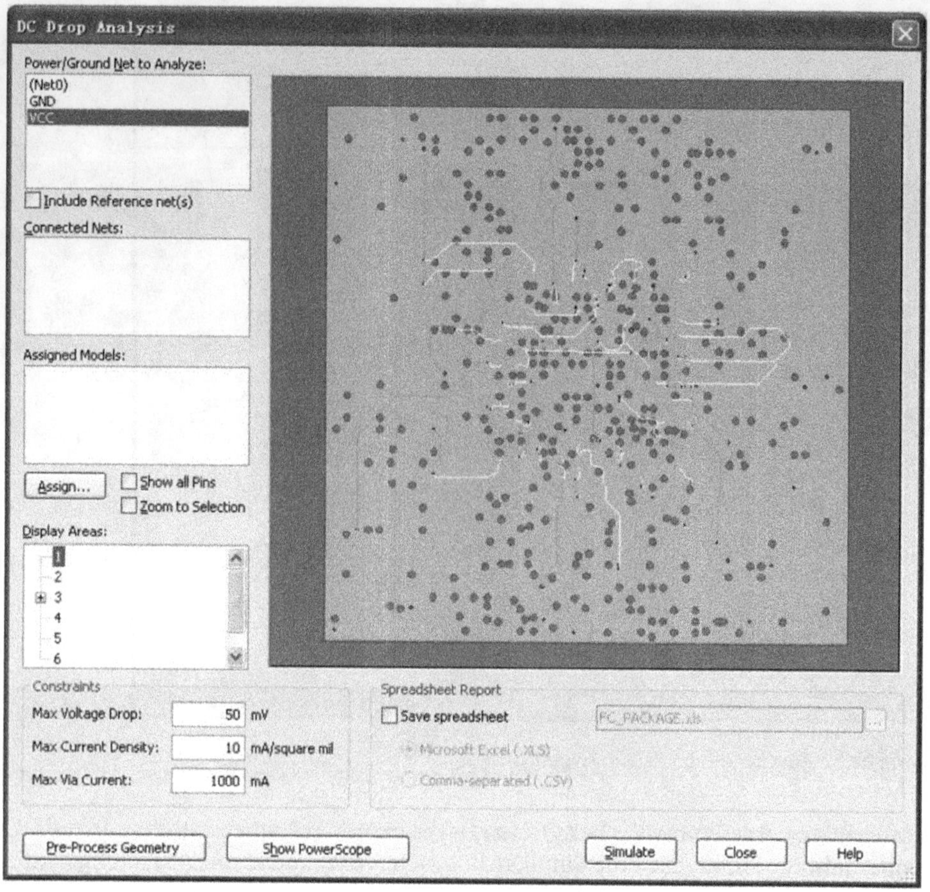

Figure 21.23 DC drop analysis interface.

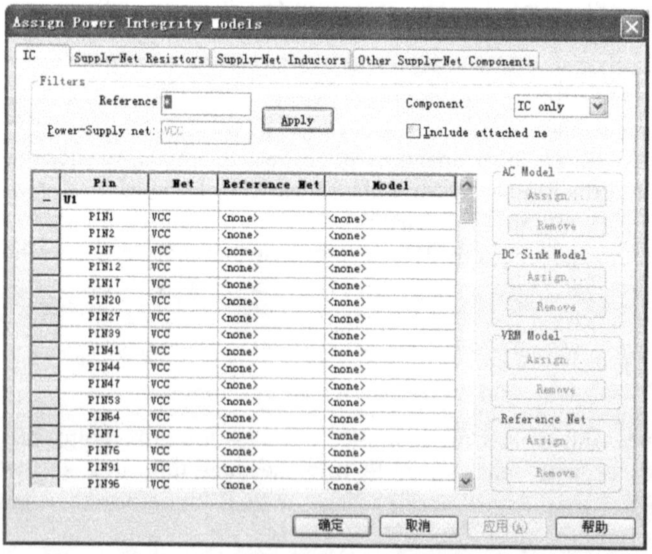

Figure 21.24 Model assignment window of PI analysis.

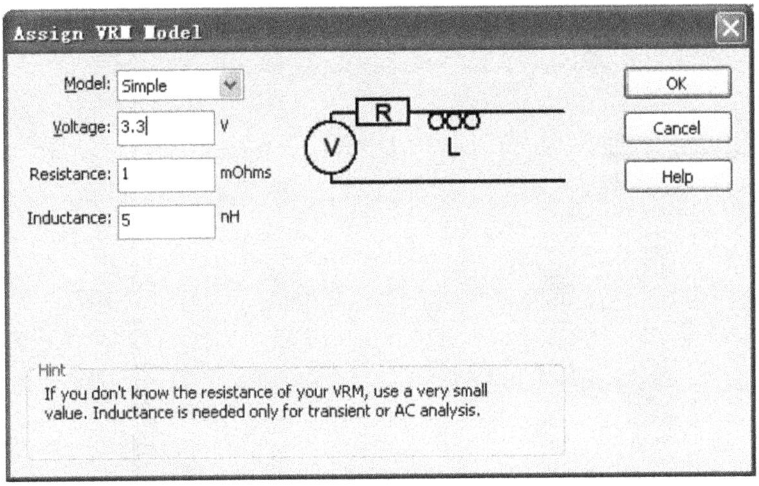

Figure 21.25 Assign VRM Model window.

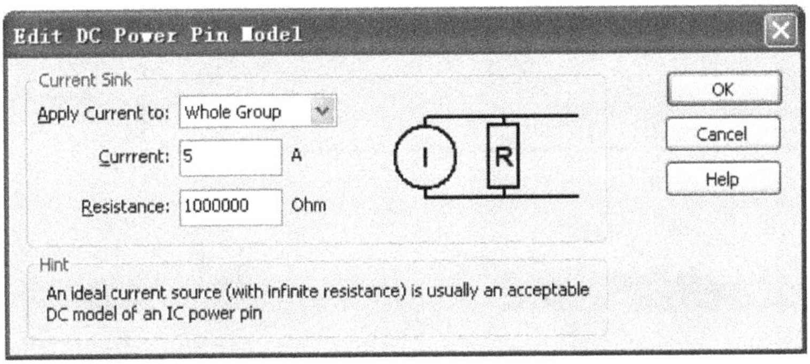

Figure 21.26 Assign DC Sink model.

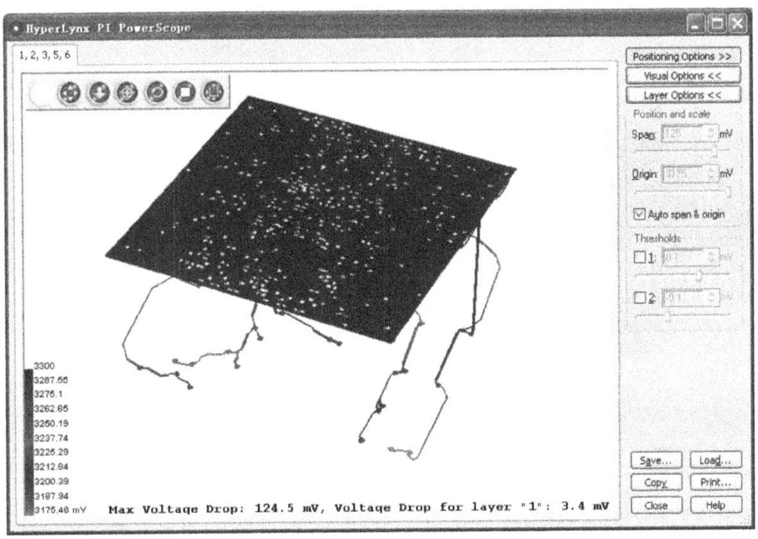

Figure 21.27 DC voltage drop 3D graphics display window.

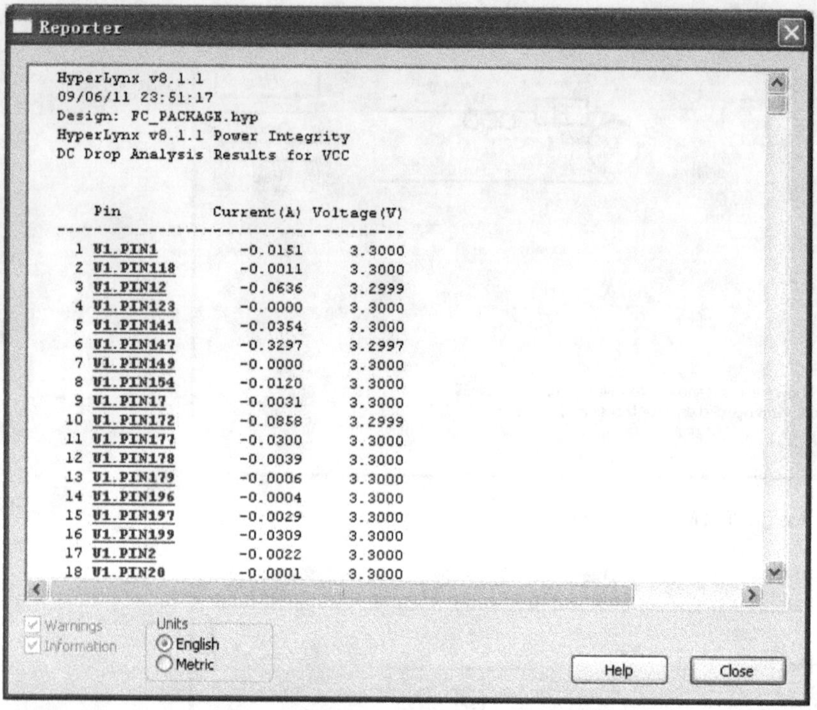

Figure 21.28 DC drop report window.

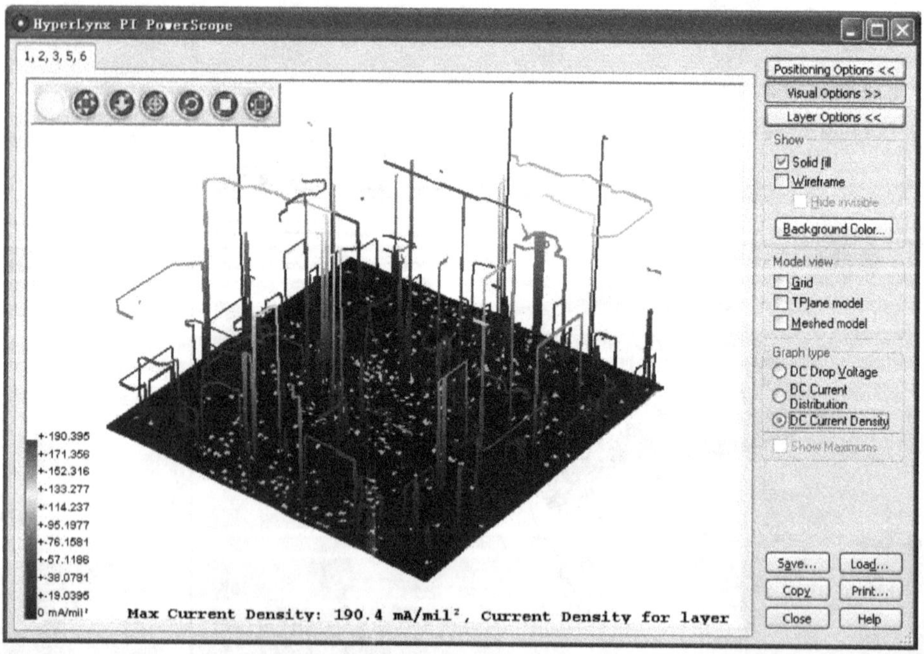

Figure 21.29 3D graphic of current density.

21.4.1 HyperLynx Thermal Introduction

HyperLynx Thermal can carry out accurate and reliable analysis of SiP's heating effect. Using the determination of SiP substrate temperature, temperature gradient and temperature of IC chips, designers can easily find potential heat dissipation and thermal reliability problems in SiP design.

HyperLynx Thermal uses the local variable step finite differential algorithm; compared with the traditional FEM method (Finite Element Method), the calculation speed is greatly increased.

To solve problems of heat conduction, convection and radiation, HyperLynx Thermal can establish airflow and thermal field models, and consider whether the device has a heat sink, radiator, fan and thermal pad, etc.

HyperLynx Thermal can analyze multiple layers and irregular-shaped substrates, and mixed circuit or sub-circuit boards. Substrate can be cooled using edge cooling in a closed space or using forced convection cooling in an open system; the gas flow can be a natural or forced convection; a closed system can be cooled by a heat exchanger.

HyperLynx Thermal modeling can consider the effects of gravity, gas pressure and gas flow direction, and can also consider anisotropic trace; IC devices can be connected with a heat sink, heat pipe, chip fan and heat conduction, etc.

A HyperLynx Thermal simulation screenshot is shown in Figure 21.30.

1) Components library

   HyperLynx Thermal provides two component libraries: work library and master library.

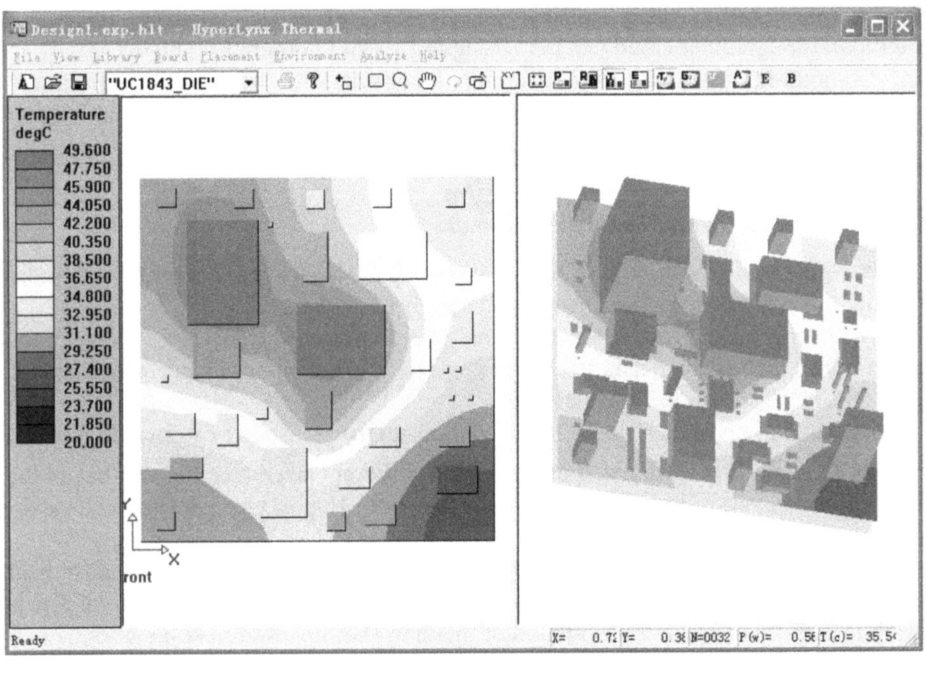

Figure 21.30 HyperLynx Thermal screenshot.

The work library can be generated by the ECAD interface, and includes all components on printed circuit boards. The master library contains more than 2500 defined components, and allows users to expand this infinitely.

2) Components temperature
The one goal of thermal analysis is to get the temperature of the IC devices. HyperLynx Thermal calculates the average temperature of IC devices, and, if THETAjc is available, it can calculate the temperature of junctions. Component shell and junction temperatures are shown as a colored cloud, enabling the designer to quickly find the higher temperature components; the temperatures can also be output to a numerical table.
3) Excess temperature
IC device failure rate has an exponential relationship with the temperature of the IC junction, so different types of components have different limiting temperatures. HyperLynx Thermal allows the user to specify the limiting temperature and to use the graphical display to monitor the difference between the excess temperature and the limiting temperature of IC devices.
4) Temperature cloud picture
HyperLynx Thermal's temperature cloud picture reveals the phenomenon of heat conduction in SiP substrate. Due to the fact that thermal expansion is proportional to temperature, a high temperature area may swell or warp, resulting in broken connections. HyperLynx Thermal can quickly identify the heat point on SiP substrate at design stage, using simulation.
5) Temperature gradient cloud
HyperLynx Thermal can output a temperature gradient cloud for the whole SiP substrate. Because of thermal expansion, high thermal stress can be caused in high gradient areas, with cracking and warping often occurring in high stress areas. The temperature gradient cloud can help designers to find potential problems.

Comparing HyperLynx Thermal simulation results with the results of experimental measures and infrared scanning, HyperLynx Thermal simulation has higher accuracy, more than 90%, and sometimes even reaching up to 95%.

21.4.2 HyperLynx Thermal Simulation Example

The above-mentioned flip chip design is exported to HyperLynx Thermal for simulation, and the analysis process is introduced simply here.

In Expedition select from the menu Analysis→Export to HyperLynx Thermal; the system transfers the design data to HyperLynx Thermal and opens HyperLynx Thermal automatically, as shown in Figure 21.31.

In HyperLynx Thermal, click the Next Side/Layer button ; we can switch between different layer displays. By looking at the data for different layers, we can see that all the pads, traces and vias in Expedition were transferred to HyperLynx Thermal accurately, as shown in Figure 21.31.

The layout data is quite important for heat transfer in the substrate, because heat transfer coefficients are very different in metal and electric materials; the metal distribution in each layer and the proportion of metal in the substrate have a big impact on the accuracy of thermal analysis.

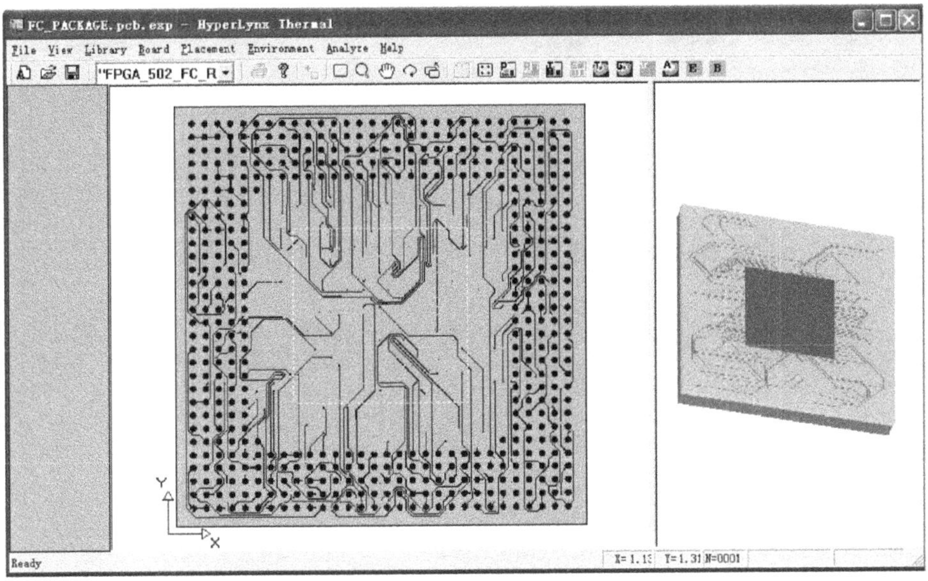

Figure 21.31 Transfer data from Expedition to HyperLynx Thermal.

Using a one-key command, we can accurately transfer design data from Expedition to HyperLynx Thermal; this is the advantage of the Mentor unified design and simulation platform.

1) Environment conditions definition
   In HyperLynx Thermal, before thermal analysis we must first define the environment conditions, such as air temperature, air pressure, gravity, humidity and air velocity, and the casing-related settings, as shown in Figure 21.32.
2) Component properties setting
   After environment condition definition, we need to set component properties, including component name, position and power factor (ratio of actual working power consumption with maximum power consumption), etc.

   Double-click the left mouse button on any component; the Component Properties window will pop up; the basic parameters in this window are inherited from Expedition, and the designer only needs to input the power-scaling factor according to the ratio of component power in actual work and the component maximum power.

   Then click the Edit this part button; the Edit part window pops up. For detailed settings for this component, see Figure 21.33. In this window, we can set up pin parameters and the gap between component and substrate, gap conductivity, component power consumption, etc.

   Here, we set the power dissipation to 5 W; other properties keep the default values, as shown in Figure 21.33. If we have set the related properties such as power consumption at the time of component-library creation, these properties will automatically transfer from Expedition to HyperLynx Thermal.

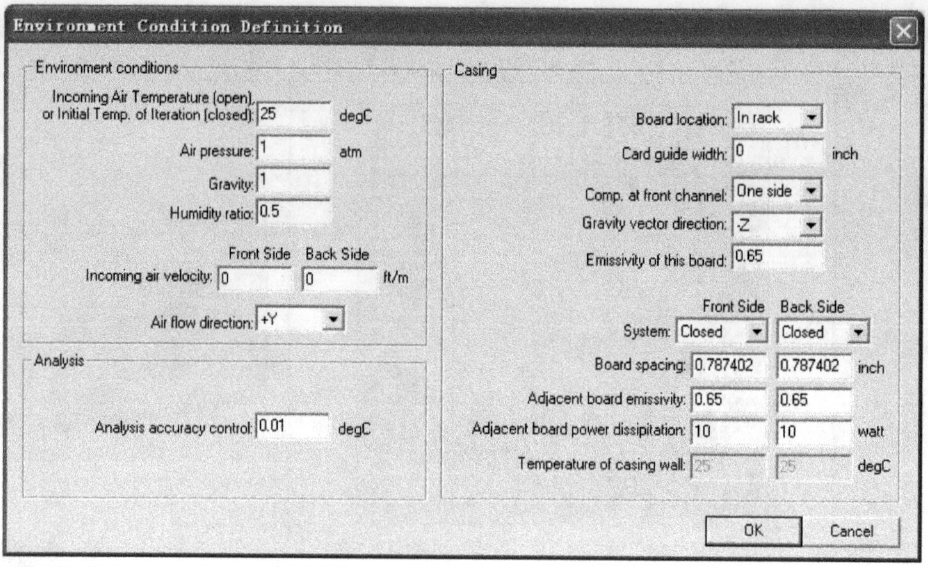

Figure 21.32 Environment condition definition.

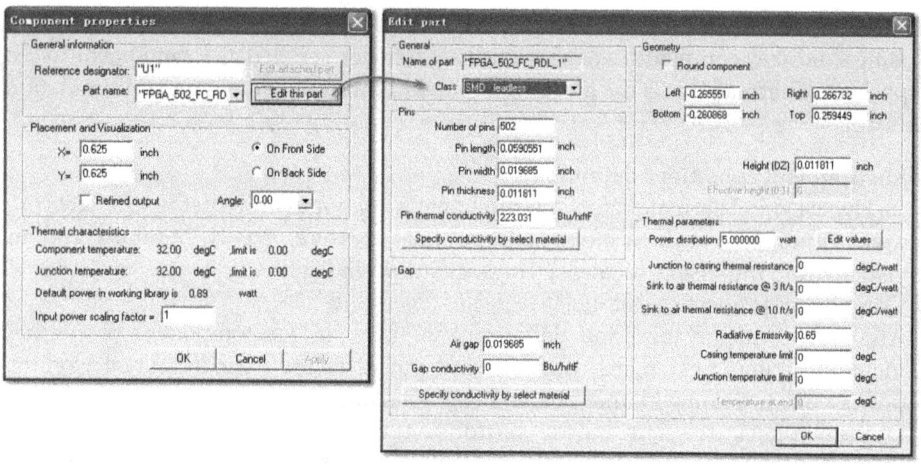

Figure 21.33 Component property setting.

3) Board properties definition

After the completion of component property setting, we start substrate board properties definition, such as board size, layers, thickness, thermal conductivity of conductor, VFM (Volume Fraction of Metal) settings, and temperature limiting. In the transfer of data from Expedition to HyperLynx Thermal, the board size, layers, thickness and VFM will automatically transfer from design to the thermal analysis environment.

The thermal conductivity of a conductor needs to be chosen depending on the conductor material used in the substrate board. For example, if the design adopts

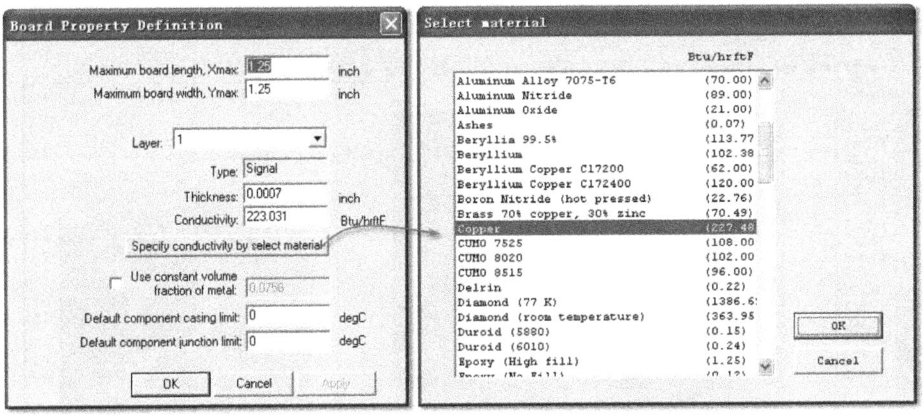

Figure 21.34 Substrate board property definition.

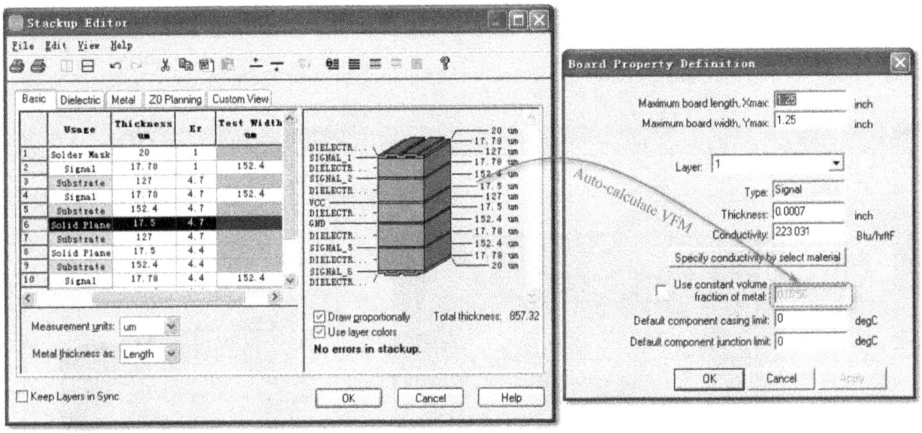

Figure 21.35 The VFM value when VCC/GND copper-layer thickness set to 0.5 Oz.

copper, click the "Specify conductivity by select material" button and select copper in the material list, as shown in Figure 21.34.

Here we need to emphasize that VFM has a big influence on substrate heat transfer; if the thermal analysis software cannot correctly import VFM data, designers need to estimate based on the actual design, which will probably cause larger tolerance. The transfer of data from Expedition to HyperLynx Thermal ensures that the VFM can be transferred accurately, so as to ensure the accuracy of thermal analysis. This can be explained by the following example.

In Expedition CES, set VCC and GND layer thickness to 17.5 μm, that is, copper thickness of 0.5 Oz, and then export the data from design to HyperLynx Thermal; we can see the VFM value is 0.0756, as shown in Figure 21.35.

After the parameters have been set, click the Run Analysis button ; after thermal analysis, the highest substrate temperature is 154 °C, as shown in Figure 21.36.

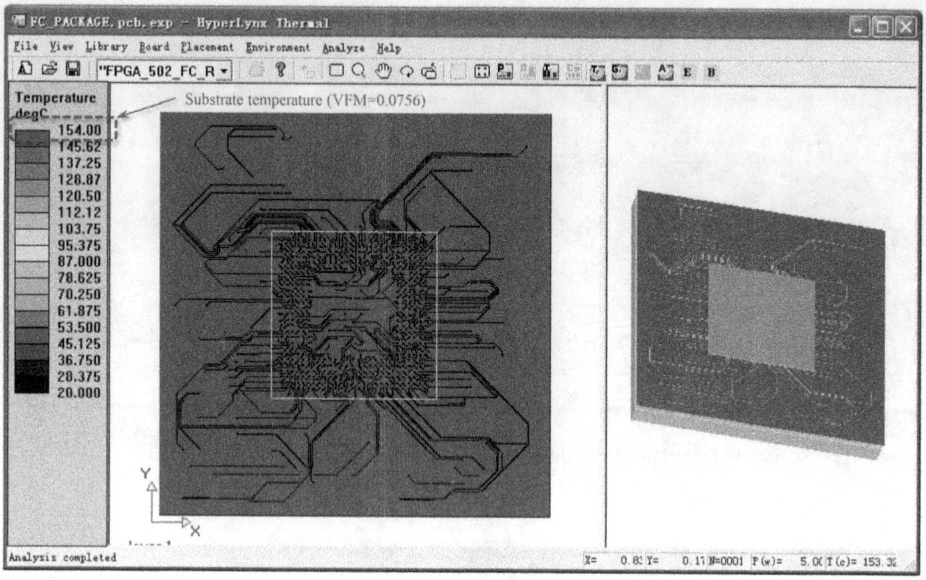

Figure 21.36 The highest substrate temperature is 154°C, when VFM=0.0756.

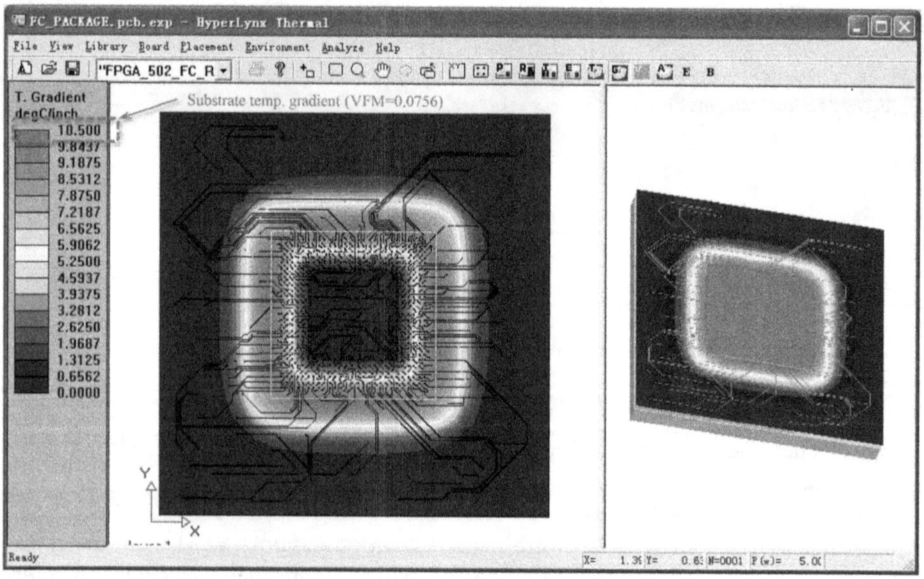

Figure 21.37 The maximum temperature gradient of the substrate is 10.5°C/inch when VFM=0.0756.

Click the Thermal gradient button to switch to temperature gradient; we can see that the maximum temperature gradient of the substrate is 10.5 °C/inch, as shown in Figure 21.37.

In Expedition CES, set VCC and GND layer thickness to 35 μm, namely the copper thickness is 1 Oz, and then export design to HyperLynx Thermal; we can see that the VFM value changes to 0.1076, as shown in Figure 21.38.

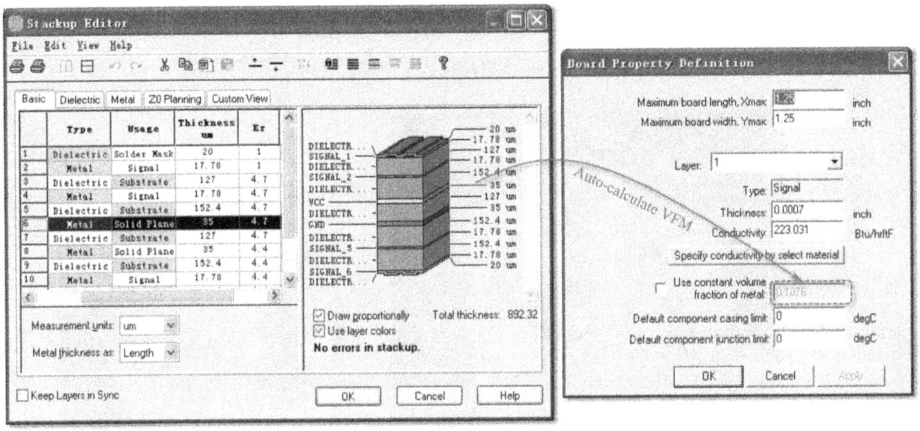

Figure 21.38 The VFM value when VCC/GND copper-layer thickness set to 1 Oz.

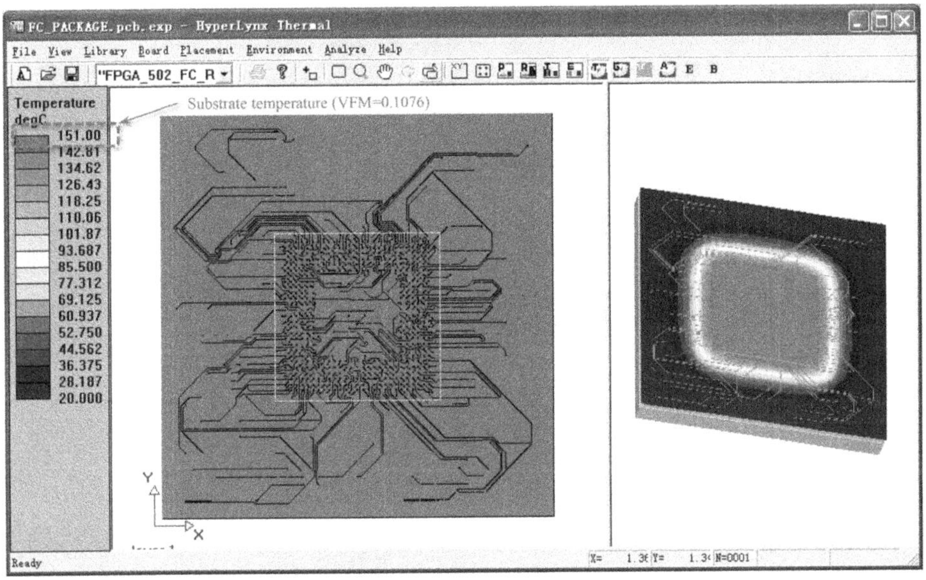

Figure 21.39 The highest substrate temperature is 151 °C when VFM=0.1076.

It can be seen that the VFM value is directly affected by copper thickness. Apart from VFM change, the other parameters remain exactly the same. Next click the Run Analysis button to run thermal analysis; we can see that the highest temperature of the substrate is 151 °C, as shown in Figure 21.39.

Click the Thermal gradient button to switch to temperature gradient; we can see that the maximum temperature gradient of the substrate is 8.3 °C/ inch, as shown in Figure 21.40.

In many cases, components are not run at maximum power consumption. For example, if the component's maximum power consumption is 5 W, in practical work

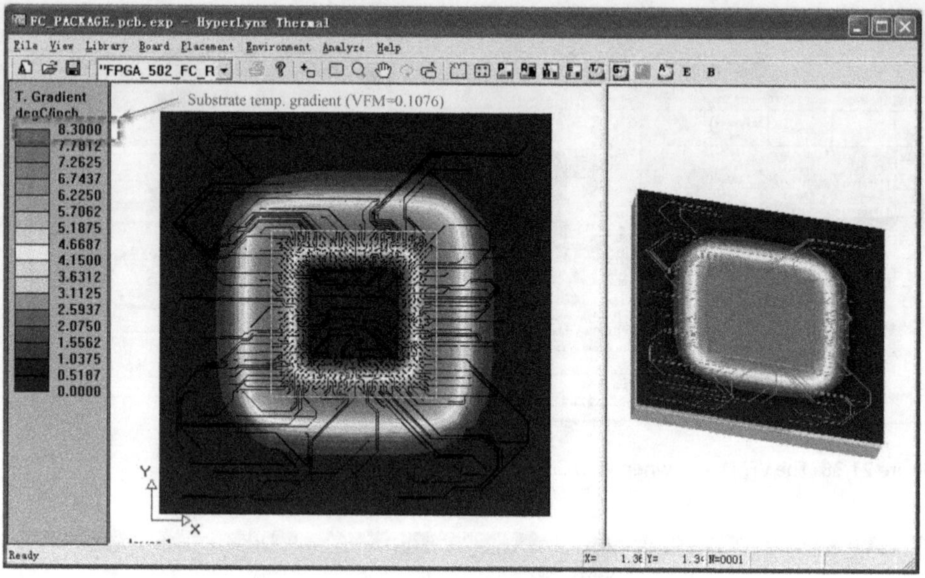

Figure 21.40 The temperature gradient is 8.3 °C/inch when VFM=0.1076.

the power consumption is often less than this; if the chip works at about 25% of maximum power consumption, the power factor can be changed from 1 to 0.25, as shown in Figure 21.41.

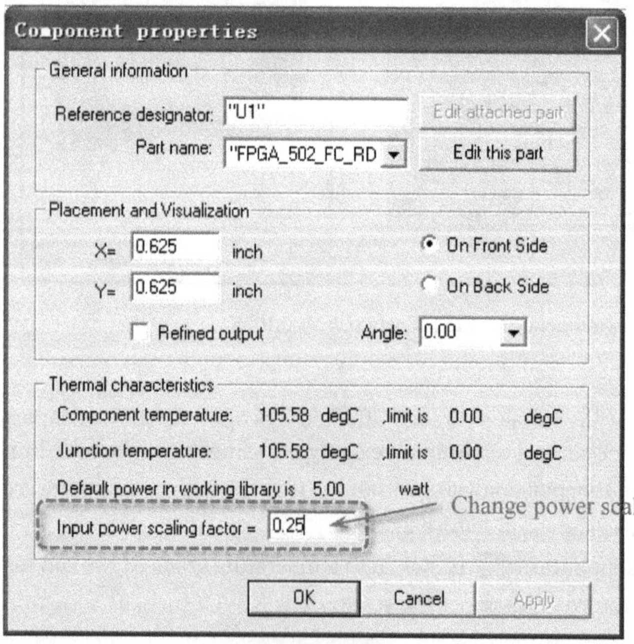

Figure 21.41 Change power-scaling factor to 0.25.

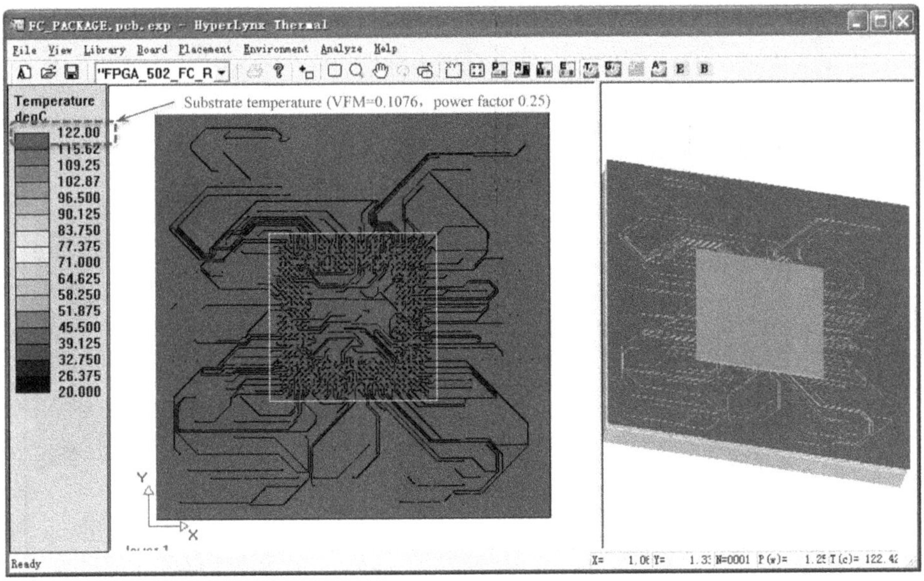

Figure 21.42 The highest temperature is 122 °C when run with 25% power consumption.

Other parameters remain the same as above; click the Analysis button to run thermal analysis again; this time the highest substrate temperature is 122 °C, as shown in Figure 21.42.

Click the Thermal gradient button to switch to temperature gradient; we can see that the substrate maximum temperature gradient has changed to 1.3 °C/inch, as shown in Figure 21.43.

If the temperature doesn't meet the design requirements, we can use other auxiliary cooling methods to lower the temperature of the substrate and components, such as adding boundary conditions, heat sinks, heat pipes, cooling screws, etc.

Below, by using the heat-dissipation method boundary condition as an example, we simulate the influence of package boundary cooling on the chip temperature.

Select from the menu Environment→Boundary, and draw along the flip chip substrate from the upper-left corner to the upper-right corner; the system pops up a Boundary Condition Definition window, as shown in Figure 21.44. Set "Thermal resistance of wedge lock at edge" to 0.3, and "Temperature of sink at edge" to 60 °C.

After thermal boundary condition definition, leave the other parameters unchanged; click the Analysis button to run thermal analysis; we get highest substrate temperature of 61.8 °C, as shown in Figure 21.45.

Click the Thermal gradient button to switch to temperature gradient; we can see that the substrate maximum temperature gradient has changed to 6.5 °C/inch, as shown in Figure 21.46.

By adding thermal boundary conditions and then checking the results of thermal analysis to determine whether this method meets the needs of a real project, it is possible to prove that a method is feasible. Otherwise, the designer needs to adopt

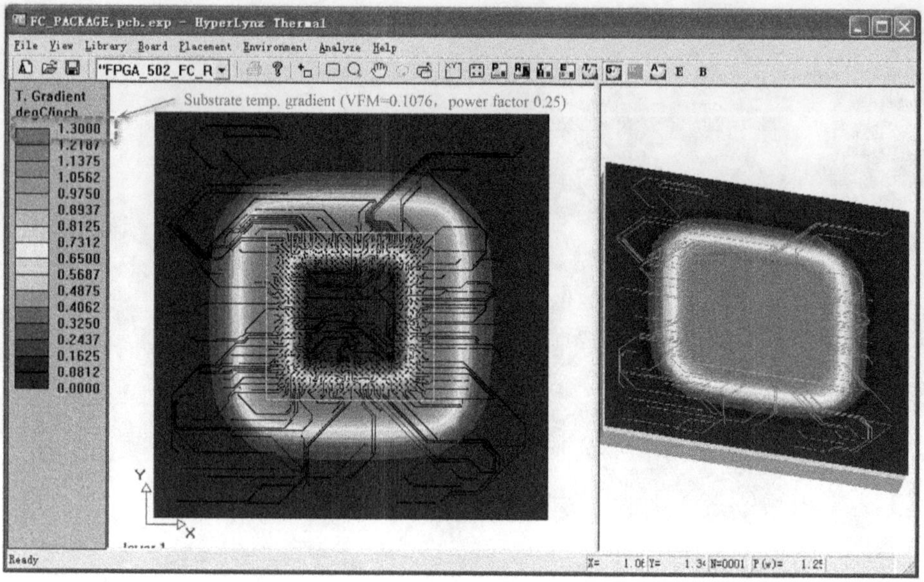

Figure 21.43 The maximum temperature gradient is 1.3 °C/inch when run with 25% power consumption.

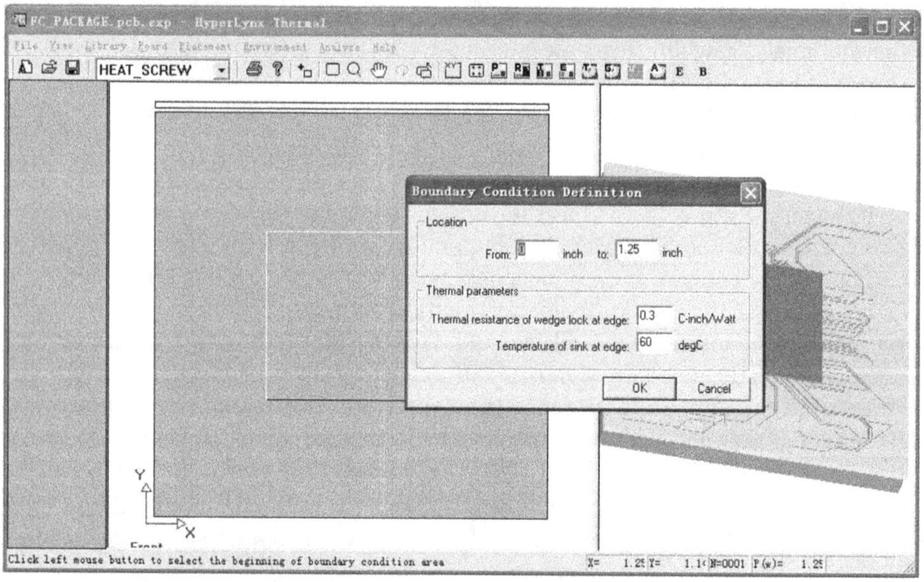

Figure 21.44 Boundary Condition Definition.

other cooling methods, such as heat sinks, heat pipes, cooling screws, etc., and then run analysis in HyperLynx Thermal to verify their feasibility.

Because of space limitations, this book does not describe those methods in detail; readers can verify through practice.

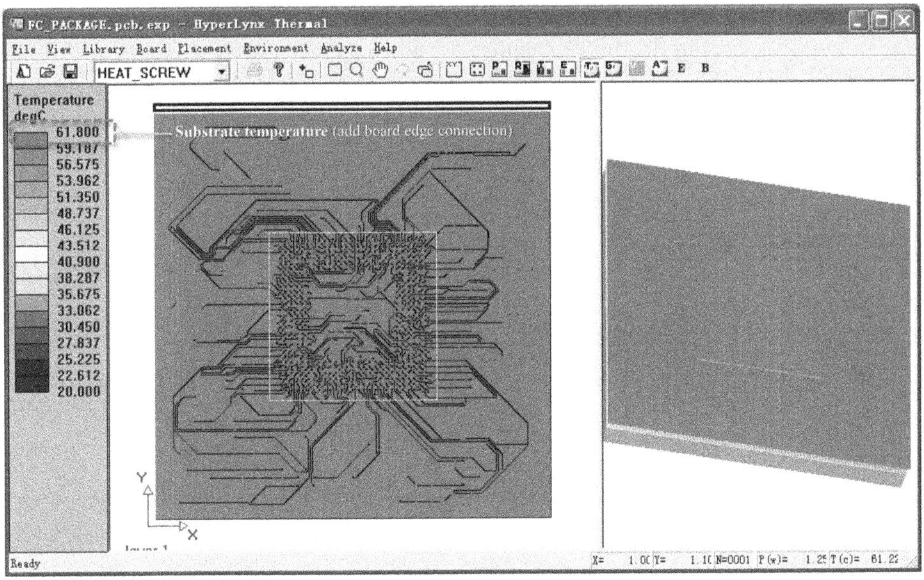

Figure 21.45 Substrate temperature after adding boundary condition.

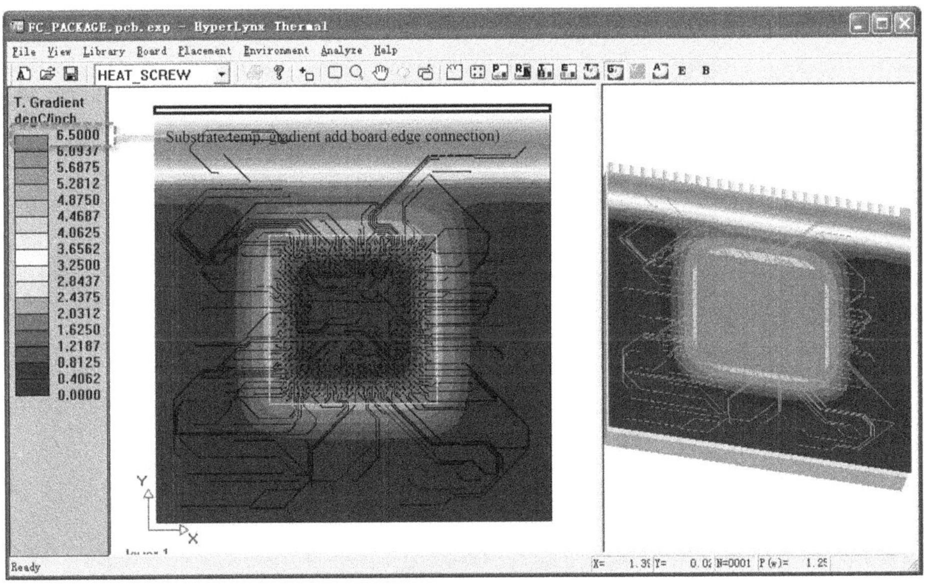

Figure 21.46 Temperature gradient after adding boundary condition.

21.5 EMI/EMC Analysis

21.5.1 HyperLynx DRC Introduction

HyperLynx DRC (its old name was Quiet Expert) is Mentor high-performance EMI/EMC expert system, which uses the research achievements of the University of

Missouri EMC laboratory to provide electromagnetic analysis of various interference sources, and, depending on the relevant conditions, to calculate the electromagnetic radiation.

HyperLynx DRC incorporates an extensible design rules checker and allows users to add their own rules; the designer can use different EMI/EMC standards to check the SiP substrate design.

HyperLynx DRC can provide rising, falling, frequency cycle or duty adjustable square waveform, and, using the calculation and simulation of signals in a time domain, analyze the features of transmission lines and gates which influence system sequence and signal quality.

Compositing component gate features and transmission line features such as delay, impedance, etc. allows for an integrated simulation, provides engineers with high-speed circuit design simulation methods, and allow users to change the types of components and parameters in order to achieve the ideal signal quality.

The HyperLynx DRC verification process is shown in Figure 21.47.

HyperLynx DRC provides the designer with a powerful EMI/EMC verification method, allowing for inspection of incorrect or dangerous routing structures that might result in EMI/EMC problems for board-level design. HyperLynx DRC can not only indicate the location of errors by highlighting but also can report the reasons for problems and provide solutions. It can be applied in the whole process of layout design, even for finished routing parts only. It is very convenient for the designer to get timely feedback during design.

The easy-to-use model can benefit all design team members; it is the most effective way to find potential EMI/EMC problems early in SiP substrate design.

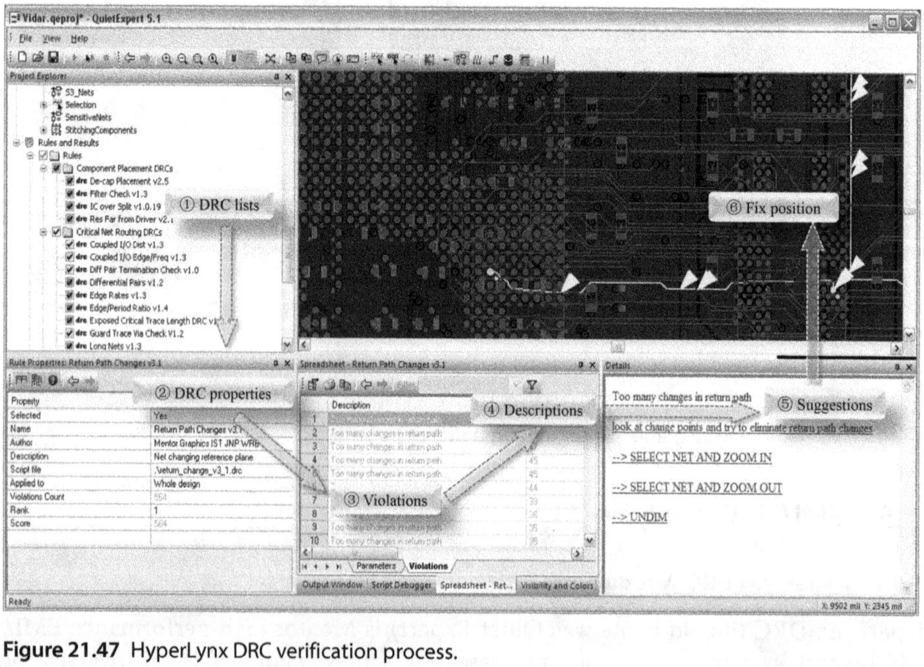

Figure 21.47 HyperLynx DRC verification process.

It is based on an embedded design rules checker, which is the core of HyperLynx DRC. DRC analysis is based on using physical and electrical parameters to validate the design and determine problem areas; this tool allows the designer to comprehensively analyze the EMI/EMC. A powerful API allows the creation of complex design rule checking, including the physical parameters which can change with frequency.

In actual design, high-frequency nodes should be routed according to strict constraints, rather than lowering the frequency or relaxing the constraints. EMI/EMC-DRC rules can also be parameterized, based on the type of design or product line.

21.5.2 HyperLynx DRC Example

As above, we use the flip chip design to introduce the HyperLynx DRC process.

In Expedition select from the menu Analysis→Export and Launch HyperLynx DRC, the system automatically transfers data to HyperLynx DRC and opens the tool, as shown in Figure 21.48.

We can see from Figure 21.48 that layers, components, electrical nets, physical nets, net classes and other design elements can be transferred accurately from Expedition to HyperLynx DRC.

In analysis setup, the designer needs to specify the related models for EMI/EMC analysis, such as component models, package models, etc. It is important to note that in HyperLynx DRC some check rules need model support, while others don't need model support.

The check rules are the core of HyperLynx DRC; at present HyperLynx DRC has 25 embedded rules, which are divided into five categories, as shown in Figure 21.49.

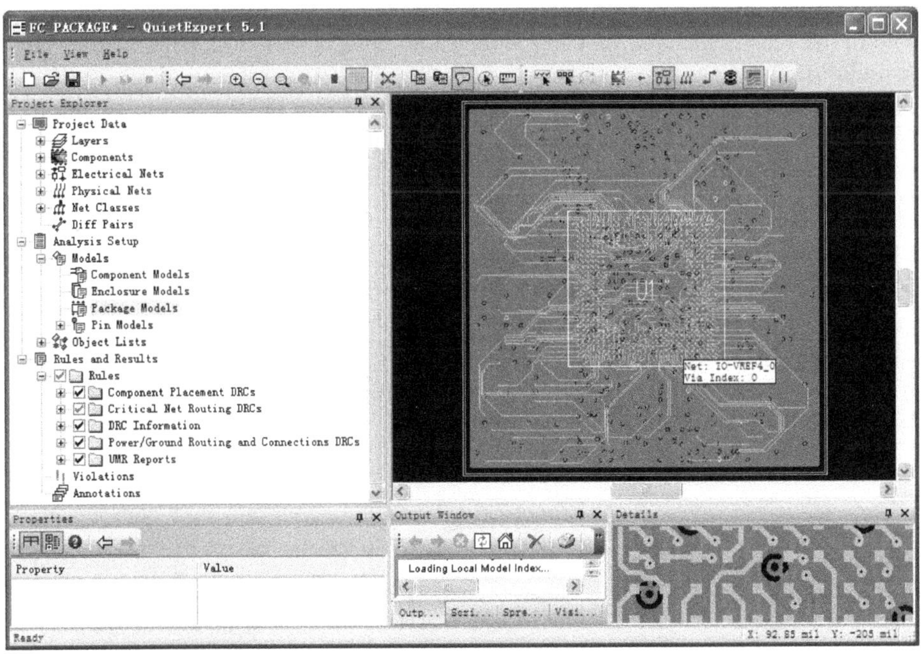

Figure 21.48 Transfer data from Expedition to HyperLynx DRC.

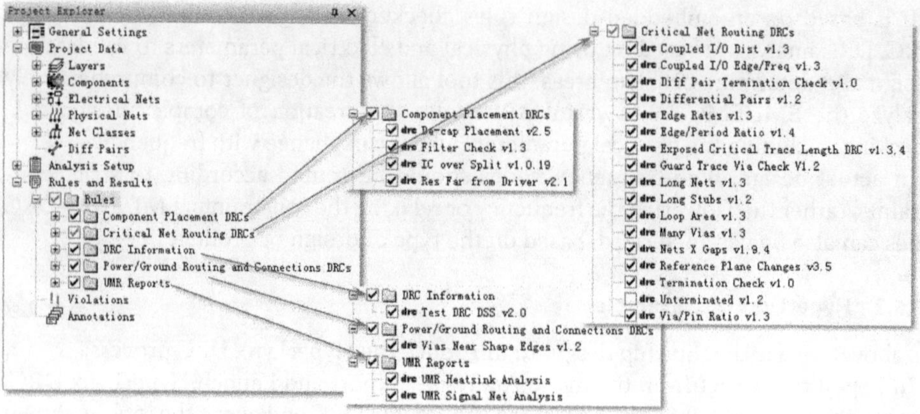

Figure 21.49 DRC rules in HyperLynx DRC.

Depending on the check rules, HyperLynx DRC will check all components, nets, power and GND one by one, like an experienced verification expert. After the completion of inspection, if the design contains unreasonable phenomena, HyperLynx DRC gives warnings and suggests amendments. In an actual project, the designer can choose all rules or some of the rules.

After setup, click the Execute Rules button ▶ to run DRC checking. On completion of DRC checking, we can see that some check boxes are red and some are green; red indicates that the DRC rule check has failed and green indicates that the DRC rule check has passed, as shown in Figure 21.50.

In this project, there are a total of eight kinds of DRC violation; these rule checks have failed. In the Violations item, we can see the eight kinds of violation. Expand any violation and the nets and components involved in the violation are shown.

For example, in the Microstrip Segments violation, we can see the nets in violation of this rule, and the system gives a detailed description of the violation. As shown in

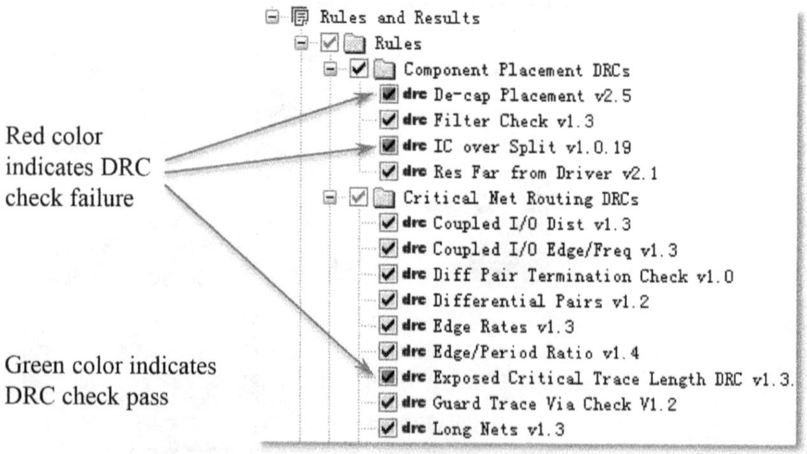

Figure 21.50 The color indicates DRC checking fail or pass.

Figure 21.51, the length of exposed segments routed on net DONE is greater than 25% of the total length, which is not allowed for critical nets.

Use the mouse to click any of the violations and the view window automatically switches to the appropriate size, highlights the net violation, and shows detailed information on the net violation in the details area at the bottom of the window; see Figure 21.52.

It is important to note that not all violations need to be changed; depending on the actual project conditions, the designer can modify the feasible design violations so as to optimize the EMI/EMC performance of this project.

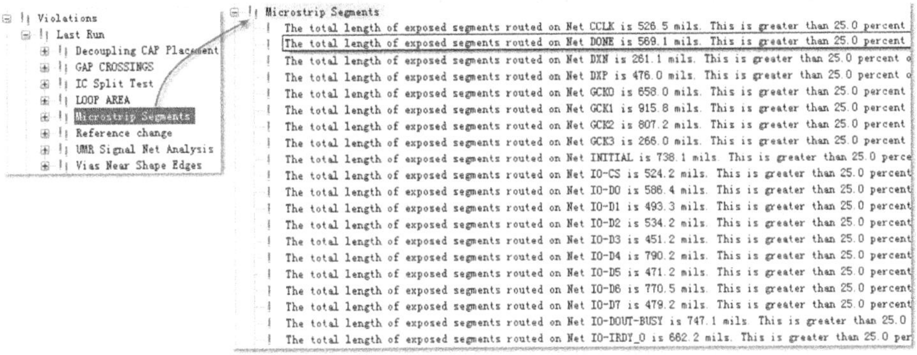

Figure 21.51 Violation display.

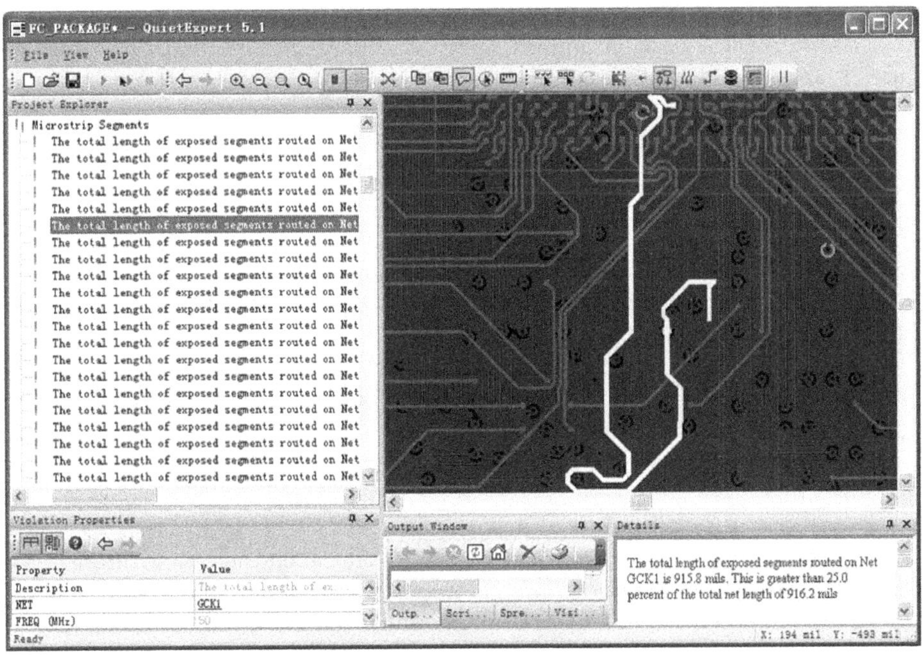

Figure 21.52 The violation net is highlighted.

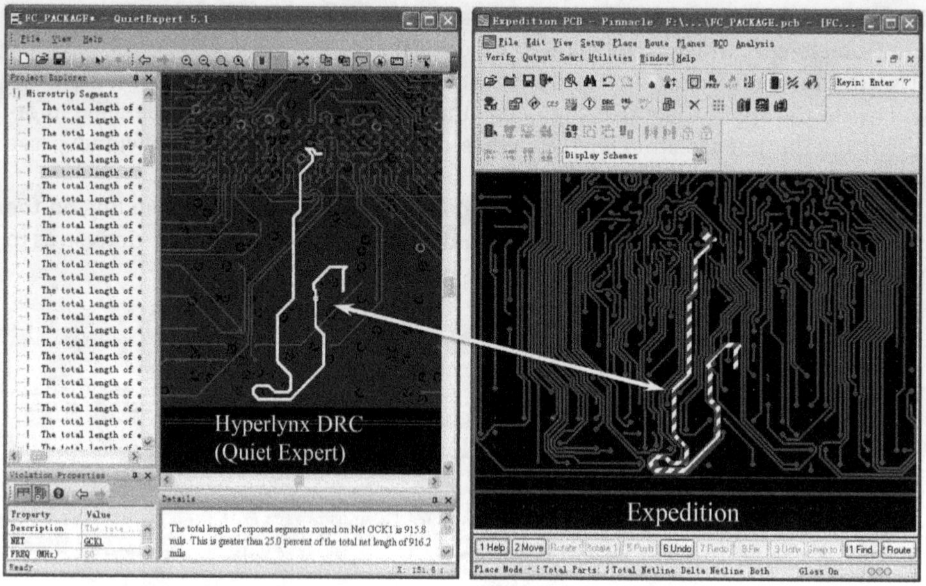

Figure 21.53 HyperLynx DRC cross probe with Expedition.

Click on the Enable Cross Probing button; HyperLynx DRC can cross probe interactively with Expedition. In HyperLynx DRC select any net; this net will also be selected in Expedition and highlighted, indicating that the data are updated in real-time between the two tools. This greatly increases the convenience of interactive checking, and also facilitates the designer in making timely changes in Expedition, as shown in Figure 21.53.

21.6 Mixed-Signal Simulation Introduction

1) HyperLynx Analog – mixed-signal simulation tool based on DxDesigner
   HyperLynx Analog is based on the DxDesigner environment. In DxDesigner if the components are attached to the correct model, we can directly start HyperLynx Analog for simulation.

   HyperLynx Analog supports analog, digital and mixed-signal simulation, providing the designer with a comprehensive simulation environment. HyperLynx supports DC, AC, time-domain and frequency-domain simulation, and parametric sweep, temperature sweep, Monte Carlo, worst case and other advanced analysis methods.

   HyperLynx Analog simulation is based on DxDesigner; in DxDesigner if the HyperLynx Analog license is checked we can invoke simulation, as shown in Figure 21.54.

   HyperLynx Analog supports multiple language simulation models, based on the embedded ADMS hybrid simulation engine which can support multiple languages, including industry standard languages such as Eldo/SPICE, Verilog and VHDL, and the latest mixed-simulation languages, Verilog-AMS, VHDL-AMS, etc.

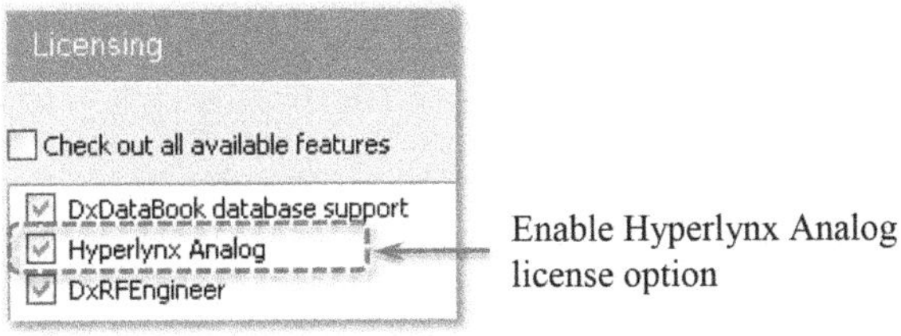

Figure 21.54 Enable HyperLynx Analog license in DxDesigner.

HyperLynx Analog has a unique hybrid simulation function; for analog circuits it can use the SPICE model, for digital circuits, it can directly use VHDL language.

Without any model transformation, the hybrid simulation function can be conveniently implemented. Figure 21.55 is the HyperLynx Analog simulation tools interface.

HyperLynx Analog supports various analysis methods, such as AC analysis, DC analysis, transient analysis, frequency-domain analysis, time-domain analysis, etc., and can be extended to advanced analysis including parametric sweep, temperature sweep, Monte Carlo analysis, worst case analysis, etc., to ensure design quality and stability.

It can fully consider discrete circuits and components, to ensure the reliability of the product.

Using a dedicated EZwave tool to display, measure and calculate the simulation waveform, various complex types of waveforms can be shown and measured. At the same time waveforms can be calculated with many kinds of functional operation to create new waveforms.

Figure 21.56 shows all kinds of waveforms obtained in HyperLynx Analog simulation.

2) The necessity to use simulation tools

For various reasons, such as the manufacturing process and conditions of use, electronic products are inevitably influenced by factors such as random disturbance, which make the circuit present a variety of errors between theoretical and actual data.

For example, in actual electronic circuits, component parameter value is not necessarily equal to nominal value, but may be a random number within the tolerance range of nominal values.

In the process of circuit design, we often use the nominal value of component parameters to analyze and design. Therefore, performance differences necessarily exist between circuit design and the actual product.

When electronic products are actually used, the difference will increase with time and environmental temperature. In serious cases, the deviation is likely to reach a level affecting the normal use of the electronic product, resulting in function failure.

In addition, the parameter tolerances of components are directly related to the cost of electronic products. Usually, the higher precision the components, the smaller the error range, and the higher the price; considering the factor of economy, the designer does not always choose high-precision components.

To sum up, the performance and price of electronic products are closely related to the parameters and tolerance of the components adopted.

Fully considering this factor, the use of a variety of simulation aids is necessary for the design of highly reliable, highly competitive electronic products.

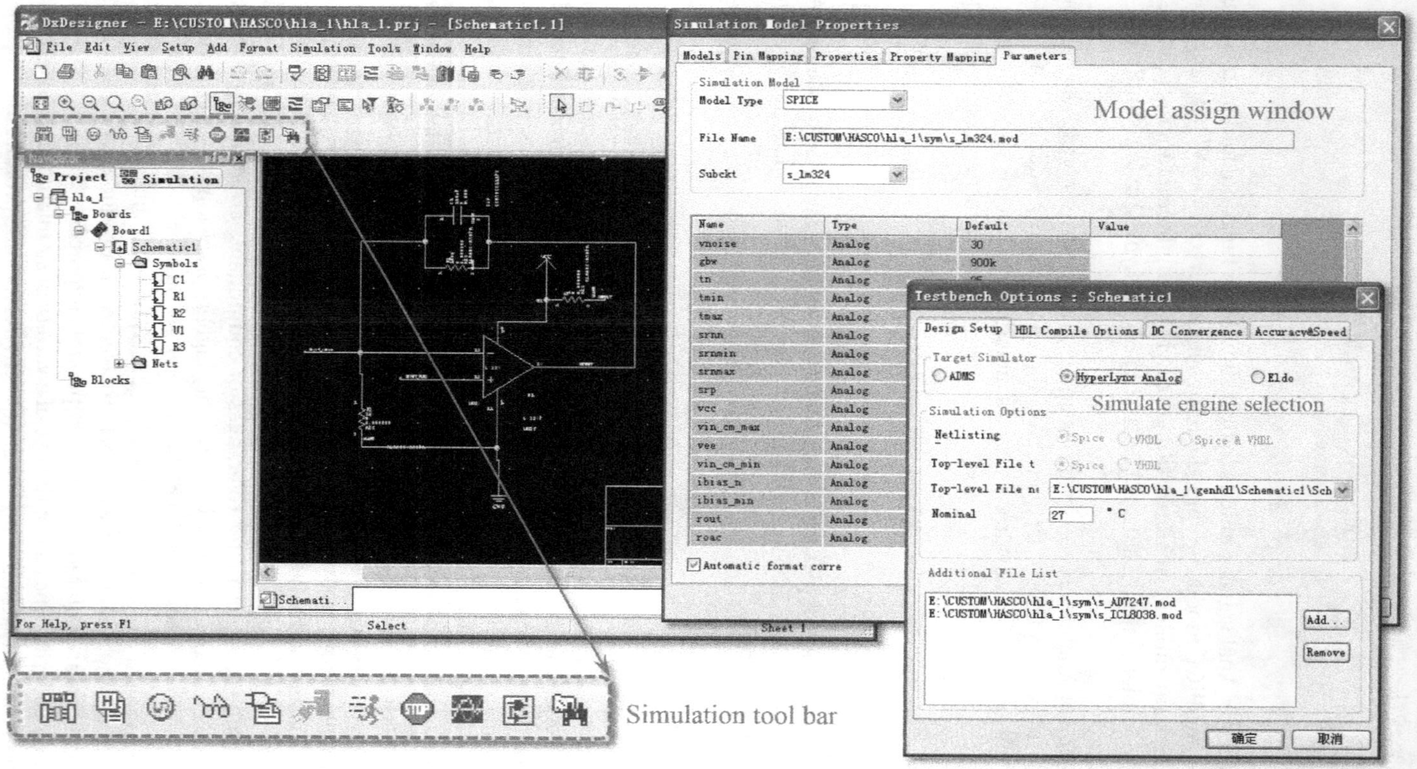

Figure 21.55 HyperLynx Analog simulation interface.

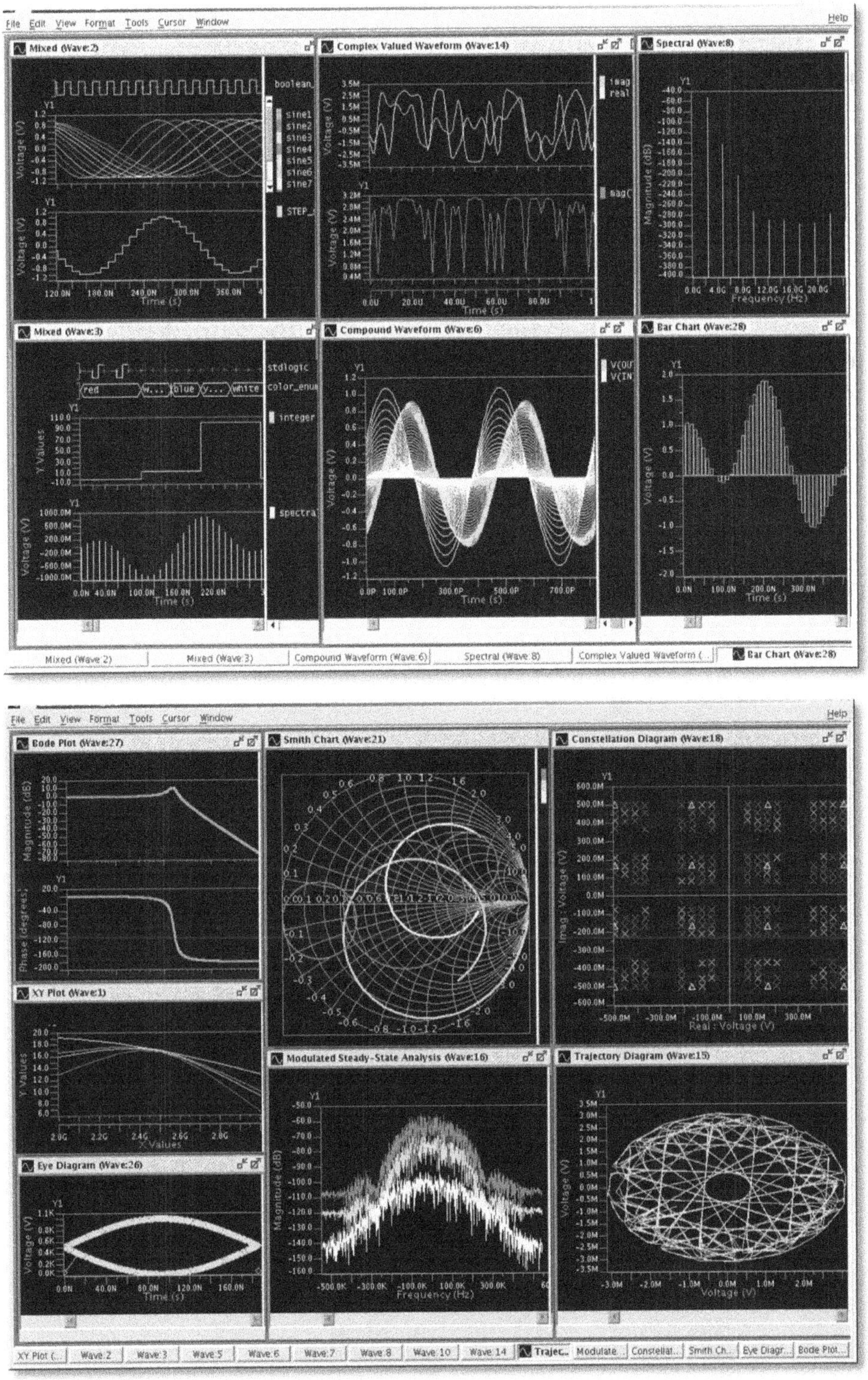

Figure 21.56 The waveforms of HyperLynx Analog simulation.

Reference Materials

1) Wire Bonding and Cavities User Guide
2) RF in DxDesigner® – Expedition® Flow User Manual
3) RF Design Tools User and Reference Manual Supplement User Manual
4) Expedition® PCB User Guide
5) Constraint Editor System™ (CES) User Manual
6) 3D PCB Viewer User Guide
7) DxDesigner® Symbol Editor
8) Cell Editor User Guide
9) Getting Started with DxDesigner® in the Expedition® Enterprise Flow
10) DxDesigner® Reference Manual
11) DxDesigner® User Guide
12) Expedition® PCB Getting Started
13) Getting Started with HyperLynx® Analog
14) XtremePCB™ User Guide
15) Stackup Editor User Guide
16) BoardSim® User Guide
17) Quiet™ Expert User Manual
18) HyperLynx Thermal User Manual
19) Expedition PCB basic training material
20) Some pictures come from Per Viklund's and Alex Roos's technical materials
21) Some pictures and information are taken from the Internet

SiP System-in-Package Design and Simulation: Mentor EE Flow Advanced Design Guide,
First Edition. Suny Li (Li Yang).

Postscript and Thanks

The Chinese version of this book *SiP – System in Package Design and Simulation* took more than one year to write, and was published in 2012.

It has taken about two years to translate the book from Chinese to English and to update some of the contents.

Because my job is busy at normal times, this book was basically written at weekends and in the middle of the night. I often continued to write until 1 a.m. or 2 a.m.

It took me a lot of energy to write this book. I often felt burnt out and regretted the huge energy it required. Midway, I found it difficult to continue, and wanted to give up, but eventually I resolved to keep going.

The book is finished and ready for publication, and I want to thank many people for their support.

First, I want to thank my SiP/MCM/package customers, and the people engaged in the SiP/MCM or package design industry. It was their various needs in actual projects that gave me insight into SiP design requirements and the actual needs of SiP projects, which gave me the motivation to write this book. The book is intended as a practical reference for readers.

Special thanks to my company AcconSys (Beijing) Technology Co., Ltd for its support. It is only with my colleagues' support and encouragement that I have completed this book.

Heartfelt thanks to my Mentor colleagues Per Viklund and Alex Roos; they have provided so much powerful technical support to me. With their help, I solved many difficult SiP design problems and widened design thinking.

I also would like to express thanks to my wife MS Yangyan; because the writing was so hard, and I so often needed to work late, it affected the whole family, but she made no complaint. In addition she supported me with ideas, and undertook the typesetting of the first draft, which made me write in a more orderly fashion, and which, to a certain extent, improved the quality of the book.

Finally, I want to thank PHEI Vice Executive Chief MS Zhao Lisong's for her strong support, and senior editor MS Chai Yan for her hard work. The fact that this book can be published is attributable to them.

The book describes the SiP design flow in detail, but the simulation of SiP used is relatively simple; depending on the demands of readers and the needs of SiP technology development, in subsequent versions, I will strengthen the simulation scenario.

SiP System-in-Package Design and Simulation: Mentor EE Flow Advanced Design Guide,
First Edition. Suny Li (Li Yang).

Finally, I hope that this book can become a useful reference book for the engineers and technical personnel who are interested in SiP and related technologies.

Author: Suny Li (Li Yang)
Chinese version, March 2012, Beijing
English version, July 2016, Beijing

Index

SiP System-in-Package Design and Simulation: Mentor EE Flow Advanced Design Guide,
First Edition. Suny Li (Li Yang).

www.ingramcontent.com/pod-product-compliance
Lightning Source LLC
Chambersburg PA
CBHW080731021025
32954CB00034BA/240
* 9 7 8 1 1 1 9 0 4 5 9 3 9 *